SPRINGER HANDBOOK OF
AUDITORY RESEARCH

Series Editors: Richard R. Fay and Arthur N. Popper

Springer
New York
Berlin
Heidelberg
Barcelona
Hong Kong
London
Milan
Paris
Singapore
Tokyo

SPRINGER HANDBOOK OF AUDITORY RESEARCH

Donata Oertel
Richard R. Fay
Arthur N. Popper
Editors

Integrative Functions in the Mammalian Auditory Pathway

With 122 Illustrations

Springer

Donata Oertel
Department of Physiology
University of Wisconsin Medical School
Madison, WI 53706, USA

Arthur N. Popper
Department of Biology and
Neuroscience and
Cognitive Science Program
University of Maryland
College Park, MD 20742-4415, USA

Richard R. Fay
Department of Psychology and
Parmly Hearing Institute
Loyola University of Chicago
Chicago, IL 60626, USA

Series Editors: Richard R. Fay and Arthur N. Popper

Cover illustration: Three-dimensional schematic diagram of the innervation of the medial superior olive in the cat. The complete figure appears on p. 121 of the text.

Library of Congress Cataloging-in-Publication Data
Integrative functions in the mammalian auditory pathway editors, Donata Oertel,
    Richard R. Fay, Arthur N. Popper.
        p.   ;   cm.—(Springer handbook of auditory research; 15)
        Includes bibliographical reference and index.
        ISBN 0-387-98903-X (hc : alk. paper)
        1. Auditory pathways.   2. Auditory cortex.   I. Oertel, Donata.   II. Fay, Richard R.
    III. Popper, Arthur N.   IV. Series.
        [DNLM: 1. Auditory Pathways—physiology. WV 272 [605 2001]
    QP361.I5635   2001
    573.8'9179—dc21   2001032810

Printed on acid-free paper.

Production managed by Terry Kornak; manufacturing supervised by Jeffrey Taub.
Typeset by Best-set Typesetter Ltd., Hong Kong.
Printed and bound by Sheridan Book Associates, Ann Arbor, MI.
Printed in the United States of America.

9 8 7 6 5 4 3 2 1

ISBN 0-387-98903-X   SPIN 10738451

Springer-Verlag   New York Berlin Heidelberg
A member of BertelsmannSpringer Science+Business Media GmbH

We dedicate this volume to our colleague, Joseph E. Hind. Joe was instrumental in making some of the key observations that underlie many of the conclusions in this volume. He made use of one of the earliest lab computers in the country to document how neurons in the early stages of the auditory pathway encode phase. The series of papers published in the late 1960s with Jerzy Rose, John Brugge, and David Anderson elucidated the ability of auditory nerve fibers and their targets in the brainstem to fire in phase with sound, a code that is now known to underlie our judgment of pitch and localization of low-frequency sounds in the horizontal plane. In the 1980s Joe led the effort, with Joe Chan and Alan Musicant, to measure the head and pinna-related spectral distortions that underlie the ability of mammals to localize sounds monaurally and in the vertical plane. Joe's quiet modesty has made it easy to overlook his important contributions. In addition to being a superb scientist and engineer, Joe is an exceptionally kind, thoughtful, and generous colleague. He has been an inspiration to those privileged to work alongside of him.

Series Preface

The *Springer Handbook of Auditory Research* presents a series of comprehensive and synthetic reviews of the fundamental topics in modern auditory research. The volumes are aimed at all individuals with interests in hearing research including advanced graduate students, post-doctoral researchers, and clinical investigators. The volumes are intended to introduce new investigators to important aspects of hearing science and to help established investigators to better understand the fundamental theories and data in fields of hearing that they may not normally follow closely.

Each volume is intended to present a particular topic comprehensively, and each chapter will serve as a synthetic overview and guide to the literature. As such, the chapters present neither exhaustive data reviews nor original research that has not yet appeared in peer-reviewed journals. The volumes focus on topics that have developed a solid data and conceptual foundation rather than on those for which a literature is only beginning to develop. New research areas will be covered on a timely basis in the series as they begin to mature.

Each volume in the series consists of five to eight substantial chapters on a particular topic. In some cases, the topics will be ones of traditional interest for which there is a substantial body of data and theory, such as auditory neuroanatomy (Vol. 1) and neurophysiology (Vol. 2). Other volumes in the series will deal with topics that have begun to mature more recently, such as development, plasticity, and computational models of neural processing. In many cases, the series editors will be joined by a co-editor having special expertise in the topic of the volume.

Richard R. Fay, Chicago, IL
Arthur N. Popper, College Park, MD

Preface

The task of extracting biologically useful information from the representation of sound that is provided by the cochlea requires rapid neuronal computation. Even a casual consideration of the speed, accuracy, and effortlessness with which most people can turn toward a speaker and understand speech shows how rapidly and efficiently that computation is performed by mammals. The tonotopic array of auditory nerve fibers signals the physical characteristics of sounds. The spatial pattern of activation of the array of auditory nerve fibers is indicative of the spectrum of sounds. Details concerning the precise timing of the onset and of phase are carried in the temporal firing patterns of individual auditory nerve fibers with a resolution in the hundreds of microseconds. This volume is concerned with how the neuronal circuits of mammals extract the origin and meaning from sounds.

This volume builds on and extends information summarized in the first volumes of the *Springer Handbook of Auditory Research* series, *The Mammalian Auditory Pathway: Neuroanatomy* (Volume 1) and *The Mammalian Auditory Pathway: Neurophysiology* (Volume 2). While the emphasis of the first two volumes was on providing descriptions of the structure and function of auditory pathways, the present volume seeks to explain how these pathways lead to an animal's ability to localize and interpret sounds.

Chapter 1 by Oertel introduces the volume and points out that for all its complexity, the auditory system accomplishes two tasks, localization and interpretation of sounds. In Chapter 2, Smith and Spirou provide a summary of what has been learned about the auditory pathways from the cochlea to the cerebral cortex. This chapter allows readers to place the role of particular groups of cells discussed in later chapters into the context of the auditory system as a whole. Chapter 3 by Trussell summarizes a series of findings, not covered in previous volumes, that show that neurons in the brainstem of vertebrates that carry auditory information have biophysical specializations for preserving and conveying information contained in the temporal firing patterns of neurons.

Chapters 4, 5, and 6 review what is known about three of the parallel pathways from the cochlear nuclei to the inferior colliculus through the brainstem. Yin, in Chapter 4, reviews the compelling evidence that the pathway through the ventral cochlear nucleus and medial superior olive extracts interaural timing information for localization in the horizontal plane. In Chapter 5, Young and Davis examine a pathway through the dorsal cochlear nucleus to the inferior colliculus that has some of the characteristics that would be expected from a pathway that uses spectral information for localizing in the vertical plane. Chapter 6 by Oertel and Wickesberg indicates that a pathway through the cochlear nucleus and through the ventral and intermediate nuclei of the lateral lemniscus about which much less is known carries information about patterns of firing that are useful for interpreting sounds.

Casseday, Fremouw, and Covey discuss the inferior colliculus in Chapter 7. This is the site at which the parallel pathways of the brainstem converge. The chapter points out that the convergence of inputs allows neurons in the inferior colliculus to encode higher order features of sounds.

Chapters 8 and 9 examine how sound localization and feature detection are encoded in neurons of the auditory cortex. Chapter 8 by Middlebrooks, Xu, Furukawa, and Mickey examines what is known about how the location of sound sources is reflected in responses of cortical neurons while Nelken (Chapter 9) examines how features are encoded in Chapter 9. These chapters include an examination of what it means for neurons to encode information in the context of localizing sound sources and in recognizing features.

One purpose of this volume has been to expand and update material presented in the first two volumes of this series. In addition, material in these chapters is complemented by chapters in other volumes in the series. Sound localization, which is discussed in this volume from the viewpoint of central processing by Middlebrooks et al. and by Yin, is considered in terms of behavioral capabilities by Brown in Volume 4 (*Comparative Hearing: Mammals*) and Wightman and Kistler in Volume 3 (*Human Psychoacoustics*). Several different aspects of CNS processing of signals in the auditory system of a highly specialized group of mammals, the bats, are considered by Covey and Casseday (lower brainstem), Pollak and Park (inferior colliculus), thalamus (Wenstrup), and auditory cortex (O'Neill) in Volume 5 of the series (*Hearing by Bats*). Additional comparisons between the mammalian system discussed in this volume and the auditory systems of other animals can be gleaned from the chapter on neural processing in insect auditory systems by Pollack in Volume 10 (*Comparative Hearing: Insects*), in fish and amphibians in chapters by McCormick and by Feng and Schellart in Volume 11 (*Comparative Hearing: Fish and Amphibians*), and in reptiles and birds in a chapter by Carr and Code in Volume 13 (*Comparative Hearing: Birds and Reptiles*).

Donata Oertel, Madison, WI
Arthur N. Popper, College Park, MD
Richard R. Fay, Chicago, II
April, 2001

Contents

Contributors

John H. Casseday
Department of Psychology, University of Washington, Seattle, WA 98195, USA

Ellen Covey
Department of Psychology, University of Washington, Seattle, WA 98195, USA

Kevin A. Davis
Department of Biomedical Engineering, Johns Hopkins University, Baltimore, MD 21205, USA

Thane Fremouw
Department of Psychology, University of Washington, Seattle, WA 98195, USA

Shigeto Furukawa
Kresge Hearing Research Institute, University of Michigan, Ann Arbor, MI 48109-0506, USA

Brian J. Mickey
Kresge Hearing Research Institute, University of Michigan, Ann Arbor, MI 48109-0506, USA

John C. Middlebrooks
Kresge Hearing Research Institute, University of Michigan, Ann Arbor, MI 48109-0506, USA

Israel Nelken
Department of Physiology, Hebrew University, Hadassah Medical School, Jerusalem 91120 Israel

Donata Oertel
Department Physiology, University of Wisconsin Medical School, Madison, WI 53706, USA

Philip H. Smith
Department of Anatomy, University of Wisconsin Medical School, Madison, WI 53706, USA

George Spirou
Department of Otolaryngology, West Virginia Medical School, Morgantown, WV 26506, USA

Laurence O. Trussell
Oregon Hearing Research Center and Vollum Institute,Oregon Health Sciences University, Portland OR 97201, USA

Robert E. Wickesberg
Department of Psychology, University of Illinois at Urbana-Champaign, Champaign, IL 61820, USA

Tom C.T. Yin
Department of Physiology, University of Wisconsin Medical School, Madison, WI 53706, USA

Eric D. Young
Department of Biomedical Engineering, Johns Hopkins University, Baltimore, MD 21205, USA

Li Xu
Kresge Hearing Research Institute, University of Michigan, Ann Arbor, MI 48109-0506, USA

1
Introduction

DONATA OERTEL

The biological importance of the auditory system of vertebrates lies in its ability to provide animals with information about where sounds arise and what the sounds mean. The auditory system can alert animals to the presence of danger or prey in the dark and around corners. In many animals, sounds also serve as a basis for communication. The ear transduces the mechanical energy in sound into electrical signals that provide the brain with an ongoing representation of the physical characteristics of sounds arriving at the two ears. The task of extracting useful information from the representation of the physical characteristics of sound, the location of its source, and its meaning, is a complex one to which mammals have devoted a significant proportion of their brains. This volume summarizes what is known about how this complex task is subdivided in mammals, and what is known about the roles of the individual brain nuclei in localizing and interpreting sounds.

The present volume builds on and extends information summarized in Volumes 1 and 2 of the SHAR series concerning the anatomy and physiology of the auditory pathway (Popper and Fay 1992; Webster et al. 1992). Readers are referred to those volumes for descriptions of the structure and function of auditory pathway. The present volume summarizes what has been learned about how some of the components of the auditory pathway contribute to an animal's ability to localize and interpret sounds, and updates and integrates the material from the two earlier volumes.

Chapter 2, by Smith and Spirou, presents an overview of what is known about the connections in the mammalian auditory pathway. It is noteworthy that the neuronal circuits are intricate and vary between species. The task of understanding the mammalian auditory pathway is made more complicated, and also more interesting, because mammals vary widely in the way they use sound stimuli and that their neuronal circuits differ accordingly. Small mammals with small heads, for example, have ears that are close together. For these animals, it is more difficult to make use of interaural time cues for localization in the horizontal plane. Thus, many small mammals, such as mice and rats, have relatively poor hearing at the lower frequencies, and the neuronal circuitry in the medial superior olivary nuclei that makes use of interaural phase differences for sound localization is small. Other small mammals, however, such as the least weasel and the gerbil, overcome those physical limitations with specialized cochleas and large medial superior olivary nuclei. Studies of bats have contributed substantially to the understanding of the mammalian auditory system. The auditory systems of bats follow the general mammalian pattern but some are hypertrophied and specialized for echolocation. These specializations, which meet very specific functional demands, make the bat auditory system relatively more accessible to study and easier to interpret. Understanding the general mammalian pattern and the general strategies used by the auditory system to acquire biologically significant information thus requires the elucidation of neuronal circuits in multiple species with a variety of biological needs.

One of the advances in anatomical studies of the past decade is an increased appreciation of the existence of feedback loops. It is natural to consider a sensory pathway (a pathway that feeds information to the central nervous system) as generally ascending. The earliest descriptions of the auditory pathway correspondingly emphasized the series of neuronal connections from the cochlea to the auditory cortex. Chapter 2 and Chapter 7 (Casseday, Fremouw and Covey) show that at every stage of the auditory pathway, neurons are subject to descending influences. The functional implications of these descending connections are just beginning to be understood.

The past decade has also brought an increased appreciation of the mechanisms that enable neurons in the auditory brainstem nuclei to convey timing information. Chapter 3, by Trussell, summarizes a series of findings, not covered in previous volumes, showing that vertebrate brainstem neurons that carry auditory information have biophysical specializations for preserving and conveying information contained in the neurons' temporal firing patterns. Variations on a theme of specializations have been documented in neurons in the cochlear nuclei, medial nucleus of the trapezoid body, medial superior olive, and ventral nucleus of the lateral lemniscus. These neurons are endowed with unusual combinations of voltage-sensitive potassium channels that enable them to respond rapidly and precisely. Neurons in the auditory pathways of the brainstem also have neurotransmitter receptors that are an order of magnitude faster than those in adjacent, nonauditory regions of the brain. That timing information underlies the ability of mammals to localize sounds in the horizontal plane (Chapter 4, Yin, and Chapter 7, Casseday, Fremouw and Covey) and probably also contributes to the recognition of patterns (Chapter 6, Oertel and Wickesberg).

It is useful to consider the simple scheme that is presented in Fig. 1.1 as an intellectual framework for understanding the role of particular neuronal circuits in the biological functions of localizing and interpreting sounds. Chapters 4 through 9 review what is known about how the various stages

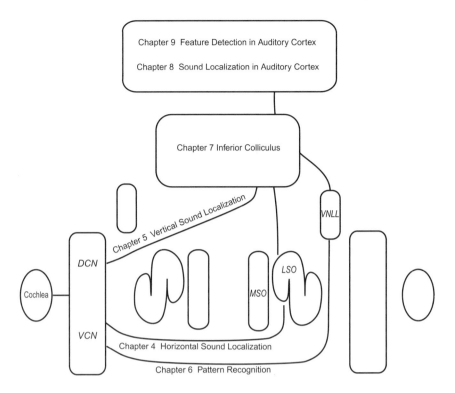

Chapter 9 Feature Detection in Auditory Cortex

Chapter 8 Sound Localization in Auditory Cortex

Chapter 7 Inferior Colliculus

Chapter 5 Vertical Sound Localization

VNLL

DCN

LSO

Cochlea

MSO

VCN

Chapter 4 Horizontal Sound Localization

Chapter 6 Pattern Recognition

Chapter 3 Biophysical Specializations in the Brain Stem

Chapter 2 Anatomical Connections in Auditory Pathways

FIGURE 1.1. This volume reviews what is known about how the auditory system allows animals to localize sound sources and to recognize the meaning in sounds. Chapter 2 reviews what is known about the anatomical connections along the auditory pathway with emphasis on recent findings. The auditory system has not only ascending components that feed sensory information to the cortex but also has descending connections at many stages. Chapter 3 describes biophysical specializations in some of the cells of the brainstem nuclei. Chapters 4–6 review functions that are associated with particular structures and pathways as illustrated schematically. Pathways through the ventral cochlear nucleus (VCN) and through the medial (MSO) and lateral (LSO) superior olivary nuclei are involved in the localization of sound in the horizontal plane. Pathways through the dorsal cochlear nucleus (DCN) contribute to the localization of sound in the vertical plane. Pathways through the VCN and ventral nuclei of the lateral lemniscus (VNLL) probably contribute to the recognition of temporal and spectral patterns in sounds. These pathways converge in the inferior colliculus and ascend to the auditory cortex.

of the auditory pathway contribute to the localization and interpretation of sounds. Those chapters make clear that while some issues are well understood, others are not. Chapter 4 reviews the compelling evidence that the pathway through the ventral cochlear nucleus and medial superior olive extracts interaural timing and interaural level information for localization of sounds in the horizontal plane. Chapter 5, by Young and Davis, examines a pathway through the dorsal cochlear nucleus to the inferior colliculus that has some of the characteristics that are expected from a pathway that uses spectral information for localizing sound sources in the vertical plane. Chapter 6 reviews circumstantial evidence that pathways through the cochlear nucleus and through the ventral nuclei of the lateral lemniscus might carry and convey information that contribute to the ability of animals to interpret and understand sounds. Chapter 7 concerns the inferior colliculus, the site where the parallel pathways of the brainstem converge. The convergence of inputs allows neurons in the inferior colliculus to encode higher order features of sounds. Chapters 8 and 9 examine how sound localization and feature detection are encoded in neurons of the auditory cortex. Chapter 8, by Middlebrooks, reviews what is known about how the location of sound sources is reflected in responses of cortical neurons. Chapter 9, by Nelken, reviews recent experiments that give clues of how features are encoded. These chapters include an examination of what it means for neurons to "encode" information in the context of localizing sound sources and in recognizing features. While some studies have emphasized the existence of maps in the systematic encoding of information, others have focused on what information is carried by individual neurons.

It is important to keep in mind that the scheme presented in Figure 1.1 is overly simplistic. Chapter 2 reviews recent anatomical studies which reveal the rich and complex network of connections in the mammalian auditory pathway. In addition to the ascending connections of arrays of principal neurons that form the ascending pathways, interneurons and axon collaterals of projecting neurons form interconnections within and between nuclei and descending projections amplify or suppress sensory input. The intricacy in anatomical connections is mirrored in the function of neurons and neuronal circuits whose activity shapes inputs from the cochlea. The individual neurons and the nuclei that form the various stages of the auditory pathway function not independently but in the context of their ascending, horizontal, and descending inputs. Assigning functions to particular pathways is also simplistic in that some of the functions we consider clearly involve several steps at several levels of the auditory pathway. For example, the localization of sound sources requires not only the computation of interaural time delays in the medial superior olive (Chapter 4) and the computation of spectral distortions in the dorsal cochlear nucleus (Chapter 5) but it also requires the suppression of echoes which may occur in several brainstem auditory nuclei. The recognition of patterns takes place in many steps. One of these steps involves the shaping of excitation from the cochlear nuclei and from the medial and lateral superior olivary nuclei by inhibition from the ventral nuclei of the lateral lemniscus (Chapter 6). The recognition of patterns and features undoubtedly also involves the inferior colliculus, thalamus and cortex (Chapter 9).

This volume reflects how an understanding of the auditory system arises from the combined observations from diverse approaches on various species, and how that understanding is still fragmented and incomplete. A general picture emerges in which the separate parallel pathways through the brainstem auditory nuclei simultaneously extract information about the location and spectrum of a sound source, and about patterns of synchronous firing in auditory nerve fibers and convey that information to the inferior colliculus. In the inferior colliculus, localization and pattern recognition are combined. As signals pass from the inferior colliculus through the thalamus to the cortex, neuronal transformations reveal more and more complex features in sound stimuli. The cellular basis for the transformations at the higher levels of the auditory pathway is not yet well understood, however.

The neuronal circuit through bushy cells to the principal cells of the medial superior olive provides a rare and compelling example of how individual neurons are specialized to perform an identifiable integrative functions. Chapters 3 and 4 show how this specialized pathway provides animals with a measure of the difference in the time of arrival of sounds to the two ears and thus with a report of the location of the sound source in the horizontal plane.

This volume also leaves important fundamental questions unanswered. Little is understood about the mechanisms whereby spatial and temporal patterns in sounds such as those in speech are detected and extracted. Mammals differ in the frequency range of what they detect in their acoustic environments and in what information they extract from environmental sounds yet they use an auditory system that has a common pattern. To what extent are the auditory pathways shaped by experience? As human beings, most of us are interested in how the auditory system enables us to understand speech and how we can restore these functions to people with hearing losses. Our ability to focus on individual speakers at a party and listen to one conversation amidst the noisy conversations of others depends on our ability to focus on the speech emanating from particular locations and reveals that localization and feature detection are interconnected. In future studies it will be important to determine which pathways are critical for understanding speech, both in quiet and in noise, so that hearing aids and cochlear implants can be designed to stimulate those pathways optimally.

References

Webster DB, Popper AN, Fay RR (Editors) (1992) The Mammalian Auditory Pathway: Neuroanatomy, Springer Verlag, New York.

Popper AN, Fay RR (Editors) (1992) The Mammalian Auditory Pathway: Neurophysiology, Springer Verlag, New York.

2
From the Cochlea to the Cortex and Back

PHILIP H. SMITH and GEORGE A. SPIROU

1. Introduction

Among the many unique features of the auditory system one of the more notable is the large number of processing centers (cell groups) interposed between the system's periphery and its cortex. Our main purpose in this chapter is to build upon the first 2 volumes in this series (Popper and Fay 1992; Webster et al. 1992) by highlighting progress made over the last decade in understanding these structures of the auditory pathway, especially when this information can be related to functional hypotheses about particular cell groups or neural circuits.

The fundamental organizing principle of the auditory system is tonotopy, which derives from the smooth mapping of frequency sensitivity along the length of the cochlea. In many regions of the system other functional attributes are superimposed on this tonotopy. Such structural heterogeneity superimposed on a tonotopic representation begins with the presence of two hair cell populations in the cochlea. Three rows of outer hair cells (OHCs; for abbreviations throughout this chapter, see Table 2.1) receive sparse afferent innervation from the peripheral processes (radial fibers) of spiral ganglion cells classified as type II and are directly contacted by efferent fibers of the medial olivocochlear system (MOC). This is in contrast to the single row of inner hair cells (IHCs), which receive dense afferent innervation from the radial fibers of type I ganglion cells (Fig. 2.1, 95% of total). It is this type I ganglion cell peripheral process that is contacted by efferent fibers of the lateral olivocochlear system (LOC). The MOC system, which affects the micromechanics of the organ of Corti through its action on the OHCs as measured by modulation of otoacoustic emissions, represents a class of brainstem neurons that are distinct from the LOC/IHC system. As we will see below, we have learned much during the past decade about the type I auditory nerve (AN) fibers innervating the IHC and about the MOC providing feedback from the brain to the OHCs. In contrast, we still know little about the type II AN fibers innervating OHCs and the LOCs affecting the innervation of IHCs.

TABLE 2.1. Abbreviations

AAF	anterior auditory field
AC	auditory cortex
ACh	acetylcholine
AEC	anterior ectosylvian cortex
AES	anterior ectosylvian sulcus
AEV	ventral bank of the anterior extosylvian cortex
AHP	afterhyperpolarization
AI	primary auditory cortex
AL	anterolateral amygdala
AMPA	α-amino-3-hydroxy-5-methyl-4-isoxazole propionic acid
AN	auditory nerve
AVCN	anteroventral cochlear nucleus
BIC	brachium of the inferior colliculus
CB	calbindin
CGRP	calcitonin gene-related peptide
CN	cochlear nucleus
CNQX	6-cyano-7-nitroquinoxalin-2,3-dione
CR	conditioned response
CS	conditioned stimulus
CSD	current source density
DCN	dorsal cochlear nucleus
DNLL	dorsal nucleus of the lateral lemniscus
EC	external cortex of the IC
DMPO	dorsomedial periolivary nucleus
GAD	glutamic acid decarboxylase
GABA	γ-aminobutyric acid
IHC	inner hair cell
IB	intrinsic bursting response
IC	inferior colliculus
ICC	central nucleus of the inferior colliculus
ICX	external nucleus of the inferior colliculus
LNTB	lateral nucleus of the trapezoid body
LSO	lateral superior olive
LTP	long-term potentiation
MGB	medial geniculate body
MGD	dorsal division of the medial geniculate
MGM	medial division of the medial geniculate
MGV	ventral division of the medial geniculate
MNTB	medial nucleus of the trapezoid body
MOC	medial olivocochlear system
MSO	medial superior olive
NBIC	nucleus of the brachium of the inferior colliculus
NMDA	N-methyl-D-aspartate
OHC	outer hair cell
PET	positron emission tomography
PIN	posterior interlaminar nucleus
Po	posterior thalamic nuclei
PON	periolivary nuclei
PP	peripeduncular nucleus
PV	parvalbumin
PVCN	posteroventral cochlear nucleus
RS	regular spiking response

TABLE 2.1. *Continued*

SC	superior colliculus
SG	suprageniculate nucleus
SK	Ca^{+2}-activated potassium channel
SOC	superior olivary complex
SPN	superior paraolivary nucleus
SR	spontaneous rate
TRN	thalamic reticular nucleus
US	unconditioned stimulus
VCN	ventral cochlear nucleus
VNLL	ventral nucleus of the lateral lemniscus
VNTB	ventral nucleus of the trapezoid body
5-HT	serotonin

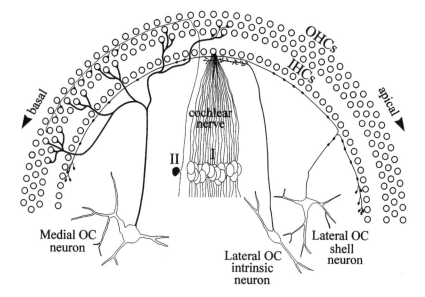

FIGURE 2.1. Output from and input to the cochlea (figure provided by W. Bruce Warr). Two types of auditory nerve fibers, I and II, innervate hair cells. Type I fibers innervate inner hair cells (IHC), and type II fibers innervate outer hair cells (OHC). The two fiber types combine to form the cochlear division of cranial nerve VIII. Three types of neurons in the brainstem provide the olivocochlear (OC) innervation of the hair cell populations. Medial OC (MOC) neurons are located in periolivary nuclei medial to the MSO. In several species, axons of MOC neurons directly contact multiple clusters of OHCs over a wide area of cochlea. Two populations of LOC neurons, located within the LSO (intrinsic neuron) and along the exterior of the LSO (shell neuron), differ in the length of cochlear distance along which they contact type I ganglion cell dendrites. Intrinsic or core neurons innervate a focal group of radial dendrites of the type I AN fibers beneath inner hair cells. In contrast, shell or paraLSO axons bifurcate and innervate fibers beneath large stretches of IHCs. (For abbreviations, also see Table 2.1.)

While their peripheral processes innervate the hair cell receptors, the central processes of type I and II ganglion cells form the auditory nerve: the conduit for carrying information from the cochlea to the brain. AN fibers arising from a single type I ganglion cell terminate solely in the cochlear nucleus (CN) in a variety of nerve terminal morphologies. These range from one of the largest terminals in the brain, the end bulb of Held, located at the rostral tip of the ascending branch, to more typical bouton terminals along the distal portions of the descending branch, to the smallest terminals on the octopus cells of the posteroventral cochlear nucleus (PVCN). In addition to their varied anatomical features, several important biophysical specializations have developed at these synapses. Details of these specializations in the cochlear nucleus and at other brainstem stations and their functional significance are described in Chapter 3. The fibers bifurcate at their entry point into the CN, which splits the ventral CN into anterior (AVCN) and PVCN divisions. The third subdivision, the dorsal cochlear nucleus (DCN), sits atop the PVCN. The different terminal morphologies are associated with different postsynaptic cell types that are spatially segregated in the CN. In the ventral CN, spherical bushy cells are rostral, globular bushy cells are in the middle near the bifurcation zone of AN fibers, multipolar cells are concentrated in the caudal half of the nucleus, and octopus cells are located caudally.

An emerging concept of the 1990's was the VCN core, comprised of the cell types mentioned above which are mostly excitatory (Fig. 2.2). Besides the specialized synaptic properties noted above, other unusual membrane properties that have been developed by VCN core neurons will be described in Chapter 3. Surrounding this core of large neurons is a rind of small, mostly multipolar cells partially embedded in a granule cell lamina that covers most boundaries of the nucleus. The dorsal cochlear nucleus has a layered structure with considerable resemblance to the cerebellum. It's specialized structure, cell types, and potential functional role in audition will be treated in detail in Chapter 5.

Spherical and globular bushy cells are contacted by large end bulb and smaller end bulb terminals, respectively, and provide excitatory drive to the principal cell groups, or core, of the superior olivary complex (Fig. 2.2, SOC). These cell groups, the medial nucleus of the trapezoid body (MNTB), the medial superior olive (MSO), and the lateral superior olive (LSO), are tied into networks subserving the beginning stages of sound localization and will be treated in detail in Chapter 4. Surrounding the core SOC nuclei are peri-olivary nuclei (PON). These neurons, like the rind of the ventral cochlear nucleus, are mostly inhibitory and have been grouped into as many as 11 nuclei. Most prominent among the PON are the superior paraolivary nucleus (SPN), which in rodents rivals or exceeds the primary cell groups in size, the lateral nucleus of the trapezoid body (LNTB) and the ventral nucleus of the trapezoid body (VNTB). Other PON are named based on their spatial relationship to the MSO, which occupies the center of the SOC

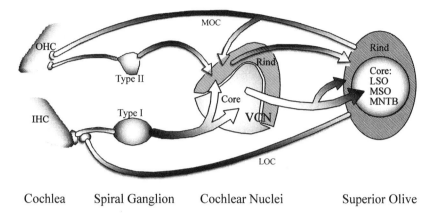

Cochlea   Spiral Ganglion   Cochlear Nuclei   Superior Olive

FIGURE 2.2. Core and rind structure of the lower auditory system. Core neurons innervate both core and rind regions of the next station along the auditory pathway. The core pathway emerges from type I spiral ganglion cells, whose dendrites contact IHCs and axons innervate both core and rind regions of the ventral cochlear nucleus (VCN). Core neurons of the VCN innervate the principal nuclei of the superior olivary complex, which we represent as its core, and periolivary nuclei, which we represent as its rind. VCN rind cells contact lateral and medial olivocochlear (LOC, MOC) neurons, located mostly in the rind of the SOC. MOC cells innervate the VCN rind en route to the cochlea, creating a feedback loop hypothesized to provide an absolute scale for activation of VCN rind cells in relation to sound intensity. (For further abbreviations, see Table 2.1.)

in coronal sections: ventromedial, dorsolateral, dorsomedial, rostral, posterior and posteroventral periolivary nuclei. As we describe below, the LNTB and VNTB, joined in some species by the SPN, contribute significant descending inputs to the CN and all three cell groups send projections forward along the pathway. Of particular note are the MOCs, located in the VNTB, MNTB and SPN, which contact outer hair cells. Their axonal trajectory is dorsal out of the SOC crossing the midline at the floor of the fourth ventricle. This anatomical arrangement has proven convenient for stimulation, recording, and lesion studies of this system. LOC neurons are located within and dorsal to the LSO and project via small fibers more diffusely in a lateral direction prior to joining the MOC fiber bundle.

The output cells of the cochlear nucleus mentioned above, and those of the olivary complex, also provide an ascending influence on the next two stations, the nuclei of the lateral lemnisci and the inferior colliculus (IC). In Chapters 5, 6, and 7 the details of these output pathways and how they influence and shape the operation of these two stations are described and will not be dealt with here.

2. Cochlea to Cortex and Back   11

2.1 The Cochlea: The Output from the Cochlea

The review by Ryugo (1992) earlier in this series provides an excellent description of auditory nerve structure and function, and includes the classical papers on this topic.

The cochlea houses one row of inner hair cells and three rows of outer hair cells. Each population of hair cells contacts the peripheral processes, called radial fibers, of a separate population of spiral ganglion cells. The ganglion cell central processes form the auditory portion of cranial nerve VIII, the vestibulocochlear nerve, which enters and terminates entoto in the cochlear nucleus. The large majority of the ganglion cells are classified as type I. Because of their large diameter, type I axons have been well characterized physiologically, using both extra- and intraaxonal recording methods. The radial fibers and central projections of type I axons have been labeled using tract tracing molecules. The radial fiber contacts one inner hair cell in most species. In cats, high spontaneous rate (SR) fibers contact the lateral side of the inner hair cell; low and medium SR fibers contact the hair cell's modiolar side. High SR neurons are located dorsally in the spiral ganglion and low to medium SR neurons are located ventrally (Liberman 1982; Leake and Snyder 1989; Liberman et al. 1990; Kawase and Liberman 1992; Leake et al. 1992; Merchan-Perez and Liberman 1996; Tsuji and Liberman 1997). Therefore, it appears that each hair cell drives both low, medium, and high SR spiral ganglion cells that have distinctly different thresholds and which, only as a population, can account for the dynamic range of audition. Hair cell and auditory nerve biologists have not investigated the mechanisms by which one hair cell can generate a high SR and particular threshold in one AN fiber and a low SR and different threshold in another. Such a revelation is of fundamental importance to understanding basic elements of auditory perception, such as intensity discrimination.

Evidence indicates that the neurotransmitter glutamate is released by hair cells to activate AN fibers and by AN fibers to activate cochlear nucleus cells (see below). In a guinea pig preparation, application of glutamate in vitro induced calcium currents (Harada et al. 1994) and action potentials (Ehrenberger and Felix 1991) in spiral ganglion cells. In human cochlear sections monoclonal antibodies to glutamate were seen adjacent to outer and inner hair cells (Nordang et al. 2000). Aspartate is released into cochlear fluids in response to sound and may also function in hair cell to AN synaptic transmission (Jager et al. 1998).

The large diameter type I axons head centrally to the CN nucleus where they bifurcate, giving rise to an ascending branch that heads into and innervates AVCN cells and a descending branch that innervates PVCN and DCN cells. The projection pattern of AN fibers to the cat DCN and VCN have been described in detail (Fekete et al. 1984; Rouiller et al. 1986; Ryugo and Rouiller 1988; Sento and Ryugo 1989; Liberman 1991a, 1993) and there are some spontaneous rate dependent differences. In the AVCN, low SRs have the greatest number of terminals and contact the most cells, and high SR fibers the least. The small cell cap or rind region of the VCN is innervated mostly by low and medium SRs. Within the core region, two of the principal cell types, AVCN multipolar cells and globular bushy cells, are contacted mostly by low-medium and high SR fibers, respectively.

Not all AN descending branches reach the DCN. Those that do don't reach into the molecular layer and superficial granule cell areas. Electron microscopy studies indicate that the descending AN fiber branches don't synapse on deep DCN giant cells and fusiform cell bodies in the fusiform cell layer. AN fiber branches do synapse on multipolar, globular, granule (in deeper layers), and octopus cell bodies, with the largest terminals on globular and the smallest on octopus cells. Ryugo and May (1993) looked more specifically at axons to the cat DCN and found terminals in layer III (deep DCN) arranged tonotopically with the axon collateral fields flattened in the "transstrial axis." Snyder et al. (1997) did multiple focal cochlear injections and looked at sheets of type I axons. When comparing the features of axonal sheets produced in the CN (length = rostrocaudal dimension, width = mediolateral dimension, thickness = dorsoventral dimension) they noted that AVCN sheets were thicker than PVCN or DCN but two sheets were separated by larger distances. They also noted that midfrequency lamina were larger in width and length. Similar innervation patterns were found in guinea pigs (Tsuji and Liberman 1997) indicating that there may be a fundamental organizing principal.

The best-studied AN terminal is the endbulb: the specialized terminals that contact bushy cells of the AVCN. Using a measure called the form factor (endbulb silhouette area/perimeter), Sento and Ryugo (1989) noted that high SR terminals had more complex shapes than low to medium SR terminals and that SBCs are often contacted by two endbulbs with similar form factors. In a subsequent paper (Ryugo et al. 1996) it was reported that high SR terminals had larger mitochondria, smaller but more numerous synaptic specializations, more synaptic vesicles and greater curvature. Serial recontruction of single globular bushy cells indicated that there are from four to fourteen calyces on each globular bushy cell (Spirou 1999).

What is the neurotransmitter for the auditory nerve? Using glutamate and glutamine antibodies, Hackney et al. (1996) showed that the antibody label was over "primary" terminals, indicating that AN terminals are glutamatergic. Whole cell patch recordings from rat bushy cells in vitro revealed that a large N-methyl-D-aspartate (NMDA) component is largely replaced by an 2-amino-3-hydroxy-5-methyl-4-isoxazole propionic acid (AMPA) component during development (Bellingham et al. 1998). The remaining NMDA component may undergo a subunit composition change. The subunit composition of AMPA receptors appears to be mainly GluR3 and GluR4 according to electron microscopic studies (Wang et al. 1998) of the rat endbulb/bushy cell terminal. Additional evidence from physiological data (Ferragamo et al. 1998) indicated that the auditory nerve inputs to mouse T stellate cells mostly use AMPA receptors as well.

What are some outstanding issues regarding AN fibers and their innervation of the CN? One pertains to the ganglion cells that innervate single inner hair cells. Depending on species, several type I ganglion cells innervate a given hair cell (Fig. 2.1). It is not known why this occurs or whether these ganglion cells from a given hair cell converge or diverge in the cochlear nucleus. It would be of considerable interest to label all of the auditory nerve inputs to a globular bushy cell and then determine if they belonged to spiral ganglion cells that innervated the same hair cell. A transynaptic, viral marker that labeled all postsynaptic elements when injected into a single hair cell, or that labeled all presynaptic elements when injected into a globular bushy cell, would be two useful approaches to this question.

Type II auditory nerve cells are poorly understood because, in contrast to type I neurons, their cell bodies and nerve terminals are small and their axons are thin and unmyelinated. Type II peripheral processes travel basally from their cell body to innervate multiple outer hair cells in all three rows (Fig. 2.1) and may give off collaterals that innervate the non-neuronal cochlear Deiters and Hensons cells (Burgess et al. 1997; Fechner et al. 1998). Small extracellular HRP injections into the rodent cochlea (Brown et al. 1988a; Brown and Ledwith 1990; Berglund and Brown 1994; Berglund et al. 1996; Hurd et al. 1999) revealed that type II axons enter the CN, bifurcate, and follow a course similar to type I axons except that they rarely enter the DCN and their axonal swellings are often in CN areas that have high granule cell concentrations. Type II axons in cat followed a similar course with few branches, many en passant swellings but not many synapses (Ryugo et al. 1991; Morgan et al. 1994). Determining when and to what auditory stimuli these type II fibers respond and characterizing the function of their input to the cochlear nucleus are issues of considerable interest. No one has ever recorded from, labeled, and recovered one of these fibers. Only one third of spiral ganglion cells antidromically activated from the AN root and presumed to be type IIs responded to noise (Brown 1994), while the others didn't respond to tones or noise. One hypothesis as to their function will be described below.

2.2 The Olivocochlear System: The Input to the Cochlea

Basic organizational features of the olivocochlear system were nicely reviewed by Warr in an earlier publication in this series (Warr 1992). Two separate systems, designated lateral (LOC) and medial (MOC) olivocochlear systems based on the cell body location relative to the medial superior olive, comprise this feedback population whose axons project to different hair cell populations in the inner ear (Fig. 2.1). In the cochlea, LOC neuron terminals end primarily on peripheral processes of Type I spiral ganglion cells immediately below the inner hair cells. The terminals of the MOC neurons end directly on outer hair cells. During the past decade, roles for the olivocochlear system in development and frequency-dependent protection from loud sounds have been added to the long-standing hypothesized role in preserving dynamic range in noisy environments (Liberman 1991b; Liberman and Gao 1995; Rajan 1995, 1996; Walsh et al. 1998; Vetter et al. 1999). New diagnostic tests of peripheral hearing have emerged that are based on anatomical demonstrations of a crossed pathway whereby sounds in one ear can modulate cochlear function in the contralateral ear by olivocochlear activation. All of these effects are attributable to the MOC system; therefore the function of the LOC system remains elusive.

2.2.1 The Lateral Olivocochlear System

As earlier work suggested (Brown 1987), there are two populations of LOC neurons (Fig. 2.1): one located within the LSO or its hilus (intrinsic or core neurons), and a second group of larger cells located mostly dorsal to the LSO outside its fiber capsule (para-LSO or shell neurons) (Campbell and Henson 1988; Vetter and Mugnaini 1992; Warr et al. 1997; Azeredo et al. 1999). Most LOC cell axons project to the ipsilateral cochlea although the relative percentages of ipsilateral versus contralateral varies between species. On its course to the cochlea the axon heads through the CN, giving off few or no collaterals depending on species, and into the vestibular division of the vestibulocochlear nerve. Core LOC neurons innervate a focal group of radial dendrites of the type I AN fibers beneath inner hair cells. In contrast, para-LSO axons bifurcate and innervate fibers beneath large stretches of IHCs. Most LOC neurons innervate the cochlea exclusively, but some also send a projection to the inferior colliculus (Riemann and Reuss 1998). The primary neurotransmitter for the LOC neurons is acetylcholine (ACh). The type I ganglion cells, on which these LOC terminals are found, express $\alpha6$, $\alpha7$, and $\beta2$ receptor subunits (Elgoyhen et al. 1994; Morley et al. 1998), so the LOC terminals are probably acting on receptors containing one or more of these subunits. The only receptor subunits expressed by outer hair cells (which receive efferents from the MOC system) are the $\alpha9$ and $\alpha10$ subunits (Elgoyhen et al. 1994; Vetter et al. 1999; Elgoyhen 2001). Knockout mice that lack the $\alpha9$ subunit show a complete absence of the classical effects of electrically shocking the OCB, leaving the functional effects of the LOC a mystery. The synaptic terminal location on peripheral process of the type I spiral ganglion cells indicates that LOC terminals might modulate the spontaneous and/or driven firing patterns of type I AN fibers. Other neurotransmitters detected in LOC neurons include γ-amino butyric acid (GABA), dopamine, calcitonin gene-related peptide (CGRP) and CGRP-like neuropeptides, dynorphins, and enkephalins (Eybalin 1993). GABA-immunoreactive LOC neurons can also contact peripheral processes of type II ganglion cells (Sobkowicz et al. 1998). Warr et al. (1999)

present arguments that the shell neurons are the source of the LOC terminals that take up GABA.

Not much is known about the LOC cell inputs. Modern tract tracing methods indicate a dendritic input from stellate cells in the ipsilateral PVCN (Thompson and Thompson 1991), noradrenergic and serotonin (5-HT) inputs (Woods and Azeredo 1999), and a projection from the IC (Vetter et al. 1993) and auditory cortex (Feliciano and Thompson 1995).

Just as elusive as the LOC cell anatomical features are their physiological characteristics. No recordings from LOC axons or cells have been reported in vivo, so the auditory response features are unknown. Recent in vitro experiments from rat LSO (Adam et al. 1999) showed that cells anatomically resembling LOC neurons had onset responses to depolarizing currents. In contrast, Fugino et al. (1997) reported that neonatal rat MOC and LOC neurons both showed tonic spike responses. Different animal ages and recording techniques (sharp versus patch electrodes) probably account for these discrepancies. Thus, a breakthrough in the function of the LOC system in the cochlea and experiments to determine if the two proposed cell populations within the LOC system are performing different tasks would be of major importance.

2.2.2 The Medial Olivocochlear System

The neuron of the MOC are situated medial to the MSO, are larger than most LOC cells and, in most species, are scattered in the middle and rostral periolivary regions of the SOC. In mouse, rat, and guinea pig, MOC cells reside primarily in a periolivary nucleus: the VNTB (Aschoff and Ostwald 1987; Campbell and Henson 1988; Vetter and Mugnaini 1992), while in chinchilla they are found predominantly in the dorsomedial periolivary nucleus (DMPO) (Azeredo et al. 1999). Direct effects of MOC stimulation on outer hair cells, monitored by intracellular recording, show that the normally excitatory neurotransmitter ACh elicits an inhibitory synaptic event. This occurs because calcium (Ca^{+2}) influx opens a Ca^{+2}-activated potassium (SK) channel that hyperpolarizes the cell (Oliver et al. 2000). A subsequent reduction in basilar membrane motion occurs that alters AN fiber activity and reduces the dynamic range of hearing.

MOC axons emerge from the SOC dorsally, decussate in the fourth ventricle floor, and join LOC axons to become part of the vestibular nerve. As MOC axons exit the brainstem, collateral branches are sent to CN areas that contain high granule cell concentrations and to the inferior vestibular nucleus (Benson et al. 1996; Benson and Brown 1996). At the midline decussation and in the vestibular nerve, these large axons are accessible for recording, stimulation, and blockade; hence we understand the MOC much better than the LOC system.

Several studies by Brown and colleagues (Brown et al. 1988b; Benson and Brown 1990; Benson et al. 1996) indicated that MOC axon terminals in the

CN can be on multipolar cell dendrites that are also receiving inputs from unmyelinated type II spiral ganglion cell axons. Low spontaneous rate type I AN fibers also synapse preferentially in these regions. Benson and Brown, in their 1990 paper, proposed a function for these MOC terminals. They thought that the decreased output of the cochlea and AN fibers, which was due to the activation of the MOC system's action on the outer hair cells, might be compensated for in these stellate cells by an increased excitatory drive from the MOC collaterals. Thus, these cells would respond at a given sound level even when the cochlea had been turned down by the MOC system. In vivo recordings from guinea pig MOC axons (Brown et al. 1998a) revealed that they respond better to noise than tones, have wide dynamic ranges, and could participate in intensity encoding at medium/high sound levels. Retrograde labeling of MOC neurons combined with anterograde labeling of PVCN neurons indicates that ascending MOC cell inputs are primarily from stellate cells in the contralateral cochlear nucleus (Thompson and Thompson 1991), many of which are in the marginal shell region (Ye et al. 2000). These stellate cells have wide dynamic ranges (Ghoshal and Kim, 1996) and innervate MOC (as well as LOC) neurons. Therefore, stellate cells of the CN rind may determine the wide dynamic range of MOC neurons and complete the ascending branch of the intensity encoding circuit called the "feedback gain control system" (Ye et al. 2000). If CN rind stellate cells are encoding absolute intensity, then one would expect significant interactions with other auditory brainstem and midbrain neurons.

Rat MOC neurons receive descending inputs from the central nucleus of the IC (ICC) and the external cortex of the IC (EC) (Caicedo and Herbert 1993), as well as the auditory cortex. Labeled corticofugal terminals were seen apposed to backfilled MOC neurons in the VNTB (Mulders and Robertson 2000) indicating that the cortex can influence auditory function as far peripherally as the cochlea. Retrograde labeling of MOC cells from the cochlea, combined with anterograde labeling of the outputs of the ICC, showed that the IC probably directly influences MOC cells (Thompson and Thompson 1993; Vetter et al. 1993). Recent reports give physiological evidence of this IC influence (Ota and Dolan 1999; Scates et al. 1999). Electrical IC stimulation results in reduction of the cochlear compound action potential in a frequency specific manner (Ota and Dolan 1999) and alters distortion product otoacoustic emissions (Scates et al. 1999), most likely via the MOC pathway. Immunohistochemical studies showed that MOC cells also get several other inputs which, like LOC cells, include a seretoninergic source (Thompson and Thompson 1995). In vitro recordings in slices (Wang and Robertson 1997b, 1998a) revealed that norepinephrine and substance-P influence the responses of these MOC cells.

Because of the size of their axons and cell bodies, a lot of physiological data has been collected from positively identified members of this cell population in vivo and in vitro. In vivo recordings from cat and guinea pig (Robertson and Gummer 1985; Liberman and Brown 1986; Liberman 1988;

Brown 1989) showed that MOC cells are monaurally driven (usually by the same ear that they innervate), display regular spiking, and have high thresholds. The cells that innervate the same ear that drives them are usually located in the brainstem contralateral to that ear. Those cells that innervate the opposite ear that drives them are usually located ipsilateral to that ear. Although MOC neurons are usually driven only monaurally, noise or tones to the opposite ear can facilitate the response. Subsequent to this group of studies, a report by Brown and others (1998b) demonstrated that MOC neurons could change their acoustic responsiveness based on their stimulus history. This finding may explain why electrical or sound-evoked MOC neuron activation can exert protective effects against subsequent exposure to loud sound, especially at high frequencies (Canlon et al. 1988; Rajan and Johnstone 1988; Liberman 1991b; Liberman and Gao 1995; Rajan 1995, 1996; Kujawa and Liberman, 1997). Most likely several cochlear mechanisms are at play, potentially including second messenger systems (Sridhar et al. 1995) and the synthesis of heat shock proteins (Yoshida et al. 1999). These recent experiments showed that stressful situations increased the amount of mRNA in the cochlea (for the synthesis of heat shock proteins) and were correlated with subsequent cochlear protection from noise damage (Yoshida et al. 1999).

Olivocochlear neurons reach the cochlea prior to maturation of cells in the organ of Corti and thus may also be important in cochlear development. An intact olivocochlear system during development is crucial for establishing sensitivity and tuning of AN fibers (Walsh et al. 1998). Removing innervation by making a genetic knockout of the α9 cholinergic receptor (postsynaptic to MOC neurons) has less severe effects than transection of fibers (Vetter et al. 1999), probably because neurotransmission per se is not the crucial factor in normal development.

The function of the MOC system in humans has been studied using two methods, one involving otoacoustic emissions. Otoacoustic emissions are spontaneous (SOAE) or evoked (EOAE) sounds originating in the inner ear presumably generated by the motile activity of outer hair cells. The click-evoked EOAE and SOAE amplitude can be reduced by either electrical or sound stimulation of the contralateral ear (Collet et al. 1990). These effects are eliminated when the vestibular nerve, the olivocochlear system pathway to the cochlear, is cut (Williams et al. 1994; Giraud et al. 1997a, b), indicating that the cut axons of the activated MOC cells projecting to the opposite ear lose their influence. Thus, auditory or electrical stimulation of the contralateral ear provides a means of investigating the uncrossed MOC system. These findings have been verified and extended in animals. May et al. (1995) showed that bilateral OCB section reduced a cat's ability to detect intensity changes of midfrequency (8 kHz) signals in noisy backgrounds. Hienz et al. (1998) lesioned cat olivocochlear axons in the floor of the fourth ventricle and saw evidence that one MOC function may be to "enhance speech processing in noise."

2.3 Deafness and Cochlear Connexin

During the last decade, a greater understanding has emerged of the potential role of the cochlear protein Connexin in congenital hearing loss. Genetic abnormalities are estimated to account for about 50% of all severe or profound childhood deafness (Morton 1991). Although genetically induced deafness can be one of several symptoms making up a syndrome (syndromic deafness), in more than two thirds of cases it occurs as a single symptom (nonsyndromic hearing loss). A large number of identified mutant genes can give rise to the syndromic or nonsyndromic forms (Keats and Berlin 1999; Steel and Bussoli 1999; Sundstrom et al. 1999; Rabionet et al. 2000). Mutations in one gene in particular, the autosomal (on chromosome 13) recessive (deleterious only in homozygotes) gene *GJB2*, may account for over half of the genetic, nonsyndromic childhood deafnesses. This gene, that encodes the production of the gap junction protein connexin26, was found to be mutated in members of several families with deafness (Kelsell et al. 1997; Cohn and Kelley 1999). Connexins are proteins that form transmembrane channels called gap junctions connecting the interiors of neighboring cells. In the cochlea connexin26 and other connexin forms are widely expressed (Kikuchi et al. 1995; Kelsell et al. 1997; Lautermann et al. 1998, 1999; Xia et al. 1999; Frenz and Van De Water 2000) and associated with gap junctions. Both nonsensory epithelial cells and connective tissue cells of the cochlea *but not* the sensory hair cells were connected by gap junctions. Why would a malfunction in gap junctions between nontransducing cells of the cochlea generate deafness? Kikuchi and others (1995) suggested that the serial arrangement of the junctions between the cochlear supporting cells is a way for potassium ions to recirculate back into the endolymph. Potassium is kept at high concentrations in the scala media endolymph by the stria vascularis and is used by hair cells to generate a transducer current after which it is removed from the hair cells through their basolateral membrane. The gap junction network between various cochlear supporting cells might provide a pathway for this "spilled" potassium to be recycled back to the stria where it could be resecreted into the endolymph. Malfunction and/or malformation of connexin through a genetic mutation could upset this potassium homeostatic mechanism to retard cochlear output.

3. The Periolivary Nuclei

New information on the major brainstem nuclei (LSO, MSO, MNTB) will be described in Chapters 3 and 4 and, for those nuclei, the reader is referred to those chapters.

Surrounding the major auditory brainstem nuclei are a set of nuclei collectively designated as the periolivary nuclei (PON). Numbering up to eleven in some schemes, the boundaries of some of these cell groups are less well defined that those of the major nuclei. Most PON cells are GABA and/or glycine immunoreactive, and a subset of cells is cholinergic and forms the MOC system. Roughly one fourth of the ascending input to the IC and three fourths of the descending inputs to the cochlear nucleus from the SOC originate from the PON. Recent tract tracing studies have revealed that PON also innervated the principal nuclei of the SOC. Different PON (and cell types within individual PON) probably serve unique functions in hearing based on their differences in anatomical, biochemical and physiological features. Similar to CN organization, the PON constitute the rind of the SOC and, as in the CN, many cells have inhibitory actions on their targets.

3.1 The Ventral Nucleus of the Trapezoid Body

The VNTB is a heterogeneous collection of cell types embedded within trapezoid body axons ventral and medial to the MSO and below the MNTB. The rodent VNTB extends lateral to the MSO. There is anatomical evidence for a tonotopic, lateral to medial arrangement of low to high frequencies (Caicedo and Herbert 1993; Warr and Beck 1996). Tonotopy is not revealed using *cfos* expression (Saint Marie et al. 1999), so perhaps not all VNTB cell types conform to this organizational pattern. VNTB cell outputs influence several upstream and downstream auditory structures. Histological techniques suggest that VNTB cells are GABA and/or glycinergic or cholinergic, with the latter forming an important component of the MOC system. Intracellular labeling (Robertson 1996) reveals multipolar cells that can have thin, smooth tapering or thick, long, spiny dendrites.

Ascending VNTB inputs arise from the contralateral CN (Spirou et al. 1990; Kuwabara et al. 1991; Smith et al. 1991, 1993; Thompson 1998) in the form of stellate and bushy cell collaterals that are probably excitatory (Smith and Rhode 1989; Sanes 1990; Banks and Smith 1992; Wu and Kelly 1992). Stimulation of the midline trapezoid body in slices, presumably activating both stellate and bushy cell axons, elicits glutamatergic epsps in VNTB cells (Robertson, 1996) that were blocked with the AMPA channel antagonist 6-cyano-7-nitroquinoxalin-2,3-dione (CNQX). Shock-induced ipsps indicate that either there are also inhibitory inputs to VNTB or that the excitatory input activates inhibitory VNTB cells with local collaterals. VNTB cells are also sensitive to noradrenaline and 5-HT, and the peptides substance P and enkephalin, but sources of these inputs are unknown (Wang and Robertson 1997a,b, 1998a,b; Wynne and Robertson, 1997).

The VNTB is the preferred territory for descending projections to the superior olive, receiving inputs from the IC, medial geniculate, and auditory cortex (Feliciano and Thompson 1995; Kuwabara 1999; Schofield and Cant 1999; Mulders and Robertson 2000). These projections terminate primarily in the ventral half of the nucleus and define a dorsoventral axis of organization within the VNTB. The VNTB, along with the rostral PON, also serves as a preferential site for the termination of IC axons, which contact cells that project to one or both cochlear nuclei (Schofield and Cant 1999). The effect of descending projections on cellular activity within the VNTB remains unexplored.

In addition to housing members of the MOC system that project to both CN and cochlea, the rat VNTB contains populations of cells that innervate both CN, ipsilateral IC, contralateral LSO, and, via intrinsic collaterals, the VNTB itself (Sherriff and Henderson 1994; Robertson 1996; Warr and Beck 1996; Ostapoff et al. 1997b). The locations of cell populations that project to different sites are segregated but overlapping in the nucleus. Cells projecting to cochlea and IC are located more dorsally than cells projecting to CN or those sending their axons within the SOC (Warr and Beck 1996). VNTB cells projecting to IC could be glycinergic (Saint Marie and Baker 1990), while those using either of three neurotransmitter systems (acetylcoholine, GABA, and glycine) project to the cochlear nucleus. It is not known which CN cells are affected by these neurotransmitters (Sherriff and Henderson 1994; Ostapoff et al. 1997a). Acetylcholine, perhaps released by VNTB cell terminals, can alter the spontaneous firing rates of several CN cell types (Chen et al. 1998; Fujino and Oertel 2000).

VNTB cells show different anatomical and intrinsic membrane features that may correlate with their axonal output. Intracellular recordings from rat VNTB cells (Robertson, 1996; Wang and Robertson 1997b, 1998a) have distinguished three kinds of action potential afterhyperpolarization waveforms (AHP1, 2, and 3) that were correlated with cell morphology. Based on axonal trajectories, Robertson (1996) surmised that AHP 1 cells may be those projecting locally to other SOC nuclei like the LSO, while AHP2 may contain those VNTB cells projecting to the cochlea or IC, and, finally, AHP3 cells may be those heading to the CN. In summary, VNTB cells with different innervation targets may be anatomically, neurochemically, and physiologically distinct, but the functional dissection of this nucleus is far from complete.

3.2 The Lateral Nucleus of the Trapezoid Body

The LNTB is ventral to the LSO and lateral to the MSO. Based on shared input from primarily ipsilateral VCN core neurons (Tolbert et al. 1982; Spirou et al. 1990, 1995; Smith et al. 1991) the LNTB may be parceled into three subregions (Spirou and Berrebi 1996), although one subregion the posteroventral LNTB (pvLNTB) is considered by some to be a separate nucleus (posteroventral PON, Warr 1969; Schofield and Cant 1992). Excitatory CN inputs are balanced by likely inhibitory inputs from VNTB neurons (Warr and Beck 1996) and local LNTB cell collaterals (Spirou, unpublished). Most cat LNTB cells are glycine immunoreactive (nearly one half co-localize glycine and GABA) and a small percentage, found mostly near the LSO ventral hilus, are immunonegative. Cells of the pvLNTB resemble MNTB principal cells in their morphology, responses to intracellular current injection, and glycine immunoreactivity. In addition, they are contacted by large nerve terminals that probably arise from globular bushy cells, although large terminals are found throughout the LNTB (Smith et al. 1991; Spirou et al. 1995, 1998). The remaining LNTB cells are multipolar in shape and constitute the main LNTB subnucleus (mLNTB). mLNTB neurons and some posteroventral PON cells in rodents provide the predominant descending inhibitory projections to all subdivisions of the CN (Schwartz 1992; Spirou et al. 1995; Ostapoff et al. 1997a; Thompson and Schofield 2000) and are probably the first descending projections that are activated by sound. The LNTB, like the VNTB, has intrinsic projections to the SOC. A substantial projection is made to the ipsilateral MSO (Cant and Hyson 1992). Electrical activation of the lateral trapezoid body induces glycinergic inhibitory synaptic activity in MSO and LSO cells, presumably due to LNTB inputs (Grothe and Sanes 1993; Wu and Kelly 1994). Many of these cells may project to both MSO and LSO, where they terminate in perisomatic endings (Kuwabara and Zook 1992). Thus both the LSO and MSO, in addition to receiving contralateral inhibition via MNTB cells, may also be receiving an ipsilateral inhibitory input by way of LNTB cells. The role of ipsilateral inhibition in the function of these cell groups has yet to be explored. LNTB cells make a minimal contribution to the ascending auditory pathway with only sparse projections to the IC (Schofield and Cant 1992), and none of these originate in the pvLNTB (Adams 1983). The IC, in turn, sends very few descending projections to the LNTB. Therefore, anatomical properties support physiological descriptions of this nucleus as being monaurally activated (Guinan et al. 1972a,b; Tsuchitani 1977), then providing rapid feedback inhibition to the CN (Spirou et al. 1995).

3.3 The Superior Paraolivary Nucleus

SPN is a large, well-defined periolivary nucleus sitting dorsal and medial to the MSO and dorsal to the MNTB. It is especially well developed in rodents and guinea pigs (Ostapoff et al. 1990; Saint Marie and Baker 1990; Saldana and Berrebi 2000) and is thought to be the homolog of the less well developed DMPO in other mammals (Morest 1968). The significance of this over development in rodents is unknown. SPN neurons are driven by octopus cells, which form large terminals, and stellate cells, that form small terminals, of the contralateral CN (Schofield 1995) and are contacted by substance P immunoreactive terminals of unknown origin (Reuss et al. 1999). Intrinsic projections from other SOC cells arise from presumably excitatory MSO cell collaterals and tonotopic inhibitory projections arise from MNTB cells (Kuwabara and Zook 1991, 1999; Banks and Smith 1992; Sommer et al. 1993). Therefore, in addition to the direct excitatory input from octopus and stellate cells in the contralateral CN, the activity of spherical bushy cells bilaterally would excite SPN cells through their connections with the MSO. The activity in globular bushy cells also located in the contralateral cochlear nucleus would inhibit SPN through their activation of MNTB.

Cell morphology is variable in guinea pigs and gerbils, but correlates with axonal projection patterns. Neurons forming descending projections to the CN tend to be smaller and rounder than those whose axons ascend to the IC, and only a small percentage of cells project to both territories. Connections to CN and IC are primarily ipsilateral and are both glycinergic and GABAergic, arising from distinct classes of cells (Helfert et al. 1989; Ostapoff et al. 1990, 1997a; Schofield and Cant 1992). The IC projection is tonotopic and widespread, innervating central nucleus as well as the dorsal and external cortices (Gonzalez-Hernandez et al. 1996; Kelly et al. 1998; Cant et al. 1999; Fuentes et al. 1999). In most species, the SPN contains MOC neurons (Aschoff and Ostwald 1987; Campbell and Henson 1988; Vetter and Mugnaini 1992; Azeredo et al. 1999) although there is one conflicting report (Riemann and Reuss 1998). In rats, the neurons are homogeneous with dendritic trees flattened along the tonotopic axes. All SPN neurons in this species project to the ipsilateral IC but not the CN or cochlea, are GABAergic, and deliver local collaterals within the SPN (Kulesza and Berrebi 1999, 2000; Kulesza et al. 2000; Saldana and Berrebi 2000).

Some data (from rodents) are available on the physiology of these cells in vivo and their responses to sound (Finlayson and Adam 1997; Dehmel et al. 1999, Behrend et al. 2000). Cells could be monaural, primarily from the contralateral ear, or binaural. Binaural units showed little sensitivity to interaural time difference cues and low best-frequency units showed little phase locking. The function of the large GABAergic SPN outputs to either the CN or IC is unknown and the physiological features of these cells require further characterization.

4. The Nucleus of the Brachium of the Inferior Colliculus

Details the anatomy, physiology and possible functional role of cells comprising the nuclei of the lateral lemniscus and the IC, together with their afferent and efferent connections, are covered in Chapters 5, 6, and 7.

The IC is a major center along the ascending pathway. Axons running between the IC and auditory thalamus (medial geniculate, MGB) are collectively known as the brachium or arm of the IC. Cells composing the nucleus of the brachium of the inferior colliculus (NBIC) lie within this bundle. In 1985, Kelly and Judge noted that complete bilateral destruction of the MGB or auditory cortex had little effect on a rat's ability to perform a sound localization task. When the lesion included the brachium of IC, a severe deficit in performance occurred indicating that the NBIC might play an important role in sound localization. As described below, this role might be in helping to create the neural representation of auditory space in the mammalian superior colliculus (SC).

In the cat SC, Middlebrooks and Knudsen (1984) concluded that if one measured the point in space that elicits a cell's maximum firing rate, then there is a class of cells whose locations map out a orderly physical representation of auditory space (Fig. 2.3). A similar map is found in the barn owl optic tectum, the bird SC homologue (Knudsen 1982). In this system, the source of this map is the external nucleus of the inferior colliculus (ICX), the site of a similar map, and the only major source of auditory input to the SC (Knudsen and Konishi 1978; Knudsen and Knudsen 1983). In mammals, this does not seem to be the case. Although a topographic rep-

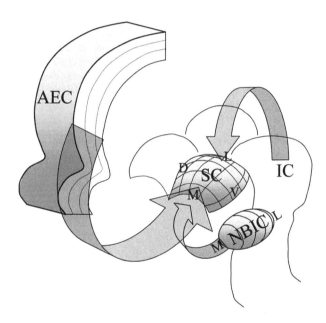

FIGURE 2.3. Potential major sources of auditory inputs to the space map in the superior colliculus. The external nucleus of the inferior colliculus (IC) has topographic representation of space and provides an input to the superior colliculus (SC) but loss of this input does not affect the SC space map. The nucleus of the brachium of the IC (NBIC) also provides an input to the SC and spatially tuned cells are laid out to create a topographic ordering of azimuth from medial (M) to lateral (L) auditory space along the rostrocaudal axis of the nucleus. Loss of this area can lead to localization deficits. The anterior ectosyvian cortex (AEC) also sends a major projection to the SC. Elimination or inactivation of this input eliminates the ability of cells in the SC space map to carry out sensory integration of multimodal sensory cues arriving from the same or separate points in space. See text for details. The grid on the SC represents the cellular map of sensory space. (D, dorsal; V, ventral)

resentation of space is found in ICX (Binns et al. 1992) that provides input
to the SC, loss of this input does not affect the SC space map (Binns 1991).
Another major auditory input to the mammalian SC is the ipsilateral NBIC
(Fig. 2.3). Backfilling cells projecting to SC showed a strong, spatially
ordered, ipsilateral NBIC projection (King et al. 1998). Extracellular
recordings from NBIC cells (Aitkin and Jones 1992) revealed a large percentage of cells (compared to downstream IC and upstream MGB) with
discrete spatial receptive fields. When these spatially tuned cells were
mapped (Schnupp and King 1997), a topographic ordering of azimuth along
the rostrocaudal axis of the nucleus was noted. Azimuthal tuning was
broader and less stable than in SC cells, leading the authors to postulate a
further refinement of the NBIC input to form a SC space map. It should be
noted that Thiele et al. (1996) saw a similar NBIC to SC projection in an
echolocating bat where no SC map of space has been reported.

What auditory inputs could potentially help to shape such a topographic
map in NBIC? The IC gives rise to a strong ipsilateral and a weak contralateral NBIC input but it is not known whether this is excitatory,
inhibitory, or both (Kudo and Niimi 1980; Wenstrup et al. 1994). A
GABAergic projection from the dorsal nucleus of the lateral lemniscus
(DNLL) and a descending, excitatory cortical input has also been demonstrated (Diamond et al. 1969; Andersen et al. 1980c; Whitley and Henkel
1984; Winer et al. 1998). Nonauditory inputs may serve to modify the map
or perform different functions. Axons of cells in the accommodation region
of the SC (an area which changes the shape of the pupil) leave terminals
in NBIC (Sato and Ohtsuka 1996) and there is a descending noncholiner-
gic input from globus pallidus (most globus pallidus outputs are cholinergic) (Shinonaga et al. 1992; Shammah-Lagnado et al. 1996). Some terminals
from the cat dorsal column nuclei also end in NBIC (Bjorkeland and Boivie
1984), a finding that is consistent with a report of some NBIC cells that are
responsive to tactile as well as auditory stimuli (Blomqvist et al. 1990).

NBIC cells project to other areas in addition to the SC projection, but
the actions of these projections are unknown. Kudo et al. (1983) demonstrated heavy terminal labeling in the midbrain reticular formation and
anterior pretectal nucleus. NBIC also sends a nonGABAergic projection to
regions of the auditory thalamus (Calford and Aitkin 1983; Itoh et al. 1984;
Kudo et al. 1984; Peruzzi et al. 1997) and to layer I of auditory cortical areas
(Mitani et al. 1984, 1987; Winer 1985; Rouiller et al. 1989).

5. The Auditory Cortex and Thalamus

More so than any two structures within the classical boundaries of the auditory pathway, the highest structures, the auditory thalamus and auditory
cortex, are inextricably linked into a functional unit. Over the past decade,
a greater appreciation has developed for the concept that the communication between them is two way and extensive. Not only is the major thalamic

output directed to the cortex, but the major input to the thalamus, in terms of sheer number of synapses, is from the cortex outnumbering ascending sensory endings by four to one. A second concept gaining acceptance is that of multiple pathways in this two-way communication link that may subserve different functions. The complexity of the anatomy and physiology of this system is thoroughly presented in chapters published in earlier volumes in this series (Clarey et al. 1992; Winer 1992) and it would be impossible to review that data as well as incorporate more recent information here. Instead, we provide a brief overview of the anatomy and physiology followed by a selection of what we think are some interesting new findings.

The thalamus contains three regions known to function in hearing. First, the MGB, often designated the auditory "relay" nucleus of the dorsal thalamus. Three MGB subdivisions are distinguished based on cytoarchitecture, fiber connections, and physiology. These three subdivisions, designated ventral (MGV), dorsal (MGD) and medial (MGM), receive, modify and transfer sensory information largely to specific regions of the cerebral cortex but, as we'll see below, to other functionally important areas (e.g. amygdala) as well. The ventral division is considered the principal relay nucleus in the lemniscal or auditory core projection to the cortex. The second region in the thalamus known to function in hearing is the subregion of the posterior group of thalamic nuclei designated Po (Imig and Morel 1985a) located adjacent to the anterior and medial edge of the ventral MGB. Po is contiguous with MGV and similar in its tonotopic organization and cortical connections and is thus thought to be part of the core or lemnical pathway. Finally, the thalamic reticular nucleus (TRN), a sheet-like nucleus with cells situated lateral and anterior to the dorsal thalamus, is also known to be involved in hearing. TRN cells are all GABAergic, receive excitatory inputs from thalamocortical and corticothalamic axons and project to relay nuclei of the dorsal thalamus including the MGB. Details of what is known about the auditory sector of this nucleus are discussed in a separate section below.

5.1 The Medial Geniculate Body

The ventral division constitutes the major conduit for information traveling from IC to auditory cortex. The primary ascending input to MGV is from the IC central nucleus and it is reciprocally connected with the primary auditory cortex. Bushy, spindle shaped cells, whose tufted dendrites in some species are aligned with isofrequency laminae, comprise most of the neurons and project to the cortex with few local collaterals. These cells are complemented by a population of smaller stellate cells whose axons ramify locally and include a subset of neurons that are immunopositive for the GABA-synthesing enzyme glutamic acid decarboxylase (GAD). Relative numbers of these interneurons varies among species. Extracellular studies in cat, rodent, and monkey have shown a predominance of narrowly tuned, shorter latency responses arranged tonotopically in MGV (and Po) (Aitkin and Webster 1971, 1972; Calford and Webster 1981; Calford 1983; Allon and Yeshurun 1985; Imig and Morel 1985b; Bordi and LeDoux 1994; Edeline et al. 1999).

The dorsal division is less homogeneous and is composed of several subdivisions, each with unique anatomical and physiological features. Cells are mostly stellate shaped but tufted neurons are also present. The main ascending input is from the dorsal cortex of the IC. Generally, MGD sends its thalamocortical output to and receives its cortical input from areas caudal to the primary cortex. The same physiological studies noted above indicate that MGD cells are not tonotopically arranged. Some MGD cells, like MGV cells, are sharply tuned and have short response latencies. Many others show broad frequency tuning with longer, more variable latencies and more rapid habituation to repetitive stimuli. (Aitkin and Webster 1971; Aitkin and Prain 1974; Rodrigues-Dagaeff et al. 1989; Bordi and LeDoux 1994; Edeline et al. 1999).

The medial division is also composed of multiple subregions and contains a diverse population of cells, the most notable of which are the magnocellular neurons. The MGM is one of a group of nuclei, often referred to as paralaminar (Herkenham 1980), that is situated medial to the MGB and includes the posterior interlaminar (PIN), suprageniculate (SG), and peripeduncular (PP) nuclei (Winer and Morest 1983; Ledoux et al. 1987; Winer and Larue 1988). The MGM and associated nuclei receive IC input from the external cortex, a region whose cells can also be driven by auditory and somatosensory input (Aitkin et al. 1978), and superior colliculus (Linke et al. 1999). MGM also receives direct somesthetic inputs from the spinal cord (Ledoux et al. 1987). Other more "downstream" cells in auditory structures like the ventral nucleus of the lateral lemniscus (VNLL), DNLL and SOC also impinge on cells here (Henkel 1983; Whitley and Henkel 1984; Bajo et al. 1993; Peruzzi et al. 1997). Given this multisensory convergence, it is not surprising that paralaminar cells can be bimodal (Calford and Aitkin 1983; Benedek et al. 1997). MGM cells project rather diffusely to multiple cortical areas and receive a similar diffuse cortical input in return. As we will see below, an important feature of this thalamocortical MGM projection is that much of it is to cortical layer 1, as compared to layers 3 and 4 for MGD and/or MGV cells. MGM cells also project to other areas including the caudate putamen and amygdala, a connection that, as described below, may be vital for emotional attachments to auditory stimuli. Another interesting feature of the MGM is the plastic nature of its responses, also described below.

5.2 The Auditory Cortex

The auditory cortex (AC) occupies much of the superior aspect of the temporal lobe. In the two species where considerable work has been done, cat and monkey, the AC may be partitioned into core, belt and parabelt regions (Fig. 2.4). Preliminary evidence suggests two streams of information flow within these regions: a dorsal stream for spatial information processing and a ventral stream for pattern (speech, vocalization) processing. Both streams lead out of the temporal cortex to higher order cortical areas (Kaas and Hackett 1998; Rauschecker 1998; Kaas et al. 1999). Physiological mapping of the core or primary cortical area, reveals multiple, cochleotopically orga-

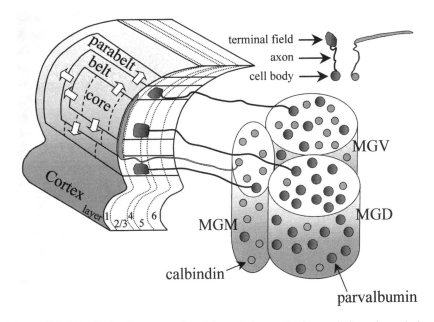

FIGURE 2.4. The thalamic core and matrix and the cortical core, belt and parabelt. Schematic representation of the auditory thalamus and its input to the auditory cortex. Calbindin-labeled cells are thought to form a background "matrix" in the thalamus and are distributed throughout all MGB divisions. The output of these diffusely innervates cortical layer 1 paying little attention to cortical borders. Parvalbumin-labeled cells form the thalamic "core" and are most concentrated in MGV. Core cells project onto the middle layers (3 and 4) of one or a few cortical fields in a precise tonotopic fashion. The auditory cortex may be partitioned into core, belt and parabelt areas. The core is subdivided into multiple independent cochleotopically organized fields (dashed lines). Surrounding the core is the belt region, with several subdivisions (dashed lines) that receive their main inputs from primary cortical areas and nonlemniscal thalamus. The surrounding parabelt region may also have subdivisions that get little input from the core regions or lemniscal thalamus, but a significant input from the belt region and extralemniscal thalamis. There would seem to be a progression of information flow and/or processing from the core area outward through belt and parabelt regions (arrows) and beyond. See text for details. (Numbers in the cortex represent the various cortical layers. For abbreviations, see Table 2.1.)

nized fields. Each of these multiple fields receives different blends of thalamocortical input (Andersen et al. 1980b; Imig and Morel 1984; Morel et al. 1993; Pandya 1995) but, in general, the core area receives its primary thalamic input from MGV. Although some thalamocortical cells innervate multiple primary cortical areas (Imig and Morel 1983), these areas are independent from one another in that destruction of a region containing one tonotopic maps does not compromise the tone-evoked responses of other (Rauschecker et al. 1997).

Surrounding the core is the belt region. Several histological features distinguish this region, and it receives its main inputs not from the MGV but from primary cortical areas and nonlemniscal thalamus (i.e. MGD and MGM). Several belt region subdivisions have been distinguished and named based on their relative locations, and many of them appear to be tonotopically arranged. In monkeys, aspiration lesions of one core area, which had little effect on the pure tone responses of cells in another core area, abolished pure tone responses in one of the belt regions (Rauschecker et al. 1997). This result, combined with connectional data and the fact that first spike latencies tend to be longer in belt regions, indicates that the belt regions represent a second processing stage in the core-belt-parabelt sequence. Adjacent and peripheral to the belt region is the parabelt. This region gets little input from the core regions or lemniscal thalamus but receives a significant input from the belt region and extralemniscal MGD, MGM and paralaminar nuclei (Hackett et al. 1998a,b). So, there seems to be a general progression from the inner core area of the auditory cortex outward through belt and parabelt.

5.3 Ascending Inputs to the Thalamus

As described above, while some of the ascending auditory inputs to the MGB originate from subcollicular structures, most arise from the IC. A feature distinguishing this ascending input from that of other sensory systems is that it contains both an excitatory glutamatergic and an inhibitory GABAergic component. Ascending thalamic inputs to visual and somatosensory thalamus are strictly excitatory. This may belie a different information processing strategy in the auditory thalamus.

Combined tract tracing and immunolabeling studies have revealed details about the inhibitory projection from IC to MGB. Both major IC cell types, disc and stellate, are retrogradely labeled from the medial geniculate (Oliver 1984). The first indication that an inhibitory projection existed was the finding that some disc and stellate cells were immunolabeled using GAD and/or GABA antibodies (Oliver et al. 1994). Likewise, biocytin injections into IC or its brachium (BIC) labeled two types of terminal boutons in the MGB that may represent excitatory and inhibitory inputs (Pallas and Sur 1994; Wenstrup et al. 1994; Malmierca et al. 1997). Colocalization of retrograde label from the MGB and GABA immunoreactivity in individual IC cells revealed definitively that part of the IC to MGB projection was inhibitory (Winer et al. 1996). Similarly, an excitatory projection was indicated by the retrograde labeling of IC cells from tritiated aspartate injections (thought to be taken up by glutamatergic neurons) into the MGB (Saint Marie 1996). Thus the evidence strongly supported the existence of both glutamatergic and GABAergic IC projections to MGB. Saint Marie et al. (1997) noted that many of the largest BIC axons were GABA immunoreactive and postulated a rapid feedforward colliculothalamic inhibition. Combined labeling of collicular inputs with GABA antibodies revealed that both GABAergic and nonGABAergic synapses terminated on the dendritic trees of cells in MGV and MGD (Bartlett et al. 2000). In vitro intracellular recordings confirmed that GABAergic ipsps were elicited in MGB cells by stimulating IC axons and could arrive before epsps (Peruzzi et al. 1997; Bartlett and Smith 1999). Brain slice and in vivo experiments (Hu et al. 1994; Bartlett and Smith 1999; Webber et al. 1999) verified that the excitatory events elicited by IC inputs were glutamatergic and utlilized both NMDA and nonNMDA receptors. The functional necessity for combined excitatory/inhibitory inputs from the IC to the auditory thalamus is, at present, unknown. In a later section, we discuss potential reasons for such an input.

5.4 Local Inhibition from the Thalamic Reticular Nucleus

The thalamic reticular nucleus is a sheet of GABAergic cells situated along the rostral and lateral surface of the dorsal thalamus (Fig. 2.5). Thalamocortical and corticothalamic axons pass through the TRN giving it a reticulated appearance. Cells in different TRN sectors are devoted to different sensory modalities. Each sector receives most of its inputs from cortical and thalamic areas of that sensory modality yet sends its axons only to the region of its thalamic input. Both thalamocortical axons and those corticothalamic axons of layer 6 pyramidal cells give off collaterals in the TRN that make excitatory glutamatergic synaptic contacts (Fig. 2.5), operating via both ionotropic and metabotropic glutamate receptors (Ohishi et al. 1993a,b; Cox and Sherman 1999), with TRN cells (Jones, 1975a, 1985; Ohara and Lieberman 1985; Williamson et al. 1993; Bourassa and Deschenes 1995). Thus, when viewed from the thalamus, the TRN acts as an inhibitory feedback and, when viewed from the cortex, it acts as a cortical inhibitory feedforward input to the thalamus. These collaterals and their terminals are spatially restricted and maintain a topographic order as they pass through the TRN (Guillery et al. 1998). Likewise, TRN cells innervate their thalamic region in a topographic fashion. TRN axons deliver local collaterals to other TRN cells (Yen et al. 1985; Spreafico et al. 1988; Liu et al. 1995) en route to the thalamus. One study indicates that rat TRN cells do not form connections with the thalamocortical cells from which they receive their

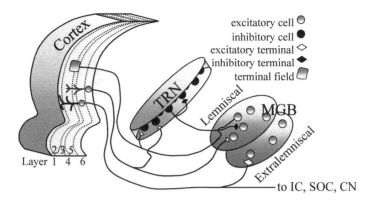

FIGURE 2.5. Interactions of the thalamus and cortex. Schematic representation of the connections between thalamus and cortex. Cells in cortical layers 5 and 6 project to the thalamus. Layer 5 projections are excitatory collaterals of corticofugal axons heading to other downstream structures. Their thalamic terminals are large and are found in extralemniscal thalamus. Layer 6 projections give off collaterals to TRN cells before innervating both lemniscal and extralemniscal thalamus (corticothalamic innervation of extralemniscal thalamus not shown for simplicity). Their thalamic terminals are small. GABAergic TRN cells also receive collaterals from other TRN cells and excitatory collaterals from thalamocortical cell axons heading for the cortex. Their main axon heads to thalamus to innervate both lemniscal and extralemniscal regions (TRN innervation of extralemnical thalamus not shown). See text for details. (For abbreviations, see Table 2.1.)

synaptic input (Pinault and Deschenes 1998). Rather, they project to nearby cells and thus may act to generate lateral inhibition in these cells.

The auditory sector of the TRN is located in the caudoventral region of the nucleus (Jones 1975a; Shosaku and Sumitomo 1983). Cells here are topographically organized according to the MGB region and/or nuclear subdivision they innervate (Rouiller et al. 1985; Conley et al. 1991; Crabtree 1998). Recordings from auditory TRN cells have shown that they are responsive primarily to auditory stimuli or to shock stimuli applied to the IC or the auditory cortex (Shosaku and Sumitomo 1983; Simm et al. 1990; Villa 1990). Some cells at the sector margins are also responsive to a second sensory modality. TRN axon terminals have been demonstrated on MGB cells (Montero 1983) and, in slices, act via GABA subfamily receptor types A and B on MGD, MGV, and MGM cells (Bartlett and Smith 1999). Stimulation of the TRN auditory region suppresses MGB cell spontaneous activity as well as their response to auditory stimuli or shock stimuli applied to the IC (Shosaku and Sumitomo 1983).

TRN cells may also play a role in synchronized MGB cell activity. TRN feedback in other sensory systems has been implicated in the global thalamic synchronization that occurs during slow wave phases of sleep and in some forms of epilepsy (Steriade et al. 1986, 1993; Buzsaki et al. 1990; Steriade, 1998). TRN feedback may also act during the waking state to selectively enhance sensory information transfer through certain thalamo-cortical cells (Warren and Jones 1994) and may play a role in cortical arousal during the attentive state (Macdonald et al. 1998). A recent inter-esting paper noted that electrical stimulation of rat auditory TRN evoked focal gamma band oscillations (40 Hz oscillatory electrical activity) in audi-tory cortex (Macdonald et al. 1998). Previous studies in humans (Llinas and Ribary 1993; Joliot et al. 1994) showed that this activity reflects the cognitive processing of auditory stimuli and the "temporal binding" of sensory stimuli into a single experience. Temporal binding simply means that sensory events occurring during a particular time will be consciously bound together as a single experience and that the 40 Hz oscillation is "required" for this to occur. Previously, using multielectrode mapping of cortical activity, Barth and MacDonald (1996) noted that gamma oscilla-tions could occur spontaneously or be evoked in AC by auditory stimuli. Further, this study showed that the oscillations could be modulated by the auditory thalamus. Electrical stimulation of the MGV or the MGD inhib-ited this 40 Hz cortical response while stimulation more medially, in the MGM, elicited it. The finding that stimulation of the TRN can elicit, pre-sumably through its connection with MGM, oscillations in the same area of auditory cortex suggests a functional role for TRN in cognition.

5.5 Functional States of the Thalamus

Early brain slice work (Jahnsen and Llinas 1984a,b) demonstrated that thal-amocortical cells can be put into two different response modes by chang-ing their membrane potential. These responses are called the "burst" and "tonic" modes and depend upon the activation or inactivation of a voltage-sensitive Ca^{2+} conductance. At around -70 mV, a depolarization will acti-vate this Ca^{2+} conductance generating a large depolarization and 3–5 action potentials. At -50 to -60 mV this conductance is inactivated and the same depolarization will elicit a train of single action potentials. Therefore, spike activity patterns transmitted from the MGB to the cortex would differ depending on the membrane potential at the arrival time of a synaptic inputs generated by auditory stimuli. Recordings from rat MGB cells (Hu 1995; Tennigkeit 1997; Bartlett and Smith 1999, 2000) have confirmed that dorsal and ventral MGB cells and some MGM cells show burst and tonic modes. It has been proposed that MGD cells exhibit sluggish, poorly timed responses to sound because they sit in burst mode, while the brisk sound evoked activity of MGV cells reflects their tonic mode status (Hu 1995). Conflicting evidence (Bartlett and Smith 1999) indicates that this may not be the case. The ascending IC inhibition, described above, could play a role in shaping MGB cell response features as it was noted that the ipsp could arrive prior to the epsp (Bartlett and Smith 1999). Inhibition could also potentially modulate an MGB cell's transition between firing modes and may operate in a spectrally dependent fashion. For example, bicuculline, the GABA antagonist, broadened tuning curves in the mustached bat (Suga et al. 1997), indicating the presence of broadband or cross frequency inhibitory projections.

In the visual thalamus, thalamocortical neurons can exhibit rhythmic, synchronized bursts during slow wave sleep (Livingstone and Hubel 1981; Steriade et al. 1993). A hypothesis was developed that these cells in awake animals were in the tonic mode and would reliably transmit sensory information while cells in slow wave sleep mode were hyperpolarized and in burst mode and would less reliably transfer the content and timing of sensory information to cortex. This hypothesis has fallen from favor, since recent finding showed that visual thalamic cells in the alert, awake cat also respond in a burst fashion (Guido and Weyand 1995; Reinagel et al. 1999). The current thinking is that cells in burst mode are better able to detect near threshold visual stimuli. Because they are hyperpolarized, spontaneous spike activity is reduced and a small synaptic input can activate the Ca^{2+} conductance.

5.6 The Role of the Thalamus in Dyslexia

Dyslexia and other language related disorders are characterized by a difficulty, displayed by certain people of otherwise normal intelligence, in learning to read and in performing other language-related tasks (Tallal 1980). Evidence indicates that this is not due to specific visual system deficits, but rather results from a deficit in the language processing system (Shaywitz 1996, 1998). More specifically, dyslexics have difficulty in processing phonemes, the linguistic units making up both written words and speech. This deficit may be related to an inability to process rapid frequency transitions or brief, rapid successive auditory stimuli (Fitch et al. 1997; Wright et al. 1997; Nagarajan et al. 1999) that are pervasive in speech. This inability is especially evident in children with learning and attention problems such as specific language impairment (Tallal and Piercy 1973; Tallal and Stark 1981; Merzenich et al. 1996). Adaptive training exercises using synthetic speech with reduced, rapid transitions have improved comprehension in such children (Tallal and Piercy 1973; Tallal and Stark 1981; Merzenich et al. 1996; Tallal et al. 1996; Nagarajan et al. 1998). Some evidence indicates that the problem may be related to processing deficits occurring in the information transfer between the MGB and cortical levels.

Evidence from the postmortem examination of the brains of dyslexics (Galaburda et al. 1994) showed MGB cell anomalies leading to speculation that this anatomical abnormality might be related to the auditory processing deficits. Kraus et al. (1996), using an evoked response feature called mismatch negativity (MMN), noted that children with learning problems who could not discriminate certain rapid auditory changes (like those in speech) also showed a decreased MMN amplitude. Mismatch negativity is the evoked response generated by an oddball stimulus amidst a series of stimuli that may provide a measure of the ability to discriminate auditory stimuli. Kraus and colleagues (Kraus et al. 1994a,b; King et al. 1995) demonstrated that cells in the nonprimary portions (nonMGV) of the auditory thalamus contribute to the generation of this evoked event. In a recent report (Herman et al. 1997), rats with similar thalamic abnormalities, induced by freeze injury to the neocortex, showed behavioral deficits in fast auditory processing. Thus, a link was made between abnormalities in the ability to process rapid acoustic transitions and structural differences in MGB and its targets. What might be the problem in the MGB? One possibility might be that the normal set points of resting membrane potentials, perhaps those in the MGD or the MGM, are altered and their mode of firing is changed. Perhaps in dyslexics these cells are in the incorrect firing mode and don't evoke as large a MMN response or, more importantly, do not transmit the appropriate information about speech for cortical interpretation.

5.7 Thalamocortical Projections. At Least Two Separate Systems?

The majority of cells in the MGB form projections to the cortex. This thalamocortical projection is not homogeneous in that cells from different MGB regions send axons to different auditory cortical areas (see Winer 1992 and below). An early report also suggested that it is also not homogeneous in terms of the cortical layers innervated (Imig and Morel 1983). Recent studies of the thalamocortical terminals and the neurochemical features of thalamocortical neurons has provided a finer grain of detail to the general principal that the MGV projects to primary cortex and that the MGD and the MGM project to belt and parabelt regions. As the following evidence indicates, one cell population innervates primarily the middle cortical layers in a tonotopic fashion. A second population innervates primarily layer 1 in a more diffuse, unconstrained fashion, and, in some species, perhaps a third population innervates both. These inputs may have significantly different roles in influencing cortical function.

Retrograde tracer injected into either layer 1 or 4 of cat primary (core) and nonprimary (belt and/or parabelt) auditory cortical areas labeled 2 sets of cells (Mitani et al. 1984, 1987; Niimi et al. 1984). Cells with smaller somata, mostly in the MGD and the MGM but scattered through all the major MGB subdivisions and associated paralaminar nuclei, sent inputs to layer 1. Layer 4 inputs to these same cortical areas arose from larger cells mostly in MGV, MGD, and MGM. These two sets of cells are also distinguishable by their immunoreactivity to antibodies against the calcium binding proteins parvalbumin (PV) or calbindin (CB). Injections into layer 1 or 4 of the monkey or rabbit primary auditory cortex backfilled either calbindin or parvalbumin-positive MGB cells respectively (Hashikawa et al. 1991; de Venecia et al. 1995, 1998). In the rat MGB, the distribution of CB staining is similar, but there appears to be no PV-immunoreactive cells (Celio 1990; Friauf 1994). Based on these observations, a "core and matrix" organization (Fig. 2.4) has been proposed for the monkey thalamus (Hashikawa et al. 1991, 1995; Molinari et al. 1995; Jones, 1998a,b). CB-labeled cells form a background matrix: they do not respect MGB divisional borders and are distributed throughout all MGB divisions including the associated paralaminar nuclei. The inputs to these cells are more diffuse, arising primarily from nonlemniscal sources. The output of these cells, which projects to layer 1 of auditory cortex and is also diffuse, is not constrained by regional cortical borders. The PV-labeled cells form the core, are confined to only certain MGB divisions and are highly concentrated in the MGV where CB cells are few. Inputs to the core cells arise from tonotopically organized lemniscal divisions of the auditory system (i.e. central nucleus of the IC) that contain cells with well-defined receptive fields. Likewise, the PV cells project onto one or a few cortical fields in a precise tonotopic fashion.

How do individual thalamocortical axons project onto the cortical layers? Interpretation of these results from gross injection studies is complicated by the fact, described above, that most MGB regions don't contain a pure population of PV or CB immunoreactive cells. Injections into the MGV (McMullen and de Venecia 1993; Hashikawa et al. 1995) stain PV cell axons that form elongated patches of terminals along isofrequency contours in layers 3 and 4 of primary cortical areas (Fig. 2.4) and are interspersed with terminal-sparse regions. Injections into the dorsal MGB, presumably filling the PV cell axons, labeled patches in nonprimary cortical areas primarily in layer 3 (Fig. 2.4). Injections into the medial magnocellular division that labeled presumed PV cell axons showed that these axons innervated layer 3 and 4 in both primary and non-primary cortical areas. Where individual axons were traced (Hashikawa et al. 1995; Cetas et al. 1999) single PV axons innervated one or multiple patches in layers 3 and 4. Axons going to layer 1, presumed to arise from CB cells, had a widespread projection over several cortical fields with single axons running long distances parallel to the cortical surface (Fig. 2.4). In the rabbit, a third type of axon was described (Cetas et al. 1999) that arose from PV positive MGV cells, innervated layers 3 and 4 and sent collaterals to layer 1.

What are the functions of these multiple thalamocortical inputs? The terminations of the PV-positive core cells in layers 3 and 4 presumably relay precise sensory information to the cortex as part of the pathway whose function forms the basis for the perception of auditory events. Jones (1998a,b) has speculated that the diffuse CB cell system projecting to layer 1, through its lack of respect for cortical boundaries, may engage multiple cortical areas at times when it is necessary to bind the many aspects of a sensory experience together.

5.8 Learning, Memory, and Fear and the Medial Geniculate Body

As described above, the MGM and associated nuclei receive convergent inputs from auditory, somatosensory, and visual structures and can respond to one or more of these multiple modalities. Early reports indicated that the medial division is also unique in its ability to show physiological plasticity or long term potentiation (LTP) during classical conditioning paradigms that pair a somatosensory with an auditory stimulus (Ryugo and Weinberger 1978; Gerren and Weinberger 1983). Since then, this region has been implicated in the learning-induced expression of the conditioned fear response to auditory stimuli, via the amygdala (LeDoux 1995; Ledoux and Muller 1997), and the expression of long term plasticity in the auditory cortex as well (Weinberger et al. 1995; Weinberger 1998).

One basic form of learning is the conditioned fear response, which occurs when an emotionally neutral auditory stimulus (conditioned stimulus, CS) is paired with an aversive, "fear-evoking" stimulus (unconditioned stimulus, US). After a number of pairings the neutral auditory stimulus acquires the capacity, *on its own*, to evoke the defensive or fear response (conditioned response, CR). This fear response includes changes in behavioral and autonomic responses. Ryugo and Weinberger (1978) reported that pairing a foot shock (the US) with a noise burst (the CS) greatly enhanced pupillary dilation (the CR) to subsequent presentation of the noise burst alone. Multiunit recordings during this paradigm revealed that only MGM cells increased their responses to the conditioned stimulus. These increases developed in parallel with the increased pupillary dilation, as noise was being paired with the shock, and persisted when noise was then presented alone. Subsequently, electrical stimulation of the brachium of the IC was shown to increase the amplitude and decrease the latency of MGM-evoked responses for over an hour (Gerren and Weinberger 1983). These two studies show that rapid, persistent changes occur in MGM during behavioral conditioning that might result from long-term potentiation of its inputs. More recent evidence also implicated MGM as a part of a learning circuit. McEchron and others (1996) used a classical conditioning paradigm, pairing an auditory stimuli (the CS) with a corneal air puff (the US). After training, stimulation of the brachium, but not inputs from the superior colliculus, generated enhanced responses for over an hour. Thus, the path carrying auditory CS information to MGM neurons increased in strength as the result of associative conditioning with an acoustic CS.

NMDA receptors have been shown elsewhere to be involved in such long-term plastic changes (Bliss and Collingridge 1993). Webber and co-workers (1999) noted that NMDA and nonNMDA receptor antagonists affected the in vivo responses of rabbit MGM cells to shocks of the brachium, implying that such mechanisms might be at work in this system. However, NMDA receptors are also present on MGV and MGD cells (no plasticity has been observed here) and are activated by IC inputs (Hu et al. 1994; Bartlett and Smith 1999). It would be of interest to see if brief high frequency stimulation of the IC input to MGM, in the presence of NMDA antagonists, would still generate enhanced responses with decreased latency and increase amplitude.

Over the past fifteen years, Ledoux and his colleagues have elucidated the interaction between the MGM and the lateral nucleus of the amygdala and shown it to be a vital part of the pathway responsible for the behavioral and autonomic responses elicited by the conditioned fear response (LeDoux et al. 1988; LeDoux 1995; Ledoux and Muller 1997). The role of the MGB was originally noted when, after lesioning its amygdalar connection, the ability to develop a conditioned fear response to auditory stimuli was disrupted, despite the fact that visual conditioned responses were unaffected (LeDoux et al. 1986). Subsequent lesion, tracer, and recording experiments verified that the MGM, (together with the PIN and SG) are the interface between anterolateral amygdala (AL) and the ascending auditory system (Clugnet et al. 1990; LeDoux et al. 1990a,b). Terminals of the MGB input were found to be excitatory, using both AMPA and NMDA glutamate receptors (LeDoux et al. 1991; Li et al. 1995; Farb and LeDoux 1997). Experiments also revealed that rapid electrical stimulation of the MGB potentiated the MGB/amygdala connection (Clugnet and LeDoux 1990) and produced enhanced AL cell responses to auditory stimuli (Rogan and LeDoux 1995). This short latency AL cell response to auditory stimuli was also enhanced after fear conditioning and the enhancement was disrupted by NMDA receptor blockade in the amygdala (Quirk et al. 1995; Gewirtz and Davis 1997). Thus amygdalar responses to auditory inputs can be enhanced either by MGB stimulation or by prior conditioning of the auditory stimulus, presumably through NMDA receptor activation. Finally, Rogan and others (1997) reported that amygdalar response changes closely parallel the acquisition of CS-elicited fear behavior, are enduring, and do not occur if the conditioned and unconditioned stimuli remain unpaired. Similar results were documented in slices (McKernan and Shinnick-Gallagher 1997) where MGB synaptic inputs to amygdala were strengthened, by a presynaptic mechanism, in previously conditioned rats, as compared to normals. This synaptic strengthening only occurred when the conditioned and unconditioned stimuli were paired. Thus, LTP-like associative processes occur during fear conditioning and may underlie the long-term associative plasticity that constitutes memory of the conditioning experience.

Recently, a debate has arisen (Cahill et al. 1999; Fanselow and LeDoux 1999; Maren 1999) over whether the plasiticity or "learning" that underlies this phenomenon occurs within the AL or somewhere upstream, for example in the MGB where, as described earlier, LTP-like events can also occur. Regardless of the outcome of the debate, this pathway will continue to be a fascinating component of the auditory system's connection to a behavioral output.

5.9 Differences in the Middle (Thalamocortical Input) Layers of Primary Auditory Cortex

Early investigators of cortical cytoarchitecture described sensory cortical areas as "granulous" in nature and designated these areas as "koniocortical." The designation of koniocortex is based on several cytoarchitectonic features including the dense packing of granule cells in the granular layer (layer 4) and supragranular layers (layers 2 and 3), and a consequent reduction of pyramidal cells there (von Economo 1929). Studies in the primary visual and somatosensory cortices show a unique form of excitatory interneuron, the spiny stellate cell, populating layer 4 to give it a granular appearance and replacing the pyramidal cell as the dominant excitatory neuron (Jones 1975b; Lund 1984; Martin and Whitteridge 1984; Simons and Woolsey 1984). In visual and somatosensory cortices, these spiny stellate cells form an important link between thalamus and cortex. Studies in visual and somatosensory cortices (LeVay 1973; LeVay and Gilbert 1976; White 1978; White and Rock 1979, 1980) reveal that the spiny stellate cells are the major layer 4 recipient of thalamocortical axon terminals arising from primary thalamic nuclei. Evidence also suggests that one of the main responsibilities of spiny stellates is to transfer this thalamic input to other, primarily superficial, cortical layers (Gilbert and Wiesel 1983). The thalamocortical termination in layer 4 is quite restricted. Individual thalamocortical axons from X and Y cells in cat visual thalamus end almost exclusively in different regions of layer 4 with only a very minor spillover into the deepest part of layer 3 (LeVay and Gilbert 1976; Ferster and LeVay 1978). Similar results have been noted for the monkey magno- and parvocellular pathways (Freund et al. 1989; Katz et al. 1989). This restriction does not seem to hold in the auditory cortex where tracer injections that centered on the cat and rabbit ventral MGB (Winer 1992; McMullen and de Venecia 1993; Huang and Winer 2000) labeled thalamocortical axons that terminated broadly in layers 3 and 4 with single axons often going to multiple patches in these layers (Cetas et al. 1999). Comparison of current source density (CSD) studies from the auditory and visual systems also provide evidence for this input difference. In one study (Steinschneider et al. 1992), where click stimuli evoked responses throughout the depth of macaque primary auditory cortex (AI), the CSD response looks very different from similar recordings taken from monkey and cat visual cortex (Mitzdorf and Singer 1979; Mitzdorf 1985, 1991). In auditory cortex, the first major current sink, corresponding to monosynaptic activation of cells by thalamocortical axons, extends deep into layer 3. In visual cortex, this sink stops abruptly at the top of layer 4.

The second major difference that may distinguish the auditory cortex is the prodominant excitatory cell population in layer 4. In visual and somatosensory systems, both in vivo and in vitro experiments have verified that the spiny stellate cell is the major excitatory cell type in layer 4. These studies have also added a great deal of information about the anatomy and physiology of these cells and how they fit into the circuitry of their regions (Gilbert 1983; Martin and Whitteridge 1984; Katz et al. 1989; Anderson et al. 1993; Hirsch 1995; Woolsey 1996; Callaway 1998; Fleidervish et al. 1998; Hirsch et al. 1998; Somogyi et al. 1998). The same cannot be said for the auditory cortex where, beyond the initial golgi studies, little is known about layer 4. Four papers have reported on the combined anatomy and physiology of cat auditory cortical cells in vivo (Mitani and Shimokouchi 1985; Mitani et al. 1985; Ojima et al. 1991, 1992), and one paper reported on cells in layer 4 from slices taken from developing rats (Metherate and Aramakis 1999). Although early studies of the primary auditory cortical areas of many species including human (Von Economo and Koskinas 1925), monkey (Walker 1937), and cat (Bremer and Dow 1939) described this primary sensory area as granulous or koniocortical, a subsequent study in cat (Rose 1949) questioned the granular nature of this region. Rose noted that layer 4 is not heavily granular and thus, by strict definition, cat auditory cortex is not koniocortical. A review of the subsequent literature gives an indication that Rose's basic observation may underlie a fundamental difference between layer 4 of the auditory cortex of cat and other species in terms of the excitatory cell types found there. Golgi studies of primary auditory cortex in several species including humans indicate that spiny stellate cells are not highly concentrated in layer 4 (McMullen and Glaser 1982; Winer 1984; Meyer et al. 1989; Fitzpatrick and Henson 1994). These same studies also show that, in contrast to the symmetric dendritic trees of visual and somatosensory spiny stellates, there is great variety in the rarely observed auditory spiny stellate dendritic pattern. Intracellular recording and labeling studies from cat auditory cortical slices (Smith and Populin 1999) also suggests that nonpyramidal spiny cells are rare in layer 4 and do not resemble spiny stellates in visual or somatosensory cortex in terms of dendritic configuration or axonal targets. If spiny stellate cells are not the primary excitatory cell type within layer 4, what is? Several reports show that pyramidal cells populate layer 4 in several species (McMullen and Glaser 1982; Meyer et al. 1989; Fitzpatrick and Henson 1994). In contrast, pyramidal cells are reported to be absent in cat auditory cortex (Winer 1984). Here, the primary cell type within layer 4 was reported to be the "medium-sized tufted neuron" whose features resembled inhibitory rather than excitatory neurons. Smith and Populin (1999) reported that there was no obvious break in the population of labeled pyramidal cells at depths corresponding to layer 4, indicating that there may be pyramidal cells in this layer in cat.

These two pieces of evidence, that the thalamocortical input innervates middle layers 3 and 4 and that the major excitatory cell type in these layers is pyramidal, indicate a fundamental difference between auditory cortical and other sensory cortical areas. In this light, it is interesting to note that (1) in the visual system binocularity first occurs in the cortex and one of the postulated functions of the mostly monocular spiny stellate cells is to converge and confer binocularity onto cells in more superficial cortical layer, and (2) in the auditory system, binaurality first occurs at the brainstem level and in this system cortical spiny stellate cells are rare.

5.10 The Dorsal Zone of Cortical Field AI

AI is one of multiple tonotopically arranged areas of the core region of several species including cat. Evidence suggests that the dorsal aspect of cat AI differs from the rest of AI in terms of its thalamic input, its cell's response properties, and perhaps its function (Middlebrooks and Zook 1983; Sutter and Schreiner 1991; He and Hashikawa 1998). This area, originally referred to as the "dorsal zone" (Middlebrooks and Zook 1983), may be part of AI (Andersen et al. 1980a) or a functionally distinct region (Middlebrooks and Zook 1983; He et al. 1997; He and Hashikawa 1998). Retrograde tracing studies (Middlebrooks and Zook 1983) revealed that the thalamic inputs to the dorsal zone are from the dorsocaudal aspect of the MGV, the MGD, and the Po. Besides this unique thalamic input, distinctive cortical inputs have also been noted (Ojima and He 1997; He and Hashikawa 1998). Different labels injected side by side in separate AI frequency regions labeled terminals patches with considerable overlap in the dorsal zone, suggesting an overlapping of frequency inputs to single cells. Cortical inputs from other primary and secondary fields were also seen.

In addition to input differences, a fairly large percentage of cells in the dorsal zone of AI differ in several of their response features including frequency specificity and/or binaural response. Cells were more broadly tuned, were typically excited by both ears and had higher thresholds and longer latencies than in the rest of AI. A larger high frequency representation was also seen and the AI isofrequency contours turned caudally as they entered the region (Middlebrooks and Zook 1983). Another interesting difference of some dorsal zone cells was their multipeaked tuning curves (Sutter and Schreiner 1991). Such units could show decreased response latencies and thresholds to two-tone combinations, a potentially important feature for detecting complex signals containing harmonics, formants, or resonances. Whether the converging AI inputs are involved in shaping these multipeaked curves is not known. Similar cells are found in specific regions of the bat cortex (Suga et al. 1979, 1983) and are thought to be involved in echolocation. Another set of cells in this same area of cat cortex have an equally interesting set of responses described as "duration tuning" (He et al. 1997). These long latency neurons could show several variations of duration tuning described as long pass, short pass, or band pass/duration sensitive. Several units responding only at the stimulus offset (Off-units) were also duration sensitive, showing either an increasing offset discharge with increasing stimulus duration or an offset discharge peaking at an intermediate stimulus duration. Like the cells displaying multipeaked tuning curves described above, no mechanisms underlying such behavior have been reported. Duration-tuned neurons emerge at lower levels of the auditory system as well, in the IC and MGB (Casseday et al. 1994; Ehrlich et al. 1997). Intracellular recordings from the bat IC revealed that duration-tuned, offset unit responses occurred when a sustained inhibitory input, that caused a rebound response at stimulus offset, combined with a transient delayed excitatory response. Duration tuning occurred only at specific stimulus durations when the offset rebound and the delayed excitation occurred coincidentally. A similar mechanism may be at work in the offset units of the dorsal zone or may simply reflect their ascending input. Interesting topics for future investigation would be (1) to record intracellularly from dorsal zone cells showing these response features and test the hypothesis that offset cells may be duration tuned because of a coincidence of a rebound response from inhibition and a long latency excitatory response and (2) to record intracellularly from the nonoffset, duration-tuned population and test the hypothesis that some unique synaptic events or intrinsic features can account for the temporal summation or suppression mechanisms proposed to be responsible for the various categories of tuning.

5.11 The Anterior Ectosylvian Cortex: Forming Unique Associations

In the cat, an area of association cortex defined by the anterior ectosylvian sulcus and known as the anterior ectosylvian cortex (AEC) is of considerable interest (Rauschecker 1995, 1996, 1997; Stein 1998) for several reasons including: (1) its participation in eye movements, (2) its role in shaping the response features of cells in the multimodal space map located in the superior colliculus, and (3) the developmental plasticity that this region has been shown to exhibit. Situated between auditory cortical fields AI and AII caudally and the anterior auditory field (AAF) and somatosensory cortical areas rostrally, the AEC has regions devoted primarily to each of three sensory modalities. Visual cells cluster on the ventral bank of the sulcus (AEV) (Mucke et al. 1982; Olson and Graybiel 1987), with somatosensatory cells on the anterior two thirds of the dorsal bank (S4) (Clemo and Stein 1982) and auditory cells on the banks and fundus of the posterior three fourths of the sulcus (AES) (Clarey and Irvine 1986, 1990a). About 60% of AEC cells are unimodal while the remaining 40%, lying mostly at the subregion boundaries, are bi- or trimodal (Jiang et al. 1994). Multimodal cells have their receptive fields in spatial register, i.e., the visual and auditory receptive fields are in the same spatial location (Wallace et al. 1993). AES cells have broad tuning to auditory stimuli with multiple response peaks, are most commonly onset, and are usually driven by both ears (Clarey and Irvine 1986, 1990a). Retrograde tracers (Clarey and Irvine 1990b) revealed that auditory inputs to the AES arise from the MGM and suprageniculate nucleus, an area where cells can show a similar physiology (Benedek et al. 1996; Clasca et al. 1997).

In cat, the AEC is the only cortical region with a robust input to the SC (Meredith and Clemo 1989). It is in the SC where multimodal cells receive a convergence of auditory, visual, and/or somatosensory inputs from a given region of space and form a cellular map of sensory space (Fig. 2.3). Curiously, the AEC multimodal cells do not project to the SC to help form this map. Electrical stimulation of the SC (Wallace et al. 1993) antidromically activated unimodal and not multimodal AEC cells. Furthermore, during AEC inactivation cells in the SC could still respond to unimodal stimuli, indicating that the AEC is not directly responsible for these responses. It also indicated that SC maps may arise from other sources of inputs including the nucleus of the brachium of the IC described above. The striking change noted during AEC inactivation was that SC cells no longer integrated multisensory inputs and animals could no longer use these multisensory cues to localize objects in space (Wallace and Stein 1994; Wilkinson et al. 1996). Several studies have reported that when SC cells are presented with two sensory cues, e.g., auditory and visual, from the same spatial location they show "response enhancement." That is, their response is not merely an addition of the two responses but an enhanced response (King and Palmer 1985; Frens et al. 1995; Wilkinson et al. 1996). If the two cues are not at the same spatial location, a response to a single modality can show "response depression" when both stimuli are presented at the same time (Meredith and Stein 1996; Kadunce et al. 1997). It is this multisensory integration that is eliminated when the AEC is inactivated (Wallace and Stein 1994). Response enhancement or depression also carries over to behavioral studies where two sensory stimuli with similar spatial locations give enhanced behavioral orienting responses, while disparate stimuli give suppressed responses. Inactivation of the AEC had no effect on orientation to unimodal cues but eliminated these enhanced or depressed behavioral responses (Wilkinson et al. 1996).

The multisensory nature of the AEC has also made it amenable to experiments that have yielded information regarding cortical plasticity (Rauschecker 1995, 1999), a topic that will be elucidated in the next section. In cats with eyelids sutured at birth, the responses of the majority of cells in the AEC visual region were taken over by auditory inputs (Rauschecker and Korte 1993). Recordings from AEC cells in these cats, when compared to normal cats, showed a higher precentage of directionally tuned cells that were more sharply tuned. Behaviorally, the lid-sutured cats were better than control animals at sound localization tasks (Rauschecker 1995). Whether this cortical takeover resulted from the strengthening of pre-existing connections or from the growth of new connections is unknown. These results indicate that cortex can modify it's connectivity to suit the needs of its situation in a fashion that favorably alters the behavior of the animal.

5.12 Cortical Plasticity and Long-term Potentiation in the Auditory Cortex

A large body of literature has shown that sensory cortical areas, including auditory cortex, can display plasticity during development and in the mature animal (Buonomano and Merzenich 1998; Klintsova and Greenough 1999; Rauschecker 1999). Plasticity has been described at several functional levels. These levels range from the extreme case where auditory cortical areas are usurped by other sensory modalities, to the moderate case where there is a reorganization of cortical frequency maps due to perturbations of sensory inputs (like partial peripheral denervation), and finally to the more "simplistic" changes in synaptic strength at the single cell level due to LTP-inducing shock protocols.

Some recent papers have highlighted the extreme flexibility of auditory cortex under certain circumstances. Using positron emission tomography (PET, an imaging technique using a cyclotron to measure short lived radioactive substances), Nishimura and co-workers (1999) noted that secondary cortical areas in congenitally deaf patients were activated not by auditory stimuli presented using cochlear implants, but by visual signals generated by sign language. In contrast, primary cortical areas were activated by the cochlear implant but not the sign language. This indicates that, in humans deafened before acquiring language, primary cortex is hard wired while secondary areas can be usurped by other stimulus modalities. In another set of reports, it appears that even the sensory function of the primary auditory cortex can be usurped if changes occur before thalamocortical connections are made (Sharma et al. 2000; von Melchner et al. 2000). In these experiments, when normal IC inputs to the MGB were replaced (during development) by optic nerve inputs, a map of visual space arose in primary auditory cortex with cells organized into a visual orientation map similar to that seen in the visual cortex. Behavioral experiments indicated that this connection could actually mediate visually directed behavior.

Less extreme examples of cortical reorganization have also been described where sensory cortical areas, including auditory cortex, can reorganize their maps of the sensory sheet both during development and in the mature animal. Recordings from auditory cortex of several species showed that small cochlear lesions could cause an expansion of frequency regions adjacent to lesioned frequency regions (Robertson and Irvine 1989; Rajan et al. 1993; Schwaber et al. 1993; Rajan and Irvine 1998). Training is also capable of generating map alterations in that frequencies used to train monkeys to perform a task were subsequently over represented in the cortex (Recanzone et al. 1993). The mechanisms leading to this reorganization are unknown; however, stimulation of the cholinergic input from the nucleus basalis (part of the basal forebrain) paired with auditory stimulation can also alter the cortical map to favor the paired frequencies (Kilgard and Merzenich 1998). Another recent paper reported a stimulus-induced change in the auditory cortical map (Klinke et al. 1999). In deaf white kittens, where the auditory pathway is intact despite a degenerated organ of Corti, chronic cochlear stimulation enhanced the amount of cortical area responding to stimulation. Hoping to restore some aspects of normal central auditory pathway maturation, a similar strategy of attempting to preserve or expand elements of the central auditory system, using cochlear implants for electrical stimulation, has been initiated in very young children with hearing loss. Whatever mechanism is involved, such stimulation ultimately leads to improved language acquisition and speech performance (Shepherd et al. 1997; Robinson 1998; Waltzman and Cohen 1998; Lenarz 1999; Ponton et al. 1999).

Plasticity has also been demonstrated at the cellular level. Pairing suprathreshold depolarization with a simultaneously presented tone, enhanced the subsequent responses of some cortical cells to that paired tone (Cruikshank and Weinberger 1996). In a related experiment, pairing a tone with nucleus basalis stimulation also produced long-term changes in cortical cell receptive fields such that they selectively increased their response to the paired tone (Bakin and Weinberger 1996). Plasticity, in the form of LTP, has also been demonstrated in slices of auditory cortex, but its function in the above-mentioned in vivo plastic events is not completely understood. In a series of papers, Kudoh and Shibuki (1994, 1996, 1997a,b) induced LTP in auditory cortical pyramidal cells of layers 2 and 3 by tetanic stimulation of the white matter. When compared with responses from visual cortical cells, auditory cortical LTP was much greater. LTP induced by postsynaptic cell depolarization while stimulating the white matter induced comparable LTP in auditory and cortical cells. The authors concluded, based on this and other evidence, that the enhanced LTP was primarily due to a greater density of local horizontal axon collaterals in the supragranular layers of auditory cortex. Such a large collateral system caused a greater level of depolarization, a greater activation of NMDA channels, and subsequently, a greater influx of calcium ions leading to LTP. Buonomano (1999) also noted that LTP could be induced in pyramidal cells within layers 2 and 3 by depolarizing the cells to spike above threshold while stimulating the synaptic inputs from the same layer. No one has looked at the thalamocortical input to the auditory cortex and its ability to generate LTP. The white matter stimulation in the LTP experiments described above should not be taken as an indication of such an effect since many inputs are stimulated in such a paradigm. In the somatosensory system, stimulation of thalamocortical synapses can generate LTP in the spiny stellate cells of layer 4 of the developing animal but the ability disappears in adults (Crair and Malenka 1995). This may be related to the fact that, during development, a significant proportion of thalamacocortical synapses are functionally "silent" under normal stimulus conditions because they contain only NMDA receptors (Isaac et al. 1997). Strong stimulation elicits enough depolarization to activate these voltage sensitive NMDA receptors and generate LTP. As development progresses, there is a decrease in the NMDA receptor mediated currents and an increase in AMPA mediated transmission. Thus, despite the fact that NMDA receptors still mediate some of the synaptic events (Crair and Malenka 1995; Gil and Amitai 1996; Isaac et al. 1997), LTP can not be induced by tetanic stimuli unless GABA inhibition is blocked or the spiny stellate cell depolarized.

5.13 The Descending Cortical System

Auditory cortex projects to several downstream regions as far peripherally as the cochlear nucleus. Most notable of these are the MGB, the IC, the nucleus sagulum, the ventral nucleus of the trapezoid body, the lateral superior olive, and the cochlear nucleus. Strong evidence from the visual, somatosensory, and auditory systems indicates that there are two populations of cells in primary sensory cortex sending axons back to thalamus (Sherman and Guillery 1996). The endings of these two axons have the following characteristics: (1) they are excitatory and terminate at different locations in the sensory thalamus, (2) they synapse at different dendritic locations on the thalamic neurons innervated, (3) they are drastically different in size, (4) they may have different effects on the thalamic cells that they innervate, and finally (5) they may have uniquely different functions/roles.

5.13.1 The Small Terminal Corticothalamic Input from Layer 6 Pyramidal Cells

Retrograde labeling of the primary or lemniscal areas of sensory thalamus (Gilbert and Kelley 1975; Lund et al. 1975; Kelly and Wong 1981; Wong and Kelly 1981) or anterograde labeling of specific cortical layers (Hoogland et al. 1987, 1991; Rouiller and Welker 1991; Ojima 1994; Bourassa and Deschenes 1995; Bourassa et al. 1995) indicate that thalamic cells receive inputs from layer 6 pyramidal cells (Fig. 2.5). Reconstruction of these axons showed that they first send collaterals to innervate the GABAergic TRN nucleus, then branch and terminate in the primary or lemniscal thalamus. In the thalamus, these axons give rise to the small, excitatory terminals synapsing on the distal dendrites of thalamic neurons. The layer 6 corticothalamic cells intermingle in their layer with other pyramidal cells projecting intracortically to higher order auditory cortical areas, to contralateral auditory cortex, or elsewhere (Kelly and Wong 1981; Games and Winer 1988; Bourassa and Deschenes 1995; Bourassa et al. 1995).

The function of this small terminal thalamic input is not completely understood. In the visual system, repetitive stimulation of this cortical input activates both NMDA and nonNMDA ionotropic glutamate receptors as well as a metabotropic glutamate receptor that depolarizes the cell by blocking a potassium conductance via a second messenger pathway

(McCormick and von Krosigk 1992). This depolarization lasts tens of seconds after stimulus offset, leading the authors to speculate that it could play a role in how the thalamic neuron processes its ascending sensory information. As described above, membrane potential changes alter thalamic neuron response features. Thalamocortical cells can "sit" in two different firing modes—when hyperpolarized they are in burst mode and ascending inputs can be amplified by the activation of a low threshold Ca^{2+} conductance that elicits a Ca^{2+} spike and a burst of high frequency sodium spikes. When depolarized, the Ca^{2+} conductance is inactivated and the cell responds only with a single, well timed spike. Thus, metabotropic receptor activation and depolarization could alter the throughput capabilities of thalamus cells. Evidence from slices (Bartlett and Smith 1999a) showed that repetitive simulation of the small terminal corticothalamic input elicited ionotropic synaptic events and activated metabotropic receptors in MGB cells, resulting in depolarizations lasting for tens of seconds.

Several in vivo studies have looked at the cortical influence on MGB cells. Although most effects that are elicited by cortical cooling or stimulation are probably mediated through the small terminal corticothalamic input, some of the effects could be the indirect result of cortical inputs to other downstream areas i.e., the IC or CN. Early studies using electrical stimulation of the cortex reported mostly inhibitory effects in the MGB (Aitkin and Dunlop 1969; Amato et al. 1969). Cortical cooling elicited both facilitatory and inhibitory effects (Ryugo and Weinberger 1976; Orman and Humphrey 1981; Villa et al. 1991, 1999). In these experiments, the precise location of the cortical stimulation or cooling relative to frequency maps was not controlled. When specific frequency regions of the bat cortex were inactivated with focal lidocaine injections, cells in the MGB that were tuned to the same frequency decreased their auditory responses without shifting their frequency response curves (Zhang and Suga 1997). In contrast, MGB cells that were tuned to a different frequency than those at the cortical inactivation site showed an increased response, and their favored frequency shifted toward that of the inactivated cortical cells. This suggests that the cortical input enhances the response of MGB cells with the same preferred frequency but suppresses the response of cells with different preferred frequencies. A similar on-CF (frequency with the lowest threshold) facilitation of MGB cell responses by electrical stimulation of the cat cortex was reported (He 1997).

Several questions arise regarding the small terminal corticothalamic system: (1) When are these layer 6 pyramidal cells active i.e., what auditory stimuli turn them on? (2) Do their terminals end on thalamocortical neurons and interneurons? (3) Do the GABAergic TRN cells receiving inputs from a layer 6 cell project to and inhibit the MGB cell that is receiving the same or a different layer 6 input? (4) Are the layer 6 inputs capable, under normal conditions, of activating the metabotropic glutamate receptor and does this lead to a change in the MGB cell's response to it's auditory input?

5.13.2 The Large Terminal Corticothalamic Input from Layer 5 Pyramidal Cells

The same tracer studies referred to above indicated that large layer 5 pyramidal cells also project to the thalamus (Fig. 2.5). Unlike the small terminal axonal projection of layer 6 pyramidals, the axons of these cells were larger in diameter and did not give off collaterals to the TRN. In addition, layer 5 pyramidal cell axons were long range corticofugal axons that did not completely terminate in the thalamus. Instead, each of these axons gave off a collateral to the thalamus. Finally, these collaterals did not terminate in the primary thalamic region but made contacts on cells in the "association" or nonlemniscal thalamus. These terminals appeared as very large swellings. In the somatosensory system, these large terminals formed "glomerulus-like" terminals that made multiple synaptic contacts on more proximal dendrites of thalamic neurons (Hoogland et al. 1991). Data from the rat auditory thalamus, also indicated that there is a population of cortical axons giving rise to large swellings in nonlemniscal regions of MGD and MGM (Rouiller and Welker 1991; Ojima 1994) that are glomerular-like and synapse on proximal dendrites (Bartlett et al. 2000).

In the visual system, corticothalamic layer 5 pyramidal cells are a unique set of cells that respond to depolarization with an intrinsic burst (IB—an initial "intrinsic burst" of spikes followed by very regular single spikes at a lower rate). In contrast, other layer 5 pyramidal cells projecting to the contralateral cortex show the RS (an initial "rapid spike" rate followed by firing at a low rate with progressively increasing rate adaptation) type of response (Kasper et al. 1994). In the cat auditory cortex, two sets of layer 5 cells have been described, including one with large cell bodies located in superficial layer 5 that send their axons toward the thalamus (Ojima et al. 1992). A recent report also describes two anatomically and physiologically distinct sets of layer 5 pyramidal cells in the rat auditory cortex and provides evidence that the IB category cells project to the MGB (Hefti and Smith 2000). Other evidence indicates that these layer 5 bursters also project to the IC (Games and Winer 1988; Ojima 1994) and cochlear nucleus (Weedman and Ryugo 1996a,b). No evidence is available to indicate how these giant terminals influence their postsynaptic targets, but recently a very intriguing idea has been set forth. It proposes that the large terminal projection provides a pathway for information transfer between primary and higher order sensory cortical areas, not directly but *back through* the thalamus (Guillery 1995). There is no evidence that these large terminals end on thalamocortical cells that project to higher order cortices, but it is certainly the case that the many cells in regions that receive large terminal input project to higher order cortical areas. There is also evidence of a very strong influence of this layer 5 projection on higher order thalamic nuclei in visual and somatosensory systems. Cells in these thalamic centers stop responding to the external sensory environment when their cortical input is turned off by cooling or other methods (Bender 1983; Diamond et al. 1992). Such data implies that these nonlemniscal thalamic nuclei are primarily driven not by ascending sensory input but by their *descending cortical input* and may act as a potentially very active and important cortical communication link.

5.13.3 The Corticocollicular Input

Layer 5 pyramidal cells from all areas of the auditory cortex combine to send an organized excitatory projection to the inferior colliculus that may be the same population projecting to the MGB (Feliciano and Potashner 1995). Much of this projection terminates primarily in the external and dorsal cortical areas of the bilateral IC (Andersen et al. 1980b; Faye-Lund and Osen 1985; Games and Winer 1988; Herbert et al. 1991; Saldana et al. 1996; Winer et al. 1998). The terminals are usually large (Ojima 1994) but it is not known what cell types they influence. Several studies have looked at the influence of this input (Amato et al. 1969; Mitani et al. 1983; Syka and Popelar 1984; Sun et al. 1989; Zhang and Suga 1997; Jen et al. 1998; Torterolo et al. 1998; Jen and Zhang 1999). In most of these studies, the best frequency of cells in the stimulated cortical area was not matched to the best frequency of the cells recorded from in the IC. Activation or suppression of cortical activity could have either a facilitative or suppressive effect on the spontaneous or acoustically driven responses of cells that were primarily located in the ICC. In one study, (Zhang and Suga, 1997) attention was paid to the CF of the cortical region. As described above for their results on MGB cells, similar results from IC cells suggest that the cortical input may enhance the responses of IC cells with the same preferred frequency but suppress the response of IC cells with different preferred frequencies.

5.13.4 Other Corticofugal Outputs

Inputs to the olivary complex and/or olivocochlear cells were described above. In the cochlear nucleus, layer 5 pyramidal cells have terminal fields in the granule cell domain of the ipsilateral cochlear nucleus where their major targets appear to be the granule cell dendrite (Weedman and Ryugo 1996a, b). Whether this layer 5 population of pyramidal cells is the same population that innervates the MGB and the IC remains to be determined.

References

Adam TJ, Schwarz DWF, Finlayson PG (1999) Firing properties of chopper and delay neurons in the lateral superior olive of the rat. Exp Brain Res 124:489–502.

Adams JC (1983) Cytology of periolivary cells and the organization of their projections in the cat. J Comp Neurol 215:275–289.

Aitkin L, Jones R (1992) Azimuthal processing in the posterior auditory thalamus of cats. Neurosci Lett 142:81–84.

Aitkin LM, Dunlop CW (1969) Inhibition in the medial geniculate body of the cat. Exp Brain Res 7:68–83.

Aitkin LM, Prain SM (1974) Medial geniculate body: unit responses in the awake cat. J Neurophysiol 37:512–521.

Aitkin LM, Webster WR (1971) Tonotopic organization in the medial geniculate body of the cat. Brain Res 26:402–405.

Aitkin LM, Webster WR (1972) Medial geniculate body of the cat: organization and responses to tonal stimuli of neurons in ventral division. J Neurophysiol 35:365–380.

Aitkin LM, Dickhaus H, Schult W, Zimmermann M (1978) External nucleus of inferior colliculus: auditory and spinal somatosensory afferents and their interactions. J Neurophysiol 41:837–847.

Allon N, Yeshurun Y (1985) Functional organization of the medial geniculate body's subdivisions of the awake squirrel monkey. Brain Res 360:75–82.

Amato G, La Grutta V, Enia F (1969) The control exerted by the auditory cortex on the activity of the medial geniculate body and inferior colliculus. Arch Sci Biol 53:291–313.

Andersen RA, Knight PL, Merzenich MM (1980a) The thalamocortical and corticothalamic connections of AI, AII, and the anterior auditory field (AAF) in the cat: evidence for two largely segregated systems of connections. J Comp Neurol 194:663–701.

Andersen RA, Roth GL, Aitkin LM, Merzenich MM (1980b) The efferent projections of the central nucleus and the pericentral nucleus of the inferior colliculus in the cat. J Comp Neurol 194:649–662.

Andersen RA, Snyder RL, Merzenich MM (1980c) The topographic organization of corticocollicular projections from physiologically identified loci in the AI, AII, and anterior auditory cortical fields of the cat. J Comp Neurol 191:479–494.

Anderson JC, Martin KA, Whitteridge D (1993) Form, function, and intracortical projections of neurons in the striate cortex of the monkey *Macacus nemestrinus*. Cereb Cortex 3:412–420.

Aschoff A, Ostwald J (1987) Different origins of cochlear efferents in some bat species, rats, and guinea pigs. J Comp Neurol 264:56–72.

Azeredo WJ, Kliment ML, Morley BJ, Relkin E, Slepecky NB, Sterns A, Warr WB, Weekly JM, Woods CI (1999) Olivocochlear neurons in the chinchilla: a retrograde fluorescent labelling study. Hear Res 134:57–70.

Bajo VM, Merchan MA, Lopez DE, Rouiller EM (1993) Neuronal morphology and efferent projections of the dorsal nucleus of the lateral lemniscus in the rat. J Comp Neurol 334:241–262.

Bakin JS, Weinberger NM (1996) Induction of a physiological memory in the cerebral cortex by stimulation of the nucleus basalis. Proc Natl Acad Sci U S A 93: 11219–11224.

Banks MI, Smith PH (1992) Intracellular recordings from neurobiotin-labeled cells in brain slices of the rat medial nucleus of the trapezoid body. J Neurosci 12: 2819–2837.

Barth DS, MacDonald KD (1996) Thalamic modulation of high-frequency oscillating potentials in auditory cortex. Nature 383:78–81.

Bartlett EL, Smith PH (1999) Anatomic, intrinsic, and synaptic properties of dorsal and ventral division neurons in rat medial geniculate body. J Neurophysiol 81: 1999–2016.

Bartlett EL, Stark JM, Guillery RW, Smith PH (2000) A comparison of the fine structure of cortical and collicular terminals in the rat medial geniculate body. Neurosci, 100:811–828.

Behrend O, Brand A, Bruetel G, Grothe B (2000) Temporal processing in the gerbil superior olivary complex. Assoc Res Otolaryngol. 23:33.

Bellingham MC, Lim R, Walmsley B (1998) Developmental changes in EPSC quantal size and quantal content at a central glutamatergic synapse in rat. J Physiol (Lond) 511:861–869.

Bender DB (1983) Visual activation of neurons in the primate pulvinar depends on cortex but not colliculus. Brain Res 279:258–261.

Benedek G, Fischer-Szatmari L, Kovacs G, Perenyi J, Katoh YY (1996) Visual, somatosensory and auditory modality properties along the feline suprageniculate-anterior ectosylvian sulcus/insular pathway. Prog Brain Res 112:325–334.

Benedek G, Pereny J, Kovacs G, Fischer-Szatmari L, Katoh YY (1997) Visual, somatosensory, auditory and nociceptive modality properties in the feline suprageniculate nucleus. Neuroscience 78:179–189.

Benson TE, Berglund AM, Brown MC (1996) Synaptic input to cochlear nucleus dendrites that receive medial olivocochlear synapses. J Comp Neurol 365:27–41.

Benson TE, Brown MC (1990) Synapses formed by olivocochlear axon branches in the mouse cochlear nucleus. J Comp Neurol 295:52–70.

Benson TE, Brown MC (1996) Synapses from medial olivocochlear branches in the inferior vestibular nucleus. J Comp Neurol 372:176–188.

Berglund AM, Benson TE, Brown MC (1996) Synapses from labeled type II axons in the mouse cochlear nucleus. Hear Res 94:31–46.

Berglund AM, Brown MC (1994) Central trajectories of type II spiral ganglion cells from various cochlear regions in mice. Hear Res 75:121–130.

Binns KE (1991) In the guinea pig the central representation of auditory space in the superior colliculus and external nucleus of the inferior colliculus are independent. Br J Audiol 25:55.

Binns KE, Grant S, Withington DJ, Keating MJ (1992) A topographic representation of auditory space in the external nucleus of the inferior colliculus of the guinea-pig. Brain Res 589:231–242.

Bjorkeland M, Boivie J (1984) An anatomical study of the projections from the dorsal column nuclei to the midbrain in cat. Anat Embryol 170:29–43.

Bliss TV, Collingridge GL (1993) A synaptic model of memory: long-term potentiation in the hippocampus. Nature 361:31–39.

Blomqvist A, Danielsson I, Norrsell U (1990) The somatosensory intercollicular nucleus of the cat's mesencephalon. J Physiol (Lond) 429:191–203.

Bordi F, LeDoux JE (1994) Response properties of single units in areas of rat auditory thalamus that project to the amygdala. I. Acoustic discharge patterns and frequency receptive fields. Exp Brain Res 98:261–274.

Bourassa J, Deschenes M (1995) Corticothalamic projections from the primary visual cortex in rats: a single fiber study using biocytin as an anterograde tracer. Neuroscience 66:253–263.

Bourassa J, Pinault D, Deschenes M (1995) Corticothalamic projections from the cortical barrel field to the somatosensory thalamus in rats: a single-fibre study using biocytin as an anterograde tracer. Eur J Neurosci 7:19–30.

Bremer F, Dow RS (1939) The acoustic area of the cerebral cortex in the cat: A combined oscillographic and cytoarchitectonic study. J Neurophysiol 2:308–318.

Brown MC (1987) Morphology of labeled efferent fibers in the guinea pig cochlea. J Comp Neurol 260:605–618.

Brown MC (1989) Morphology and response properties of single olivocochlear fibers in the guinea pig. Hear Res 40:93–109.

Brown MC (1994) Antidromic responses of single units from the spiral ganglion. J Neurophysiol 71:1835–1847.

Brown MC, Ledwith JVd (1990) Projections of thin (type-II) and thick (type-I) auditory-nerve fibers into the cochlear nucleus of the mouse. Hear Res 49: 105–118.

Brown MC, Berglund AM, Kiang NY, Ryugo DK (1988a) Central trajectories of type II spiral ganglion neurons. J Comp Neurol 278:581–590.

Brown MC, Liberman MC, Benson TE, Ryugo DK (1988b) Brainstem branches from olivocochlear axons in cats and rodents. J Comp Neurol 278:591–603.

Brown MC, Kujawa SG, Duca ML (1998a) Single olivocochlear neurons in the guinea pig. I. Binaural facilitation of responses to high-level noise. J Neurophysiol 79:3077–3087.

Brown MC, Kujawa SG, Liberman MC (1998b) Single olivocochlear neurons in the guinea pig—II—Response plasticity due to noise conditioning. J Neurophysiol 79:3088–3097.

Buonomano DV (1999) Distinct functional types of associative long-term potentiation in neocortical and hippocampal pyramidal neurons. J Neurosci 19:6748–6754.

Buonomano DV, Merzenich MM (1998) Cortical plasticity: from synapses to maps. Annu Rev Neurosci 21:149–186.

Burgess BJ, Adams JC, Nadol JB, Jr. (1997) Morphologic evidence for innervation of Deiters' and Hensen's cells in the guinea pig. Hear Res 108:74–82.

Buzsaki G, Smith A, Berger S, Fisher LJ, Gage FH (1990) Petit mal epilepsy and parkinsonian tremor: hypothesis of a common pacemaker. Neuroscience 36:1–14.

Cahill L, Weinberger NM, Roozendaal B, McGaugh JL (1999) Is the amygdala a locus of "conditioned fear"? Some questions and caveats. Neuron 23:227–228.

Caicedo A, Herbert H (1993) Topography of descending projections from the inferior colliculus to auditory brainstem nuclei in the rat. J Comp Neurol 328:377–392.

Calford MB (1983) The parcellation of the medial geniculate body of the cat defined by the auditory response properties of single units. J Neurosci 3:2350–2364.

Calford MB, Aitkin LM (1983) Ascending projections to the medial geniculate body of the cat: evidence for multiple, parallel auditory pathways through thalamus. J Neurosci 3:2365–2380.

Calford MB, Webster WR (1981) Auditory representation within principal division of cat medial geniculate body: an electrophysiology study. J Neurophysiol 45: 1013–1028.

Callaway EM (1998) Local circuits in primary visual cortex of the macaque monkey. Annu Rev Neurosci 21:47–74.

Campbell JP, Henson MM (1988) Olivocochlear neurons in the brainstem of the mouse. Hear Res 35:271–274.

Canlon B, Borg E, Flock A (1988) Protection against noise trauma by pre-exposure to a low level acoustic stimulus. Hear Res 34:197–200.

Cant NB, Hyson RL (1992) Projections from the lateral nucleus of the trapezoid body to the medial superior olivary nucleus in the gerbil. Hear Res 58:26–34.

Cant NB, Benson CG, Schofield BR (1999) Projections from the superior paraolivary nucleus to the inferior colliculus in guinea pigs and gerbils. Assoc Res Otolaryngol 22:89.

Casseday JH, Ehrlich D, Covey E (1994) Neural tuning for sound duration: role of inhibitory mechanisms in the inferior colliculus. Science 264:847–850.

Celio MR (1990) Calbindin D-28k and parvalbumin in the rat nervous system. Neuroscience 35:375–475.

Cetas JS, de Venecia RK, McMullen NT (1999) Thalamocortical afferents of Lorente de No: medial geniculate axons that project to primary auditory cortex have collateral branches to layer I. Brain Res 830:203–208.

Chen K, Waller HJ, Godfrey DA (1998) Effects of endogenous acetylcholine on spontaneous activity in rat dorsal cochlear nucleus slices. Brain Res 783:219–226.

Clarey JC, Irvine DR (1986) Auditory response properties of neurons in the anterior ectosylvian sulcus of the cat. Brain Res 386:12–19.

Clarey JC, Irvine DR (1990a) The anterior ectosylvian sulcal auditory field in the cat: I. An electrophysiological study of its relationship to surrounding auditory cortical fields. J Comp Neurol 301:289–303.

Clarey JC, Irvine DR (1990b) The anterior ectosylvian sulcal auditory field in the cat: II. A horseradish peroxidase study of its thalamic and cortical connections. J Comp Neurol 301:304–324.

Clarey JC, Barone P, Imig TJ (1992) Physiology of thalamus and cortex. In: Popper AN, Fay RR (eds) The mammalian auditory pathway: Neurophysiology pp. 232–334. New York: Springer-Verlag.

Clasca F, Llamas A, Reinoso-Suarez F (1997) Insular cortex and neighboring fields in the cat: a redefinition based on cortical microarchitecture and connections with the thalamus. J Comp Neurol 384:456–482.

Clemo HR, Stein BE (1982) Somatosensory cortex: a 'new' somatotopic representation. Brain Res 235:162–168.

Clugnet MC, LeDoux JE (1990) Synaptic plasticity in fear conditioning circuits: induction of LTP in the lateral nucleus of the amygdala by stimulation of the medial geniculate body. J Neurosci 10:2818–2824.

Clugnet MC, LeDoux JE, Morrison SF (1990) Unit responses evoked in the amygdala and striatum by electrical stimulation of the medial geniculate body. J Neurosci 10:1055–1061.

Cohn ES, Kelley PM (1999) Clinical phenotype and mutations in connexin 26 (DFNB1/GJB2), the most common cause of childhood hearing loss. Am J Med Genet 89:130–136.

Collet L, Kemp DT, Veuillet E, Duclaux R, Moulin A, Morgon A (1990) Effect of contralateral auditory stimuli on active cochlear micro-mechanical properties in human subjects. Hear Res 43:251–261.

Conley M, Kupersmith AC, Diamond IT (1991) The organization of projections from subdivisions of the auditory cortex and thalamus to the auditory sector of the thalamic reticular nucleus in Galago. Eur J Neurosci 3:1089–1103.

Cox CL, Sherman SM (1999) Glutamate inhibits thalamic reticular neurons. J Neurosci 19:6694–6699.

Crabtree JW (1998) Organization in the auditory sector of the cat's thalamic reticular nucleus. J Comp Neurol 390:167–182.

Crair MC, Malenka RC (1995) A critical period for long-term potentiation at thalamocortical synapses. Nature 375:325–328.

Cruikshank SJ, Weinberger NM (1996) Receptive-field plasticity in the adult auditory cortex induced by Hebbian covariance. J Neurosci 16:861–875.

De Venecia RK, Smelser CB, Lossman SD, McMullen NT (1995) Complementary expression of parvalbumin and calbindin D-28k delineates subdivisions of the rabbit medial geniculate body. J Comp Neurol 359:595–612.

De Venecia RK, Smelser CB, McMullen NT (1998) Parvalbumin is expressed in a reciprocal circuit linking the medial geniculate body and auditory neocortex in the rabbit. J Comp Neurol 400:349–362.

Dehmel S, Doerrscheidt GJ, Reubsamen R (1999) Electrophysiological characterization of neurons in the superior paraolivary nucleus of the gerbil (Meriones unguiculatus). Assoc Res Otolaryngol 22:94.

Diamond IT, Jones EG, Powell TP (1969) The projection of the auditory cortex upon the diencephalon and brainstem in the cat. Brain Res 15:305–340.

Diamond ME, Armstrong-James M, Budway MJ, Ebner FF (1992) Somatic sensory responses in the rostral sector of the posterior group (POm) and in the ventral posterior medial nucleus (VPM) of the rat thalamus: dependence on the barrel field cortex. J Comp Neurol 319:66–84.

Edeline JM, Manunta Y, Nodal FR, Bajo VM (1999) Do auditory responses recorded from awake animals reflect the anatomical parcellation of the auditory thalamus? Hear Res 131:135–152.

Ehrenberger K, Felix D (1991) Glutamate receptors in afferent cochlear neurotransmission in guinea pigs. Hear Res 52:73–80.

Ehrlich D, Casseday JH, Covey E (1997) Neural tuning to sound duration in the inferior colliculus of the big brown bat, Eptesicus fuscus. J Neurophysiol 77:2360–2372.

Elgoyhen AB (2001) Cloning and functional properties of hair cell nACHRs. Assoc Res Otolaryngol. 24:138.

Elgoyhen AB, Johnson DS, Boulter J, Vetter DE, Heinemann S (1994) Alpha 9: an acetylcholine receptor with novel pharmacological properties expressed in rat cochlear hair cells. Cell 79:705–715.

Eybalin M (1993) Neurotransmitters and neuromodulators of the mammalian cochlea. Physiol Rev 73:309–373.

Fanselow MS, LeDoux JE (1999) Why we think plasticity underlying Pavlovian fear conditioning occurs in the basolateral amygdala. Neuron 23:229–232.

Farb CR, LeDoux JE (1997) NMDA and AMPA receptors in the lateral nucleus of the amygdala are postsynaptic to auditory thalamic afferents. Synapse 27:106–121.

Faye-Lund H, Osen KK (1985) Anatomy of the inferior colliculus in rat. Anat Embryol 171:1–20.

Fechner FP, Burgess BJ, Adams JC, Liberman MC, Nadol JB, Jr. (1998) Dense innervation of Deiters' and Hensen's cells persists after chronic deefferentation of guinea pig cochleas. J Comp Neurol 400:299–309.

Fekete DM, Rouiller EM, Liberman MC, Ryugo DK (1984) The central projections of intracellularly labeled auditory nerve fibers in cats. J Comp Neurol 229:432–450.

Feliciano M, Potashner SJ (1995) Evidence for a glutamatergic pathway from the guinea pig auditory cortex to the inferior colliculus. J Neurochem 65:1348–1357.

Feliciano M, Thompson AM (1995) Descending auditory cortical projections to midbrain and brainstem auditory structures in the cat. Assoc Res Otolaryngol 18:163.

Ferragamo MJ, Golding NL, Oertel D (1998) Synaptic inputs to stellate cells in the ventral cochlear nucleus. J Neurophysiol 79:51–63.

Ferster D, LeVay S (1978) The axonal arborizations of lateral geniculate neurons in the striate cortex of the cat. J Comp Neurol 182:923–944.

Finlayson PG, Adam TJ (1997) Excitatory and inhibitory response adaptation in the superior olive complex affects binaural acoustic processing. Hear Res 103:1–18.

Fitch RH, Miller S, Tallal P (1997) Neurobiology of speech perception. Annu Rev Neurosci 20:331–353.

Fitzpatrick DC, Henson OW, Jr. (1994) Cell types in the mustached bat auditory cortex. Brain Behav Evol 43:79–91.

Fleidervish IA, Binshtok AM, Gutnick MJ (1998) Functionally distinct NMDA receptors mediate horizontal connectivity within layer 4 of mouse barrel cortex. Neuron 21:1055–1065.

Frens MA, Van Opstal AJ, Van der Willigen RF (1995) Spatial and temporal factors determine auditory-visual interactions in human saccadic eye movements. Percept Psychophys 57:802–816.

Frenz CM, Van De Water TR (2000) Immunolocalization of connexin 26 in the developing mouse cochlea. Brain Res Rev 32:177–180.

Freund TF, Martin KA, Soltesz I, Somogyi P, Whitteridge D (1989) Arborisation pattern and postsynaptic targets of physiologically identified thalamocortical afferents in striate cortex of the macaque monkey. J Comp Neurol 289:315–336.

Friauf E (1994) Distribution of calcium-binding protein calbindin-D28k in the auditory system of adult and developing rats. J Comp Neurol 349:193–211.

Fuentes V, Berrebi AS, Saldana E (1999) Trajectory, morphology, and distribution of axons of the superior paraolivary nucleus (SPON) that innervate the inferior colliculus in the rat. Assoc Res Otolaryngol 22:222.

Fujino K, Oertel D (2000) Cholinergic modulation of stellate cells in ventral cochlear nucleus in mice. Assoc Res Otolaryngol 23:33.

Fujino K, Koyano K, Ohmori H (1997) Lateral and medial olivocochlear neurons have distinct electrophysiological properties in the rat brain slice. J Neurophysiol 77:2788–2804.

Galaburda AM, Menard MT, Rosen GD (1994) Evidence for aberrant auditory anatomy in developmental dyslexia. Proc Natl Acad Sci USA 91:8010–8013.

Games KD, Winer JA (1988) Layer V in rat auditory cortex: projections to the inferior colliculus and contralateral cortex. Hear Res 34:1–25.

Gerren RA, Weinberger NM (1983) Long term potentiation in the magnocellular medial geniculate nucleus of the anesthetized cat. Brain Res 265:138–142.

Gewirtz JC, Davis M (1997) Second-order fear conditioning prevented by blocking NMDA receptors in amygdala. Nature 388:471–474.

Ghoshal S, Kim DO (1996) Marginal shell of the anteroventral cochlear nucleus: intensity coding in single units of the unanesthetized, decerebrate cat. Neurosci Lett 205:71–74.

Gil Z, Amitai Y (1996) Adult thalamocortical transmission involves both NMDA and non-NMDA receptors. J Neurophysiol 76:2547–2554.

Gilbert CD, Kelly JP (1975) The projection of cells in different layers of the cat's visual cortex. J Comp Neurol 163:81–105.

Gilbert CD (1983) Microcircuitry of the visual cortex. Annu Rev Neurosci 6: 217–247.

Gilbert CD, Wiesel TN (1983) Functional organization of the visual cortex. Prog Brain Res 58:209–218.

Giraud AL, Garnier S, Micheyl C, Lina G, Chays A, Chery-Croze S (1997a) Auditory efferents involved in speech-in-noise intelligibility. Neuroreport 8:1779–1783.

Giraud AL, Wable J, Chays A, Collet L, Chery-Croze S (1997b) Influence of contralateral noise on distortion product latency in humans: is the medial olivocochlear efferent system involved? J Acoust Soc Am 102:2219–2227.

Gonzalez-Hernandez T, Mantolansa-Mmiento B, Gonzalez-Gonzalez B, Perez-Gonzalez H (1996) Sources of Gabaergic input to the inferior colliculus of the rat. J Comp Neurol 372:309–326.

Grothe B, Sanes DH (1993) Bilateral inhibition by glycinergic afferents in the medial superior olive. J Neurophysiol 69:1192–1196.

Guido W, Weyand T (1995) Burst responses in thalamic relay cells of the awake behaving cat. J Neurophysiol 74:1782–1786.

Guillery RW (1995) Anatomical evidence concerning the role of the thalamus in corticocortical communication: a brief review. J Anat 187:583–592.

Guillery RW, Feig SL, Lozsadi DA (1998) Paying attention to the thalamic reticular nucleus. Trends Neurosci 21:28–32.

Guinan JJJr, Guinan SS, Norris BE (1972a) Single auditory units in the superior olivary complex. Responses to sounds and classifications based on physiological properties. Int J Neurosci 4:101–120.

Guinan JJJr, Norris BE, Guinan SS (1972b) Single auditory units in the superior olivary complex: II. Location of unit categories and tonotopic organization. Int J Neurosci 4:147–166.

Hackett TA, Stepniewska I, Kaas JH (1998a) Subdivisions of auditory cortex and ipsilateral cortical connections of the parabelt auditory cortex in macaque monkeys. J Comp Neurol 394:475–495.

Hackett TA, Stepniewska I, Kaas JH (1998b) Thalamocortical connections of the parabelt auditory cortex in macaque monkeys. J Comp Neurol 400:271–286.

Hackney CM, Osen KK, Ottersen OP, Storm-Mathisen J, Manjaly G (1996) Immunocytochemical evidence that glutamate is a neurotransmitter in the cochlear nerve: a quantitative study in the guinea-pig anteroventral cochlear nucleus. Eur J Neurosci 8:79–91.

Harada N, Han DY, Komeda M, Yamashita T (1994) Glutamate-induced intracellular Ca2+ elevation in isolated spiral ganglion cells of the guinea pig cochlea. Acta Otolaryngol 114:609–612.

Hashikawa T, Rausell E, Molinari M, Jones EG (1991) Parvalbumin- and calbindin-containing neurons in the monkey medial geniculate complex: differential distribution and cortical layer specific projections. Brain Res 544:335–341.

Hashikawa T, Molinari M, Rausell E, Jones EG (1995) Patchy and laminar terminations of medial geniculate axons in monkey auditory cortex. J Comp Neurol 362:195–208.

He J (1997) Modulatory effects of regional cortical activation on the onset responses of the cat medial geniculate neurons. J Neurophysiol 77:896–908.

He J, Hashikawa T (1998) Connections of the dorsal zone of cat auditory cortex. J Comp Neurol 400:334–348.

He J, Hashikawa T, Ojima H, Kinouchi Y (1997) Temporal integration and duration tuning in the dorsal zone of cat auditory cortex. J Neurosci 17:2615–2625.

Hefti BJ, Smith PH (2000) Anatomy, physiology and synaptic responses of layer V cells in rat auditory cortex: characterization and study of inhibition through intracellular GABA$_A$ blockade. J Neurophysiol 83:2626–2638.

Helfert RH, Bonneau JM, Wenthold RJ, Altschuler RA (1989) GABA and glycine immunoreactivity in the guinea pig superior olivary complex. Brain Res 501: 269–286.

Henkel CK (1983) Evidence of sub-collicular auditory projections to the medial geniculate nucleus in the cat: an autoradiographic and horseradish peroxidase study. Brain Res 259:21–30.

Herbert H, Aschoff A, Ostwald J (1991) Topography of projections from the auditory cortex to the inferior colliculus in the rat. J Comp Neurol 304:103–122.

Herkenham M (1980) Laminar organization of thalamic projections to the rat neocortex. Science 207:532–535.

Herman AE, Galaburda AM, Fitch RH, Carter AR, Rosen GD (1997) Cerebral microgyria, thalamic cell size and auditory temporal processing in male and female rats. Cereb Cortex 7:453–464.

Hienz RD, Stiles P, May BJ (1998) Effects of bilateral olivocochlear lesions on vowel formant discrimination in cats. Hear Res 116:10–20.

Hirsch JA (1995) Synaptic integration in layer IV of the ferret striate cortex. J Physiol (Lond) 483:183–199.

Hirsch JA, Alonso JM, Reid RC, Martinez LM (1998) Synaptic integration in striate cortical simple cells. J Neurosci 18:9517–9528.

Hoogland PV, Welker E, Van der Loos H (1987) Organization of the projections from barrel cortex to thalamus in mice studied with *Phaseolus vulgaris*-leucoagglutinin and HRP. Exp Brain Res 68:73–87.

Hoogland PV, Wouterlood FG, Welker E, Van der Loos H (1991) Ultrastructure of giant and small thalamic terminals of cortical origin: a study of the projections from the barrel cortex in mice using *Phaseolus vulgaris* leuco-agglutinin (PHA-L). Exp Brain Res 87:159–172.

Hu B (1995) Cellular basis of temporal synaptic signalling: an in vitro electrophysiological study in rat auditory thalamus. J Physiol (Lond) 483:167–182.

Hu B, Senatorov V, Mooney D (1994) Lemnical and non-lemniscal synaptic transmission in rat auditory thalamus. J Physiol (Lond) 479:217–231.

Huang CL, Winer JA (2000) Auditory thalamocortical projections in the cat: laminar and areal patterns of input. J Comp Neurol 427:302–331.

Hurd LB, Hutson KA, Morest DK (1999) Cochlear nerve projections to the small cell shell of the cochlear nucleus: The neuroanatomy of extremely thin sensory axons. Synapse 33:83–117.

Imig TJ, Morel A (1983) Organization of the thalamocortical auditory system in the cat. Annu Rev Neurosci 6:95–120.

Imig TJ, Morel A (1984) Topographic and cytoarchitectonic organization of thalamic neurons related to their targets in low-, middle-, and high-frequency representations in cat auditory cortex. J Comp Neurol 227:511–539.

Imig TJ, Morel A (1985a) Tonotopic organization in lateral part of posterior group of thalamic nuclei in the cat. J Neurophysiol 53:836–851.

Imig TJ, Morel A (1985b) Tonotopic organization in ventral nucleus of medial geniculate body in the cat. J Neurophysiol 53:309–340.

Isaac JT, Crair MC, Nicoll RA, Malenka RC (1997) Silent synapses during development of thalamocortical inputs. Neuron 18:269–280.

Itoh K, Kaneko T, Kudo M, Mizuno N (1984) The intercollicular region in the cat: a possible relay in the parallel somatosensory pathways from the dorsal column nuclei to the posterior complex of the thalamus. Brain Res 308:166–171.

Jager W, Goiny M, Herrera-Marschitz M, Flock A, Hokfelt T, Brundin L (1998) Sound-evoked efflux of excitatory amino acids in the guinea-pig cochlea in vitro. Exp Brain Res 121:425–432.

Jahnsen H, Llinas R (1984a) Electrophysiological properties of guinea-pig thalamic neurones: an in vitro study. J Physiol (Lond) 349:205–226.

Jahnsen H, Llinas R (1984b) Ionic basis for the electro-responsiveness and oscillatory properties of guinea-pig thalamic neurones in vitro. J Physiol (Lond) 349: 227–247.

Jen PHS, Chen QC, Sun XD (1998) Corticofugal regulation of auditory sensitivity in the bat inferior colliculus. J Comp Physiol 183:683–697.

Jen PHS, Zhang JP (1999) Corticofugal regulation of excitatory and inhibitory frequency tuning curves of bat inferior collicular neurons. Brain Res. 841:184–188.

Jiang H, Lepore F, Ptito M, Guillemot JP (1994) Sensory modality distribution in the anterior ectosylvian cortex (AEC) of cats. Exp Brain Res 97:404–414.

Joliot M, Ribary U, Llinas R (1994) Human oscillatory brain activity near 40 Hz coexists with cognitive temporal binding. Proc Natl Acad Sci USA 91:11748–11751.

Jones EG (1975a) Some aspects of the organization of the thalamic reticular complex. J Comp Neurol 162:285–308.

Jones EG (1975b) Varieties and distribution of non-pyramidal cells in the somatic sensory cortex of the squirrel monkey. J Comp Neurol 160:205–267.

Jones EG (1985) The thalamus. New York: Plenum Press. 425–442.

Jones EG (1998a) A new view of specific and nonspecific thalamocortical connections. Adv Neurol 77:49–73.

Jones EG (1998b) Viewpoint: the core and matrix of thalamic organization. Neuroscience 85:331–345.

Kaas JH, Hackett TA (1998) Subdivisions of auditory cortex and levels of processing in primates. Audiol Neurootol 3:73–85.

Kaas JH, Hackett TA, Tramo MJ (1999) Auditory processing in primate cerebral cortex. Current Opin Neurobiol 9:164–170.

Kadunce DC, Vaughan JW, Wallace MT, Benedek G, Stein BE (1997) Mechanisms of within- and cross-modality suppression in the superior colliculus. J Neurophysiol 78:2834–2847.

Kasper EM, Larkman AU, Lubke J, Blakemore C (1994) Pyramidal neurons in layer 5 of the rat visual cortex. I. Correlation among cell morphology, intrinsic electrophysiological properties, and axon targets. J Comp Neurol 339:459–474.

Katz LC, Gilbert CD, Wiesel TN (1989) Local circuits and ocular dominance columns in monkey striate cortex. J Neurosci 9:1389–1399.

Kawase T, Liberman MC (1992) Spatial organization of the auditory nerve according to spontaneous discharge rate. J Comp Neurol 319:312–318.

Keats BJ, Berlin CI (1999) Genomics and hearing impairment. Genome Res 9:7–16.

Kelly JB, Judge PW (1985) Effects of medial geniculate lesions on sound localization by the rat. J Neurophysiol 53:361–372.

Kelly JB, Liscum A, Vanadel B, Ito M (1998) Projections from the superior olive and lateral lemniscus to tonotopic regions of the rats inferior colliculus. Hear Res 116:43–54.

Kelly JP, Wong D (1981) Laminar connections of the cat's auditory cortex. Brain Res 212:1–15.

Kelsell DP, Dunlop J, Stevens HP, Lench NJ, Liang JN, Parry G, Mueller RF, Leigh IM (1997) Connexin 26 mutations in hereditary non-syndromic sensorineural deafness. Nature 387:80–83.

Kikuchi T, Kimura RS, Paul DL, Adams JC (1995) Gap junctions in the rat cochlea: immunohistochemical and ultrastructural analysis. Anat Embryol 191:101–118.

Kilgard MP, Merzenich MM (1998) Plasticity of temporal information processing in the primary auditory cortex. Nat Neurosci 1:727–731.

King AJ, Jiang ZD, Moore DR (1998) Auditory brainstem projections to the ferret superior colliculus: anatomical contribution to the neural coding of sound azimuth. J Comp Neurol 390:342–365.

King AJ, Palmer AR (1985) Integration of visual and auditory information in bimodal neurones in the guinea-pig superior colliculus. Exp Brain Res 60:492–500.

King C, McGee T, Rubel EW, Nicol T, Kraus N (1995) Acoustic features and acoustic changes are represented by different central pathways. Hear Res 85:45–52.

Klinke R, Kral A, Heid S, Tillein J, Hartmann R (1999) Recruitment of the auditory cortex in congenitally deaf cats by long-term cochlear electrostimulation. Science 285:1729–1733.

Klintsova AY, Greenough WT (1999) Synaptic plasticity in cortical systems. Curr Opin Neurobiol 9:203–208.

Knudsen EI (1982) Auditory and visual maps of space in the optic tectum of the owl. J Neurosci 2:1177–1194.

Knudsen EI, Knudsen PF (1983) Space-mapped auditory projections from the inferior colliculus to the optic tectum in the barn owl (Tyto alba). J Comp Neurol 218:187–196.

Knudsen EI, Konishi M (1978) Space and frequency are represented separately in auditory midbrain of the owl. J Neurophysiol 41:870–884.

Kraus N, McGee T, Carrell T, King C, Littman T, Nicol T (1994a) Discrimination of speech-like contrasts in the auditory thalamus and cortex. J Acoust Soc Am 96: 2758–2768.

Kraus N, McGee T, Littman T, Nicol T, King C (1994b) Nonprimary auditory thalamic representation of acoustic change. J Neurophysiol 72:1270–1277.

Kraus N, McGee T, Carrell T, Zecker S, Nicol T, Koch D (1996) Auditory neurophysiologic responses and discrimination deficits in children with learning problems. Science 273:971–973.

Kudo M, Itoh K, Kawamura S, Mizuno N (1983) Direct projections to the pretectum and the midbrain reticular formation from auditory relay nuclei in the lower brainstem of the cat. Brain Res 288:13–19.

Kudo M, Niimi K (1980) Ascending projections of the inferior colliculus in the cat: an autoradiographic study. J Comp Neurol 191:545–556.

Kudo M, Tashiro T, Higo S, Matsuyama T, Kawamura S (1984) Ascending projections from the nucleus of the brachium of the inferior colliculus in the cat. Exp Brain Res 54:203–211.

Kudoh M, Shibuki K (1994) Long-term potentiation in the auditory cortex of adult rats. Neurosci Lett 171:21–23.

Kudoh M, Shibuki K (1996) Long-term potentiation of supragranular pyramidal outputs in the rat auditory cortex. Exp Brain Res 110:21–27.

Kudoh M, Shibuki K (1997a) Comparison of long-term potentiation between the auditory and visual cortices. Acta Otolaryngol Suppl 532:109–111.

Kudoh M, Shibuki K (1997b) Importance of polysynaptic inputs and horizontal connectivity in the generation of tetanus-induced long-term potentiation in the rat auditory cortex. J Neurosci 17:9458–9465.

Kujawa SG, Liberman MC (1997) Conditioning-related protection from acoustic injury: effects of chronic deefferentation and sham surgery. J Neurophysiol 78: 3095–3106.

Kulesza R, Holt A, Spirou G, Berrebi A (2000) Intracellular labeling of axonal collaterals of SPON neurons. Assoc Res Otolaryngol 23:37.

Kulesza RJ, Berrebi AS (1999) Distribution of GAD isoforms in the superior paraolivary nucleus (SPON) of the rat. Assoc Res Otolaryngol 22:70.

Kulesza RJ, Berrebi AS (2000) The superior paraolivary nucleus of the rat is a GABAergic nucleus. J Assoc Res Otolaryngol 1:255–269.

Kuwabara N (1999) Multiple descending projections to the ventral nucleus of the trapezoid body. Assoc Res Otolaryngol 22:148.

Kuwabara N, Zook JM (1991) Classification of the principal cells of the medial nucleus of the trapezoid body. J Comp Neurol 314:707–720.

Kuwabara N, Zook JM (1992) Projections to the medial superior olive from the medial and lateral nuclei of the trapezoid body in rodents and bats. J Comp Neurol 324:522–538.

Kuwabara N, Zook JM (1999) Local collateral projections from the medial superior olive to the superior paraolivary nucleus in the gerbil. Brain Res 846:59–71.

Kuwabara N, DiCaprio RA, Zook JM (1991) Afferents to the medial nucleus of the trapezoid body and their collateral projections. J Comp Neurol 314:684–706.

Lautermann J, Frank HG, Jahnke K, Traub O, Winterhager E (1999) Developmental expression patterns of connexin26 and-30 in the rat cochlea. Dev Genet 25: 306–311.

Lautermann J, Tencate WJF, Altenhoff P, Grummer R, Traub O, Frank HG, Jahnke K, Winterhager E (1998) Expression of the gap-junction connexins 26 and 30 in the rat cochlea. Cell Tissue Res 294:415–420.

Leake PA, Snyder RL (1989) Topographic organization of the central projections of the spiral ganglion in cats. J Comp Neurol 281:612–629.

Leake PA, Snyder RL, Merzenich MM (1992) Topographic organization of the cochlear spiral ganglion demonstrated by restricted lesions of the anteroventral cochlear nucleus. J Comp Neurol 320:468–478.

LeDoux JE (1995) Emotion: clues from the brain. Annu Rev Psychol 46:209–235.

Ledoux JE, Muller J (1997) Emotional memory and psychopathology. Phil Trans Royal Soc B: Biological Sciences 352:1719–1726.

LeDoux JE, Sakaguchi A, Iwata J, Reis DJ (1986) Interruption of projections from the medial geniculate body to an archi-neostriatal field disrupts the classical conditioning of emotional responses to acoustic stimuli. Neuroscience 17:615–627.

LeDoux JE, Ruggiero DA, Forest R, Stornetta R, Reis DJ (1987) Topographic organization of convergent projections to the thalamus from the inferior colliculus and spinal cord in the rat. J Comp Neurol 264:123–146.

LeDoux JE, Iwata J, Cicchetti P, Reis DJ (1988) Different projections of the central amygdaloid nucleus mediate autonomic and behavioral correlates of conditioned fear. J Neurosci 8:2517–2529.

LeDoux JE, Cicchetti P, Xagoraris A, Romanski LM (1990a) The lateral amygdaloid nucleus: sensory interface of the amygdala in fear conditioning. J Neurosci 10: 1062–1069.

LeDoux JE, Farb C, Ruggiero DA (1990b) Topographic organization of neurons in the acoustic thalamus that project to the amygdala. J Neurosci 10:1043–1054.

LeDoux JE, Farb CR, Milner TA (1991) Ultrastructure and synaptic associations of auditory thalamo-amygdala projections in the rat. Exp Brain Res 85:577–586.

Lenarz T (1999) Sensorineural hearing loss in children. Int J Ped Otorhinolaryngol 49:179–181.

LeVay S (1973) Synaptic patterns in the visual cortex of the cat and monkey. Electron microscopy of Golgi preparations. J Comp Neurol 150:53–85.

LeVay S, Gilbert CD (1976) Laminar patterns of geniculocortical projection in the cat. Brain Res 113:1–19.

Li XF, Phillips R, LeDoux JE (1995) NMDA and non-NMDA receptors contribute to synaptic transmission between the medial geniculate body and the lateral nucleus of the amygdala. Exp Brain Res 105:87–100.

Liberman MC (1982) Single-neuron labeling in the cat auditory nerve. Science 216:1239–1241.

Liberman MC (1988) Response properties of cochlear efferent neurons: monaural vs. binaural stimulation and the effects of noise. J Neurophysiol 60:1779–1798.

Liberman MC (1991a) Central projections of auditory-nerve fibers of differing spontaneous rate. I. Anteroventral cochlear nucleus. J Comp Neurol 313:240–258.

Liberman MC (1991b) The olivocochlear efferent bundle and susceptibility of the inner ear to acoustic injury. J Neurophysiol 65:123–132.

Liberman MC (1993) Central projections of auditory nerve fibers of differing spontaneous rate, II: Posteroventral and dorsal cochlear nuclei. J Comp Neurol 327: 17–36.

Liberman MC, Brown MC (1986) Physiology and anatomy of single olivocochlear neurons in the cat. Hear Res 24:17–36.

Liberman MC, Gao WY (1995) Chronic cochlear de-efferentation and susceptibility to permanent acoustic injury. Hear Res 90:158–168.

Liberman MC, Dodds LW, Pierce S (1990) Afferent and efferent innervation of the cat cochlea: quantitative analysis with light and electron microscopy [published erratum appears in J Comp Neurol 1991 Feb 8;304(2):341]. J Comp Neurol 301: 443–460.

Linke R, De Lima AD, Schwegler H, Pape HC (1999) Direct synaptic connections of axons from superior colliculus with identified thalamo-amygdaloid projection neurons in the rat: possible substrates of a subcortical visual pathway to the amygdala. J Comp Neurol 403:158–170.

Liu XB, Warren RA, Jones EG (1995) Synaptic distribution of afferents from reticular nucleus in ventroposterior nucleus of cat thalamus. J Comp Neurol 352:187–202.

Livingstone MS, Hubel DH (1981) Effects of sleep and arousal on the processing of visual information in the cat. Nature 291:554–561.

Llinas R, Ribary U (1993) Coherent 40-Hz oscillation characterizes dream state in humans. Proc Natl Acad Sci USA 90:2078–2081.

Lund JS (1984) Spiny stellate neurons. In: Jones EG, Peters A (eds) The cerebral cortex, pp. 255–308. New York: Plenum Press.

Lund JS, Lund RD, Hendrickson AE, Bunt AH, Fuchs AF (1975) The origin of efferent pathways from the primary visual cortex, area 17, of the macaque monkey as shown by retrograde transport of horseradish peroxidase. J Comp Neurol 164: 287–303.

Macdonald KD, Fifkova E, Jones MS, Barth DS (1998) Focal stimulation of the thalamic reticular nucleus induces focal gamma waves in cortex. J Neurophysiol 79:474–477.

Malmierca MS, Rees A, LeBeau FEN (1997) Ascending projections to the medial geniculate body from physiologically identified loci in the inferior colliculus. In: Syka J (ed) Acoustic signal processing in the central auditory system, pp. 295–302. New York: Plenum.

Maren S (1999) Neurotoxic basolateral amygdala lesions impair learning and memory but not the performance of conditional fear in rats. J Neurosci 19: 8696–8703.

Martin KA, Whitteridge D (1984) Form, function and intracortical projections of spiny neurones in the striate visual cortex of the cat. J Physiol (Lond) 353: 463–504.

May BJ, McQuone SJ, Lavoie A (1995) Effects of olivocochlear lesions on intensity discrimination in cats. Assoc Res Otolaryngol 18:581.

McCormick DA, Von Krosigk M (1992) Corticothalamic activation modulates thalamic firing through glutamate "metabotropic" receptors. Proc Natl Acad Sci USA 89:2774–2778.

McEchron MD, Green EJ, Winters RW, Nolen TG, Schneiderman N, McCabe PM (1996) Changes of synaptic efficacy in the medial geniculate nucleus as a result of auditory classical conditioning. J Neurosci 16:1273–1283.

McKernan MG, Shinnick-Gallagher P (1997) Fear conditioning induces a lasting potentiation of synaptic currents in vitro. Nature 390:607–611.

McMullen NT, De Venecia RK (1993) Thalamocortical patches in auditory neocortex. Brain Res 620:317–322.

McMullen NT, Glaser EM (1982) Morphology and laminar distribution of nonpyramidal neurons in the auditory cortex of the rabbit. J Comp Neurol 208: 85–106.

Merchan-Perez A, Liberman MC (1996) Ultrastructural differences among afferent synapses on cochlear hair cells: correlations with spontaneous discharge rate. J Comp Neurol 371:208–221.

Meredith MA, Clemo HR (1989) Auditory cortical projection from the anterior ectosylvian sulcus (Field AES) to the superior colliculus in the cat: an anatomical and electrophysiological study. J Comp Neurol 289:687–707.

Meredith MA, Stein BE (1996) Spatial determinants of multisensory integration in cat superior colliculus neurons. J Neurophysiol 75:1843–1857.

Merzenich MM, Jenkins WM, Johnston P, Schreiner C, Miller SL, Tallal P (1996) Temporal processing deficits of language-learning impaired children ameliorated by training. Science 271:77–81.

Metherate R, Aramakis VB (1999) Intrinsic electrophysiology of neurons in thalamorecipient layers of developing rat auditory cortex. Brain Research. Devel Brain Res 115:131–144.

Meyer G, Gonzalez-Hernandez TH, Ferres-Torres R (1989) The spiny stellate neurons in layer IV of the human auditory cortex. A Golgi study. Neuroscience 33:489–498.

Middlebrooks JC, Knudsen EI (1984) A neural code for auditory space in the cat's superior colliculus. J Neurosci 4:2621–2634.

Middlebrooks JC, Zook JM (1983) Intrinsic organization of the cat's medial geniculate body identified by projections to binaural response-specific bands in the primary auditory cortex. J Neurosci 3:203–224.

Mitani A, Shimokouchi M (1985) Neuronal connections in the primary auditory cortex: an electrophysiological study in the cat. J Comp Neurol 235:417–429.

Mitani A, Shimokouchi M, Nomura S (1983) Effects of stimulation of the primary auditory cortex upon colliculogeniculate neurons in the inferior colliculus of the cat. Neurosci Lett 42:185–189.

Mitani A, Itoh K, Nomura S, Kudo M, Kaneko T, Mizuno N (1984) Thalamocortical projections to layer I of the primary auditory cortex in the cat: a horseradish peroxidase study. Brain Res 310:347–350.

Mitani A, Shimokouchi M, Itoh K, Nomura S, Kudo M, Mizuno N (1985) Morphology and laminar organization of electrophysiologically identified neurons in the primary auditory cortex in the cat. J Comp Neurol 235:430–447.

Mitani A, Itoh K, Mizuno N (1987) Distribution and size of thalamic neurons projecting to layer I of the auditory cortical fields of the cat compared to those projecting to layer IV. J Comp Neurol 257:105–121.

Mitzdorf U (1985) Current source-density method and application in cat cerebral cortex: investigation of evoked potentials and EEG phenomena. Physiol Rev 65: 37–100.

Mitzdorf U (1991) Physiological sources of evoked potentials. Electroenceph Clin Neurophysiol Suppl 42:47–57.

Mitzdorf U, Singer W (1979) Excitatory synaptic ensemble properties in the visual cortex of the macaque monkey: a current source density analysis of electrically evoked potentials. J Comp Neurol 187:71–83.

Molinari M, Dell'Anna ME, Rausell E, Leggio MG, Hashikawa T, Jones EG (1995) Auditory thalamocortical pathways defined in monkeys by calcium-binding protein immunoreactivity. J Comp Neurol 362:171–194.

Montero VM (1983) Ultrastructural identification of axon terminals from the thalamic reticular nucleus in the medial geniculate body of the rat: An EM autoradiogrtaphic study. Exp Brain Res 51:338–342.

Morel A, Garraghty PE, Kaas JH (1993) Tonotopic organization, architectonic fields, and connections of auditory cortex in macaque monkeys. J Comp Neurol 335: 437–459.

Morest DK (1968) The collataral system of the medial nucleus of the trapezoid body of the cat, its neuronal architecture and relationship to the olivo-cochlear bundle. Brain Res 9:288–311.

Morgan YV, Ryugo DK, Brown MC (1994) Central trajectories of type II (thin) fibers of the auditory nerve in cats. Hear Res 79:74–82.

Morley BJ, Li HS, Hiel H, Drescher DG, Elgoyhen AB (1998) Identification of the subunits of the nicotinic cholinergic receptors in the rat cochlea using RT-PCR and in situ hybridization. Brain Res Mol Brain Res 53:78–87.

Morton NE (1991) Genetic epidemiology of hearing impairment. Ann N Y Acad Sci 630:16–31.

Mucke L, Norita M, Benedek G, Creutzfeldt O (1982) Physiologic and anatomic investigation of a visual cortical area situated in the ventral bank of the anterior ectosylvian sulcus of the cat. Exp Brain Res 46:1–11.

Mulders WH, Robertson D (2000) Evidence for direct cortical innervation of medial olivocochlear neurons in rats. Hear Res 144:65–72.

Nagarajan S, Mahncke H, Salz T, Tallal P, Roberts T, Merzenich MM (1999) Cortical auditory signal processing in poor readers. Proc Nat Acad Sci USA 96:6483–6488.

Nagarajan SS, Wang X, Merzenich MM, Schreiner CE, Johnston P, Jenkins WM, Miller S, Tallal P (1998) Speech modifications algorithms used for training language learning-impaired children. IEEE Trans Rehabil Eng 6:257–268.

Niimi K, Ono K, Kusunose M (1984) Projections of the medial geniculate nucleus to layer 1 of the auditory cortex in the cat traced with horseradish peroxidase. Neurosci Lett 45:223–228.

Nishimura H, Hashikawa K, Doi K, Iwaki T, Watanabe Y, Kusuoka H, Nishimura T, Kubo T (1999) Sign language "heard" in the auditory cortex. Nature 397:116.

Nordang L, Cestreicher E, Arnold W, Anniko, M (2000) Glutamate is the afferent neurotransmitter in the human cochlea. Acta Otolaryngol 120:359–362.

Ohara PT, Lieberman AR (1985) The thalamic reticular nucleus of the adult rat: experimental anatomical studies. J Neurocytol 14:365–411.

Ohishi H, Shigemoto R, Nakanishi S, Mizuno N (1993a) Distribution of the messenger RNA for a metabotropic glutamate receptor, mGluR2, in the central nervous system of the rat. Neuroscience 53:1009–1018.

Ohishi H, Shigemoto R, Nakanishi S, Mizuno N (1993b) Distribution of the mRNA for a metabotropic glutamate receptor (mGluR3) in the rat brain: an in situ hybridization study. J Comp Neurol 335:252–266.

Ojima H (1994) Terminal morphology and distribution of corticothalamic fibers originating from layers 5 and 6 of cat primary auditory cortex. Cereb Cortex 4:646–663.

Ojima H, He JF (1997) Cortical convergence originating from domains representing different frequencies in the cat AI. Acta Otolaryngolog Suppl 532:126–128.

Ojima H, Honda CN, Jones EG (1991) Patterns of axon collateralization of identified supragranular pyramidal neurons in the cat auditory cortex. Cereb Cortex 1:80–94.

Ojima H, Honda CN, Jones EG (1992) Characteristics of intracellularly injected infragranular pyramidal neurons in cat primary auditory cortex. Cereb Cortex 2: 197–216.

Oliver DL (1984) Neuron types in the central nucleus of the inferior colliculus that project to the medial geniculate body. Neuroscience 11:409–424.

Oliver DL, Winer JA, Beckius GE, Saint Marie RL (1994) Morphology of GABAergic neurons in the inferior colliculus of the cat. J Comp Neurol 340:27–42.

Oliver D, Klonker N, Schuck J, Baukrowitz T, Ruppersberg JP, Fakler (2000) Gating of Ca2+-activated K+ channels controls fast inhibitory synaptic transmission at auditory outer hair cells. Neuron 26:595–601.

Olson CR, Graybiel AM (1987) Ectosylvian visual area of the cat: location, retinotopic organization, and connections. J Comp Neurol 261:277–294.

Orman SS, Humphrey GL (1981) Effects of changes in cortical arousal and of auditory cortex cooling on neuronal activity in the medial geniculate body. Exp Brain Res 42:475–482.

Ostapoff EM, Morest DK, Potashner SJ (1990) Uptake and retrograde transport of [³H]GABA from the cochlear nucleus to the superior olive in the guinea pig. J Chem Neuroanat 3:285–295.

Ostapoff EM, Benson CG, Saint Marie RL (1997a) GABA- and glycine-immunoreactive projections from the superior olivary complex to the cochlear nucleus in guinea pig. J Comp Neurol 381:500–512.

Ostapoff EM, Benson CG, Saintmarie RL (1997b) Gaba- and glycine-immunoreactive projections from the superior olivary complex to the cochlear nucleus in guinea pig. J Comp Neurol 381:500–512.

Ota Y, Dolan DF (1999) Localized electrical stimulation of the inferior colliculus produces frequency specific reductions in the cochlear whole-nerve action potential. Assoc Res Otolaryngol 22:210.

Pallas SL, Sur M (1994) Morphology of retinal axon arbors induced to arborize in a novel target, the medial geniculate nucleus. II. Comparison with axons from the inferior colliculus. J Comp Neurol 349:363–376.

Pandya DN (1995) Anatomy of the auditory cortex. Rev Neurol (Paris) 151:486–494.

Peruzzi D, Bartlett E, Smith PH, Oliver DL (1997) A monosynaptic GABAergic input from the inferior colliculus to the medial geniculate body in rat. J Neurosci 17:3766–3777.

Pinault D, Deschenes M (1998) Anatomical evidence for a mechanism of lateral inhibition in the rat thalamus. Eur J Neurosci 10:3462–3469.

Ponton CW, Moore JK, Eggermont JJ (1999) Prolonged deafness limits auditory system developmental plasticity: evidence from an evoked potentials study in children with cochlear implants. Scand Audiol 28:13–22.

Popper AN, Fay RR (Eds) (1992) Springer Handbook of Auditory Research. New York: Springer-Verlag.

Quirk GJ, Repa C, LeDoux JE (1995) Fear conditioning enhances short-latency auditory responses of lateral amygdala neurons: parallel recordings in the freely behaving rat. Neuron 15:1029–1039.

Rabionet R, Gasparini P, Estivill X (2000) Molecular genetics of hearing impairment due to mutations in gap junction genes encoding beta connexins. Hum Mutat 16:190–202.

Rajan R (1995) Involvement of cochlear efferent pathways in protective effects elicited with binaural loud sound exposure in cats. J Neurophysiol 74:582–597.

Rajan R (1996) Additivity of loud-sound-induced threshold losses in the cat under conditions of active or inactive cochlear efferent-mediated protection. J Neurophysiol 75:1601–1618.

Rajan R, Irvine DR (1998) Neuronal responses across cortical field A1 in plasticity induced by peripheral auditory organ damage. Audiol Neurootol 3:123–144.

Rajan R, Johnstone BM (1988) Binaural acoustic stimulation exercises protective effects at the cochlea that mimic the effects of electrical stimulation of an auditory efferent pathway. Brain Res 459:241–255.

Rajan R, Irvine DR, Wise LZ, Heil P (1993) Effect of unilateral partial cochlear lesions in adult cats on the representation of lesioned and unlesioned cochleas in primary auditory cortex. J Comp Neurol 338:17–49.

Rauschecker JP (1995) Developmental plasticity and memory. Behav Brain Res 66:7–12.

Rauschecker JP (1996) Substitution of visual by auditory inputs in the cat's anterior ectosylvian cortex. Prog Brain Res 112:313–323.

Rauschecker JP (1997) Processing of complex sounds in the auditory cortex of cat, monkey, and man. Acta Otolaryngolog Suppl 532:34–38.

Rauschecker JP (1998) Parallel processing in the auditory cortex of primates. Audiol Neurootol 3:86–103.

Rauschecker JP (1999) Auditory cortical plasticity: a comparison with other sensory systems. Trends Neurosci 22:74–80.

Rauschecker JP, Korte M (1993) Auditory compensation for early blindness in cat cerebral cortex. J Neurosci 13:4538–4548.

Rauschecker JP, Tian B, Pons T, Mishkin M (1997) Serial and parallel processing in rhesus monkey auditory cortex. J Comp Neurol 382:89–103.

Recanzone GH, Schreiner CE, Merzenich MM (1993) Plasticity in the frequency representation of primary auditory cortex following discrimination training in adult owl monkeys. J Neurosci 13:87–103.

Reinagel P, Godwin D, Sherman SM, Koch C (1999) Encoding of visual information by LGN bursts. J Neurophysiol 81:2558–2569.

Reuss S, Disque-Kaiser U, De Liz S, Ruffer M, Riemann R (1999) Immunfluorescence study of neuropeptides in identified neurons of the rat auditory superior olivary complex. Cell Tis Res 297:13–21.

Riemann R, Reuss S (1998) Projection neurons in the superior olivary complex of the rat auditory brainstem: a double retrograde tracing study. J of Otorhinolaryngol 60:278–282.

Robertson D (1996) Physiology and morphology of cells in the ventral nucleus of the trapezoid body and rostral periolivary regions of the rat superior olivary complex studied in slices. Aud Neurosci 2:15–32.

Robertson D, Gummer M (1985) Physiological and morphological characterization of efferent neurones in the guinea pig cochlea. Hear Res 20:63–77.

Robertson D, Irvine DR (1989) Plasticity of frequency organization in auditory cortex of guinea pigs with partial unilateral deafness. J Comp Neurol 282:456–471.

Robinson K (1998) Implications of developmental plasticity for the language acquisition of deaf children with cochlear implants. Int J Pediatr Otorhinolaryngol 46:71–80.

Rodrigues-Dagaeff C, Simm G, De Ribaupierre Y, Villa A, De Ribaupierre F, Rouiller EM (1989) Functional organization of the ventral division of the medial geniculate body of the cat: evidence for a rostro-caudal gradient of response properties and cortical projections. Hear Res 39:103–125.

Rogan MT, LeDoux JE (1995) LTP is accompanied by commensurate enhancement of auditory-evoked responses in a fear conditioning circuit. Neuron 15:127–136.

Rogan MT, Staubli UV, LeDoux JE (1997) AMPA receptor facilitation accelerates fear learning without altering the level of conditioned fear acquired. J Neurosci 17:5928–5935.

Rose JE (1949) The cellular structure of the auditory region of the cat. J Comp Neurol 91:409–440.

Rouiller EM, Welker E (1991) Morphology of corticothalamic terminals arising from the auditory cortex of the rat: a Phaseolus vulgaris-leucoagglutinin (PHA-L) tracing study. Hear Res 56:179–190.

Rouiller EM, Colomb E, Capt M, De Ribaupierre F (1985) Projections of the reticular complex of the thalamus onto physiologically characterized regions of the medial geniculate body. Neurosci Lett 53:227–232.

Rouiller EM, Cronin-Schreiber R, Fekete DM, Ryugo DK (1986) The central projections of intracellularly labeled auditory nerve fibers in cats: an analysis of terminal morphology. J Comp Neurol 249:261–278.

Rouiller EM, Hornung JP, De Ribaupierre F (1989) Extrathalamic ascending projections to physiologically identified fields of the cat auditory cortex. Hear Res 40:233–246.

Ryugo DK (1992) The auditory nerve: Peripheral innervation, cell body morphology and central projections. In: Webster DB, Popper AN, Fay RR (eds) The mammalian auditory pathway: Neuroanatomy, pp. 23–65. New York: Springer-Verlag.

Ryugo DK, May SK (1993) The projections of intracellularly labeled auditory nerve fibers to the dorsal cochlear nucleus of cats. J Comp Neurol 329:20–35.

Ryugo DK, Rouiller EM (1988) Central projections of intracellularly labeled auditory nerve fibers in cats: morphometric correlations with physiological properties. J Comp Neurol 271:130–142.

Ryugo DK, Weinberger NM (1976) Corticofugal modulation of the medial geniculate body. Exp Neurol 51:377–391.

Ryugo DK, Weinberger NM (1978) Differential plasticity of morphologically distinct neuron populations in the medical geniculate body of the cat during classical conditioning. Behav Biol 22:275–301.

Ryugo DK, Dodds LW, Benson TE, Kiang NY (1991) Unmyelinated axons of the auditory nerve in cats. J Comp Neurol 308:209–223.

Ryugo DK, Wu MM, Pongstaporn T (1996) Activity-related features of synapse morphology: a study of endbulbs of held. J Comp Neurol 365:141–158.

Saint Marie RL (1996) Glutamatergic connections of the auditory midbrain: selective uptake and axonal transport of D-[^{3}H]aspartate. J Comp Neurol 373:255–270.

Saint Marie RL, Baker RA (1990) Neurotransmitter-specific uptake and retrograde transport of [^{3}H]glycine from the inferior colliculus by ipsilateral projections of the superior olivary complex and nuclei of the lateral lemniscus. Brain Res 524:244–253.

Saint Marie RL, Stanforth DA, Jubelier EM (1997) Substrate for rapid feedforward inhibition of the auditory forebrain. Brain Res 765:173–176.

Saint Marie RL, Luo L, Ryan AF (1999) Effects of stimulus frequency and intensity on c-fos mRNA expression in the adult rat auditory brainstem. J Comp Neurol 404:258–270.

Saldana E, Berrebi AS (2000) Anisotropic organization of the rat superior paraolivary nucleus. Anat Embryol 202:265–279.

Saldana E, Feliciano M, Mugnaini E (1996) Distribution of descending projections from primary auditory neocortex to inferior colliculus mimics the topography of intracollicular projections. J Comp Neurol 371:15–40.

Sanes DH (1990) An in vitro analysis of sound localization mechanisms in the gerbil lateral superior olive. J Neurosci 10:3494–3506.

Sato A, Ohtsuka K (1996) Projection from the accommodation-related area in the superior colliculus of the cat. J Comp Neurol 367:465–476.

Scates KW, Woods CI, Azeredo WJ (1999) Inferior colliculus stimulation and changes in 2f(1)-f(2) distortion product otoacoustic emissions in the rat. Hear Res 128:51–60.

Schnupp JWH, King AJ (1997) Coding for auditory space in the nucleus of the brachium of the inferior colliculus in the ferret. J Neurophysiol 78:2717–2731.

Schofield BR (1995) Projections from the cochlear nucleus to the superior paraolivary nucleus in guinea pigs. J Comp Neurol 360:135–149.

Schofield BR, Cant NB (1992) Organization of the superior olivary complex in the guinea pig: II. Patterns of projection from the periolivary nuclei to the inferior colliculus. J Comp Neurol 317:438–455.

Schofield BR, Cant NB (1999) Descending auditory pathways: Projections from the inferior colliculus contact superior olivary cells that project bilaterally to the cochlear nuclei. J Comp Neurol 409:210–223.

Schwaber MK, Garraghty PE, Kaas JH (1993) Neuroplasticity of the adult primate auditory cortex following cochlear hearing loss. Am J Otol 14:252–258.

Schwartz IR (1992) The superior olivary complex and lateral lemnisal nuclei. In: Webster DB, Popper AN, Fay RR (eds) The mammalian auditory pathway: Neuroanatomy, pp. 117–167. New York: Springer-Verlag.

Sento S, Ryugo DK (1989) Endbulbs of held and spherical bushy cells in cats: morphological correlates with physiological properties. J Comp Neurol 280:553–562.

Shammah-Lagnado SJ, Alheid GF, Heimer L (1996) Efferent connections of the caudal part of the globus pallidus in the rat. J Comp Neurol 376:489–507.

Sharma J, Angelucci A, Sur M (2000) Induction of visual orientation modules in auditory cortex. Nature 404:841–847.

Shaywitz SE (1996) Dyslexia. Sci Am 275:98–104.

Shaywitz SE (1998) Dyslexia. N Engl J Med 338:307–312.

Shepherd RK, Hartmann R, Heid S, Hardie N, Klinke R (1997) The central auditory system and auditory deprivation: experience with cochlear implants in the congenitally deaf. Acta Otolaryngol Suppl 532:28–33.

Sherman SM, Guillery RW (1996) Functional organization of thalamocortical relays. J Neurophysiol 76:1367–1395.

Sherriff FE, Henderson Z (1994) Cholinergic neurons in the ventral trapezoid nucleus project to the cochlear nuclei in the rat. Neuroscience 58:627–633.

Shinonaga Y, Takada M, Ogawa-Meguro R, Ikai Y, Mizuno N (1992) Direct projections from the globus pallidus to the midbrain and pons in the cat. Neurosci Lett 135:179–183.

Shosaku A, Sumitomo I (1983) Auditory neurons in the rat thalamic reticular nucleus. Exp Brain Res 49:432–442.

Simm GM, de Ribaupierre F, de Ribaupierre Y, Rouiller EM (1990) Discharge properties of single units in auditory part of reticular nucleus of thalamus in cat. J Neurophysiol 63:1010–1021.

Simons DJ, Woolsey TA (1984) Morphology of Golgi-Cox-impregnated barrel neurons in rat SmI cortex. J Comp Neurol 230:119–132.

Smith PH, Populin LC (2001) Fundamental differences in the thalamocortical recipient layer of the cat auditory and visual cortices. J Comp Neurol 436:508–519.

Smith PH, Rhode WS (1989) Structural and functional properties distinguish two types of multipolar cells in the ventral cochlear nucleus. J Comp Neurol 282:595–616.

Smith PH, Joris PX, Carney LH, Yin TC (1991) Projections of physiologically characterized globular bushy cell axons from the cochlear nucleus of the cat. J Comp Neurol 304:387–407.

Smith PH, Joris PX, Yin TC (1993) Projections of physiologically characterized spherical bushy cell axons from the cochlear nucleus of the cat: evidence for delay lines to the medial superior olive. J Comp Neurol 331:245–260.

Snyder RL, Leake PA, Hradek GT (1997) Quantitative analysis of spiral ganglion projections to the cat cochlear nucleus. J Comp Neurol 379:133–149.

Sobkowicz HM, Slapnick SM, Nitecka LM, August BK (1998) Tunnel crossing fibers and their synaptic connections within the inner hair cell region in the Organ of Corti in the maturing mouse. Anat Embryol 198:353–370.

Sommer I, Lingenhohl K, Friauf E (1993) Principal cells of the rat medial nucleus of the trapezoid body: an intracellular in vivo study of their physiology and morphology. Exp Brain Res 95:223–239.

Somogyi P, Tamas G, Lujan R, Buhl EH (1998) Salient features of synaptic organisation in the cerebral cortex. Brain Res Rev 26:113–135.

Spirou GA (1999) Convergence of auditory nerve projections onto globular bushy cells. Assoc Res Otolaryngol 22:147.

Spirou GA, Berrebi AS (1996) Organization of ventrolateral periolivary cells of the cat superior olive as revealed by PEP-19 immunocytochemistry and Nissl stain. J Comp Neurol 368:100–120.

Spirou GA, Brownell WE, Zidanic M (1990) Recordings from cat trapezoid body and HRP labeling of globular bushy cell axons. J Neurophysiol 63:1169–1190.

Spirou GA, Walker MP, Berrebi AS (1995) Connectivity of the lateral nucleus of the trapezoid body in cats. Assoc Res Otolaryngol 18:155.

Spirou GA, Rowland KC, Berrebi AS (1998) Ultrastructure of neurons and large synaptic terminals in the lateral nucleus of the trapezoid body of the cat. J Comp Neurol 398:257–272.

Spreafico R, de Curtis M, Frassoni C, Avanzini G (1988) Electrophysiological characteristics of morphologically identified reticular thalamic neurons from rat slices. Neuroscience 27:629–638.

Sridhar TS, Liberman MC, Brown MC, Sewell WF (1995) A novel cholinergic "slow effect" of efferent stimulation on cochlear potentials in the guinea pig. J Neurosci 15:3667–3678.

Steel KP, Bussoli TJ (1999) Deafness genes—expressions of surprise. Trends in Genet 15:207–211.

Stein BE (1998) Neural mechanisms for synthesizing sensory information and producing adaptive behaviors. Exp Brain Res 123:124–135.

Steinschneider M, Tenke CE, Schroeder CE, Javitt DC, Simpson GV, Arezzo JC, Vaughan HG, Jr. (1992) Cellular generators of the cortical auditory evoked potential initial component. Electroenceph Clin Neurophysiol 84:196–200.

Steriade M (1998) Corticothalamic networks, oscillations, and plasticity. Adv Neurol 77:105–134.

Steriade M, Domich L, Oakson G (1986) Reticularis thalami neurons revisited: activity changes during shifts in states of vigilance. J Neurosci 6:68–81.

Steriade M, McCormick DA, Sejnowski TJ (1993) Thalamocortical oscillations in the sleeping and aroused brain. Science 262:679–685.

Suga N, O'Neill WE, Manabe T (1979) Harmonic-sensitive neurons in the auditory cortex of the mustache bat. Science 203:270–274.

Suga N, O'Neill WE, Kujirai K, Manabe T (1983) Specificity of combination-sensitive neurons for processing of complex biosonar signals in auditory cortex of the mustached bat. J Neurophysiol 49:1573–1626.

Suga N, Zhang Y, Yan J (1997) Sharpening of frequency tuning by inhibition in the thalamic auditory nucleus of the mustached bat. J Neurophysiol 77:2098–2114.

Sun XD, Jen PH, Sun DX, Zhang SF (1989) Corticofugal influences on the responses of bat inferior collicular neurons to sound stimulation. Brain Res 495:1–8.

Sundstrom RA, Van Laer L, Van Camp G, Smith RJH (1999) Autosomal recessive nonsyndromic hearing loss. Am J Med Genet 89:123–129.

Sutter ML, Schreiner CE (1991) Physiology and topography of neurons with multipeaked tuning curves in cat primary auditory cortex. J Neurophysiol 65:1207–1226.

Syka J, Popelar J (1984) Inferior colliculus in the rat: neuronal responses to stimulation of the auditory cortex. Neurosci Lett 51:235–240.

Tallal P (1980) Language disabilities in children: a perceptual or linguistic deficit? J Pediatr Psychol 5:127–140.

Tallal P, Piercy M (1973) Developmental aphasia: impaired rate of non-verbal processing as a function of sensory modality. Neuropsychol 11:389–398.

Tallal P, Stark RE (1981) Speech acoustic-cue discrimination abilities of normally developing and language-impaired children. J Acoust Soc Am 69:568–574.

Tallal P, Miller SL, Bedi G, Byma G, Wang X, Nagarajan SS, Schreiner C, Jenkins WM, Merzenich MM (1996) Language comprehension in language-learning impaired children improved with acoustically modified speech. Science 271:81–84.

Tennigkeit F, Puil E, Schwarz DW (1997) Firing modes and membrane properties in lemniscal auditory thalamus. Acta Otolaryngol 117:254–257.

Thiele A, Rubsamen R, Hoffmann KP (1996) Anatomical and physiological investigation of auditory input to the superior colliculus of the echolocating megachiropteran bat *Rousettus aegyptiacus*. Exp Brain Res 112:223–236.

Thompson AM (1998) Heterogeneous projections of the cat posteroventral cochlear nucleus. J Comp Neurol 390:439–453.

Thompson AM, Thompson GC (1991) Posteroventral cochlear nucleus projections to olivocochlear neurons. J Comp Neurol 303:267–285.

Thompson AM, Thompson GC (1993) Relationship of descending inferior colliculus projections to olivocochlear neurons. J Comp Neurol 335:402–412.

Thompson AM, Thompson GC (1995) Light microscopic evidence of serotoninergic projections to olivocochlear neurons in the bush baby (*Otolemur garnettii*). Brain Res 695:263–266.

Thompson AM, Schofield BR (2000) Afferent projections of the superior olivary complex. Microsc Res Tech 51:330–354.

Tolbert LP, Morest DK, Yurgelun-Todd DA (1982) The neuronal architecture of the anteroventral cochlear nucleus of the cat in the region of the cochlear nerve root: horseradish peroxidase labelling of identified cell types. Neuroscience 7:3031–3052.

Torterolo P, Zurita P, Pedemonte M, Velluti RA (1998) Auditory cortical efferent actions upon inferior colliculus unitary activity in the guinea pig. Neurosci Lett 249:172–176.

Tsuchitani C (1977) Functional organization of lateral cell groups of cat superior olivary complex. J Neurophysiol 40:296–318.

Tsuji J, Liberman MC (1997) Intracellular labeling of auditory nerve fibers in guinea pig: central and peripheral projections. J Comp Neurol 381:188–202.

Vetter DE, Liberman MC, Mann J, Barhanin J, Boulter J, Brown MC, Saffiote-Kolman J, Heinemann SF, Elgoyhen AB (1999) Role of alpha9 nicotinic ACh receptor subunits in the development and function of cochlear efferent innervation. Neuron 23:93–103.

Vetter DE, Mugnaini E (1992) Distribution and dendritic features of three groups of rat olivocochlear neurons. A study with two retrograde cholera toxin tracers. Anat Embryol 185:1–16.

Vetter DE, Saldana E, Mugnaini E (1993) Input from the inferior colliculus to medial olivocochlear neurons in the rat: a double label study with PHA-L and cholera toxin. Hear Res 70:173–186.

Villa AE (1990) Physiological differentiation within the auditory part of the thalamic reticular nucleus of the cat. Brain Res Brain Res Rev 15:25–40.

Villa AE, Rouiller EM, Simm GM, Zurita P, de Ribaupierre Y, de Ribaupierre F (1991) Corticofugal modulation of the information processing in the auditory thalamus of the cat. Exp Brain Res 86:506–517.

Villa AEP, Tetko IV, Dutoit P, De Ribaupierre Y, De Ribaupierre F (1999) Corticofugal modulation of functional connectivity within the auditory thalamus of rat, guinea pig and cat revealed by cooling deactivation. J Neurosci Methods 86:161–178.

Von Economo C (1929) The cytoarchitecture of the human cerebral cortex. London: Oxford University Press.

Von Economo C, Koskinas GN (1925) Die cytoarchitektonik der hirnrinde des erwachsesen menchen. Berlin: Springer.

Von Melchner L, Pallas SL, Sur M (2000) Visual behaviour mediated by retinal projections directed to the auditory pathway. Nature 404:871–876.

Walker AE (1937) The projection of the medial geniculate body to the cerebral cortex in the macaque monkey. J Anat 71:319–331.

Wallace MT, Stein BE (1994) Cross-modal synthesis in the midbrain depends on input from cortex. J Neurophysiol 71:429–432.

Wallace MT, Meredith MA, Stein BE (1993) Converging influences from visual, auditory, and somatosensory cortices onto output neurons of the superior colliculus. J Neurophysiol 69:1797–1809.

Walsh EJ, McGee J, McFadden SL, Liberman MC (1998) Long-term effects of sectioning the olivocochlear bundle in neonatal cats. J Neurosci 18:3859–3869.

Waltzman SB, Cohen NL (1998) Cochlear implantation in children younger than 2 years old. Am J Otol 19:158–162.

Wang X, Robertson D (1997a) Effects of bioamines and peptides on neurones in the ventral nucleus of trapezoid body and rostral periolivary regions of the rat superior olivary complex: an in vitro investigation. Hear Res 106:20–28.

Wang X, Robertson D (1997b) Two types of actions of norepinephrine on identified auditory efferent neurons in rat brainstem slices. J Neurophysiol 78:1800–1810.

Wang X, Robertson D (1998a) Substance P-induced inward current in identified auditory efferent neurons in rat brainstem slices. J Neurophysiol 80:218–229.

Wang X, Robertson D (1998b) Substance P-sensitive neurones in the rat auditory brainstem: possible relationship to medial olivocochlear neurones. Hear Res 116:86–98.

Wang YX, Wenthold RJ, Ottersen OP, Petralia RS (1998) Endbulb synapses in the anteroventral cochlear nucleus express a specific subset of AMPA-type glutamate receptor subunits. J Neurosci 18:1148–1160.

Warr WB (1969) Fiber degeneration following lesions in the posteroventral cochlear nucleus of the cat. Exp Neurol 23:140–155.

Warr WB (1992) Organization of the olivocochlear efferent systems in mammals. In: Webster DB, Popper AN, Fay RR (eds) The auditory mammalian pathway: Neuroanatomy, pp. 410–448. New York: Springer-Verlag.

Warr WB, Beck JE (1996) Multiple projections from the ventral nucleus of the trapezoid body in the rat. Hear Res 93:83–101.

Warr WB, Boche JB, Neely ST (1997) Efferent innervation of the inner hair cell region—origins and terminations of two lateral olivocochlear systems. Hear Res 108:89–111.

Warr WB, Gratton MA, Boche JEB (1999) Fine structure of labeled lateral efferent axons in the rat cochlea. Assoc Res Otolaryngol 22:211.

Warren RA, Jones EG (1994) Glutamate activation of cat thalamic reticular nucleus: effects on response properties of ventroposterior neurons. Exp Brain Res 100:215–226.

Webber TJ, Green EJ, Winters RW, Schneiderman N, McCabe PM (1999) Contribution of NMDA and non-NMDA receptors to synaptic transmission from the brachium of the inferior colliculus to the medial subdivision of the medial geniculate nucleus in the rabbit. Exp Brain Res 124:295–303.

Webster DB, Popper AN, Fay RR (Eds) (1992) Springer Handbook of Auditory Research. New York: Springer-Verlag.

Weedman DL, Ryugo DK (1996a) Projections from auditory cortex to the cochlear nucleus in rats: synapses on granule cell dendrites. J Comp Neurol 371:311–324.

Weedman DL, Ryugo DK (1996b) Pyramidal cells in primary auditory cortex project to cochlear nucleus in rat. Brain Res 706:97–102.

Weinberger NM (1998) Physiological memory in primary auditory cortex—characteristics and mechanisms. Neurobiol Learn Mem 70:226–251.

Weinberger NM, Javid R, Lepan B (1995) Heterosynaptic long-term facilitation of sensory-evoked responses in the auditory cortex by stimulation of the magnocellular medial geniculate body in guinea pigs. Behav Neurosci 109:10–17.

Wenstrup JJ, Larue DT, Winer JA (1994) Projections of physiologically defined subdivisions of the inferior colliculus in the mustached bat: targets in the medial geniculate body and extrathalamic nuclei. J Comp Neurol 346:207–236.

White EL (1978) Identified neurons in mouse SmI cortex which are postsynaptic to thalamocortical axon terminals: a combined Golgi-electron microscopic and degeneration study. J Comp Neurol 181:627–661.

White EL, Rock MP (1979) Distribution of thalamic input to different dendrites of a spiny stellate cell in mouse sensorimotor cortex. Neurosci Lett 15:115–119.

White EL, Rock MP (1980) Three-dimensional aspects and synaptic relationships of a Golgi-impregnated spiny stellate cell reconstructed from serial thin sections. J Neurocytol 9:615–636.

Whitley JM, Henkel CK (1984) Topographical organization of the inferior collicular projection and other connections of the ventral nucleus of the lateral lemniscus in the cat. J Comp Neurol 229:257–270.

Wilkinson LK, Meredith MA, Stein BE (1996) The role of anterior ectosylvian cortex in cross-modality orientation and approach behavior. Exp Brain Res 112: 1–10.

Williams EA, Brookes GB, Prasher DK (1994) Effects of olivocochlear bundle section on otoacoustic emissions in humans: efferent effects in comparison with control subjects. Acta Otolaryngol (Stockh) 114:121–129.

Williamson AM, Ohara PT, Ralston HJd (1993) Electron microscopic evidence that cortical terminals make direct contact onto cells of the thalamic reticular nucleus in the monkey. Brain Res 631:175–179.

Winer JA (1984) Anatomy of layer IV in cat primary auditory cortex (AI). J Comp Neurol 224:535–567.

Winer JA (1985) The medial geniculate body of the cat. Advances in Anatomy, Embryology & Cell Biology 86:1–97.

Winer JA (1992) Thalamus and cortex. In: Webster DB, Popper AN, Fay RF (eds) The mammalian auditory pathway: Neuroanatomy, pp. 222–409. New York: Springer-Verlag.

Winer JA, Larue DT (1988) Anatomy of glutamic acid decarboxylase immunoreactive neurons and axons in the rat medial geniculate body. J Comp Neurol 278:47–68.

Winer JA, Morest DK (1983) The medial division of the medial geniculate body of the cat: implications for thalamic organization. J Neurosci 3:2629–2651.

Winer JA, Saint Marie RL, Larue DT, Oliver DL (1996) GABAergic feedforward projections from the inferior colliculus to the medial geniculate body. Proc Natl Acad Sci USA 93:8005–8010.

Winer JA, Larue DT, Diehl JJ, Hefti BJ (1998) Auditory cortical projections to the cat inferior colliculus. J Comp Neurol 400:147–174.

Wong D, Kelly JP (1981) Differentially projecting cells in individual layers of auditory cortex: a double labeling study. Brain Res 230:362–366.

Woods CI, Azeredo WJ (1999) Noradrenergic and serotonergic projections to the superior olive: potential for modulation of olivocochlear neurons. Brain Res 836: 9–18.

Woolsey TA (1996) Barrels: 25 years later. Somatosens Mot Res 13:181–186.

Wright BA, Lombardino LJ, King WM, Puranik CS, Leonard CM, Merzenich MM (1997) Deficits in auditory temporal and spectral resolution in language-impaired children. Nature 387:176–178.

Wu SH, Kelly JB (1992) Synaptic pharmacology of the superior olivary complex studied in mouse brain slice. J Neurosci 12:3084–3097.

Wu SH, Kelly JB (1994) Physiological evidence for ipsilateral inhibition in the lateral superior olive: synaptic responses in mouse brain slice. Hear Res 73:57–64.

Wynne B, Robertson D (1997) Somatostatin and substance P-like immunoreactivity in the auditory brainstem of the adult rat. J Chem Neuroanat 12:259–266.

Xia AP, Kikuchi T, Hozawa K, Katori Y, Takasaka T (1999) Expression of the connexin 26 and Na, K-ATPase in the developing mouse cochlear lateral wall: functional applications. Brain Res 846:106–111.

Ye Y, Machado DG, Kim DO (2000) Projection of the marginal shell of the anteroventral cochlear nucleus to olivocochlear neurons in the cat. J Comp Neurol 420:137–148.

Yen CT, Conley M, Hendry SH, Jones EG (1985) The morphology of physiologically identified GABAergic neurons in the somatic sensory part of the thalamic reticular nucleus in the cat. J Neurosci 5:2254–2268.

Yoshida N, Kristiansen A, Liberman MC (1999) Heat stress and protection from permanent acoustic injury in mice. J Neurosci 19:10116–10124.

Zhang Y, Suga N (1997) Corticofugal amplification of subcortical responses to single tone stimuli in the mustached bat. J Neurophysiol 78:3489–3492.

3
Cellular Mechanisms for Information Coding in Auditory Brainstem Nuclei

LAURENCE O. TRUSSELL

1. Introduction and Overview

The brainstem auditory nuclei carry out a wide variety of transformations of the signals carried by the auditory nerve. Although basic frequency and intensity information is first encoded in the cochlea, brainstem circuitry must perform further neural definitions and refinements of these parameters, as well as integrate the cues necessary for the localization of sounds in space. Each of these aspects is associated not just with certain cell types, morphologies, and synaptic connections, but with cells having characteristic electrical response profiles. Such response properties are an outcome of the complement of ion channels that the cells possess and of the dynamic properties of the synapses through which cells communicate.

The link between the properties of channels and synapses and the higher-order functions of circuits is not simply correlative. Rather, we are beginning to understand the function of cellular and membrane properties in terms of the relationship between a given sound stimulus and the responses of the auditory nerve and different brainstem neurons in vivo. The cell types of the cochlear nuclei and superior olivary complex are numerous, and the full spectrum of their responses and functions is not yet clear. However, some key cell types have been well described and serve as excellent illustrations of the basic electrical themes. Auditory nerve fibers respond to simple acoustic stimuli with two general response profiles (Fig. 3.1; see Ruggero 1992). For low-frequency stimuli, nerve fibers fire action potentials or spikes in a phase-locked manner, i.e., with a spike occurring most often with a certain phase relationship to the sound stimulus. For high-frequency stimuli, responses show an initial peak at the onset of the sound, and then a rapid decline in firing rate down to a steady-state level of random (not phase-locked) activity. These responses are documented using a poststimulus-time histogram (PSTH), in which the time of occurrence of spikes during repeated presentations of a stimulus is recorded (Fig. 3.1). Such histograms illustrate at a glance the temporal relationship between the sound stimulus and the firing of the neuron. Recordings made from

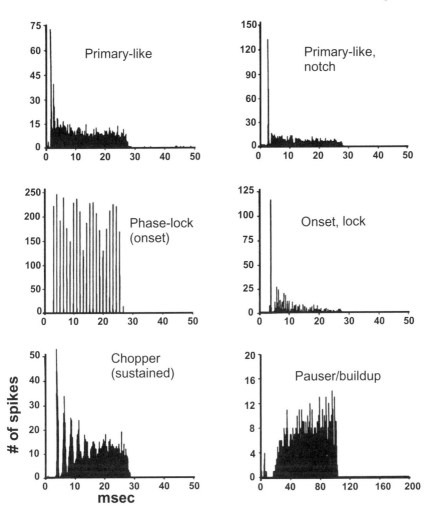

FIGURE 3.1. Summary of poststimulus-time histograms discussed in this chapter. Panels adapted from Rhode and Greenberg (1992).

neurons in the auditory brainstem show characteristic levels of transformation of the primary afferent (auditory nerve) response pattern, and indeed these transformations are used in part to define the cell type. For example, postsynaptic neurons in the ventral cochlear nucleus (VCN) may respond almost identically to the activity patterns of the auditory nerve fibers; such responses are termed "primary-like" or PL (Pfeiffer 1966; Rhode 1986). Alternatively, neurons may show subtle or major deviations from this pattern. The goal of this chapter is explore how, at a membrane level, these firing patterns in postsynaptic cells might arise during synaptic activity. In the process, we will outline basic concepts in the physiology of ion channels and synapses as they relate to the function of auditory neurons.

2. Techniques

The earliest studies of single neurons in the brainstem used extracellular recording techniques, which allowed investigators to associate a stereotyped pattern of firing of action potentials with a particular sound stimulus. This approach led to the definition of cell types by response profile, such as PL, "chopper", or "onset" cells (Fig. 3.1). Further definition was made by associating the recording with a particular subregion of the brainstem, through the use of histological tracking methods. However, additional characterization of the properties of individual cells required the use of intracellular recording techniques, which permit both higher-level resolution of electrical activity and the injection of dye into the cells to reveal dendritic branching patterns, axonal projections, and synaptic distribution (Oertel 1983; Smith and Rhode 1987, 1989). Such studies, although very difficult, have provided critical information about the relationships between PSTH patterns, cell types, and the membrane potential. Most recently, the patch clamp technique has been applied to single-cell analysis because of the comparative ease with which recordings can be obtained and because of the ability to voltage clamp neurons and monitor ionic currents (Hamill et al. 1981). Such recordings require the use of brain slice preparations which allow one to change the composition of the extracellular solution and thus better define the ionic species giving rise to a particular current. With this information, the ionic selectivity and the pharmacological sensitivity of an ion channel can be accurately defined. Drugs identified in this way can then be used to determine the contribution of the channel to the firing properties of the neurons. Individual neurons may also be labeled, using dyes in the patch pipettes to provide morphological identification of the neuron after the voltage or current clamp recordings are complete (Fig. 3.2). A drawback of the patch clamp technique is that it is difficult to apply in the in vivo setting, although some success has been reported (Covey et al. 1996). Further development of these and other in vivo recording techniques will be key to progress in the field.

Several further refinements and issues deserve mention. Electrical activity arising in dendrites and axons present a problem for electrophysiological analysis, in that voltage responses are distorted by the cable properties of the membranes. Conventional voltage clamp of such structures is impossible, and the currents generated in poorly clamped cells inaccurately reflect the properties of the ion channels (Spruston et al. 1994). One solution is to use neurons physically isolated from brain slices, generally after application of a cocktail of proteolytic enzymes. By cleaving off the axon and most of the dendritic arbor, the cell membrane potential is better controlled during

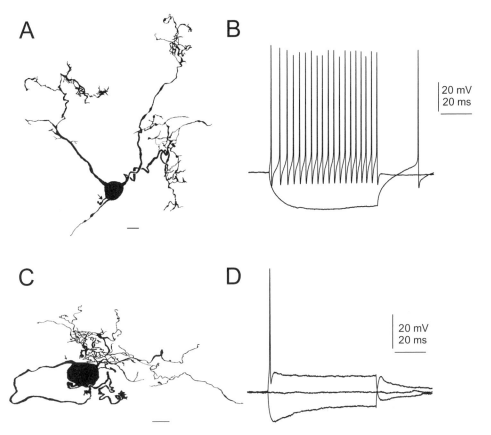

FIGURE 3.2. Stellate and bushy cells of the mouse ventral cochlear nucleus in a brain slice preparation. A: camera lucida drawing of a stellate cell labeled using biocytin following patch-clamp recording. Arrow indicates the beginning of the axon. Calibration bar 10 μm. B: response of a stellate cell to current steps of −200 and +200 pA. C: camera lucida drawing of a biocytin-labeled bushy cell, with the axon marked by arrow, and calibration bar of 10 μm. D: response of a bushy cell to current steps of −400, 0 and +800 pA.

voltage clamp experiments. Remarkably, such isolated cells often retain the same response profiles observed in the intact neurons (Manis and Marx 1991). However, isolation also removes synaptic contacts, and so issues related to transmission and circuitry cannot be studied. Another variation that has been used in studies of auditory neurons is the excised patch technique, in which a small region of cell membrane is transferred to a separate set of bath solutions. This has allowed the detailed characterization of receptors mediating synaptic transmission in the auditory system and of voltage-gated channels that shape the response to synaptic stimuli (Raman et al. 1994; Kanold and Manis 1999).

3. The Synaptic Potential

Before examining how different inventories of ion channels are used to generate a variety of response profiles, we will first discuss general properties of synapses in the auditory brainstem, as these properties may be common to diverse cell types.

3.1 Excitatory Transmission

Rapid excitatory postsynaptic potentials (EPSPs) in the auditory pathway, such as those mediating auditory nerve signaling, are generated by ionotropic glutamate receptors (reviewed by Parks 2000). Glutamate receptors fall into three major categories: the AMPA (α-amino-3-hydroxy-5-methyl-4-isoxazolepropionic acid) receptor, the kainate receptor, and the NMDA (N-methyl-D-aspartate) receptor (reviewed by Hollmann 1999). While there is evidence for all three receptor subtypes in the cochlear nuclei and superior olivary complex, the weight of evidence indicates that the AMPA receptor plays the major role in fast transmission. For example, quinoxaline-derived antagonists, which act at AMPA and kainate receptors, completely block transmission mediated by auditory nerve fibers in the cochlear nuclei or by axons of bushy cells in the medial nucleus of the trapezoid body (MNTB) (Wu and Kelly 1992b; Zhang and Trussell 1994b; Isaacson and Walmsley 1995). The noncompetitive antagonist of AMPA receptors GYKI-52466 also blocks auditory nerve transmission (T. Otis and L. Trussell, unpublished observations), and synaptic responses on bushy cells and MNTB neurons are modulated by cyclothiazide, a drug selective for the AMPA receptor (Trussell et al. 1993; Wu and Borst 1999). NMDA receptors are also activated by the excitatory neurotransmitter of auditory nerve fibers (Zhang and Trussell 1994a; Isaacson and Walmsley 1995; Ferragamo et al. 1998), but the role of these receptors in auditory transmission is less clear, perhaps providing slow, stable depolarization or the mediation of long-term plasticity. In bushy cells and MNTB neurons, the expression of NMDA receptors subsides during development (Bellingham et al. 1998; Taschenberger and von Gersdorff 2000). It remains even less clear what the role of kainate receptors is in the auditory system (Petralia et al. 1996). These receptors have been proposed to regulate transmitter release presynaptically in other brain regions (Rodriguez-Moreno et al. 1997), but such analyses have not yet been extended to auditory synapses. Regarding the identity of the transmitter itself, available evidence suggests that it is glutamate. Although both glutamate and aspartate are commonly found in nerve terminals, only glutamate is an effective agonist for both AMPA and NMDA receptors (Patneau and Mayer 1990). There is evidence, however, that a novel and as yet unidentified compound may be the excitatory transmitter, at least at the hair cell synapse (Sewell and Mroz 1990).

The AMPA receptor is composed of four or five protein subunits. These may be generated by 1–4 gene products, termed Glutamate Receptor (GluR) 1–4 or A–D. Each of these subunits may be processed in critical ways, either by alternative splicing or by RNA editing. Although these issues are well reviewed elsewhere (Hollmann 1999), they are worth summarizing here. Alternative splicing occurs mainly through a 38-amino acid cassette, generating two subunit isoforms, flip or flop (Sommer et al. 1990). A biophysical consequence of splicing is that it determines the gating kinetics of ion channels composed of such subunits (Mosbacher et al. 1994). Channel gating kinetics influence the time course of voltage-clamped synaptic currents and of the responses to brief or long pulses of glutamate applied directly to receptor-rich membrane patches. As shown in Fig. 3.3, such brief pulses produce a response that terminates quickly, and the decay of such a response is called "deactivation". A long pulse results in receptor desensitization; the onset of this process is also a key parameter in kinetic analysis (Fig. 3.3). For GluR-3 and -4, splicing is particularly effective, so that receptors containing flop subunits show desensitization time constants of less than 1 ms, and deactivation times of under 0.5 ms at room tempera-

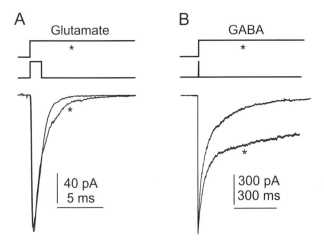

FIGURE 3.3. Response of short or long pulses of 1 mM glutamate (A) or 1 mM GABA (B) to an outside-out patch excised from a neuron in nucleus magnocellularis. The 2 upper traces indicate the time course of the transmitter application, while the lowest traces show the current responses. The asterisk marks the longer pulse and its response. In A, the response is mediated largely by AMPA receptors, whose deactivation and desensitization occur within milliseconds. Note that the desensitizing response (*) is slower than the deactivating response, reflecting different kinetic processes of the AMPA receptor. In B, GABA application activates GABAA receptors, whose kinetics of deactivation and desensitization are far slower than those of the AMPA receptors. The slower time course of these responses accounts for the slower time course of the GABAergic IPSC.

ture (Mosbacher et al. 1994). As shown below, these features impart significant functional advantages to some auditory synapses.

Another pertinent modification to the AMPA receptor subunit is the alteration of the coding sequence for one of the pore-lining amino acids, at which the code for glutamine is replaced by that of arginine (Hollmann 1999). Such editing is nearly 100% complete for GluR-2 and is rare in the other AMPA receptor subunits. The presence of GluR-2 in an AMPA receptor alters dramatically two features related to ion permeation. Such receptors have a low permeability to Ca^{2+} and to polyamines (Hume et al. 1991; Bowie and Mayer 1995). Ca^{2+} activates certain types of ion channels, drives cellular metabolic processes leading to long-term alterations in electrical responsiveness or the biochemical state of the neuron and may also lead to cell death. How these reactions to the transmitter may play out in auditory function is not clear.

Molecular biological and immunohistochemical studies have revealed the presence of GluR-3 and -4 in many neurons of the mammalian and avian auditory brainstem and inferior colliculus, particularly in cells which only weakly express GluR-1 and -2 (Rubio and Wenthold 1997; Wang et al. 1998; Caicedo and Eybalin 1999; Ravindranathan et al. 2000). Reverse transcriptase polymerase chain reaction (RT-PCR) indicated further that most subunits in the avian nucleus magnocellularis (NM), angularis (NA), and laminaris (NL), and the rat MNTB were of the flop isoform (Geiger et al. 1995; Ravindranathan et al. 2000). The functional implications of this expression pattern is well supported by physiological studies, as described below.

3.1.1 Time Course of the Synaptic Response

How do receptor properties work with other cellular parameters to shape the EPSP? The synaptic potential is generated by the charging effect of the excitatory synaptic current (i.e, the EPSC) on the membrane capacitance. For synapses with very rapid EPSCs, the decay of the EPSP will be determined by the rate at which the membrane discharges, the membrane time constant. The time constant in turn is dependent on the membrane conductance imparted by the ion channels that are active near the resting potential of the cell. Thus, the duration of the AMPA receptor EPSP will be limited by intrinsic membrane properties. By contrast, the duration of NMDA EPSPs are limited by the sluggish NMDA EPSC, which is generally longer than the membrane time constant (Lester et al. 1990; see Section 4).

The duration of the EPSP has important implications for how neurons integrate synaptic signals. Brief synaptic potentials summate only when several synapses fire very close in time; in this way, the output of the postsynaptic cell depends on synchronous activity of presynaptic fibers. For neurons in the timing pathways of acoustic processing, the timing of activity of the inputs fibers is locked to the phase of the sound stimulus. Thus, convergence of weak, brief synaptic potentials provides a simple mechanism for ensuring that the timing information is well preserved through sequential synaptic levels (Joris et al. 1994). When synaptic inputs are large, such as in MNTB neurons, brevity is also critical. As the synapses fire at relatively high rates, EPSPs must be brief enough that sequential events do not overlap and distort the timing of action potential generation in the postsynaptic cell.

3.1.2 Synaptic Activation of Glutamate Receptors

The simplest picture for the molecular dynamics of a typical fast excitatory synaptic current begins with the release of the contents of a glutamate-filled vesicle. The transmitter then binds and activates a limited number of post-synaptic receptors opposite the active zone, producing a so-called minia-ture excitatory postsynaptic current (mEPSC). Glutamate binding to receptors is weak, i.e. it binds with low affinity. The remaining transmitter that has not bound diffuses away quickly, probably within a few hundred microseconds. The short lifetime of the bound state and of available trans-mitter ensure that the transmitter response is terminated quickly. In the brief period that glutamate is still bound, the receptor protein swings between channel open and closed conformations. The frequency with which such transitions occur also contributes to the speed of the overall synaptic response (Trussell 1998).

mEPSCs thus provide a convenient monitor of the overall effects of receptor affinity and channel gating kinetics. Measurements of mEPSC decay times in different cell types reveal a striking trend: the fastest decays of any synaptic current in the brain are found in the central auditory neurons, with typical values of between 400–500 µs at room temperature (Trussell 1998). Where examined, these synaptic currents also show high Ca^{2+} permeability and high sensitivity to polyamine channel blockers (Otis et al. 1995; Zhou et al. 1995; Gardner et al. 1999). Together these observa-tions are consistent with the expression of receptors composed largely of GluR-3*flop* or -4*flop*. Direct measurement of receptor deactivation and desensitization supports this conclusion. Deactivation time constants mea-sured in patches are identical to those of the mEPSCs (Otis et al. 1996b; Gardner et al. 1999; Gardner et al. 2001). These deactivation constants and the desensitization time constants are identical to those of GluR-3*flop* and -4*flop* measured in heterologous expression systems (Gardner et al. 2001). Thus, synaptic transmission at many auditory synapses depends upon selec-tive expression of one or two specialized receptor subunits.

Recent studies suggest that expression of the fast-gating subunits may be a consequence of cellular interactions during the development of synaptic contacts. Auditory nerve fibers branch upon entering the brain and inner-vate a wide variety of cell types: at least two in birds and seven in mammals

(Warchol and Dallos 1990; Cant 1992). In every case examined, the cells postsynaptic to auditory nerve fibers produce the fast-gating subunits and, in all but one case, express little GluR-2 (Raman et al. 1994; Rubio and Wenthold 1997; Wang et al. 1998; Ravindranathan et al. 2000; Gardner et al. 2001). Moreover, developmental analysis of the receptors indicates that channel gating in AMPA receptors is slow, and sensitivity to polyamines is low, at the time of innervation (Lawrence and Trussell 2000). Finally, isolation of neurons in cell culture prior to innervation prevents the change in channel kinetics (Lawrence and Trussell 2000). These results suggest that some molecular features of the auditory system arise through interactions between pre- and postsynaptic cells.

Besides receptor properties, additional factors also shape the EPSC. Because auditory synapses generally release a large number of vesicles needed to ensure a suprathreshold postsynaptic response, the time period over which the vesicles fuse substantially broadens the EPSC. At endbulb and calyceal synapses, release of vesicles requires several hundred microseconds after arrival of an action potential at the nerve terminal. This period, convolved with the duration of the mEPSC, results in an EPSC that lasts about 1 ms (Isaacson and Walmsley 1995; Borst and Sakmann 1996). Additional slower phases of the EPSC are attributed to the gradual clearance of low levels of residual transmitter (Otis et al. 1996a).

3.2 Inhibition

The major inhibitory neurotransmitters of the auditory system brainstem nuclei are GABA and glycine (Sato et al. 2000). The GABA$_A$ subtype of GABA receptors and the glycine receptors activate chloride-selective channels whose activation may have a variety of effects. When the reversal potential for chloride ions (E_{Cl}) is negative to the resting potential, the transmitter produces a hyperpolarizing inhibitory postsynaptic potential (IPSP). By shunting excitatory currents and bringing the membrane potential away from threshold, the IPSP curtails excitatory signaling. Examples of this effect in the auditory system include glycinergic transmission in the VCN, Lateral superior olive (LSO), the medial superior olive (MSO), and MNTB (Banks and Smith 1992; Grothe and Sanes 1993; Kotak et al. 1998; Wu and Fu 1998; Smith et al. 2000). In the cochlear nuclei, it has been proposed that inhibition accounts for the refinement of the range of frequencies resulting in excitation or of the responsiveness to tones in the presence of background noise (Caspary et al. 1993; Evans and Zhao 1993). Accordingly, frequency response curves (tuning curves) may be modified in vivo by local application of antagonists of GABA and glycine receptors (Caspary et al. 1993; Evans and Zhao 1993; Caspary et al. 1994). However, the precise neural circuitry that modifies tuning characteristics has not been defined; potential sources of such inhibition exist throughout the superior olivary complex and in the cochlear nuclei themselves (Ostapoff et al. 1990; Saint Marie et al. 1991; Wickesberg et al. 1991; Ferragamo et al. 1998).

Inhibitory mechanisms play a clearer role in the convergence of excitation and inhibition on neurons of the LSO. Here, the timing of convergence of the two signals is believed to be critical for the encoding of interaural sound level differences such that sounds louder at the contralateral ear will suppress the activity of neurons excited by weaker ipsilateral inputs (Tsuchitani and Boudreau 1967). Another key feature of this inhibition is that the IPSPs are comparatively brief (on the order of a few ms); this brevity is presumably required for the precise temporal convergence of ipsi- and contralateral signals (Wu and Kelly 1992a). Glycinergic signals are also seen in bushy cells of the ventral cochlear nucleus, and are thought to be mediated by tuberculoventral cells of the dorsal cochlear nucleus (Wu and Oertel 1986; Wickesberg et al. 1991). These inputs, in which the inhibitory neurons have a similar characteristic frequency as the bushy cell, have been proposed to play a role in the suppression of echos generated within the ear (Wickesberg and Oertel 1990). Again, rapid synaptic signaling and precise convergence would be essential to this mechanism.

However, this picture of inhibitory function as mediated by precise convergence of depolarizing and hyperpolarizing signals is not universal. Early in development of the LSO, the reversal potential for chloride ions is depolarizing, and the inhibitory terminals from the MNTB release GABA as well as glycine (Sanes and Friauf 2000). Due to their strong depolarizing action, these transmitters have excitatory actions in the developing brain whose functions may be related to the activation of voltage-activated Ca^{2+} channels and Ca^{2+}-dependent synapse stabilization (Sanes and Friauf 2000). In some regions, GABA and glycine are depolarizing even in the mature animal. In the dorsal cochlear nucleus, for example, glycinergic cartwheel cells provide depolarizing IPSPs onto other cartwheel cells and hyperpolarizing IPSPs onto fusiform neurons (Golding and Oertel 1996, 1997). The effect of the depolarizing IPSP is apparently complex, such that when the cartwheel cell is relatively quiescent, the IPSP will facilitate effects of excitatory signals, while during more intense periods of activity, the IPSP is inhibitory to action potential generation. Generally, inhibition produced by depolarizing IPSPs is interpreted in terms of the shunting effects of the glycinergic or GABAergic conductance, which drives the membrane potential below spike threshold. Developmentally stable depolarizing IPSPs are also evident in the avian bushy cells of NM, as described below.

Another intriguing variable in the IPSPs of auditory neurons is their time course. While glycinergic IPSPs are generally of short duration, GABAergic signals are slower to decay, often outlasting the EPSP by tens of milliseconds, a result of the slow gating kinetics of GABA receptors (Fig. 3.3). In mammals, there are few direct demonstrations of GABAergic IPSPs in the cochlear nuclei and olivary complex (Ferragamo et al. 1998; Lim et al.

2000), but in birds GABA is the major inhibitory transmitter. Neurons of the avian NM and NL, which receive depolarizing GABAergic input from the superior olive (Hyson et al. 1995), respond to trains of stimuli delivered to the GABAergic fibers with depolarizing plateau potentials of about 10 mV. After the period of stimulation, IPSPs decay gradually back to baseline over 100 ms (Yang et al. 1999; Lu and Trussell 2000). Depolarizing inhibition shortens the membrane time constant, thus ensuring that only perfectly convergent EPSPs summate to threshold (Funabiki et al. 1998). The mechanism for the generation of protracted inhibition appears to involve a combination of pre- and postsynaptic factors (Lu and Trussell 2000). Slow IPSPs, due to the slower gating kinetics of the GABA receptor, summate during high frequency activity to produce a relatively smooth plateau. Presynaptically, however, the timing of the response of the synapses to presynaptic action potentials appears to become disrupted, and GABA is released at random intervals during a train of regular stimuli. This desynchronization is due to Ca^{2+} accumulation in the terminal, the dynamics of vesicle depletion and replenishment, and the facilitation of vesicle fusion probability. The slow decline of these processes after cessation of presynaptic action potential activity delays the termination of the IPSP (Lu and Trussell 2000). Such late inhibitions have been suggested to play computationally significant roles in sound localization in the inferior colliculus (Klug et al. 1999).

3.3 Presynaptic Regulation of Synaptic Transmission

Synaptic potentials vary in amplitude during repetitive synaptic stimulation. Some variation is random from event to event, a consequence of the probabilistic nature of transmitter release (del Castillo and Katz 1954). Some variation, however, is use-dependent, resulting from simple forms of short-term synaptic plasticity, facilitation, and depression (Zucker 1989). These are synapse-specific features whose significance for sensory processing are only just being appreciated as brain slice and voltage clamp methods are applied to auditory synapses. As noted above, facilitation and depression also play a key role in shaping inhibitory plateaus in NM. In large endbulb and calyceal synapses, transmitter is released with relatively high probability, ensuring a rapid approach to action potential threshold. With continuous activity at high rates, however, these terminals undergo synaptic depression, with EPSPs and EPSCs falling to a smaller, steady-state level depending on stimulus frequency, temperature, and age of the animal (Wu and Oertel 1987; Brenowitz et al. 1998; Taschenberger and von Gersdorff 2000). A use-dependent reduction in release is understandable in terms of a simple model of the synapse. Each synaptic site has a limited number of vesicles available for release (the readily releasable pool) and these are replenished at a finite rate, usually with a time course of at least tens of milliseconds (Dittman and Regehr 1998; Wang and Kaczmarek 1998). This model has also been invoked to account for adaptation of spike rate during tones in auditory nerve fibers (Geisler and Greenberg 1986; Moser and Beutner 2000). Besides depletion of vesicles, other pre- and postsynaptic factors may also contribute to lack of responsiveness of synapses (Otis et al. 1996a; Wu and Borst 1999; Waldeck et al. 2000). Nevertheless, despite massive reductions over time in the number of vesicles that are released per stimulus, mature terminals are still able to generate suprathreshold EPSPs due to the large number of release sites that remain active. For example, at MNTB calyces, EPSPs remain suprathreshold for at least eight to ten stimuli at 600 Hz (Taschenberger and von Gersdorff 2000). For multiply-innervated bushy cells, depression at each synapse may be profound but is compensated by the larger number of synapses contributing to the depolarization.

Release at calyceal and endbulb terminals may also be regulated by activation of presynaptic receptors. For example, metabotropic glutamate, the $GABA_B$-subtype of GABA receptor, and adenosine receptors inhibit release through inhibition of presynaptic Ca^{2+} channels (Barnes-Davies and Forsythe 1995; Takahashi et al. 1996; Brenowitz et al. 1998). However, although these receptor systems are capable of dramatically altering the size of EPSCs elicited at low frequencies, their function in physiological contexts remains obscure. In chick bushy cells (NM), it has been suggested that $GABA_B$ receptors may act to reduce synaptic depression since lowering transmitter release appears to lessen postsynaptic receptor desensitization. Still, it has not yet been shown that such receptors are activated in vivo at these synapses. Lim et al. (2000) found that synaptic strength at glycinergic synapses could be enhanced by antagonists of $GABA_B$ receptors, suggesting that GABA, perhaps co-released with glycine, inhibits release of glycine. More recently, glycine receptors themselves have been demonstrated on MNTB calyces. Activation of these receptors appears to enhance EPSPs through an unusual mechanism in which opening of glycine-gated chloride channels in the nerve terminal depolarizes the synapse, resulting in an elevation of presynaptic Ca^{2+} and a facilitation of release (Turecek and Trussell 2001). These glycine receptors may be activated following stimulation of nearby glycinergic fibers, implying that glycine must effectively diffuse from glycinergic boutons to excitatory terminals.

4. Intrinsic Properties of Neurons

As we have discussed, the response of a neuron to synaptic stimuli depends on the characteristics of the synaptic receptors and of the ion channels they activate. The outcome of this change in membrane conductance, from the perspective of the output of the neuron, is related to the inventory of voltage-gated ion channels, the electrotonic structure of the neuron, the number and location of synapses, and the temporal structure of their activity. Gathering information on all of these factors in one experimental system borders on impossibility. Thus, to describe how these factors converge to generate a basic stereotypical response, we must draw upon both in vivo and in vitro experimental data and also computer simulations. When based upon real biological parameters, such simulations permit the testing of concepts that significantly enhance the interpretation of data and the subsequent design of experiments.

4.1 Choppers

The responses of neurons in the cochlear nucleus to sound stimuli are characterized by the patterns observed in the PSTH, with additional information provided by other statistical analyses, such as interspike intervals and latencies. One of the major groupings, the chopper, is characterized by the presence of a regular train of action potentials whose intervals are consistent (thus the term "regular spiking") and are not related to the frequency of the sound stimulus (Rhode and Smith 1986; Rhode and Greenberg 1992). The chopping may persist during the stimulus or dampen out after a few ms, in which case the spiking becomes random from trial to trial. In response to a low-frequency sound, the choppers often show poor phase locking. In vivo studies have identified neurons showing sustained (C_S) or transient (C_T) chopping as stellate cells of the ventral cochlear nucleus (Rhode et al. 1983a). This identification of neurons is supported by in vitro studies using microelectrodes or patch pipettes. These show that neurons identified morphologically as stellate cells, respond to prolonged, square pulses of positive current with a train of regularly spaced action potentials, as shown in Figure 3.2A and B (Oertel 1983; Manis and Marx 1991; Isaacson and Walmsley 1995). Larger current steps yield a higher frequency of spikes, with a minimum to maximum frequency range of several fold. This profile of response, in its crudest form, can be mimicked with computational models that resemble the classical squid giant axon in channel composition: an inactivating Na^+ channel, a delayed rectifier K^+ channel and a nonselective leak current (Banks and Sachs 1991; Manis and Marx 1991; White et al. 1994). Although the squid axon does not behave exactly like a chopper, simple adjustments in the properties of the component ion channels, such as gating kinetics and thresholds for activation, apparently yield a model that shows many of the features of choppers observed experimentally. The intracellular studies support this assessment, since the voltage clamping of enzymatically-isolated, regular-spiking neurons exhibit a dominant outward potassium current, blocked by tetraethylammonium ions (TEA), a sodium current, and a background leakage current (Manis and Marx 1991). Based on the properties of these ionic currents, Manis and Marx (1991) were able to recontruct the basic regular-spiking response in a computer simulation. Depolarizing currents bring the cells to threshold, setting in motion the classical Hodgkin-Huxley cycle of activation, inactivation and deactivation of the Na^+ and K^+ current. When the depolarizing current is sustained, the cycle can be repeated, primarily because the channels are able to reprime themselves sufficiently after the last action potential despite the persistent depolarization provided by the stimulus. Regularity of spike interval occurs because the degree to which the channels can reprime themselves is similar from spike to spike. These results imply that the stellate cell could produce a chopping response given a basic complement of ion channels, but specifically in response to a simple, prolong stimulus.

Beyond the voltage clamp characterization described by Manis and Marx (1991), there is little information about the molecular identity of the voltage-gated channels of choppers. The key features required of the K^+ channels to facilitate chopping include (1) rapid channel kinetics, which allow channels to open and close in the time scale of an action potential, (2) a high (more depolarized) voltage "threshold" for the channels, allowing the cell to deactivate the K^+ current even during a persistent depolarizing stimulus, and (3) little or no channel inactivation, which would otherwise impede repolarization and repetitive firing later in trains of stimuli. One candidate is the Kv3.1 potassium channel subunit. When expressed in heterologous expression systems, Kv3.1 produces a K^+ current with a high threshold for activation and rapid kinetics. Repetitive spiking neocortical neurons express this channel (Martina et al. 1998; Erisir et al. 1999). However, in situ hybridization indicates that only a fraction of stellate-like cells express this channel (Perney and Kaczmarek 1997), and thus other channels with similar properties may be important in stellate neurons.

How can these cells produce a chopping response during a barrage of complex synaptic stimuli? This question has been approached both experimentally and through computational methods. Banks and Sachs (1991) developed a model of the stellate cell that incorporated the basic morphological features of the neuron, including several dendrites, an axon, and a cell body. The pattern of synaptic activity driving the cell was made to simulate the actual temporal features of auditory nerve responses to tones. Two major parameters that were examined were the effects of varying synapse location and number. When a small number of synapses (four) converged close to the cell body, the individual synaptic potentials were large and summated to variable degrees. As a result, the cell's firing pattern was highly irregular. Synapses placed on electrotonically distant dendritic locations produced synaptic response that were heavily filtered (made smaller and slower) more severely by the resistance and capacitance of the dendrites, i.e., their cable properties. Increasing the number of synaptic inputs to ten could compensate for the reduction in amplitude, producing a smooth depolarizing response at the cell body. When the action potential was generated in or near the cell body, the output of the cell was now a regular train of spikes, reminiscent of the response of the model to a simple square pulse of current. In principle, differences among cells in the regularity of spiking could reflect differences in the location of synapses; morphological data indicates that stellate cells may indeed differ in the relative position of excitatory synapses (Smith and Rhode 1989). Similar conclusions were reached using more sophisticated models based on actual morphometric data (White et al. 1994). The weakness of phase locking then would be an outcome of the rigid response profile of the cell to a stimulus and the filtering properties of the dendrites, which together would blur the temporal information contained in auditory nerve firing.

More recent brain slice studies suggest that the number of inputs to VCN stellate cells of the mouse (five) are not as many as were required in the models described above to account for the regularity of spike output (Ferragamo et al. 1998). Moreover, it appears that the actual synaptic waveform itself was more complex than that employed in the model. Excitation was mediated by a fast, AMPA receptor-dependent EPSP, but also by an NMDA receptor EPSP that lasted for over 100 ms after the synaptic stimulus was terminated. Blockade of NMDA receptors resulted a loss of firing during periods greater than a few milliseconds after the stimulus. Late, fast EPSPs were also observed, suggesting that disynaptic pathways may enhance excitation of the stellate cell following auditory nerve stimulation. These data indicate the irregular chopping resulting from a small number of auditory nerve synaptic inputs might be alleviated by additional slow or delayed excitatory events. However, it should be noted that these effects occur over a period of hundreds of milliseconds, while chopping *per se* is defined during a period of <50 ms. Thus, it remains unclear how multiple receptor subtypes and recurrent excitation may shape the earliest response to sound in choppers.

4.2 Primary-Like Responses and Phase Locking

The main features of the primary-like (PL) response are the gradual adaptation and irregularity of firing to high-frequency stimuli and phase-locking to lower-frequency inputs. A related profile is primary-like with notch (PL-N), which shows a sharp transient drop in firing after an initial well-timed spike and excellent phase locking at low frequencies. Onset-lock (O-L) units are similar to PL-N, but have a lower rate of firing late in the response to a high-frequency tone (Rhode and Greenberg 1992). What cellular properties permit such a close mimicking of the properties of the synaptic input, which is clearly required for preservation of timing information?

PL and PL-N responses are most likely generated by AVCN spherical and globular bushy cells, respectively (Rhode et al. 1983a; Rhode and Smith 1986; Smith and Rhode 1987; Smith et al. 1993). The most basic features bushy cells must have in order to account for these responses are strong, brief synaptic inputs and ion channels that allow the membrane voltage to follow closely the changes in synaptic current and to recover quickly after each action potential. The latter have been well-studied in brain slice preparations. Oertel (1983) first showed that mouse bushy cells respond to a prolonged, strong depolarizing current injection with a single action potential, a finding confirmed in bushy cells of rat and guinea pig AVCN and chick NM (Fig. 2C, D; Manis and Marx, 1991; Reyes et al. 1994; Zhang and Trussell 1994b; Isaacson and Walmsley 1995; Schwarz et al. 1998). Repetitive, brief stimuli, however, easily elicit action potentials, indicating that the cell membranes are designed for permitting a single suprathreshold response but a rapid recovery following the termination of the stimulus. Unlike choppers, even weak depolarizing currents cause a sharp decrease in membrane resistance (increase in conductance) in bushy cells, indicating that ion channels open with a very low threshold relative to the resting potential. Further studies using voltage clamp techniques support this inference and defined in bushy cells two important K^+ currents (Manis and Marx 1991; Rathouz and Trussell 1998). A low-threshold current (LTC) is slightly active at the resting potential and becomes more completely activated when the cell is depolarized by 10–15 mV. The K^+ channel producing this current is sensitive to 4-aminopyridine (4-AP) and dendrotoxin. A high-threshold current (HTC) similar to that of stellate cells activates quickly when during an action potential the membrane potential exceeds about −20 mV. Blockade of HTC by tetraethylammonium (TEA) ions results in a broadened action potential but does not initiate repetitive firing. By contrast, blockade of LTC with 4-AP or dendrotoxin transforms the neurons into repetitively spiking cells, supporting the notion that LTC may be key to distinguishing the electrical properties of the stellate and bushy cell.

How do these ionic currents shape a single synaptic response in bushy cells? With simultaneous activation of a suitable number of inputs, the inward synaptic current depolarizes the neuron to threshold and an action potential is fired. The rapid initiation of the spike is due to the large number of synaptic vesicles released and the nearness of the somatic synapses to the spike initiation zone on the axon. The cell then quickly repolarizes due to the shut down of Na^+ and K^+ channels and the abrupt decay of the synaptic current, mediated by the fast-acting AMPA receptors. Thus, the cell is then ready to respond again. Several factors may limit the likelihood of a subsequent response. Voltage-gated channels must reprime, and stimuli at too close an interval will fall within the cell's spike refractory period (i.e., when spike threshold is elevated). Synapses also have a refractoriness of their own (see section 3.3): presynaptic active zones, once they have released a vesicle, must prepare a new release-ready vesicle, and this may take at least several milliseconds. Postsynaptic receptors may be briefly desensitized by the transmitter, and require time to recover. One factor that might aid in restoration of the synapse is that transmission is mediated by a large number of release sites, not all of which release in response to any one presynaptic action potential. Thus, a moderate release

probability may aid in the ability of the synapse to transmit at high frequencies (Schneggenburger et al. 1999).

Phase locking of auditory nerve signals is never perfect, and so the arrival
of signals at the spherical bushy cell occurs with variable latencies (jitter)
(Young et al. 1988; Blackburn and Sachs 1989). Later in the response to a
tone, such jitter is even more enhanced, particularly during high-frequency
tones in which phase locking is lost. The synaptic delay, the interval between
the arrival of the presynaptic spike and the onset of the postsynaptic
response, is also a stochastic process contributing to transmission jitter
(Barrett and Stevens 1972). Spherical bushy cells receive input from only
a few auditory nerve axons, unlike globular bushy cells (Sento and Ryugo
1989; Liberman 1991), and one would expect that strength of their synapses
to be correspondingly larger. When a few, strong EPSPs occur with some
jitter, the one with the shortest latency will initiate a postsynaptic action
potential. However, when a larger number of smaller EPSPs occur, as with
globular bushy cells, they must summate until threshold is reached. Two
factors ensure that such summation does not destroy the preservation of
timing, but may actually enhance it (Joris et al. 1994). First, each EPSP is
very brief, and so they can only summate effectively if they occur with a
similar latency. Second, if the jitter is such that the summation occurs over
a longer period of time, activation of the LTC will reduce the effectiveness
of the EPSPs and threshold cannot be reached. As a result, postsynaptic
spikes will only occur when driven by summation of well-timed subthreshold EPSPs. Such a mechanism accounts for the differences in firing rate of
PL and O-L cells late in the response to a high-frequency tone (Rhode and
Smith 1986; Young et al. 1988). The "notch" on the PL-N PSTH apparently
arises from the fact that synaptic convergence effectively drives the globular bushy cell to fire with little jitter; the ensuing refractory period of the
bushy cell must therefore also occur with uniform latency. Rothman and
co-workers (1993) have explored these aspects of bushy cell activity with
an excellent computer model which accounts for many aspects of the membrane physiology.

4.3 The Medial Nucleus of the Trapezoid Body Mirrors Bushy Cell Activity

Principal cells of the MNTB show membrane properties remarkably similar
to those of bushy cells, in terms of expression of voltage- and glutamate-
evoked currents as well as the presence of a somatic glutamatergic synapse.
The MNTB receives only a single, large calyceal terminal which reliably
transmits signals from the globular bushy cell (Guinan and Li 1990; Smith
et al. 1998). Thus, the MNTB neurons show a PL-N PSTH in response to
tones, not because of convergence (as discussed above), but rather because
of the neurons' ability reproduce the activity of the globular bushy cell so
faithfully.

The MNTB has received much attention regarding the molecular identity of the K$^+$ channels responsible for the high precision in the temporal responsiveness synaptic input. As with bushy cells, dendrotoxin or 4-AP induces repetitive firing to long current steps (Banks and Smith 1992; Brew and Forsythe 1995). Under voltage clamp, these neurons exhibit a prominent LTC (Brew and Forsythe 1995). Several features of this current are very similar to those of currents produced by the Kv1.1 and/or 1.2 K$^+$ channels, which are also strongly expressed in MNTB as well as in VCN (Grigg et al. 2000). The HTC is also quite prominent in MNTB and, as with bushy cells, is sensitive to TEA. Wang and co-workers (1998) contrasted the properties of the Kv3.1 channel expressed in cell lines with the HTC of MNTB, finding a remarkable degree of similarity in their pharmacological and voltage sensitivity, and their gating kinetics. Computer simulations incorporating the LTC and HTC suggested that the HTC is necessary to permit firing at high rates by rapidly terminating the action potential (see section 4.1). Accordingly, Kv3.1 is strongly expressed in the MNTB, as well as in bushy cells (Perney et al. 1992; Perney and Kaczmarek 1997; Wang et al. 1998).

4.4 Octopus Cells and Detection of Coherent Activity

Octopus cells of the posterior ventral cochlear nucleus respond to high-frequency tones with a strong O-L response, that is, an extremely well-timed initial response followed by a low level of random activity (Rouiller and Ryugo 1984; Rhode and Smith 1986). Octopus cells phase lock (and even entrain) extremely well to intense low-frequency sounds ($\leq$1,000 kHz). This behavior represents an outcome of several morphological and electrophysiological features of these neurons. Octopus cells send their stout dendrites across bands of auditory nerve fibers, sampling a broad spectrum of acoustic best frequencies (Osen 1969; Kane 1973). These fibers (and, rarely, other octopus cells) make excitatory synapses along the dendrites; it is estimated that at least 60 auditory nerve fibers innervate each octopus cell (Golding et al. 1995). The neurons also show striking levels of expression of two types of channel: the dendrotoxin-sensitive LTC channels and the hyperpolarization-activated nonselective cation channel, I_H (Golding et al. 1995; Ferragamo and Oertel 1998; Bal and Oertel 2000; Grigg et al. 2000). As a result, octopus cells exhibit a very low input resistance and extremely short membrane time constant (200 μsec) near the resting potential. By contrast, bushy and MNTB neurons require depolarization to attain such a short time constant. As noted above, clicks or loud tones will activate a broad spectrum of auditory nerve fibers and evoke a well-timed spike in the octopus cell. However, during the later phases of high-frequency auditory nerve activity, the input activity is random and summation cannot occur. Thus, octopus cells are well suited to respond to broad-band phase-locked responses of the auditory nerve, leading to the suggestion that they may serve to detect common interspike intervals in auditory nerve fibers, necessary to encoding pitch (Golding et al. 1995).

The leakiness of the membrane at rest, conferred by I_H, and the sharp onset of LTC may work together to allow this computation in the face of very complex activity in fibers of diverse best frequency (Bal and Oertel 2000; Cai et al. 2000). For example, such a high degree of synaptic convergence would produce significant depolarizations in other neurons, simply through spontaneous fiber activity alone. Leakiness of the resting cell membrane shunts these synaptic currents and minimizes the chances of random simultaneous EPSPs reaching threshold. When a group of fibers are active nearly simultaneously, the LTC acts to assure that the response only occurs for the most well-timed inputs and that the latency to the response is short, in a way similar to that discussed above for bushy cells. As the membrane depolarizes, LTC activates, increasing the amount of current needed to drive the cell to threshold. Thus, unless the cell can be depolarized quickly, LTC prevents excitation. In this way, octopus cells perform a rate-of-rise discrimination on synaptic signals, filtering out activity that is not sufficiently coherent (Ferragamo and Oertel 1998).

4.5 Other Coincidence Detection Mechanisms

Neurons of the MSO and the avian NL are key elements in the neural mechanism of sound localization in the horizontal plane (Carr and Konishi 1990; Yin and Chan 1990; Carr 1993). MSO and NL neurons have intrinsic membrane properties similar to those of bushy cells, and use these to detect the coincident arrival of signals from ipsi- and contralateral spherical bushy cells (see Yin, Chapter 4). A high degree of bushy cell convergence and the high sensitivity to the timing of signals is likely to produce the O-L responses and strong phase locking to monaural sound measure in MSO, using the principles outlined above (Rothman et al. 1993). However, studies in the avian NL suggest that dendrites may play a unique role in enhancing the ability of the neurons to respond particularly well to the coincidence of ipsi- and contralateral signals (Agmon-Snir et al. 1998).

The chicken NL comprises a sheet of neurons with clusters of dendrites at the dorsoventral poles of each cell body (Jhaveri and Morest 1982). Neurons receive input from ipsilateral bushy cells (from NM) on their dorsal dendrites and input from contralateral bushy cells on their ventral dendrites. Both the innervation and morphology of NL are tonotopic, in that cells receiving input from high, best-frequency NM cells have short dendrites, while cells innervated by low best-frequency NM cells have long dendrites (Rubel and Parks 1975). Agmon-Snir and colleagues explored the role of this remarkable dendritic layout using computer simulations (Agmon-Snir et al. 1998). When a population of adjacent synapses are active, the depolarization they produce reduces the driving force for current through the synaptic channels. This results in a reduction in the efficacy of each synaptic input in bringing the cell body to spike threshold. When inputs are separated at different dendritic locations, this effect of nonlinear summation is reduced, and that same number of inputs are more effective. The authors found that coincidence of ipsi- and contralateral inputs was apparently improved (and convergence from same-dendrite inputs diminished) by this mechanism. Moreover, they found that the usefulness of the dendrites falls off with high frequency input, and noted that this may account for the comparative lack of dendrites in high best-frequency NL neurons.

4.6 Fine Tuning of Response Properties in Fusiform Cells

Fusiform (pyramidal) cells of the dorsal cochlear nucleus participate in circuits that integrate somatosensory inputs with auditory signals and may play an important role in sound localization based on monaural spectral cues (see Young, chapter 5). In vivo recordings indicate that fusiform cells can exhibit a variety of PSTH profiles, including chopper, onset, and build-up patterns (Pfeiffer 1966; Godfrey et al. 1975; Rhode et al. 1983b). Rhode and colleagues (1983b) suggested that such patterns may shift with membrane potential. In view of the variety of voltage-sensitive ionic conductances in fusiform cells, one might predict that small changes in the resting potential could alter the relative balance of these channels available for shaping the response to excitation. Manis (1990) tested this explicitly using brain slice recordings, finding that the pattern of spiking could be markedly altered by small shifts in the membrane potential before a depolarizing current step was delivered. With moderate resting potentials, cells showed a response profile similar to that of stellate cells, consistent with the chopping mode recorded in vivo. Preceding the excitation with a hyperpolarizing step led to a single spike, followed by a pause of over 70 ms before the chopping pattern reemerged. When the excitation was slightly weaker, no initial spike was seen. These latter two profiles are reminiscent of the pauser and build-up PSTH patterns, respectively (Fig. 3.1).

Kanold and Manis (1999) examined the ionic conductances of fusiform cells in greater detail in order to determine what channels account for the shifting response profile described above. They found that fusiform cells express two types of transient outward currents, one of which (termed I_{KIF}) showed very rapid activation and inactivation and recovery from inactivation. I_{KIF} was highly voltage sensitive, such that half the channels inactivated at -85 mV. A detailed comparison of the gating kinetics of this channel with the timing of the pauses and buildup pattern seen under current clamp recordings, led to the following scenario for how I_{KIF} bestows such plasticity in the response to acoustic stimuli. At normal resting potentials, I_{KIF} is mostly inactivated, so that depolarization results immediately in the generation of repetitive action potentials, probably using a mechanism similar to that of the stellate cells (section 4.1). However, if a fusiform neuron has been hyper-polarized long enough to reprime the I_{KIF} channels, roughly 20–30 ms, sub-sequent depolarization activates the I_{KIF} channel, thus opposing spike generation. Repetitive spiking begins as I_{KIF} gradually inactivates. Hyper-polarizations needed to restore I_{KIF} might, in principal, result from the activity of inhibitory synapses or from strong after hyperpolarization following trains of spikes (Hirsch and Oertel 1988; Golding and Oertel 1996).

5. Summary

Individual neurons in auditory brainstem nuclei respond to tones with stereotypical patterns of action potentials. These patterns, which can mimic or transform the activity pattern of the synaptic inputs, result for the characteristic morphology and physiology of pre- and postsynaptic elements. The number and location of synapses and their transmitter release dynamics may determine the shape and size of synaptic potentials. Moreover, the biophysical properties of the transmitter receptors and of voltage-gated channels also determines how the cell will respond. Among the latter, K^+ channels figure most prominently. Key issues for future studies include: (1) the molecular composition of the ion channels, especially of K^+ channels, (2) the developmental events that determine which channels are expressed, (3) the detailed circuitry of synaptic connections within the brainstem, especially contacts between excitatory cells and inhibitory cells, and (4) how such circuitry is activated in response to sound.

Acknowledgements. I thank Achim Klug and Dan Padgett for comments on the manuscript. Su Zhang and Josh Lawrence kindly provided data in Figures 2 and 3. My work is supported by NIH grants NS28901 and DC04078.

References

Agmon-Snir H, Carr CE, Rinzel J (1998) The role of dendrites in auditory coincidence detection. Nature 393:268–272.

Bal R, Oertel D (2000) Hyperpolarization-activated, mixed-cation current (I(h)) in octopus cells of the mammalian cochlear nucleus. J Neurophysiol 84:806–817.

Banks MI, Sachs MB (1991) Regularity analysis in a compartmental model of chopper units in the anteroventral cochlear nucleus. J Neurophysiol 65:606–629.

Banks MI, Smith PH (1992) Intracellular recordings from neurobiotin-labeled cells in brain slices of the rat medial nucleus of the trapezoid body. J Neurosci 12: 2819–2837.

Cai Y, McGee J, Walsh EJ (2000) Contributions of ion conductances to the onset responses of octopus cells in the ventral cochlear nucleus: simulation results. J Neurophysiol 83:301–314.

Carr CE (1993) Processing of temporal information in the brain. Annu Rev Neurosci 16:223–243.

Carr CE, Konishi M (1990) A circuit for detection of interaural time differences in the brainstem of the barn owl. J Neurosci 10:3227–3246.

Caspary DM, Palombi PS, Backoff PM, Helfert RH, Finlayson PG (1993) GABA and glycine inputs control discharge rate within the excitatory response area of primary-like and phase-locked AVCN neurons. In: Merchan MA, Jiuz JM, Godfrey DA, Mugnaini E (eds) The mammalian cochlear nuclei. Organization and function. New York: Plenum, pp. 239–252.

Caspary DM, Backoff PM, Finlayson PG, Palombi PS (1994) Inhibitory inputs modulate discharge rate within frequency receptive fields of anteroventral cochlear nucleus neurons. J Neurophysiol 72:2124–2133.

Covey E, Kauer JA, Casseday JH (1996) Whole-cell patch-clamp recording reveals subthreshold sound-evoked postsynaptic currents in the inferior colliculus of awake bats. J Neurosci 16:3009–3018.

Del Castillo J, Katz B (1954) Quantal components of the end-plate potential. J Physiol (Lond) 124:560–573.

Dittman JS, Regehr WG (1998) Calcium dependence and recovery kinetics of presynaptic depression at the climbing fiber to Purkinje cell synapse. J Neurosci 18:6147–6162.

Erisir A, Lau D, Rudy B, Leonard CS (1999) Function of specific K(+) channels in sustained high-frequency firing of fast-spiking neocortical interneurons. J Neurophysiol 82:2476–2489.

Evans EF, Zhao W (1993) Neuropharmacological and neurophysiological dissection of inhibition in the mammalian cochlear nuclei. In: Merchan MA, Jiuz JM, Godfrey DA, Mugnaini E (eds) The mammalian cochlear nuclei. Organization and function. New York: Plenum, pp. 253–266.

Ferragamo MJ, Oertel D (1998) Shaping of synaptic responses and action potentials in octopus cells. Assoc Res Otolaryngol 21:96.

Ferragamo MJ, Golding NL, Oertel D (1998) Synaptic inputs to stellate cells in the ventral cochlear nucleus. J Neurophysiol 79:51–63.

Forsythe ID, Barnes-Davies M (1993) The binaural auditory pathway: excitatory amino acid receptors mediate dual timecourse excitatory postsynaptic currents in the rat medial nucleus of the trapezoid body. Proc R Soc Lond B Biol Sci 251: 151–157.

Funabiki K, Koyano K, Ohmori H (1998) The role of GABAergic inputs for coincidence detection in the neurones of nucleus laminaris of the chick. J Physiol (Lond) 508:851–869.

Gardner SM, Trussell LO, Oertel D (1999) Time course and permeation of synaptic AMPA receptors in cochlear nuclear neurons correlate with input. J Neurosci 19:8721–8729.

Gardner SM, Trussell LO, Oertel D (2001) Comparison of AMPA receptors associated with different inputs in the cochlear nuclei. J Neurosci 21:7428–7437.

Geiger JR, Melcher T, Koh DS, Sakmann B, Seeburg PH, Jonas P, Monyer H (1995) Relative abundance of subunit mRNAs determines gating and Ca2+ permeability of AMPA receptors in principal neurons and interneurons in rat CNS. Neuron 15:193–204.

Geisler CD, Greenberg S (1986) A two-stage nonlinear cochlear model possesses automatic gain control. J Acoust Soc Am 80:1359–1363.

Godfrey DA, Kiang NY, Norris BE (1975) Single unit activity in the posteroventral cochlear nucleus of the cat. J Comp Neurol 162:247–268.

Golding NL, Oertel D (1996) Context-dependent synaptic action of glycinergic and GABAergic inputs in the dorsal cochlear nucleus. J Neurosci 16:2208–2219.

Golding NL, Oertel D (1997) Physiological identification of the targets of cartwheel cells in the dorsal cochlear nucleus. J Neurophysiol 78:248–260.

Golding NL, Robertson D, Oertel D (1995) Recordings from slices indicate that octopus cells of the cochlear nucleus detect coincident firing of auditory nerve fibers with temporal precision. J Neurosci 15:3138–3153.

Grigg JJ, Brew HM, Tempel BL (2000) Differential expression of voltage-gated potassium channel genes in auditory nuclei of the mouse brainstem. Hear Res 140:77–90.

Grothe B, Sanes DH (1993) Bilateral inhibition by glycinergic afferents in the medial superior olive. J Neurophysiol 69:1192–1196.

Guinan JJ, Jr., Li RY (1990) Signal processing in brainstem auditory neurons which receive giant endings (calyces of Held) in the medial nucleus of the trapezoid body of the cat. Hear Res 49:321–334.

Hamill OP, Marty A, Neher E, Sakmann B, Sigworth FJ (1981) Improved patch-clamp techniques for high-resolution current recording from cells and cell-free membrane patches. Pflugers Arch 391:85–100.

Hirsch JA, Oertel D (1988) Intrinsic properties of neurones in the dorsal cochlear nucleus of mice, in vitro. J Physiol (Lond) 396:535–548.

Hollmann M (1999) Structure of ionotropic glutamate receptors. In: Jonas P, Monyer H (eds) Ionotropic glutamate receptors in the CNS. Berlin: Springer-Verlag, pp. 1–78.

Hume RI, Dingledine R, Heinemann SF (1991) Identification of a site in glutamate receptor subunits that controls calcium permeability. Science 253:1028–1031.

Hyson RL, Reyes AD, Rubel EW (1995) A depolarizing inhibitory response to GABA in brainstem auditory neurons of the chick. Brain Res 677:117–126.

Isaacson JS, Walmsley B (1995) Receptors underlying excitatory synaptic transmission in slices of the rat anteroventral cochlear nucleus. J Neurophysiol 73:964–973.

Jhaveri S, Morest DK (1982) Neuronal architecture in nucleus magnocellularis of the chicken auditory system with observations on nucleus laminaris: a light and electron microscope study. Neuroscience 7:809–836.

Joris PX, Carney LH, Smith PH, Yin TC (1994) Enhancement of neural synchronization in the anteroventral cochlear nucleus. I. Responses to tones at the characteristic frequency. J Neurophysiol 71:1022–1036.

Kane EC (1973) Octopus cells in the cochlear nucleus of the cat: heterotypic synapses upon homeotypic neurons. Int J Neurosci 5:251–279.

Kanold PO, Manis PB (1999) Transient potassium currents regulate the discharge patterns of dorsal cochlear nucleus pyramidal cells. J Neurosci 19:2195–2208.

Klug A, Bauer EE, Pollak GD (1999) Multiple components of ipsilaterally evoked inhibition in the inferior colliculus. J Neurophysiol 82:593–610.

Kotak VC, Korada S, Schwartz IR, Sanes DH (1998) A developmental shift from GABAergic to glycinergic transmission in the central auditory system. J Neurosci 18:4646–4655.

Lawrence JJ, Trussell LO (2000) Long-term specification of AMPA receptor properties after synapse formation. J Neurosci 20:4864–4870.

Lester RA, Clements JD, Westbrook GL, Jahr CE (1990) Channel kinetics determine the time course of NMDA receptor-mediated synaptic currents. Nature 346:565–567.

Liberman MC (1991) Central projections of auditory-nerve fibers of differing spontaneous rate. I. Anteroventral cochlear nucleus. J Comp Neurol 313:240–258.

Lim R, Alvarez FJ, Walmsley B (2000) GABA mediates presynaptic inhibition at glycinergic synapses in a rat auditory brainstem nucleus. J Physiol 525:447–459.

Lu T, Trussell LO (2000) Inhibitory transmission mediated by asynchronous transmitter release. Neuron 26:683–694.

Manis PB (1990) Membrane properties and discharge characteristics of guinea pig dorsal cochlear nucleus neurons studied in vitro. J Neurosci 10:2338–2351.

Manis PB, Marx SO (1991) Outward currents in isolated ventral cochlear nucleus neurons. J Neurosci 11:2865–2880.

Martina M, Schultz JH, Ehmke H, Monyer H, Jonas P (1998) Functional and molecular differences between voltage-gated K+ channels of fast-spiking interneurons and pyramidal neurons of rat hippocampus. J Neurosci 18:8111–8125.

Mosbacher J, Schoepfer R, Monyer H, Burnashev N, Seeburg PH, Ruppersberg JP (1994) A molecular determinant for submillisecond desensitization in glutamate receptors. Science 266:1059–1062.

Moser T, Beutner D (2000) Kinetics of exocytosis and endocytosis at the cochlear inner hair cell afferent synapse of the mouse. Proc Natl Acad Sci U S A 97: 883–888.

Oertel D (1983) Synaptic responses and electrical properties of cells in brain slices of the mouse anteroventral cochlear nucleus. J Neurosci 3:2043–2053.

Osen KK (1969) Cytoarchitecture of the cochlear nuclei in the cat. J Comp Neurol 136:453–484.

Ostapoff EM, Morest DK, Potashner SJ (1990) Uptake and retrograde transport of [3H]GABA from the cochlear nucleus to the superior olive in the guinea pig. J Chem Neuroanat 3:285–295.

Otis T, Zhang S, Trussell LO (1996a) Direct measurement of AMPA receptor desensitization induced by glutamatergic synaptic transmission. J Neurosci 16: 7496–7504.

Otis TS, Wu YC, Trussell LO (1996b) Delayed clearance of transmitter and the role of glutamate transporters at synapses with multiple release sites. J Neurosci 16:1634–1644.

Otis TS, Raman IM, Trussell LO (1995) AMPA receptors with high Ca2+ permeability mediate synaptic transmission in the avian auditory pathway. J Physiol (Lond) 482:309–315.

Parks TN (2000) The AMPA receptors of auditory neurons. Hear Res 147:77–91.

Patneau DK, Mayer ML (1990) Structure-activity relationships for amino acid transmitter candidates acting at N-methyl-D-aspartate and quisqualate receptors. J Neurosci 10:2385–2399.

Perney TM, Kaczmarek LK (1997) Localization of a high threshold potassium channel in the rat cochlear nucleus. J Comp Neurol 386:178–202.

Perney TM, Marshall J, Martin KA, Hockfield S, Kaczmarek LK (1992) Expression of the mRNAs for the Kv3.1 potassium channel gene in the adult and developing rat brain. J Neurophysiol 68:756–766.

Petralia RS, Wang YX, Zhao HM, Wenthold RJ (1996) Ionotropic and metabotropic glutamate receptors show unique postsynaptic, presynaptic, and glial localizations in the dorsal cochlear nucleus. J Comp Neurol 372:356–383.

Pfeiffer RR (1966) Classification of response patterns of spike discharges for units in the cochlear nucleus: tone-burst stimulation. Exp Brain Res 1:220–235.

Raman IM, Zhang S, Trussell LO (1994) Pathway-specific variants of AMPA receptors and their contribution to neuronal signaling. J Neurosci 14:4998–5010.

Rathouz M, Trussell L (1998) Characterization of outward currents in neurons of the avian nucleus magnocellularis. J Neurophysiol 80:2824–2835.

Ravindranathan A, Donevan SD, Sugden SG, Greig A, Rao MS, Parks TN (2000) Contrasting molecular composition and channel properties of AMPA receptors on chick auditory and brainstem motor neurons. J Physiol (Lond) 523:667–684.

Reyes AD, Rubel EW, Spain WJ (1994) Membrane properties underlying the firing of neurons in the avian cochlear nucleus. J Neurosci 14:5352–5364.

Rhode WS, Greenberg S (1992) Physiology of the cochlear nuclei. In: Popper AN, Fay RR (eds) The Mammalian Auditory Pathway: Neurophysiology. New York: Springer-Verlag, pp. 94–152.

Rhode WS, Smith PH (1986) Encoding timing and intensity in the ventral cochlear nucleus of the cat. J Neurophysiol 56:261–286.

Rhode WS, Oertel D, Smith PH (1983a) Physiological response properties of cells labeled intracellularly with horseradish peroxidase in cat ventral cochlear nucleus. J Comp Neurol 213:448–463.

Rhode WS, Smith PH, Oertel D (1983b) Physiological response properties of cells labeled intracellularly with horseradish peroxidase in cat dorsal cochlear nucleus. J Comp Neurol 213:426–447.

Rodriguez-Moreno A, Herreras O, Lerma J (1997) Kainate receptors presynaptically downregulate GABAergic inhibition in the rat hippocampus. Neuron 19:893–901.

Rothman JS, Young ED, Manis PB (1993) Convergence of auditory nerve fibers onto bushy cells in the ventral cochlear nucleus: implications of a computational model. J Neurophysiol 70:2562–2583.

Rouiller EM, Ryugo DK (1984) Intracellular marking of physiologically characterized cells in the ventral cochlear nucleus of the cat. J Comp Neurol 225:167–186.

Rubel EW, Parks TN (1975) Organization and development of brainstem auditory nuclei of the chicken: tonotopic organization of n. magnocellularis and n. laminaris. J Comp Neurol 164:411–433.

Rubio ME, Wenthold RJ (1997) Glutamate receptors are selectively targeted to postsynaptic sites in neurons. Neuron 18:939–950.

Ruggero MA (1992) Physiology of the auditory nerve. In: Popper AN, Fay RR (eds) The Mammalian Auditory Pathway: Neurophysiology. New York: Springer-Verlag, pp. 34–93.

Saint Marie RL, Benson CG, Ostapoff EM, Morest DK (1991) Glycine immunoreactive projections from the dorsal to the anteroventral cochlear nucleus. Hear Res 51:11–28.

Sanes DH, Friauf E (2000) Development and influence of inhibition in the lateral superior olivary nucleus. Hear Res 147:46–58.

Sato K, Shiraishi S, Nakagawa H, Kuriyama H, Altschuler RA (2000) Diversity and plasticity in amino acid receptor subunits in the rat auditory brainstem. Hear Res 147:137–144.

Schneggenburger R, Meyer AC, Neher E (1999) Released fraction and total size of a pool of immediately available transmitter quanta at a calyx synapse. Neuron 23:399–409.

Schwarz DW, Tennigkeit F, Adam T, Finlayson P, Puil E (1998) Membrane properties that shape the auditory code in three nuclei of the central nervous system. J Otolaryngol 27:311–317.

Sento S, Ryugo DK (1989) Endbulbs of held and spherical bushy cells in cats: morphological correlates with physiological properties. J Comp Neurol 280:553–562.

Sewell WF, Mroz EA (1990) Purification of a low-molecular-weight excitatory substance from the inner ears of goldfish. Hear Res 50:127–137.

Smith PH, Rhode WS (1987) Characterization of HRP-labeled globular bushy cells in the cat anteroventral cochlear nucleus. J Comp Neurol 266:360–375.

Smith PH, Rhode WS (1989) Structural and functional properties distinguish two types of multipolar cells in the ventral cochlear nucleus. J Comp Neurol 282:595–616.

Smith PH, Joris PX, Yin TC (1993) Projections of physiologically characterized spherical bushy cell axons from the cochlear nucleus of the cat: evidence for delay lines to the medial superior olive. J Comp Neurol 331:245–260.

Smith PH, Joris PX, Yin TC (1998) Anatomy and physiology of principal cells of the medial nucleus of the trapezoid body (MNTB) of the cat. J Neurophysiol 79:3127–3142.

Smith AJ, Owens S, Forsythe ID (2000) Characterisation of inhibitory and excitatory postsynaptic currents of the rat medial superior olive. J Physiol 529:681–698.

Sommer B, Keinanen K, Verdoorn TA, Wisden W, Burnashev N, Herb A, Kohler M, Takagi T, Sakmann B, Seeburg PH (1990) Flip and flop: a cell-specific functional switch in glutamate-operated channels of the CNS. Science 249:1580–1585.

Spruston N, Jaffe DB, Johnston D (1994) Dendritic attenuation of synaptic potentials and currents: the role of passive membrane properties. Trends Neurosci 17:161–166.

Takahashi T, Forsythe ID, Tsujimoto T, Barnes-Davies M, Onodera K (1996) Presynaptic calcium current modulation by a metabotropic glutamate receptor. Science 274:594–597.

Taschenberger H, von Gersdorff H (2000) Fine-tuning an auditory synapse for speed and fidelity: developmental changes in presynaptic waveform, EPSC kinetics, and synaptic plasticity. J Neurosci 20:9162–9173.

Trussell L (1998) Control of time course of glutamatergic synaptic currents. Prog Brain Res 116:59–69.

Trussell LO, Zhang S, Raman IM (1993) Desensitization of AMPA receptors upon multiquantal neurotransmitter release. Neuron 10:1185–1196.

Tsuchitani C, Boudreau JC (1967) Encoding of stimulus frequency and intensity by cat superior olive S- segment cells. J Acoust Soc Am 42:794–805.

Turecek R, Trussell LO (2001) Presynaptic glycine receptors enhance transmitter release at a mammalian central synapse. Nature 411:587–590.

Waldeck RF, Pereda A, Faber DS (2000) Properties and plasticity of paired-pulse depression at a central synapse. J Neurosci 20:5312–5320.

Wang LY, Kaczmarek LK (1998) High-frequency firing helps replenish the readily releasable pool of synaptic vesicles. Nature 394:384–388.

Wang LY, Gan L, Forsythe ID, Kaczmarek LK (1998) Contribution of the Kv3.1 potassium channel to high-frequency firing in mouse auditory neurones. J Physiol (Lond) 509:183–194.

Wang YX, Wenthold RJ, Ottersen OP, Petralia RS (1998) Endbulb synapses in the anteroventral cochlear nucleus express a specific subset of AMPA-type glutamate receptor subunits. J Neurosci 18:1148–1160.

Warchol ME, Dallos P (1990) Neural coding in the chick cochlear nucleus. J Comp Physiol [A] 166:721–734.

White JA, Young ED, Manis PB (1994) The electrotonic structure of regular-spiking neurons in the ventral cochlear nucleus may determine their response properties. J Neurophysiol 71:1774–1786.

Wickesberg RE, Oertel D (1990) Delayed, frequency-specific inhibition in the cochlear nuclei of mice: a mechanism for monaural echo suppression. J Neurosci 10:1762–1768.

Wickesberg RE, Whitlon D, Oertel D (1991) Tuberculoventral neurons project to the multipolar cell area but not to the octopus cell area of the posteroventral cochlear nucleus. J Comp Neurol 313:457–468.

Wu LG, Borst JG (1999) The reduced release probability of releasable vesicles during recovery from short-term synaptic depression. Neuron 23:821–832.

Wu SH, Oertel D (1986) Inhibitory circuitry in the ventral cochlear nucleus is probably mediated by glycine. J Neurosci 6:2691–2706.

Wu SH, Oertel D (1987) Maturation of synapses and electrical properties of cells in the cochlear nuclei. Hear Res 30:99–110.

Wu SH, Kelly JB (1992a) Binaural interaction in the lateral superior olive: time difference sensitivity studied in mouse brain slice. J Neurophysiol 68:1151–1159.

Wu SH, Kelly JB (1992b) Synaptic pharmacology of the superior olivary complex studied in mouse brain slice. J Neurosci 12:3084–3097.

Wu SH, Fu XW (1998) Glutamate receptors underlying excitatory synaptic transmission in the rat's lateral superior olive studied in vitro. Hear Res 122:47–59.

Yang L, Monsivais P, Rubel EW (1999) The superior olivary nucleus and its influence on nucleus laminaris: a source of inhibitory feedback for coincidence detection in the avian auditory brainstem. J Neurosci 19:2313–2325.

Yin TC, Chan JC (1990) Interaural time sensitivity in medial superior olive of cat. J Neurophysiol 64:465–488.

Young ED, Robert JM, Shofner WP (1988) Regularity and latency of units in ventral cochlear nucleus: implications for unit classification and generation of response properties. J Neurophysiol 60:1–29.

Zhang S, Trussell LO (1994a) Voltage clamp analysis of excitatory synaptic transmission in the avian nucleus magnocellularis. J Physiol (Lond) 480:123–136.

Zhang S, Trussell LO (1994b) A characterization of excitatory postsynaptic potentials in the avian nucleus magnocellularis. J Neurophysiol 72:705–718.

Zhou N, Taylor DA, Parks TN (1995) Cobalt-permeable non-NMDA receptors in developing chick brainstem auditory nuclei. Neuroreport 6:2273–2276.

Zucker RS (1989) Short-term synaptic plasticity. Annu Rev Neurosci 12:13–31.

4
Neural Mechanisms of Encoding Binaural Localization Cues in the Auditory Brainstem

Tom C.T. Yin

1. Introduction: the Importance of Sound Localization

When an animal hears a sound in its environment, there are several important tasks that the auditory system must try to accomplish. Two major jobs are to determine *what* it was that produced the sound and *where* it comes from. Understanding how the nervous system can accomplish these tasks is a major goal of modern auditory neurobiological research. In this book, we explore what is known about these questions at several different levels of the auditory system. The purpose of this chapter is to review the anatomical and physiological mechanisms in the auditory brainstem of mammals that encode where a sound originates. Specifically, this chapter examines the two binaural localization cues: interaural time disparities (ITDs) and interaural level disparities (ILDs) (For abbreviations, see Table 1). The neural mechanisms of sound localization are of particular interest since the location of a stimulus is not represented in the sensory epithelium, as it is in the visual or somatosensory systems, but must be computed by combining input from the two ears in the central auditory system. To a large degree, we understand how these cues are encoded by single cells at this level of the auditory system. Indeed, it appears that certain cells in the auditory brainstem are highly specialized to facilitate the encoding of these cues, and more is known about the central processing of sound localization cues than of any other auditory function (e.g., pitch perception, vowel discrimination).

It is easy to appreciate the importance to an animal of identifying the location of a sound source. For example, imagine hearing a snapping twig in the middle of the woods on a dark night. The ability to ascertain whether the sound originated from the left or right side may determine whether the animal survives predation or starvation. For *Homo sapiens* living in modern society with few predators outside of our own species, sound localization is perhaps not as critical, except for occasions like an American stepping off the curb on a busy London thoroughfare and hearing the sudden blaring horn of a rapidly oncoming cab from an unexpected direction.

TABLE 4.1. Abbreviations

AM	amplitude modulated
AVCN	anteroventral cochlear nucleus
CD	characteristic delay
CF	characteristic frequency
CP	characteristic phase
DNLL	dorsal nucleus of the lateral lemniscus
EE	excitatory (contra)/excitatory (ipsi)
EI	excitatory (contra)/inhibitory (ipsi)
EM	electron micrograph
EPSP	excitatory post-synaptic potential
GABA	gamma aminobutyric acid
GBC	globular bushy cell
HRP	horseradish peroxidase
IC	inferior colliculus
ICC	central nucleus of the inferior colliculus
IE	inhibitory (contra)/excitatory (ipsi)
ILD	interaural level difference
IPD	interaural phase difference
ITD	interaural time difference
IPSP	inhibitory post-synaptic potential
LNTB	lateral uncleus of the trapezoid body
LSO	lateral superior olive
MNTB	medial nucleus of the trapezoid body
MSO	medial superior olive
NMDA	N-methyl-d-aspartate
PL	primary-like
PLn	primary-like-with-notch
PST	post-stimulus time
SBC	spherical bushy cell
SOC	superior olivary complex
SPL	sound pressure level
VNLL	ventral nucleus of the lateral lemniscus
VNTB	ventral nucleus of the trapezoid body

Despite the relatively minor importance of sound localization, per se, for the survival of humans, the neural circuitry that has evolved for localization is nonetheless significant for other tasks facing modern human beings. It is likely that the ability to detect and discriminate sounds in a noisy environment, the so-called cocktail party problem, also relies heavily on the same neural circuits. Cherry (1953) noted that one of the major factors in solving the cocktail party problem was the spatial separation of the sound sources. Related observations were made a few years earlier by Licklider (1948) and Hirsch (1948) that detection of a signal in a background of noise is much easier when the signal has a different interaural time difference than that of the noise, e.g., if the signal is inverted in phase to the two ears while the noise masker is in phase. Since interaural time differences are related to the spatial cues for azimuthal localization, this binaural masking level difference appears to be associated with the cocktail party problem.

1.1 Acoustic Cues for Sound Localization

The classical "duplex theory of sound localization" was first formalized by Thompson (1882) and Rayleigh (1907). According to this theory, the primary cues for sound localization are interaural level differences (ILDs) and interaural time differences (ITDs; for abbreviations see Table 4.1). Rayleigh recognized that substantial ILDs will be generated only at high frequencies (in humans, above 2–3 kHz) where the wavelength is short and the head can act as an effective acoustic shadow. To demonstrate the ability to encode ITDs, Rayleigh repeated an experiment by Thompson (1882) using the perception of what is now called "binaural beats," where two tones of slightly different frequencies (<5 Hz or so) are delivered separately to the two ears. He generated binaural beats by dripping wax onto one of two identical tuning forks to slightly mistune it. The tones were delivered separately to the two ears by seating the subject in one room and delivering the tones via long tubes from adjacent rooms through long tubes. Rayleigh realized that if the two mistuned tones reached the ears without interacting physically, then the perception of the beat or difference frequency must depend upon the brain's sensitivity to interaural phase differences. The weak sensation of movement of the tone within the head led Rayleigh to propose that the perception of the binaural beats was related to azimuthal sound localization. Since the beat frequency was not perceived when the tuning forks were at high frequency, Rayleigh proposed that ITDs are only useful at low frequencies. Thus, the classical "duplex" theory proposed that high frequency tones are localized by ILDs and low frequency tones by ITDs.

The binaural localization cues of ILD and ITD are the important cues for horizontal or azimuthal localization. That additional cues must be present can be easily appreciated by considering the localization of sounds along the vertical midsagittal meridian. Here interaural cues such as ITDs and ILDs are minimal, yet we know that we can discriminate a sound that is delivered from directly in front of us from one that is above or below our heads. In recent years, there has been a growing awareness of the importance of another cue for sound localization in the vertical dimension: namely, what are usually called spectral cues. The spectral cues created by the diffraction and reflection of sound by the external ears, head and shoulders are largely responsible for our ability to localize sounds along the vertical dimension (Batteau 1966; Gardner and Gardner 1973). This topic is covered more thoroughly in Young, Chapter 5.

Support for the duplex theory has come from psychophysical experiments that show decreased azimuthal acuity in the mid-frequency range where neither cue is strong (Stevens and Newman 1936; Mills 1958; Casseday and Neff 1973), and from the inability of human observers to detect ITDs in the ongoing fine structure of high frequency pure tones (Licklider et al. 1950; Zwislocki and Feldman 1956; Mills 1960). The temporal features of stimuli are encoded by the phase locking of auditory nerve fibers, but phase locking to pure tones is limited to low frequencies (<3–4 kHz in mammals, Johnson 1980) so auditory nerve fibers (and all other cells in the ascending auditory system) are unable to carry information about the phase of high frequency tones. Thus, ITDs cannot be used to localize high frequency tones. However, it is important to note that the duplex theory only holds for pure tones. Broadband high-frequency signals can be lateralized on the basis of ITDs alone (Klumpp and Eady 1956; Leakey et al. 1958; Henning 1974a, 1974b), which depends upon the ability of high-frequency cells to phase lock to the low frequency envelope of the broadband stimuli despite the inability to phase lock to the high frequency carrier (Yin et al. 1984; Batra et al. 1989a; Joris and Yin 1995).

Corresponding to the psychophysically-defined classical duplex theory, there are two parallel pathways in the auditory brainstem that are thought to encode ITDs and ILDs. The neural circuitry underlying the encoding of these binaural localization cues is well understood, and each of the cell types in the circuit has been well characterized physiologically. Indeed, this is one of the rare cases where the behavioral function of a circuit in the central nervous system is believed to be well known. Evidence for a similar dichotomy in humans has been found in recent studies of patients with multiple sclerosis who were unable to detect ITDs but had preserved ILD sensitivity, suggesting the existence in humans of parallel pathways encoding ITDs and ILDs (Levine et al. 1993; Furst et al. 2000).

Figure 1 shows the two circuits that are thought to be involved in the coding of ITDs and ILDs in the mammalian auditory brainstem. There are many parallel features to the two circuits: both involve cell groups in the superior olivary complex (SOC), both circuits receive inputs from similar cells in the anteroventral cochlear nucleus (AVCN), and both project primarily to the inferior colliculus (IC). The circuit that encodes ITDs involves cells in the medial superior olive (MSO), which receive excitation from large spherical bushy cells (SBCs) of the AVCN of both sides (Fig. 4.1A). The circuit that encodes ILDs involves cells in the lateral superior olive (LSO), which receive excitation from the small SBCs of the ipsilateral AVCN, and inhibition from the contralateral side relayed through inhibitory neurons in the medial nucleus of the trapezoid body (MNTB). The MNTB cells receive excitatory input from the globular bushy cells (GBCs) of the contralateral AVCN (Fig. 4.1C). Cells in the LSO are sensitive to the ILD of the signal: they are excited by sounds that are greater in level at the ipsilateral ear than the contralateral ear and inhibited by sounds that are greater in level at the contralateral ear (Fig. 4.1D). Thus all of the major inputs to these localization circuits in the SOC derive from the two classes of bushy cells in the AVCN which course out of the ventral acoustic stria, or trapezoid body, and project to the SOC. We shall then begin this review by an examination of the properties of the most peripheral neural elements in these circuits: the auditory nerve fibers and bushy cells of the AVCN.

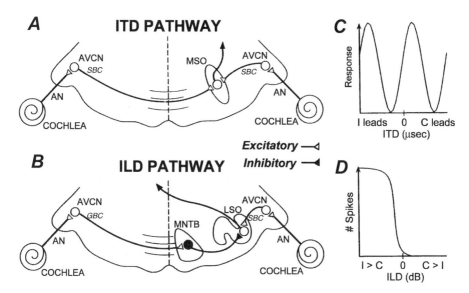

FIGURE 4.1. Schematic diagram of the two parallel pathways in the SOC for encoding ITDs (A) and ILDs (B). On the right are idealized neural responses in the MSO to variations in ITDs (C) and in the LSO to ILDs (D) of pure tones. I—ipsilateral, C—contralateral. (For further abbreviations, see Table 4.1.) (Reprinted from Neuron, Vol. 21, Joris et al., Coincidence detection in the auditory system: 50 years after Jeffress, p. 1237, Copyright 1998, with permission from Elsevier Science.)

The aim of the present chapter is to review the anatomical and physiological evidence that the two interaural cues, ILDs and ITDs, are initially encoded in the auditory brainstem nuclei of the SOC. There are several recent comprehensive reviews of these topics including earlier volumes in this series (Irvine 1986, 1992; Schwartz 1992; Kuwada et al. 1997; Joris et al. 1998). The present review will emphasize more recent work and will only highlight the older results. In addition, the review will concentrate on studies in higher mammals. This necessarily neglects a sizable and significant corpus of studies in birds (Carr and Konishi 1990). Less attention will also be paid to work done in rodents, as there appear to be fundamental differences in the organization of the SOC in rodents and in higher mammals related to the origin of the olivocochlear system, and I will not review prominent descending pathways from the SOC.

2. Encoding of Time and Intensity by the Peripheral Auditory System

Clearly, if ITD and ILD are to be useful binaural cues, the peripheral auditory system must encode the temporal and amplitude parameters of the acoustic waveform. Two characteristics of hair cells are important for this discussion. First, the basilar membrane is tonotopically organized so that each inner hair cell is maximally excited by stimulation at a characteristic frequency (CF). The tonotopicity is due to the systematic change in stiffness of the basilar membrane from base to apex that causes the maximal amplitude of vibration to be frequency dependent: high-frequency tones cause maximal deflection near the base of the basilar membrane, and low frequency tones stimulate the apex. Thus, the hair cells are tuned in frequency, and have narrow V-shaped frequency tuning curves. Second, the hair cells are polarized so that movement of the hair bundle toward the taller stereocilia is excitatory and movement in the opposite direction is inhibitory. The result is that the auditory nerve fibers innervated by the inner hair cells respond when the acoustic wave moves the hair bundle in one direction and are silent in the opposite direction. For a low-frequency sinusoidal tone, the auditory nerve fibers are then said to respond in a phase-locked fashion, as the response favors a particular phase angle. This phase locking of auditory nerve fibers is the basis for the encoding of timing information by the auditory periphery.

Since almost all auditory nerve fibers innervate a single inner hair cell, the properties of the auditory nerve fibers reflect inner hair cell responses. Figure 4.2A shows the response in the form of dot rasters of an auditory nerve fiber to many repetitions of a low frequency (340 Hz) tone. Each horizontal line represents a single repetition of the tone and each dot the time of occurrence of an action potential from the nerve fiber. The preference of the fiber to respond to a particular phase angle of the tone is seen in the vertical alignment of the dots across successive presentations of the tone. The degree of phase locking can be seen more precisely by computing a cycle ("period") histogram where the occurrence of spikes are graphed relative to the phase of one cycle of the sinusoidal tone (Fig. 4.2B) and quantified by the computing the synchronization coefficient from the vector average (Goldberg and Brown 1969). If all spikes fall into a single bin, the synchronization coefficient equals 1.0, while if all bins are equally likely, the coefficient equals 0.0. The ability of auditory nerve fibers to phase lock to tones is limited to low frequencies; in mammals, phase locking begins to decline at frequencies above about 1 kHz and disappears above 3–4 kHz (Johnson 1980; Fig. 4.3). At higher frequencies, the timing of discharges on the auditory nerve is random with respect to the stimulus phase.

As the sound pressure level (SPL) of a stimulus is increased from below to well above threshold, auditory nerve fibers increase their rate of discharge above their spontaneous level with a sigmoidal relationship to dB SPL over a relatively narrow (20–30 dB) range and then saturate at some peak discharge (Sachs and Abbas 1974). Some auditory nerve fibers have "sloping saturation" with slowing rising rates at high saturation SPLs (Fig. 4.4, left column). Since the great majority of auditory nerve fibers have sharply saturating rate-level functions with a relatively narrow range of thresholds, there arises the so-called "dynamic range problem": how does

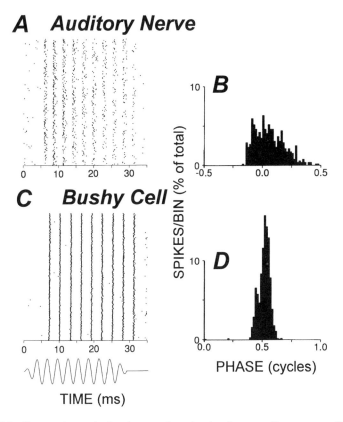

FIGURE 4.2. Comparison of stimulus synchronization in an auditory nerve fiber and AVCN bushy cell. Each dot in the rasters (left panels) indicates a spike occurrence to a short tone at the cell's best frequency (350 Hz in A,B and 340 Hz in C,D); each row of dots is the response to one of 200 repetitions of the tone. The responses are phase locked to the stimulus, which is seen in the tendency of spikes to occur at a particular phase angle by graphing the response relative to stimulus phase (B, D). Spikes in the bushy cell are temporally less dispersed than in the auditory nerve and occur in each stimulus cycle, whereas cycles are often skipped in the nerve. From Joris et al. (1998).

the auditory system achieve its wide psychophysical dynamic range of 100–120 dB when most individual fibers only vary in discharge rate over 30–40 dB? Part of the answer must lie in the different threshold ranges and in the sloping saturation responses, but the problem still does not have a satisfactory answer.

2.1 Bushy Cells of Anteroventral Cochlear Nucleus

Figure 4.1 shows that both of the interaural localization circuits receive their input from the cochlear nucleus by way of the bushy cells of the

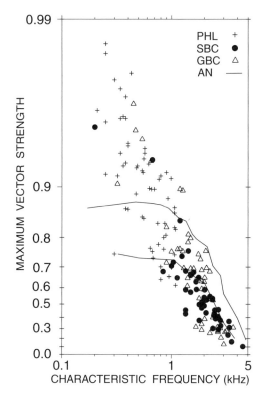

FIGURE 4.3. Maximum synchronization coefficient in AN and bushy cells plotted against CF. Solid lines: upper and lower boundaries of the range of maximum sync coefficients for AN fibers from Johnson (1980). SBC and GBC symbols indicate anatomically-identified spherical bushy cells and globular bushy cells, respectively, while PHL refers to fibers with physiological responses only. The ordinate designates (1—sync) on a logarithmic scale to provide an equal variance axis. (For abbreviations, see Table 4.1.)

AVCN. Bushy cells are only one of several cell types in the cochlear nucleus that receive direct synaptic input from auditory nerve fibers. The size and shape of the synaptic endings of auditory nerve fibers onto cells in the cochlear nucleus varies systematically, depending upon the region of the cochlear nucleus and the cell type contacted. Especially striking are the synaptic contacts between auditory nerve fibers and SBCs in the anterior AVCN, which take the form of a single or few large endbulbs of Held onto the soma of the bushy cells. More posteriorly, in the AVCN and nerve root region, GBCs receive smaller somatic terminals, modified endbulbs, from the auditory nerve fibers. Estimates based on anatomical considerations of the number of nerve fibers innervating the bushy cells range from one to two on SBCs (Brawer and Morest 1975; Cant and Morest 1984) to 19 modified endbulbs for GBCs (Liberman 1991).

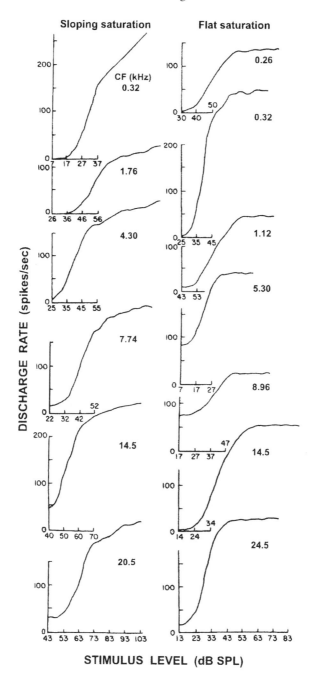

FIGURE 4.4. Rate versus level functions for 13 AN fibers from 8 cats. Stimuli were all measured at CF, which is indicated for each fiber. (Reprinted with permission from Sachs and Abbas, Rate versus level functions for auditory-nerve fibers in cats: tone-burst stimuli, Journal of the Acoustical Society of America, Vol. 56, 1974, p. 1837. Copyright 1974, Acoustical Society of America.)

2.2 Physiological Responses of Bushy Cells

The physiological response properties of the SBCs and GBCs are so similar to that of their auditory nerve inputs that they are called primary-like (PL) and primary-like-with-notch (PLn), respectively. Early studies of the bushy cells emphasized this similarity with but few exceptions. In particular, comparisons of the phase locking of cells in the AVCN with auditory nerve fibers indicated that the ability of bushy cells to phase lock was similar to or poorer than auditory nerve fibers (Lavine 1971; Goldberg and Brownell 1973; Bourk 1976; Kettner et al. 1985; Palmer et al. 1986; Blackburn and Sachs 1989; Winter and Palmer 1990).

However, recent studies using intracellular recordings and staining of single axons in the trapezoid body (Joris et al. 1994a, 1994b), and extracellular recordings in AVCN (Carney 1990) have shown some important differences in the ability of auditory nerve fibers and bushy cells to encode the temporal information in stimulus. The majority of bushy cells phase lock with considerably higher precision to low frequency tones than do auditory nerve fibers, which is clearly evident in the alignment of dot rasters (Figs. 4.2A and 4.2C) or shapes of the period histograms (Figs. 4.2B and 4.2D). This enhancement in synchronization is seen for cells of low characteristic frequency when stimulated at CF (Joris et al. 1994a), or for cells of high CF when stimulated with a low frequency tone (500 Hz) in their tail (Joris et al. 1994b). Intracellular labeling of the fibers following physiological identification showed that both GBCs and SBCs exhibited the increased synchronization (Smith et al. 1991, 1993; Joris et al. 1994a). The identification of GBCs or SBCs was made primarily on the basis of the axonal projection pattern: GBCs project to the contralateral MNTB with a large calyceal ending, while SBCs project bilaterally to the MSO (Fig. 4.1). The obvious sharpening in the period histograms of Figure 4.2D can be quantified by comparing the synchronization coefficients in Figure 4.3. Because the synchronization coefficient saturates at values greater than 0.9, the data are plotted on an expansive logarithmic scale that provides an equal variance axis (Johnson 1974). The lines represent the range of values found in the auditory nerve by Johnson (1980) as well as in our own auditory nerve recordings. Auditory nerve fibers never have synchronization coefficients greater than 0.9 while 75% of presumed bushy cells with CFs below 0.7 kHz have synchronization greater than 0.9. These results show that the temporal input provided to the binaural nuclei in the SOC by bushy cells is considerably sharper than that of auditory nerve fibers. Models of binaural processing, therefore, should use the physiological properties of bushy cells rather than auditory nerve fibers to provide accurate temporal information.

An important question is the mechanism by which the bushy cells achieve this enhanced temporal synchronization. Clearly, since the bushy cell response differs from that of the auditory nerve fiber, the large endbulb synapse does not function as a one-to-one pure relay even at low CFs where the endbulbs are largest. One way to get an enhanced synchronization is to have the bushy cell behave as a coincidence detector with a number of presynaptic auditory nerve inputs with similar CFs. If each input is sub-threshold with a fast, brief EPSP, then the bushy cell effectively behaves as a coincidence detector, requiring coincident inputs over a short time window to elicit a postsynaptic spike. Joris et al. (1994a) showed that such a simple "shot-noise" coincidence model could mimic the enhanced phase locking of bushy cells to pure tones at CF. Rothman et al. (1993) and Rothman and Young (1996) have used computer models to investigate the feasibility of various combinations of auditory nerve inputs and synaptic strengths in order to mimic the physiological responses of bushy cells using realistic auditory nerve spike trains as input. The model was adjusted to fit the response pattern, spontaneous rates, phase-locking, and regularity of discharge (the tendency of the fiber to discharge at a constant rate; Young et al. 1988). To obtain the enhanced phase locking at CF and at 500 Hz, models that utilized subthreshold inputs and required coincidence to evoke a spike were needed, but such models were also more regular than physiological responses of bushy cells. The best results were obtained with suprathreshold inputs with a tonic inhibition that functionally made the inputs subthreshold, though the results suggest that different classes of bushy cells may require different combinations of parameters.

The idea of inhibitory inputs is attractive because it is known from anatomical (Thompson et al. 1985; Smith and Rhode 1987; Wenthold et al. 1988; Saint Marie et al. 1989a; Adams and Mugnaini 1990) and intracellular physiological studies (Wu and Oertel 1986; Caspary et al. 1994) that both SBCs and GBCs have substantial synaptic inputs with the characteristics of inhibitory terminals, but there has been little physiological data supporting a role for inhibition, possibly because most experiments use anesthetized animals. Recordings from brain slices of nucleus magnocellularis in the chick (Monsivais et al. 2000), the avian homologue of the AVCN, have demonstrated prominent GABAergic inputs from the superior olive that have slow time courses which could provide the tonic inhibition in the Rothman and Young (1996) model.

As the stimulus frequency of CF tones is raised above 1 kHz, phase locking drops rapidly so that it is lower in bushy cells than in the auditory nerve above 2 kHz and becomes insignificant for all cells above 4 kHz (Fig. 4.3). At these higher frequencies, the response histograms of bushy cells take on their PL and PLn shapes for SBCs and GBCs, respectively. The large initial peak is created by the precise timing of the first spike in GBCs and the consequent alignment of the refractory period creates the notch in the PST histogram. The notch in the GBC histogram is usually only present at relatively high SPLs. The precision of the first spike in GBCs as compared to SBCs, which also receive the auditory nerve fiber input, points to some clear differences between these two classes of cells in the number and strength of their synaptic inputs and/or in the intrinsic membrane properties of their postsynaptic cell (see Trussell, Chapter 3). These high frequency SBCs and GBCs are the predominant excitatory input to the LSO and MNTB, respectively.

2.3 Projections of Bushy Cells to the Superior Olive

The classic descriptions of the projections of bushy cells from the AVCN to the SOC are shown on Figure 4.1. SBCs from the anterior AVCN (Tolbert et al. 1982; Cant and Casseday 1986) provide excitatory input bilaterally to the MSO and ipsilaterally to the LSO while the GBCs project to the contralateral MNTB, which in turn provides an inhibitory input to the LSO. Recent studies have shown more details of these projections by intracellular labeling of individual axons, immunocytochemical studies, or the use of different axonal tracers.

Using intra-axonal injections in the trapezoid body, Smith et al. (1991, 1993) were able to obtain direct evidence of the correlation between the physiological responses of individual bushy cell axons and the anatomical features of the axonal projection to the SOC. The shape of the poststimulus time histogram to CF tones was used to distinguish SBCs from GBCs: SBCs have a primary-like (PL) response whereas GBCs have a primary-like-with-notch (PLn) response at high SPLs. This physiological feature distinguishes the majority of high CF bushy cells into SBC and GBCs but does not help to separate the two classes at low CFs where phase locking obscures the notch, and the PSTs are thereby indistinguishable. Using the anatomical criteria that GBCs project to the contralateral MNTB while SBCs project bilaterally to the MSO, Smith et al. (1993) found that all GBCs crossed the midline in the ventral half of the trapezoid body while all SBCs crossed in the dorsal half, indicating that these two classes of axons are clearly physically segregated by their axonal course.

The physiological features are sometimes less reliable for differentiating these two classes of cells. For example, while no SBCs at high CFs ever exhibited a notch in the PST at high levels, a few GBCs had less well defined notches or were more onset-like with a low sustained rate even at very high SPLs. Blackburn and Sachs (1989) suggested that the use of CF tones with asynchronous starting phases could be used to distinguish low CF SBCs and GBCs, but unfortunately they had no independent confirmation of the cell type. On the other hand, Smith et al. (1993) found that such asynchronous tones did not differentiate the two classes definitively, as judged by the anatomical criteria.

2.3.1 Projections of Spherical Bushy Cells

An important question regarding the projection of SBCs to the MSO is whether these axonal projections take the form of neuronal delay lines as hypothesized by Jeffress (1948). We will postpone discussion of this question until later when looking in detail at the evidence supporting the Jeffress model.

The well known classic projections of the SBCs (studied by axonal degeneration or HRP injection) are to the ipsilateral LSO and bilaterally to the MSO (Warr 1972; Tolbert et al. 1982; Glendenning et al. 1985; Saint Marie et al. 1989b). An interesting feature of the projection to the MSO projection is that inputs from the AVCN of each side are segregated to the dendrites facing that side, so that the projection to the ipsilateral MSO ends on the lateral dendrites while the contralaterally projecting axon terminals end on the medial dendrites (Stotler 1953; Lindsey 1975; Smith et al. 1993). Shneiderman and Henkel (1985) showed that at least some of the projections to these three nuclei come from individual axons that collateralize to two or three of the nuclei by making small injections of wheat germ agglutin HRP into the low-frequency lateral limb of the LSO and tracing anterograde label into both MSO in the expected frequency representation, or vice versa. Intracellular labeling of individual SBC axons confirmed that single axons projected bilaterally to the MSO and ipsilaterally to the low-frequency limb of the LSO with additional projections to the contralateral VNTB, lateral nucleus of the trapezoid body (LNTB), and VNLL, and ipsilateral LNTB (Smith et al. 1993). However, very few high-CF PL fibers were encountered in penetrations made from the ventral surface between the two MSOs, suggesting that the projections of high frequency SBCs may be restricted or at least are biased to the ipsilateral LSO and do not include the MSO. Given the low-frequency bias of MSO and high-frequency bias of LSO (Guinan et al. 1972b), this suggestion seems reasonable.

The characteristics of synaptic terminals of SBCs in the LSO are consistent with their presumed role in mediating the excitatory response to stimulation of the ipsilateral ear in physiological experiments: immunocytochemical studies show that SBCs and their terminals in the LSO are glutamatergic or aspartergic (Glendenning et al. 1991), receptor binding suggests that the terminals are of the quisqualate type (Glendenning et al. 1991), and electron microscopic (EM) analysis of the synaptic terminals of fibers labeled with an antibody to PEP-19 (Berrebi and Spirou 1998) reveal large round vesicles with asymmetric synaptic junctions, which are likely to be the same as those found on the distal dendrites of LSO principal cells (Cant 1984; Helfert et al. 1992).

The synaptic terminals ending on MSO cells also exhibit characteristics of excitatory synapses; EM analysis revealed asymmetric terminals with large round vesicles originating from the SBCs of the AVCN (Schwartz 1972; Lindsey 1975; Kiss and Majorossy 1983), which was confirmed by EM examination of single HRP-labeled axons of SBCs that terminated in the MSO (Smith et al. 1993).

2.3.2 Projections of Globular Bushy Cells

The axons of GBCs are uniformly large in diameter and travel in the ventral half of the trapezoid body. The major target of GBCs is the contralateral MNTB where they terminate in the calyx of Held, one of the largest synaptic endings in the entire nervous system (Fig. 4.1) (Warr 1972; Friauf and Ostwald 1988; Guinan and Li 1990; Spirou et al. 1990; Kuwabara et al. 1991; Smith et al. 1991; Sommer et al. 1993). Each principal cell in the MNTB receives a single calyceal synapse which has a claw-like shape that envelopes the soma. With single fiber labeling, one can see that a single GBC axon can sometimes branch and form two calyceal endings on different MNTB cells, though this is not common (Smith et al. 1991). In recent years there has been a growing realization that the large size of the presynaptic terminal makes it a model for *in vitro* studies of an excitatory glutamatergic synapse. We will discuss these results in more detail when we get to the MNTB (section 4). In addition to the primary projection to the contralateral MNTB, some projections are evident following intracellular labeling of single GBC axons (Smith et al. 1991). On the ipsilateral side there are consistent collaterals to the posterior preolivary nucleus and LNTB, with occasional projections to the dorsolateral periolivary nucleus and the LSO. On the contralateral side, there are projections to the dorsomedial periolivary nucleus, VNLL, and to a region ventromedial to the facial nucleus.

3. The Medial Superior Olivary Circuit for Encoding Interaural Time Disparities

3.1 Anatomical Organization of the Medial Superior Olive

The basic cytoarchitectual and histological organization of the medial superior olive (MSO) is well known and has been reviewed in a previous volume in this series with particular emphasis on the comparative aspects of the SOC (Schwartz 1992), so I will only summarize the highlights and concentrate on recent work. In many mammals (cat, monkey, gerbil, rat), the MSO comprises a sheet of cells only a few cells thick in the narrowest dimension, though it can also be folded as in the dog (Goldberg and Brown 1968). The nucleus is tonotopically organized, with low frequencies represented dorsally and high frequencies ventrally (Guinan et al. 1972b). In agreement with the duplex theory regarding the importance of low frequency sounds for sound localization based on ITDs and the role of the MSO in encoding ITDs, there is an over representation of low frequencies in the MSO's tonotopic map (Guinan et al. 1972b). Generally, three cell types have been identified: prinicipal, multipolar, and marginal cells (Kiss and Majorossy 1983).

The principal cells are of particular interest since they are likely to be the coincidence detectors studied in physiological experiments. In a coronal section, the principal cells have a striking bipolar appearance, with dendrites that extend medially and laterally and receive inputs from the SBCs of the AVCN of each side: the medial dendrites from the contralateral and the lateral dendrites from the ipsilateral AVCN (Stotler 1953; Lindsey 1975). However, in horizontal or sagittal sections, the dendritic trees of the principal cells can extend over a considerable rostrocaudal extent of the nucleus (Scheibel and Scheibel 1974; Schwartz 1977). In addition, Golgi studies have shown that the terminal arbors of axons from the AVCN to the MSO can also extend for a considerable section of the rostrocaudal extent of the nucleus. These considerations will be important when we consider the Jeffress model of ITD coding later.

3.1.1 Afferents to the MSO

In addition to the input from spherical bushy cells discussed above, the afferents to the MSO have been studied using retrograde tracing following injections of HRP or other label restricted to the MSO. A difficulty in any such experiment involving the SOC is labeling axons of passage since the nuclei are embedded in the ascending axons of the trapezoid body. Nonetheless, Cant and Hyson (1992) have shown that the gerbil MSO also receives input from the LNTB and MNTB in addition to the input from the AVCN. From EM studies, it is likely that these provide inhibitory inputs to the MSO while the major spherical bushy cell input is excitatory.

EM studies of the MSO have generally found that the most common synaptic terminals found on the soma and dendrites of principal cells appear to be excitatory: they have large asymmetric profiles filled with large spherical vesicles. Since these terminals disappear following lesions of the AVCN (Clark 1969a; 1969b; Perkins 1973; Lindsey 1975; Kiss and Majorossy 1983; Schwartz 1984), they undoubtedly originate from the major SBC projection to the MSO on both sides. In addition, ultrastructural studies of the MSO have identified inhibitory-like synapses with both pleomorphic and flat vesicles (Clark 1969a, 1969b; Schwartz 1980; Kiss and Majorossy 1983; Brunso-Bechtold et al. 1994). Immunocytochemical studies have also revealed the presence of labeled perisomatic puncta in the MSO for the inhibitory neurotransmitters glycine and GABA, although the labeling is considerably less intense than in the LSO (Roberts and Ribak 1987; Glendenning and Baker 1988; Helfert et al. 1989). The source of these putative inhibitory inputs is likely to be the ipsilateral MNTB and LNTB. The MNTB is known to provide a glycinergic inhibitory input to the LSO (see below), and MNTB axons have collaterals that end in the MSO (Adams and Mugnaini 1990; Banks and Smith 1992; Kuwabara and Zook 1992; Smith et al. 1998). Direct anatomical confirmation that the synaptic terminals in the LSO of the axons of MNTB cells have flattened vesicles comes from intracellular labeling of MNTB cells with HRP or neurobiotin followed by EM analysis (Smith et al. 1998). That the ipsilateral LNTB is likely to be a source of inhibitory terminals comes from anatomical tracing studies showing retrogradely-labeled cell bodies in LNTB following small injections of HRP into the MSO (Cant and Hyson 1992). Finally, both the MNTB and LNTB show intense somatic labeling for glycine (Wenthold et al. 1987; Peyret et al. 1987; Helfert et al. 1989; Adams and Mugnaini 1990; Henkel and Brunso-Bechtold 1995; Ostapoff et al. 1997; Spirou and Berrebi 1997). Thus, there is strong anatomical and immunocytochemical evidence for inhibitory projections from the MNTB and LNTB to the MSO. We will consider the role of inhibition in the MSO later.

3.1.2 Efferent Projections of the MSO

Anatomical degeneration and tracer studies have shown that the primary projection of the MSO is to the ipsilateral dorsal nucleus of the lateral lemniscus (DNLL) and inferior colliculus (IC) with a small contralateral component (Roth et al. 1978; Adams 1979; Glendenning et al. 1981; Henkel and Spangler 1983; Aitkin and Schuck 1985; Kelly et al. 1998). Since very few MSO cells are labeled with GABA or glycine immunoreactivity (Helfert et al. 1989), and since EM autoradiographic studies show asymmetric synaptic terminals with round vesicles in the central nucleus of the inferior colliculus (ICC) (Oliver et al. 1995), it is likely that this projection is excitatory. Physiological recordings are also in accord with the conclusion that the projection is primarily ipsilateral and excitatory, since the strong bias for stimuli favoring the contralateral ear in MSO cells is largely maintained in the IC (see below). However, this conclusion is indirect, since there are no single axon labeling studies coupled with EM.

3.2 Neural Mechanisms of Coding Interaural Time Differences: the Jeffress Model

Over 50 years ago Jeffress (1948) published a short seminal paper that proposed a model for encoding ITDs. The model is remarkable in many respects. First, it was exceptionally prescient in forecasting the neural mechanisms of sound localization, particularly in light of how little was known at the time it was written. Second, it has proven to be valuable for both psychoacousticians and physiologists in generating experiments, (indeed, the paper is one of the most highly cited papers in the literature). Finally, the model forms the basis for virtually all modern psychoacoustical and physiological models of binaural processing based on ITDs. Modern consensus has placed the likely site for encoding ITDs to be in the MSO or its avian homologue of nucleus laminaris (Carr and Konishi 1990).

Despite the almost universal acceptance of the model, there are relatively few data that directly address features of the model. Partly, this is because the MSO has proven to be an exceptionally difficult nucleus from which to record single cells. The reason for the difficulty is not clear, though it is at least partly because of the narrowness of the nucleus and the presence of large field potentials in the vicinity of the MSO as a consequence of the many phase-locked afferents converging there.

There are three critical assumptions to the Jeffress model: (1) that the inputs to the binaural cells from each side are carrying accurate timing information about the acoustic stimulus; (2) that the binaural cells behave like coincidence detectors, i.e., they respond maximally when the input spikes from the two sides arrive in coincidence (the model is sometimes referred to as the coincidence model) and are therefore exquisitely sensitive to small disparities in interarrival timing; and (3) that afferents project to the binaural cells in the form of oppositely directed neuronal delay lines. The result of these three assumptions is a spatial map of ITDs across the axis of the nucleus parallel to the delay lines (Joris et al. 1998). We shall now examine each of the three assumptions in turn.

3.2.1 Assumption #1: Afferents Carry Timing Information

The first assumption has been discussed above since the afferents to the MSO from the AVCN originate from the SBCs. Figure 4.3 shows that the great majority of bushy cells with CFs <1.3 kHz have synchronization coefficients to pure tones at CF that are higher than their auditory nerve fiber inputs. Thus, the inputs to the MSO can signal timing information with high fidelity. A surprising and still inexplicable aspect of the synchrony enhancement of SBCs and GBCs described by Joris et al. (1994a) is the large number of reports (Bourk 1976, Lavine 1971, Goldberg and Brownell 1973; Kettner et al. 1985; Palmer et al. 1986; Blackburn and Sachs 1989; Van Gisbergen et al.1975; Winter and Palmer 1990) that failed to find such enhancement from recordings of low CF cells in the AVCN itself, particularly since the high synchrony and entrainment (the tendency for the cell to respond to every cycle of the tone) gives the response a unique tonal quality when heard on the audio monitor. Joris et al. (1994a,b) made recordings from the axons of AVCN cells in the trapezoid body while all other studies of the AVCN were made from the cell bodies in the AVCN. There are only a few scattered reports of high synchronization in other AVCN recordings (Rose et al. 1974; Rhode and Kettner 1987).

3.2.2 Assumption #2: Coincidence Detection in MSO Cells

The first study of coincidence detection in the MSO was the classic study of Goldberg and Brown (1969) in the dog. They showed that the interaural phase for maximal binaural stimulation could be predicted from the difference between the response phase angles to monaural stimulation of the two ears. Figure 4.5 shows a similar test from an MSO cell (Yin and Chan 1990). The cell is phase locked to monaural stimulation of each ear at the 1 kHz CF tone (Fig. 4.5B and Fig. 4.5C). The coincidence model predicts that maximal response under binaural stimulation with the same tones should occur when the responses from each ear arrive at the MSO simultaneously, i.e., when an acoustic interaural delay is introduced such that the peaks of the period histograms are in alignment. For the example, in Figure 4.5 this should occur when the ipsilateral tone is delayed by 0.14 cycles, or 140 μsec, which is the time difference between the peaks of the monaural phase-locked responses in Figures 4.5B and 4.5C. The response of the MSO cell as a function of ITD is shown in Figure 4.5A, and the peak response occurs when the ipsilateral tone is delayed by 100 μsec, close to that predicted by the coincidence model. Several studies have recorded from in and around the MSO in a number of different species of animals, and there is nearly universal agreement with the coincidence model in these studies (Moushegian et al. 1964, 1975; Crow et al. 1978; Yin and Chan 1990; Spitzer and Semple 1995; Batra et al. 1997b).

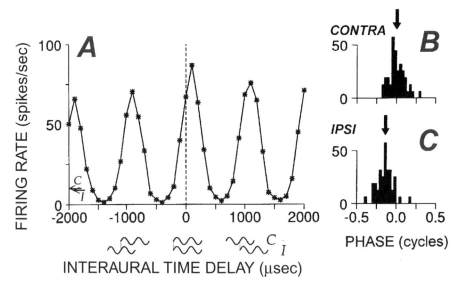

FIGURE 4.5. Sensitivity to ITD in MSO neurons. A: Responses to a binaural 1 kHz tone as a function of ITD. Positive ITDs represent a lead of the contralateral (C) relative to the ipsilateral (I) tone. B and C: Period histograms of the same cell to the same monaural ipsi- or contralateral 1 kHz tone show average response phases (arrows) that differ by 0.14 cycles, or 140 μsec. From Yin and Chan (1990).

While Goldberg and Brown only studied responses at one frequency (presumably at the CF of the cell), tests at different frequencies show that coincidence also holds across the responsive range of each cell (Yin and Chan 1990). Since most natural sounds consist of more frequency components than pure tones, the cell's responses to more than one frequency are critical to understanding its signal processing capabilities (Fig. 4.6A). Rose et al. (1966) were the first to study this issue, and they observed, from minimal experimental data (4 cells, only tested at 2 or 3 frequencies each), that cells in the IC seemed to show ITD-rate curves that had either a common peak or trough at a fixed ITD independent of stimulus frequency. They coined the phrase "characteristic delay" to denote the ITD of the common peak or trough.

The concept of the characteristic delay was widely accepted but not critically tested for almost 20 years, until Yin and Kuwada (1983b) applied a quantitative measure to test for characteristic delay in studies of interaural phase sensitivity of cells in the IC. Based on linear systems theory, Yin and Kuwada (1983b) argued that if cells show a peak in their ITD function at a common ITD regardless of the frequency of the sound, as defined by Rose et al. (1966), then the interaural phase at each frequency must be a linear function with frequency. The slope of the phase versus frequency function is the characteristic delay (CD) and corresponds to the ITD of the peak, which results from the time difference on the two sides while the phase intercept is the characteristic phase (CP), which indicates a constant phase difference between the phase-locked inputs from each side. If the inputs converging on the binaural cell are in phase, then the Jeffress model predicts that the CP = 0 (Fig. 4.6B). If the input from one side is inhibitory, as it is in the lateral superior olive (Joris and Yin 1995), then the two inputs will be 180° out of phase and the CP = 0.5. CP values between these two values indicate a constant phase difference between the two sides at a point other than the peak or trough. The advantage of using the phase/frequency relationship to determine characteristic delay is that there can be an objective measure of whether cells satisfy the linearity requirement. A technical advance in data collection has expedited CD analysis: Yin and Kuwada (1983a) showed that the use of the binaural beat stimulus, also used by Lord Rayleigh, provided a much faster means of collecting interaural phase sensitivity data. Most cells in the MSO and IC show a linear function between interaural phase and frequency, though there are differences in the distribution of CPs. In the IC, the CPs are distributed over a wide range of values (Yin and Kuwada 1983b; Kuwada et al. 1987; McAlpine et al. 1996) while in the MSO the CP values are clustered around 0.0 (Yin and Chan 1990; Spitzer and Semple 1995; Batra et al. 1997a), which suggests that MSO cells behave like classical coincidence cells while IC cells do not. McAlpine et al. (1998) suggested that the CP values between 0.0 (or 1.0) and 0.5 in the IC are achieved by converging input from the MSO and LSO in the IC. Evidence for this convergence was found by a novel and clever technique

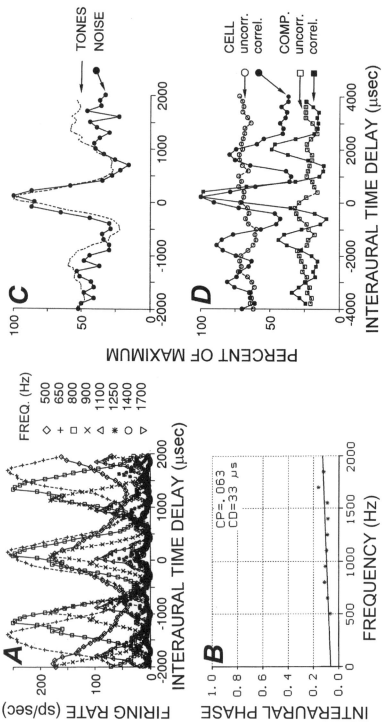

FIGURE 4.6. Interaural time delay functions for another MSO cell to tones at many different frequencies (A) and to a broadband noise stimulus (C). B: Plot of interaural phase versus frequency for the data in A shows that the function is linear with a characteristic delay (CD) of 33μsec and a characteristic phase (CP) of 0.063. C: Comparison of response to ITDs of broadband noise and the linear addition of responses in A. D: Responses of a third MSO cell were obtained to correlated (●) and uncorrelated (○) noise over a range of ITDs, as well as to these same stimuli presented monaurally. The monaural responses were then cross correlated at different time delays, for the COMP correlated (■) and uncorrelated (□) waveforms. (For further abbreviations, see Table 4.1.) From Yin and Chan (1990).

of suppressing one input with a tone at a fixed unfavorable ITD, which made the phase plots more linear and revealed pure MSO or LSO input.

In more recent studies of the MSO in which a substantial population of cells were recorded, most cells responded maximally when the stimulus to the ipsilateral ear was delayed (Yin and Chan 1990; Spitzer and Semple 1995; Batra et al. 1997a). This is reflected in the bias for CDs to be positive, meaning delays to the ipsilateral ear which, when coupled with CPs near 0, would result in peaks of the ITD functions when the ipsilateral stimulus was delayed (Yin and Chan 1990; Spitzer and Semple 1995).

By comparing the responses of a cell to ITDs of a wide band noise stimulus with those obtained at many different tones within the response area of the neuron, one can estimate the manner in which the different frequency components are summed in the MSO. Figure 4.6C shows a comparison of the composite curve, calculated by summing individual tone ITD curves, with the noise ITD curve for a cell in the MSO. The similarity in these curves suggests that MSO cells, like those in the IC (Yin et al. 1986), sum the spectral components of wide band noise stimuli in an approximately linear fashion (Yin and Chan 1990). Similar findings have been found in the IC of cats (Yin et al. 1986) and guinea pigs (McAlpine et al. 1996), but not in the barn owl (Takahashi and Konishi 1986). The noise ITD curve shown in Figure 4.6C was obtained with identical noises to the two ears. However, if two uncorrelated noises are used (one in each ear), then the noise ITD curve is flat (Fig. 4.6D, Cell, uncorr), suggesting that the cell performs an operation not unlike cross-correlation. To test the cross-correlation hypothesis, we compared the response of MSO cells to a computer cross-correlation of the input spike trains for both correlated and uncorrelated acoustic inputs (Fig. 4.6D). The spike trains in response to monaural stimulation of the ipsilateral and contralateral ears were cross correlated with a simple coincidence requirement and compared with the response obtained with binaural stimulation under the two conditions, correlated and uncorrelated inputs. Figure 4.6D shows that when the computer cross correlated inputs that were correlated, the response was similar to the MSO response to correlated inputs and likewise for uncorrelated inputs. Note that both of these tests of coincidence are only indirect: the binaural cell cross correlates the inputs arriving onto its synaptic terminals, whereas both of these tests cross correlate the *output* of the MSO cell in response to monaural stimulation. A more direct test would require cross-correlation of the intracellularly recorded postsynaptic potentials in the MSO, a procedure that has not yet been reported.

Batra et al. (1997b) have recently proposed three criteria for determining whether a binaural cell acts as a coincidence detector. Two of these criteria are similar to those discussed above. First, the frequency range over which the cell exhibits synchrony for monaural stimulation of both ears should be equal to the range over which the cell shows interaural phase sensitivity. Second, the ITD that evokes maximal discharge should be equal to the delay required to bring the ipsilateral and contralateral inputs into coincidence. Batra et al. (1997b) suggested an additional criterion based upon the synchronization coefficient: the synchrony to interaural stimulation should be equal to the product of the synchrony to monaural stimulation of each ear. The rationale for this criterion is that the process of coincidence detection is essentially the convolution of the two afferent spike trains, which predicts that the output synchronization will be the product of the input synchronizations.

3.2.3 Assumption #3: Delay Lines in the Projection of Afferents to the MSO

As discussed above, the predominant excitatory input to the MSO originates from the SBCs of the AVCN. To examine whether these afferents take the shape of systematic delay lines, Smith et al. (1993) made intraaxonal injections of biocytin into physiologically identified SBCs (i.e., those with a primary like response) as they traversed the trapezoid body. As expected from earlier anatomical studies, fibers projected bilaterally to the MSO. Since the MSO in the cat is a narrow sheet of cells which is only a few cells thick in the mediolateral dimension and the tonotopic axis runs along the dorsoventral dimension, the rostrocaudal axis is the most probable axis for representing ITDs. Therefore, the delay lines and the spatial map of ITDs would be likely to extend along the rostrocaudal axis. The Jeffress model proposes that the ipsilateral and contralateral inputs to the MSO take the form of oppositely directed delay lines. However, Smith et al. (1993) found that only the contralaterally projecting axon collateral showed the lattice-like delay line shape; 7 out of 9 axons clearly projected further caudally than rostrally, while the gradient for the other two fibers was not as clear. Ipsilaterally projecting axons did not show any systematic projection pattern.

Recently, Beckius et al. (1999) studied the same problem using the less reliable method of small injections of biotinylated dextran into the AVCN. This technique results in many axons being labeled rather than a single one, making the tracing of single fibers more problematic. They reported a wider variety of axonal projection patterns, though the overall results are similar to those found by Smith et al. (1993). Beckius et al. (1999) measured the length of axons from the injection site to each terminal and found that 6 out of 7 axons projecting to the contralateral MSO had collaterals with longer lengths innervating the caudal MSO and shorter lengths innervating the rostral MSO. In addition, those projecting ipsilaterally had a more complicated reverse, but less steep, gradient, which resembles the original Jeffress model. Thus, in the cat, there appears to be a prominent delay line on only one side with perhaps a weaker one on the other side, but the model would still function provided that the inputs from the other side all arrive at about the same time.

Interestingly, the homologous system in the bird involving afferents projecting from nucleus magnocellularis to nucleus laminaris also shows an asymmetry similar to that described by Smith et al. (1993): there is a prominent delay line pattern in the medial-to-lateral axis for the contralaterally projecting axons and no delay line in the ipsilaterally projecting axons (chick, Young and Rubel 1983; barn owl, Carr and Konishi 1990). In the barn owl, however, it is not known whether this difference in projection pattern is relevant since it is thought that the functional delay lines run across the dorsoventral width of nucleus laminaris rather than along the mediolateral dimension where the delay lines are more prominent.

3.2.4 Resultant Topographic Map of ITDs in MSO

According to the Jeffress model, if all three assumptions are met, the result should be a spatial map of ITDs along the axis of the nucleus parallel to the delay lines. In the mammalian MSO this corresponds to the rostral caudal dimension. Yin and Chan (1990) found evidence suggestive, but not definitive, of a spatial map in the MSO of the cat by combining data from several cats and looking within a narrow frequency range. When responses were restricted to a narrow band of CFs, cells located in the rostral end of the MSO tended to respond best to stimuli with ITDs near 0, while cells in the caudal MSO responded to ITDs near 400–600 μs (Fig. 4.7). The direc-

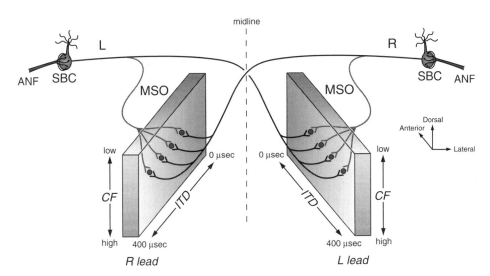

FIGURE 4.7. Three-dimensional schematic diagram of the innervation of the MSO in the cat. A single contralateral fiber and an ipsilateral fiber are shown innervating a row of MSO cells that lie in the same isofrequency lamina. The approximate ranges of the frequency and ITD axes are shown. (For abbreviations, see Table 4.1.)

tion of this spatial gradient fits the rostral-to-caudal projection of the contralateral afferents to the MSO (Smith et al. 1993; Beckius et al. 1999). The uncertainty in the spatial map in the MSO is a consequence of the difficulty in recording from the nucleus; so few cells can be recorded from an individual animal that data must be pooled from many animals to reconstruct population data. An important aspect of these data is that there is a representation of only the contralateral sound field in the MSO, which is preserved at higher levels by the predominant uncrossed projection to the IC.

3.3 Complications to the Jeffress Model

3.3.1 Nonprincipal Cells in the MSO

While the majority of cells in the MSO have the classical bipolar dendritic shape, with inputs from the ipsilateral and contralateral AVCN on the lateral and medial dendrites, respectively, there are marginal cells bordering the principal cells on both the medial and lateral sides that do not appear to receive binaural input. Moreover, physiological studies suggest the existence of a sizable population of other cells. All physiological studies of the MSO have reported the presence of a population of cells that are not binaural, though the percentage of such cells varies considerably in different studies. Goldberg and Brown (1969) found 36% of dog MSO neurons to be monaural while Yin and Chan (1990) found 18% monaural neurons in the cat MSO. A reasonable assumption, though without direct experimental support, is that the bipolar cells are the binaural, ITD-sensitive cells while the marginal cells are the monaural neurons. If these monaural cells represent a sizable proportion of the MSO, then the function of the MSO must include other roles in addition to encoding ITDs.

3.3.2 Role of the MSO in Animals with Small Heads and/or Predominantly High Frequency Hearing

If the MSO is used to encode ITDs, then it should be useful only in animals with good hearing in the frequency ranges where phase locking is possible and in animals with sufficient head size to create functional ITDs. Thus, animals with small heads and narrow range of natural ITDs are generally associated with good high-frequency hearing and are thought to rely on ILDs rather than ITDs for sound localization. For example, in the rat, an animal with a small head and predominantly high frequency hearing, Inbody and Feng (1981) found that cells were not sensitive to ongoing ITDs and were only weakly sensitive to ITDs of the stimulus onset. Masterton et al. (1975) studied the relationship between the ability to localize low- and high-frequency tones and the sizes of the MSO in four different mammalian species. All animals were able to localize high-frequency tones, but the ability to localize low-frequency tones, which was assumed to reflect the ability to detect ITDs, was correlated with the size of the MSO. However, Grothe (2000) and Grothe and Neuweiler (2000) have pointed out the interesting case of bats, with small heads and primarily high frequency hearing, some of which have a prominent MSO. Recordings from the MSO of the bat show that some cells are sensitive to ITDs in the same range as in cats: ITDs that would correspond to very coarse localization for the small head of the bat. Furthermore, in some species of bats the cells in the MSO are primarily monaural. Thus, it seems unlikely that the primary function of the MSO in these animals is to encode ITDs. A clue may lie in the discovery of a prominent inhibitory input to many of these MSO cells.

3.3.3 Role of Inhibition to the MSO

An unresolved question arises from an inconsistency between the anatomical and physiological data on the MSO in all mammals: what is the role of inhibitory inputs? As described above, there are ample anatomical data showing the presence of inhibitory inputs to the MSO, both in the form of ultrastructural data showing synapses with nonround vesicles and in the projection of glycinergic cells from the MNTB and LNTB to the MSO. Furthermore, as we shall see below, recordings from the MSO in brain slices have confirmed the existence of IPSPs evoked from stimulation of the afferent pathways to the MSO (Grothe and Sanes 1993, 1994; Smith 1995; Grothe, 2000). However, in vivo recordings from the MSO do not reveal a prominent role for inhibitory inputs; indeed, the Jeffress coincidence model in its basic form relies solely on the convergence of excitatory inputs. Initially, it was hypothesized that the inhibitory input was responsible for reducing the response of binaural, ITD-sensitive MSO cells at the unfavorable ITDs below the monaural, evoked response to either ear (Goldberg and Brown 1969; Yin and Chan 1990). However, a model of the MSO by Colburn et al. (1990) pointed out that inhibition was not needed to produce such low responses at unfavorable ITDs because even high spontaneously active auditory nerve fibers show a suppression of response during the out of phase condition, which would result in suppression in the MSO at unfavorable ITDs.

Grothe (1994) has found that many of the MSO cells in the bat receive an early excitatory input followed shortly afterward by an inhibitory one. By stimulating with amplitude modulated signals and varying the modulation frequency, Grothe (1994) found that such cells can be tuned to specific modulation frequencies. The presence of inhibition was shown by application of strychnine, which turned the response from phasic to sustained and shifted the upper frequency cutoff of the modulation transfer function to higher frequencies. Thus, in the bat at least, these cells may play a role in the processing of the amplitude modulated sounds created by the wingbeats of flying insects, rather than functioning as ITD detectors. However, these data do not address the function of inhibitory inputs within the MSOs of other mammals. One possibility is that the inhibition allows suppression of echoes for the precedence effect (Yin 1994), which has received some support from modeling studies that have incorporated inhibitory inputs (see below, Brughera et al. 1996).

3.4 Studies of the MSO in Brain Slices

In vitro studies provide a means to relate the biophysical properties of neurons with their anatomical features. Smith (1995) used intracellular recording with sharp electrodes to study cells in the MSO of the guinea pig brain and correlated their physiological responses to anatomical cell types by injecting the cells with neurobiotin. Principal cells, which correspond to the bipolar cells described in anatomical studies, responded to current injection with a single or few spikes at onset. These few spikes could be converted to a repetitive response with application of the potassium channel blocker, 4-aminopyridine (4-AP), in a manner similar to the responses of bushy cells of the AVCN (Oertel 1983) and MNTB (Banks and Smith 1992). Nonprincipal cells were multipolar in shape and responded repetitively to current injection in normal saline and near threshold (Smith 1995). Stimulation of the afferent fibers of the ipsilateral or contralateral side evoked EPSPs followed by IPSPs at higher shock strengths.

Grothe and Sanes (1993, 1994) studied the MSO of the gerbil brain slice and found physiological correlates for the anatomically-identified, glycinergic inhibitory inputs to the MSO. They showed that EPSPs and action potentials could be evoked by unilateral electrical stimulation of the ipsilateral or contralateral afferent pathways, while stimulus evoked synaptic IPSPs could also be evoked in most cells (89%), particularly at high stimulus amplitude where they could inhibit the excitatory effects. Bath application of glycine could reversibly block synaptically-evoked EPSPs and action potentials while the effects of synaptic inhibition were suppressed by application of the glycine antagonist, strychnine. With bilateral stimulation at low, subthreshold stimulus levels, coincidence could be demonstrated since action potentials could be evoked only when the two afferent pulses were within a narrow (~500 μs) temporal window, like that seen in vivo. At higher stimulus levels, the evoked spikes could be suppressed over a similar narrow range of ITDs by inhibitory inputs, which were in turn antagonized by strychnine (Grothe and Sanes 1994). The level dependency of the inhibition is rather puzzling, since the inhibitory projections to the MSO originate, in part, from the MNTB, which receives input from the large diameter GBC axons. The GBC axons would be expected to have lower thresholds to electrical stimulation than the smaller diameter axons of SBCs. Thus, these intracellular studies show that different anatomical classes of neurons in the MSO exhibit different physiological responses to current injection and revealed a prominent role for glycinergic inhibitory inputs to the MSO.

3.5 Models of the MSO

Given the difficulty in recording single units from the MSO, successful models of the nucleus become more important, but also more difficult since there are few data with which to compare model responses. Of course the most successful model of the MSO is the one by Jeffress (1948), though it is a strictly qualitative model. A more recent model (Brughera et al. 1996) used a modification of the Rothman et al. (1993) Hodgkin-Huxley model with stochastic inputs supplied by an auditory nerve model (Carney 1993) to simulate bushy cell inputs to the MSO. The MSO model had bilateral excitatory inputs along with inhibitory inputs with longer lasting conductance changes. It could simulate many tonal responses of MSO cells with or without inhibitory inputs, confirming an earlier model by Colburn et al. (1990). Unfortunately, the lack of physiological responses to the ITDs of clicks forced Brughera et al. (1996) to compare their model responses to those obtained in the IC (Carney and Yin 1989). Inhibitory inputs from more transient cell types in the AVCN (GBCs or onset cells) were the most successful in modeling the IC responses that exhibited prominent binaural inhibitory processing that mimicked the precedence effect.

A different approach to modeling led to insight into a possible role for the unique bipolar dendrites in the MSO, with inputs from each ear segregated onto each side. Agmon-Snir et al. (1998) showed that the nonlinear summation of excitatory inputs on dendrites of a neural model led to an improvement in the ITD sensitivity. The saturating nonlinearity makes the model respond better to two inputs that arrive on the dendrites of each side compared to the same two inputs arriving on one side. Since the effects of the saturating nonlinearity depend upon the size of the dendrites, which also affects the temporal sensitivity of the coincidence detection, the model predicts that, in order to optimize performance, the dendrites of high-frequency cells should be shorter than that of low-frequency cells. Such a gradient in dendritic length is seen prominently in the avian nucleus laminaris (Smith and Rubel 1979), the homologue to the MSO, and to a lesser degree in the mammalian MSO (Smith 1995).

4. Medial Nucleus of the Trapezoid Body

The other important interaural cue, interaural level disparities (ILDs), is believed to be encoded by a circuit involving the other two major nuclei in the SOC: the LSO and MNTB. Functionally, the MNTB serves as an inhibitory relay nucleus, providing a sign inversion for the signals from the contralateral ear to the LSO. Practically, the MNTB operates as much like a pure relay as any nucleus in the nervous system; its inputs are derived from the contralateral AVCN in an almost perfect one-to-one fashion, and its outputs project almost solely to the LSO. In achieving these notable features, the MNTB exhibits some of the most spectacular anatomical and physiological specializations in the brain. Yet, it can be argued that we still do not understand the precise function of these specializations.

4.1 Anatomy of the MNTB

MNTB cells lie within the axons of the trapezoid body, ventromedial to the MSO. Golgi studies of the MNTB have identified three neuron types: principal, elongate and stellate cells (Morest 1968). The somata of the principal cells are round or oval with one or two large diameter dendrites that branch profusely in a confined volume, much like the bushy cells in the AVCN (Banks and Smith 1992; Kuwubara and Zook 1991; Sommer et al. 1993; Smith et al. 1998). The other two cell types have been studied in much less detail. In Golgi preparations, it appears that only the principal cells receive the large synaptic ending known as the calyces of Held.

Immunocytochemical studies using antibodies to the inhibitory neurotransmitter glycine demonstrate that the MNTB cells are likely to be inhibitory. Several studies have shown that the MNTB and neighboring LNTB have intense staining for glycine, as well as to the calcium-binding protein calbindin (Campistron et al. 1986; Wenthold et al. 1987; Helfert et al. 1989; Saint Marie et al. 1989b; Adams and Mugnaini 1990; Bledsoe et al. 1990; Matsubara 1990; Webster et al. 1990; Henkel and Brunso-Bechtold 1995; Spirou and Berrebi 1997). Kainic acid lesions of the MNTB result in the disappearance of MNTB principal cells with a corresponding loss of labeled fibers projecting to the ipsilateral LSO (Bledsoe et al. 1990). Physiological evidence that the inhibitory input to the LSO is glycinergic is provided by the response to iontophoretic application of glycine or its antagonist, strychnine, while recording from LSO cells in vivo; glycine mimics the effect of stimulation of the contralateral ear in suppressing LSO responses, while strychnine blocks the contralateral inhibition (Moore and Caspary 1983).

4.1.1 Afferents to the MNTB

The most distinctive feature of the MNTB is the large synaptic ending known as the "calyx of Held" (Held, 1893; Ramon y Cajal 1909), which is a spectacular anatomical feature, one of the largest, if not the largest, synapses in the mammalian brain. The calyx is the synaptic terminal of GBC axons and forms a cup-like structure around the cell body of the MNTB principal cell with finger-like processes that envelope the soma. While each principal cell receives only a single calyx, single axon labeling experiments have shown that a single GBC axon often can have two (Smith et al. 1991), or even three (Kuwubara et al. 1991) calyces, so that the relationship between GBC and MNTB cell is not strictly one-to-one. EM studies of the calyx show features typical of an excitatory synapse; it contains large round vesicles and makes asymmetric synaptic contacts with the MNTB cell (Jean-Baptiste and Morest 1975; Morest 1973; Lenn and Reese 1966; Smith et al. 1998). Approximately half of the available synaptic space on the surface area of the soma is taken up by the calyx. Clearly, the calyx provides a potent excitatory drive to the MNTB.

Since the MNTB cell bodies lie within the bundles of axons making up the trapezoid body, anatomical experiments using injections of tracers or lesions will be compromised by uptake or damage to fibers coursing through the trapezoid body. Glendenning et al. (1985) circumvented this problem with a two-step experiment. First, they showed that the AVCN projected only to the contralateral, not the ipsilateral, MNTB by using anterograde transport of HRP or surgical lesions of the AVCN. Second, they injected the MNTB with HRP and showed retrogradely labeled cells in the AVCN of the contralateral side along with a smaller number of cells on the ipsilateral side. Since the anterograde experiments showed no ipsilateral projection, the conclusion was that the retrogradely labeled ipsilateral cells were filled by HRP entering through fibers of passage damaged by the micropipette, which is in accord with the conclusions of Tolbert et al. (1982).

In addition, there is evidence for inputs to the principal cell that are noncalyceal. For example, there are often collaterals that arise from the processes that make up the calyx itself, so-called calycine collaterals, that appear to innervate processes of neighboring cells. Additionally, there are also precalycine collaterals that arise from the axon. As mentioned above, only about half the synaptic space on the surface of the soma of MNTB cells is taken up by the calyx. Some of the noncalyceal endings show features of inhibitory synapses in EM material, but the origin of these endings is not known. Tract tracing studies of afferents to the MNTB have concentrated on the calyceal input from the contralateral AVCN (Warr 1972; Tolbert et al. 1982; Glendenning et al. 1985), but single axon labeling in brain slices has shown multiple sources of noncalyceal afferents to the MNTB arising from various periolivary nuclear groups (the MNTB, the dorsomedial periolivary nucleus, and the ventromedial and ventral periolivary nuclei) (Kuwubara et al. 1991).

4.1.2 Efferent Projections of the MNTB

A variety of anatomical experiments have demonstrated that the principal target of MNTB cells is the ipsilateral LSO. These include the following: classic anterograde degeneration studies following lesions placed in the MNTB (van Noort 1969; Browner and Webster 1975), anterograde tracer studies with injections of tritiated amino acid or other markers into the MNTB (Elverland 1978; Glendenning et al. 1985; Spangler et al. 1985; Henkel and Gabriele 1999), retrograde tracing with HRP injections into the LSO (Glendenning et al. 1985; Spangler et al. 1985; Henkel and Gabriele

1999), and intracellular recording and labeling of MNTB cells in brain slices (Kuwubara et al. 1991; Banks and Smith 1992; Kuwabara and Zook 1992) and in vivo (Sommer et al. 1993; Smith et al. 1998). These studies show that in addition to its primary projection to the ipsilateral LSO, there are also less consistent, minor collateral projections that end in the MSO, the DNLL, VNLL, the ventromedial and dorsomedial periolivary nucleus, and the MNTB itself. Several studies have found a spatial gradient related to the frequency representation in the projection of the MNTB to the LSO; there is a weaker projection to the lateral (low frequency) limb of the LSO than to the medial and middle limbs (Glendenning et al. 1985, 1991; Spangler et al. 1985; Sanes et al. 1987; Saint Marie et al. 1989b).

There are several distinct advantages in making intracellular recordings and labeling in vivo (Sommer et al. 1993; Smith et al. 1998): a direct correlation of the physiological responses with the anatomical features can be made in single cells, all of the collateral branches of the axon into neighboring nuclei can be followed, and thin sections can be prepared for EM for relating the ultrastructural features of the synaptic terminals with the physiology and anatomy (Smith et al. 1998). Thus, in vivo recording and labeling provide a direct demonstration that the synaptic terminals with presumed inhibitory features, (symmetrical with nonround or pleomorphic vesicles and terminations on the soma or proximal dendrites of principal cells in the LSO), originate from the MNTB (Cant 1984; Brunso-Bechtold et al. 1994).

4.2 Physiology of the MNTB

The synaptic terminal of the calyx is so large that extracellular recordings from the MNTB show a prominent prepotential which precedes each action potential by about 0.5 ms (Pfeiffer 1966; Guinan et al. 1972a). Guinan and Li (1990) systematically studied the relationship between the prepotential (which they called C1) and the action potential (or C2) evoked in MNTB principal cells. They showed that the C1 was elicited at very short latency by electrical stimulation of the trapezoid body fibers whereas one could demonstrate collision between an antidromic spike evoked by stimulation in the LSO and C2. In all cases, spontaneous or sound-evoked, C2 spikes always followed C1 spikes, suggesting a faithful one-to-one relay from the calyx to the MNTB. However, these studies found that, in some rare cases, C2 spikes could be evoked in the absence of a C1 spike, a finding that was confirmed by Smith et al. (1998). These findings suggest that the overwhelming predominant input to the MNTB is the GBC, which forms its calyx ending, but that on rare occasions there may be a noncalyceal input that can also drive the MNTB. The source of the noncalyceal input is not well established.

Since the large calyceal ending of the GBC axon provides the dominant input to the principal cells of the MNTB, it is not surprising that the response of MNTB cells resembles that of the GBCs; the response to pure tones at CF for most MNTB cells was similar to the PLn response seen in GBCs (Guinan et al. 1972a; Tsuchitani 1997; Smith et al. 1998). Despite the large, fast synapse at the calyx of Held, the timing of the first spike in MNTB cells is not as precise as that of GBCs, so that the notch in the PLn response is not as clear and reflects the jitter introduced at central synapses. Similar considerations presumably account for the loss in the upper limit of phase-locking at successive synapses in the ascending auditory system. Tsuchitani (1997) compared a number of response properties of LSO and MNTB cells. She reported that the widths of the excitatory tuning curves of LSO and MNTB cells were similar, but that the inhibitory tuning curves (derived by stimulating the ipsilateral ear at a level that elicited one spike and varying the contralateral stimulus until that spike was just inhibited) for LSO cells were narrower and had higher thresholds than for cells of the MNTB. First spike latencies of MNTB cells were slightly shorter than LSO cells. From such comparisons, Tsuchitani (1997) suggested that the inhibitory input to an individual LSO cell is provided by more than one MNTB cell.

4.2.1 Studies of MNTB in Brain Slices

The large excitatory glutamatergic synapse at the calyx of Held has attracted considerable attention in recent years with the realization by the biophysical community that the giant synapse affords a special opportunity to record from both presynaptic and postsynaptic structures in the same preparation and provides a large, fast glutamatergic synapse (Forsythe 1994; Barnes-Davies and Forsythe 1995; Borst et al. 1995). The biophysical properties of this synapse are reviewed in Trussell (Chapter 3) so will not be reiterated here.

4.3 Why is there a Calyx in the MNTB?

With the increased interest in biophysical studies of the calyx of Held in the MNTB, a natural question arises: what is the function of such a highly specialized giant synapse? Various reasons have been given for the function of the calyx: to maintain the speed of conduction (Tauschenberg and von Gerstorff 1900; Kuwubara et al. 1991; Banks and Smith 1992; Wu and Kelly 1992; Forsythe 1994; Borst et al. 1995), to preserve the precise timing of spikes by minimizing synaptic delay (Forsythe and Barnes-Davies 1993; Wu and Kelly 1994; Borst et al. 1995; Wang et al. 1998), to ensure that each presynaptic spike results in a postsynaptic spike in the MNTB (Morest 1973; Kuwubara et al. 1991; Banks and Smith 1992; Forsythe and Barnes-Davies 1993; Forsythe 1994; Wu and Kelly 1994; Barnes-Davies and Forsythe 1995;

Borst et al. 1995), to preserve high-frequency phase locking in the LSO (Koyano and Ohmori 1996; Chuhma and Ohmori 1998; Tauschenberger and von Gerstorff 2000), and to follow high rates of repetitive stimulation (Guinan and Li 1990; Wu and Kelly 1994; Borst et al. 1995; Wang et al. 1998). The question posed above arises if we make two common assumptions about the LSO-MNTB circuit: first, that the function of the LSO-MNTB circuit is to encode ILDs, and second, that the highly specialized large synaptic terminal with correspondingly large synaptic currents released at the synapse of the calyx of Held allows fast and precise timing as well as a one-to-one following of postsynaptic spikes. In other words, why is there a need for temporal precision in a circuit coding level differences rather than time differences?

The MNTB has predominantly high-frequency cells, which cannot phase lock to CF tones above 3 kHz, so preservation of phase locking or precise timing to stimulus fine structure is not essential to its role. Part of the answer may lie in the sensitivity of LSO cells to ITDs of amplitude-modulated (AM) stimuli since MNTB cells do phase lock to such signals (Joris and Yin 1995). However, this answer is unsatisfactory since the modulation transfer functions derived from phase locking to high-frequency AM stimuli in the MNTB and LSO have lower cutoff frequencies than those in the auditory nerve (Joris and Yin 1995), which in turn are almost an octave lower than the phase locking to pure tones (Joris and Yin 1992). Thus, the temporal resolution of the ITD sensitivity in the LSO is nearly two octaves slower than that seen in the ITD circuit of the MSO.

There seems no particular reason why the MNTB cells would be especially designed to follow its inputs in a one-to-one fashion or why this synapse should be concerned with one-to-one transmission from pre- to postsynaptic elements. Furthermore, firing rates in MNTB are not especially higher than in other parts of the auditory system. The most likely reason for the calyx, then, seems to be speed, to relay the inhibitory input to the LSO as quickly as possible so that it arrives nearly coincident with the excitatory input. As we will discuss below, measurements of the time of arrival of excitatory and inhibitory inputs show that the system is nearly able to achieve this coincidence notwithstanding the longer distance and extra synapse that the contralateral afferent needs to reach the LSO.

5. The Lateral Superior Olive

As mentioned earlier, cells in the LSO are binaurally activated, inhibited by stimulation of the contralateral ear, and excited by stimulation of the ipsilateral ear (Fig. 4.1). We will refer to these cells as having IE binaural interaction. This follows the convention of designating the contralateral response first followed by the ipsilateral response (Goldberg and Brown 1968) and the widespread use of the terminology EI to refer to cells that are found above the level of the superior olive and are excited by the contralateral and inhibited by the ipsilateral ear.

5.1 Anatomical Organization of the LSO

In transverse sections, the LSO of the cat has a distinctive S-shape tipped on its side, with the lateral limb extending dorsally and the medial limb ventrally. The tonotopic axis runs along the curved axis of the S, with high frequencies represented medially and low frequencies laterally (Guinan et al. 1972b; Tsuchitani 1977). Golgi studies have generally identified three major cell types: principal, marginal, and multipolar (Helfert and Schwartz 1986, 1987; Majorossy and Kiss 1990; Rietzel and Friauf 1998), with a scattering of other smaller cells. In this plane, principal cells, which make up about 75% of the population, have a distinctive bipolar or fusiform shape with dendrites extending normal to the tonotopic axis of the nucleus. However, unlike the fusiform cells in the MSO, there is no known functional difference in the afferents that project to the two sides of the principal cells; afferents to the LSO innervate the dendritic trees on both sides of the cell body (Ramon y Cajal 1909). Like cells in the MSO, both the dendrites and the afferent axons extend for a considerable length along the rostocaudal dimension of the LSO as seen in horizontal or sagittal sections (Scheibel and Scheibel 1974) so their dendritic trees will be disc-shaped and orthogonal to the medial lateral axis of the nucleus (Henkel and Brunso-Bechtold 1991).

5.1.1 Afferents to the LSO

As expected from the binaural activation of LSO, anatomical studies of its inputs show practically exclusive innervation from only two sources; injections of retrograde tracers restricted to the LSO result in labeling of cells in the ipsilateral AVCN and MNTB (Glendenning et al. 1985, 1991; Spangler et al. 1985; Cant and Casseday 1986; Henkel and Gabriele 1999). The labeled cells in the AVCN cover the entire frequency representation and are situated predominantly in the anterior subdivision, which contains primarily SBCs, with a smaller contribution from the posterior subdivision, which contains GBCs and multipolar cells (Cant and Casseday 1986). Likewise, anterograde degeneration following lesions of the spherical cell region of the AVCN shows a projection to the ipsilateral LSO (Warr 1966, 1972; Goldberg and Brown 1969; Van Noort 1969). The axons of SBCs that innervate the ipsilateral LSO are collaterals of fibers that continue across the trapezoid body (Glendenning et al. 1985; Shneiderman and Henkel 1985).

In addition to this well-documented input from the ipsilateral AVCN, there are also projections from other sources. It appears that these apparently minor projections take different forms in different species. Both anterograde degeneration and retrograde tracer studies have consistently shown a small projection from the contralateral AVCN to the high-frequency medial limb of the LSO in dog (Goldberg and Brown 1969) and cat (Warr 1972; Glendenning et al. 1985; Cant and Casseday 1986). Using crystals of dioctadecyl-tetramethylindocarbocyanine perchlorate (DiI) in the gerbil, Kil et al. (1995) have described the development and maintenance of a projection from the contralateral AVCN to the low-frequency ventrolateral limb of the LSO. Using small injections of HRP into the AVCN, Zook and Leake (1989) found a consistent anterograde projection to the expanded 60 kHz region of the contralateral LSO of the mustache bat. Most physiological recordings from the medial limb have found the classic IE binaural cells typical of the LSO, but both Joris and Yin (1995) in the cat and Kil et al. (1995) in the gerbil found a few EE cells that were located lateral to or in the lateral limb of the LSO.

The other major source of afferents to the LSO is the ipsilateral MNTB. We have already discussed above the evidence for this projection and that it provides glycinergic inhibition relayed from the contralateral GBCs.

Ultrastructural studies of the LSO support the existence of both excitatory and inhibitory inputs that are segregated onto the dendrites and somata (respectively) of LSO principal cells. Asymmetrical synaptic terminals with round vesicles tend to be found on the distal dendrites, while symmetrical terminals with flat vesicles are segregated to the region of the cell body and proximal dendrites (Cant 1984; Helfert et al. 1992; Brunso-Bechtold et al. 1994). By using postembedding immunogold histochemistry (Helfert et al. 1992), it was possible to relate the synaptic terminal morphology with putative transmitter type. The asymmetrical terminals with round vesicles were immunoreactive for glutamate while the symmetrical terminals with flat vesicles labeled intensely for glycine and γ-aminobutyricacid (GABA), confirming the excitatory and inhibitory action, respectively, of each terminal type.

As mentioned in the introduction, the organization of the olivocochlear system in the rodent appears to be different from that of carnivores and primates. In cats, the olivocochlear efferents arise from the periolivary cell groups, lying outside of the MSO, LSO, or MNTB (Warr 1992). However, in rodents, cells in the LSO itself contribute to the olivocochlear projection. There are also differences between the afferents to the LSO as mentioned above, e.g., a projection from the PVCN to the LSO, bilaterally (Thompson and Thompson 1991).

5.1.2 Efferents from the LSO

Anatomical studies using anterograde degeneration following kainic acid lesions (to circumvent damage to the many fibers of passage in the LSO), tritiated amino acid autoradiography, and retrograde transport of HRP or fluorescent dyes have shown that the LSO projects bilaterally to the IC and the DNLL (Roth et al. 1978; Glendenning and Masterton 1983; Moore 1988; Saint Marie et al. 1989b; Henkel and Brunso-Bechtold 1993; Moore et al. 1995; Kelly et al. 1998). Glendenning and Masterton (1983) have referred to this partial decussation of fibers as the acoustic chiasm, in analogy to the more well-known optic chiasm. In most studies, a large injection of HRP into the IC labels about an equal number of cells in the LSO on both sides. Anatomically, the parameters of interest for each of the ascending branches are whether the projection is excitatory or inhibitory and whether there is a topographical gradient. Glendenning and Masterton (1983) found that the lateral limb of the cat LSO had a preponderance of cells projecting ipsilaterally while the medial limb had more cells projecting contralaterally, but Henkel and Brunso-Bechtold (1993) and Moore et al. (1995) found the opposite gradient in the ferret. Using double-labeling experiments with fluorescent tracers, a small percentage of cells in the LSO (2–5%) have been found to project bilaterally (Glendenning and Masterton 1983; Saint Marie et al. 1989b; Henkel and Brunso-Bechtold 1993). Shneiderman and Henkel (1985) described a prominent banding pattern in the axonal projection to the IC, which was more prominent for the ipsilateral than the contralateral projection. The bands from a single injection in the LSO project to noncorresponding, possibly interdigitating bands in the ICs of the two sides.

A variety of methods have been used to examine the neurotransmitter and sign (excitatory or inhibitory) of the crossed and uncrossed projections from the LSO: receptor binding for the two inhibitory neurotransmitters (GABA and glycine) as well as for the excitatory amino acid neurotransmitters (glutamate, kainate, and quisqualate) (Glendenning et al. 1992); retrograde transport of tritiated GABA (Glendenning et al. 1992); immunocytochemistry using antibodies against glycine, GABA, glutamate and aspartate (Saint Marie et al. 1989b; Glendenning et al. 1992); and electron microscopic autoradiography using tritiated leucine (Oliver et al. 1995). These studies show that the projection to the ipsilateral IC from the LSO can be both inhibitory and excitatory while the contralateral projection appears to be wholly excitatory. The inhibitory projection is glycinergic, since tritiated GABA injected into the IC did not retrogradely label any cells in the LSO as it did in the DNLL whereas tritiated glycine resulted in labeled cells in the ipsilateral LSO (Glendenning et al. 1992). Furthermore, these results were confirmed with glycine and GABA immunohistochemistry. Synaptic terminals in the ICC following injections of tritiated leucine into the LSO show both symmetric synaptic junctions with pleomorphic vesicles (presumed inhibitory) and asymmetric terminals with round vesicles (presumed excitatory) in the ipsilateral ICC while the contralateral projection is almost, but not entirely, round vesicle terminals (Oliver et al. 1995). The bilateral projection of the LSO with its mixed inhibitory and excitatory components is highly unusual, and we will discuss the functional implications of the bilateral projection below.

5.2 Physiology of LSO

5.2.1 Coding of ILDs

The classical view of the LSO is that it represents the initial stage of encoding of ILDs. Most studies of the LSO in which there is histological verification of the recording site have found that most cells are IE binaural type (Boudreau and Tsuchitani 1968, 1970; Guinan et al. 1972a, 1972b; Tsuchitani 1977; Caird and Klinke 1983; Park et al. 1996, 1997; Joris and Yin 1995, 1998). Typically, the inhibitory effect of the contralateral input is established by stimulating the ipsilateral excitatory ear with an excitatory tone while varying the level of the contralateral inhibitory input (Fig. 4.8B). There is a strong and consistent tonotopical organization to the nucleus with high frequencies represented in the medial limb and progressively lower frequencies laterally (Tsuchitani and Boudreau 1966; Guinan et al. 1972b; Tsuchitani 1977). Furthermore, the representation of the cochlea in the LSO is biased towards high frequencies, in parallel with the role of the LSO in encoding ILDs and the importance of ILDs for sound localization at high frequencies (Guinan et al. 1972b).

There are two areas of disagreement regarding this uniform assessment of LSO function. First, the early classical studies suggested that the lateral, low-frequency limb of the LSO is not IE but rather contains monaural, ipsilaterally excited cells (Boudreau and Tsuchitani 1968; Guinan et al. 1972a, 1972b), but more recent studies have found the cells in the lateral limb to be IE in character like the rest of the LSO (Finlayson and Caspary 1991; Tollin et al. 2000). The reason for this discrepancy is still unresolved. Second, Brownell et al. (1979), recording in the decerebrate cat, reported a different sampling of LSO responses: fewer cells were IE, there were more EE cells (excited by both ears) and there was evidence for an inhibitory input from the ipsilateral side in the form of inhibitory sidebands.

Since the input from the contralateral ear is inhibitory onto the LSO while that from the ipsilateral ear is excitatory, it is more difficult to directly compare the properties of these inputs and their effectiveness in the LSO. Some of the most systematic measures were made in the earliest studies. Boudreau and Tsuchitani (1968) showed that the CF of the excitatory ipsilateral input was well-matched to the CF of the inhibitory contralateral input. This result was obtained by presenting a fixed-level CF tone to the ipsilateral ear and searching for the most effective inhibitory stimulus by varying the frequency and SPL of the contralateral input. Estimates of the bandwidth of the inputs are more problematic since they involve some arbitrary assumptions about how much inhibition is equivalent to an excitatory bandwidth for a given SPL above threshold; using a 25% inhibition criteria and 20 dB above threshold, Boudreau and Tsuchitani (1968) found that, for some cells, the contralateral bandwidth was wider than the ipsilateral. For other cells, the opposite was true, with no systematic trend. The shapes of the excitatory and inhibitory rate-level functions (the inhibitory ones being plotted as number of spikes inhibited for a given SPL at the excitatory ear) were similar for most cells. Caird and Klinke (1983) showed a dramatic comparison of the excitatory and inhibitory tuning curves by frequency-response plots in an LSO cell that had a high spontaneous firing rate. The two frequency response plots were nearly identical, though mirror images of each other, including inhibitory sidebands. We will discuss the relative timing of the inputs from both ears in more detail below.

5.2.2 Sensitivity to ITDs of Amplitude-Modulated Signals

Since phase locking to stimulus fine structure in auditory nerve fibers is limited to low frequencies, and the LSO is predominantly a high frequency nucleus, most cells in the LSO do not phase lock to CF tones. However, it is well known that when a temporal envelope is superimposed upon the tone, high frequency auditory nerve fibers can phase lock to low modulation frequencies (Javel 1980; Smith and Brachman 1980; Palmer 1982; Joris and Yin 1992), and that cells in the cochlear nucleus can relay, and even enhance, the coding of temporally modulated signals (Frisina et al. 1985; Wang and Sachs 1994). This suggests that temporal information about the modulating envelope, but not about the fine structure, is available to high-frequency cells in the LSO. On the other hand, low-frequency cells in the LSO receive input from low-frequency SBCs and GBCs, both of which phase lock well to CF tones so they would be expected to encode information about the fine structure of ITDs.

A considerable body of psychophysical evidence has accumulated over the past 20 years which shows that human subjects can discriminate ITDs of the low-frequency envelope of signals with high-frequency carriers (Henning 1974a, 1974b; McFadden and Pasanen 1976; Neutzel and Hafter 1976; Bernstein and Trahiotis 1985). Whether this ability to discriminate ITDs of the envelope is related to localization has been questioned since large ITDs are needed to lateralize the image in the head (Bernstein and Trahiotis 1985). Physiological correlates of the psychophysical effects were first studied in the IC (Yin et al. 1984; Batra et al. 1989). Batra et al. (1989) found that the modulation transfer function of IC cells was not as broad as the range of modulation frequencies over which the cell showed sensitivity to ITDs of the envelope. They interpreted this result to mean that the site of binaural interaction was likely to be below the IC, presumably at the level of the SOC, whose cells carry the ITD sensitivity to the IC but with degraded monaural phase locking. If this were so, then SOC cells should be sensitive to the ITDs of AM stimuli as demonstrated in the MSO (Yin and Chan 1990; Joris 1996) and in the LSO (Joris and Yin 1995, 1998; Joris 1996).

The characteristics of the ITD sensitivity of LSO cells to AM stimuli are predictable from its binaural properties as illustrated in Figure 4.8A (Joris

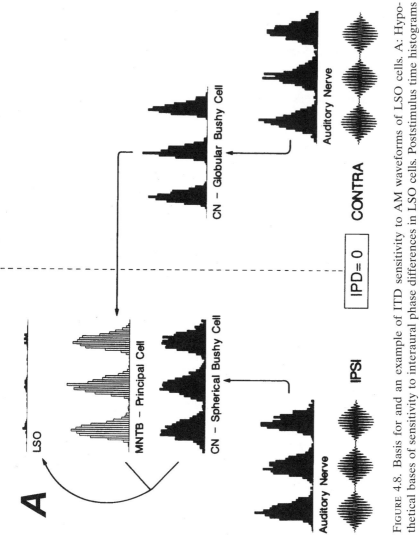

FIGURE 4.8. Basis for and an example of ITD sensitivity to AM waveforms of LSO cells. A: Hypothetical bases of sensitivity to interaural phase differences in LSO cells. Poststimulus time histograms illustrate hypothetical phase relationships for the in-phase amplitude modulated stimuli. Since the stimuli are in phase, the excitation and inhibition arrive at the LSO at the same time and effectively suppress the LSO cell. If the stimuli were out of phase (not shown), the inhibition would be out of phase with the excitation and thereby ineffective.

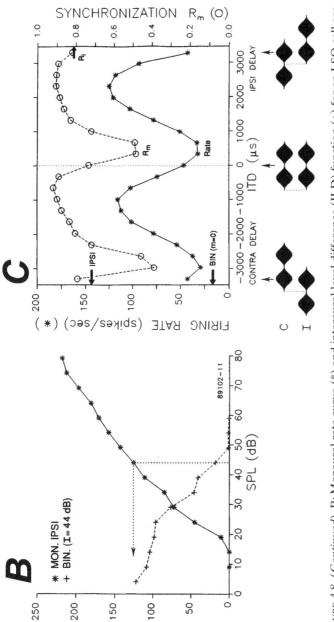

FIGURE 4.8. (*Continued*) B: Monaural rate curve (*) and interaural level difference (ILD) function (+) for an LSO cell are graphed on a common abscissa that indicates the SPL of the ipsilateral and contralateral ear, respectively. For the binaural response the abscissa is also an ILD axis, with ILD = SPL − 44 dB. Dotted line indicates the ipsilateral level chosen to obtain the ILD curve. C. Sensitivity of the cell shown in B in terms of discharge rate (*, left ordinate) and synchronization Rm to the modulation frequency (O, right ordinate) as a function of ITD. The carrier frequency was set to the CF of the cell (5.0 kHz) and the modulation frequency was 300 Hz. Arrows indicate the response rate and synchrony R to monaural ipsilateral stimulation. (For abbreviations see Table 4.1). From Joris and Yin (1995).

and Yin 1995). There are three classes of afferents in the LSO circuit: SBCs, which provide ipsilateral excitation to the LSO, GBCs, which provide the contralateral excitation to the MNTB, and MNTB cells, which inhibit the LSO. Each of these afferent populations phase locks to the AM stimuli, and the temporal modulation transfer functions to the AM stimuli of each population are similar (Joris and Yin 1998). If the ITD or, equivalently, IPD equals zero, and if there is minimal difference in the arrival time of the excitatory input from the ipsilateral side and inhibitory input from the contralateral side, then the excitatory and inhibitory phases will arrive at the LSO in phase, producing a minimal response in the LSO. On the other hand, if the stimuli are out of phase (IPD = 0.5), then the excitation and inhibition will arrive at the LSO out of phase and the inhibition will be ineffective, so that the LSO cell will respond at about the same level as it does for a monaural ipsilateral stimulus. Thus, we expect the ITD curves of LSO cells to show a minimal response, or trough, near 0 ITD and recovery close to the monaural response to the ipsilateral ear alone when the inputs are out of phase (Fig. 4.8C) (Caird and Klinke 1983; Joris and Yin 1995; Batra et al. 1997a). Such trough responses are also seen for more complex, wide-band stimuli such as noise or clicks (Caird and Klinke 1983; Joris and Yin 1998).

A limiting factor in the ability of ITDs of the envelope to serve as a potent cue for localization is the inability of LSO cells to follow higher modulation frequencies (Joris 1996). The modulation transfer function of LSO cells has a lower cutoff frequency than that of its afferents, and the discharge rate decreases dramatically for frequencies above 300 Hz. Both of these factors severely limit the sensitivity to ITDs within the physiological range.

If the latency for the arrival of ipsilateral excitation and contralateral inhibition is equal, then the trough of the ITD function would be expected to occur at 0 msec. Any differences in the timing of excitation and inhibition should be reflected in the ITD of the trough. Boudreau and Tsuchitani (1968) concluded that the latency of inhibition is as fast as the excitation because simultaneous stimulation of the contralateral ear at a higher level can inhibit the first spike excitation from the ipsilateral ear. However, since the latency of responses systematically decreases with SPL, this question can only be addressed at equal SPL, a condition not tested by Boudreau and Tsuchitani (1968).

As discussed above for the MSO, the differences in timing between the ipsilateral and contralateral inputs are best studied using the characteristic delay analysis. By varying the modulation frequency, Joris (1996) found that in response to ITDs of high-frequency, amplitude-modulated signals, LSO cells have CPs near 0.5 and CDs near 0, as expected from the IE binaural interaction. The values of CDs varied between −560 and +971 μs (mean of 202 μs) with 73% of them within the physiological range of ITDs (+/−400 μs) of the cat. The mean CD of +202 μs is rather remarkable: it implies that, on average, the inhibitory input from the contralateral side reaches the

LSO 202 μsec later than the excitatory input from the ipsilateral side. Not only does the inhibitory input have to travel farther, but it also has to traverse an extra synapse, yet there is a delay of only 0.2 ms.

Independent estimates of the time difference between ipsilateral and contralateral afferents to LSO can also be obtained from analyses of group delays measured from responses to monaural AM stimuli from the LSO. Estimates of latency derived from measurements of the monaural phase versus the modulation frequency have the advantage that they are level tolerant whereas other common measures, such as first spike latencies, are very sensitive to level (Joris and Yin 1998). Group delays measured from monaural responses to AM CF tones to the ipsilateral and contralateral ears in high frequency LSO cells varied from 4.1 to 5.6 ms, but there was a striking similarity in the delays to ipsilateral and contralateral modulation for individual LSO cells (r = 0.72). The mean difference was +182 μs, indicating that, on average, the contralateral inhibition arrived 182 μs later than the ipsilateral excitation. This estimate is close to the 202 μs difference estimated from mean CD analysis using binaural stimulation.

It is interesting to compare group delay estimates in the LSO with those obtained from the population of afferents to it (i.e., from auditory nerve fibers, GBCs, SBCs and MNTB cells). Recordings were taken near the midline in the trapezoid body, and these populations were distinguished on the basis of the physiological signatures of these cells: GBCs had PLn responses to CF tones, SBCs had PL responses, and MNTB cells had prominent prepotentials (Joris and Yin 1998). Estimates of latencies based upon group delays have the distinct advantage that they are not sensitive to stimulus level (Joris and Yin, 1998). Since latencies are affected by cochlear travel time, they also vary systematically as a function of the CF. As expected, there was a systematic increase in the latency from the auditory nerve to the bushy cells to the MNTB and finally to the LSO. To obtain a single measure of each population, Joris and Yin (1998) fit each class of cells with a power function derived from fitting auditory nerve data (Anderson et al. 1971; Smolders and Klinke 1986). Not enough points were available to fit SBCs. Figure 4.9 shows a summary of these measures and gives estimates for arrival time at each of the recording sites. Especially striking is the relatively long time delay for ipsilateral excitation of LSO cells (2.9 ms from the auditory nerve) as compared with the 3.1 ms from the contralateral side despite the additional length and extra synapse of the contralateral projection. The large caliber of GBC axons, the fast synapse in the MNTB at the calyx of Held, and their termination on the soma of LSO cells help to reduce the contralateral delay while the thin SBC axons, their sometimes circuitous route to the LSO (Smith et al. 1993), and their termination on distal dendrites of LSO cells serve to delay the ipsilateral excitation just enough that the two inputs arrive nearly coincidentally at the LSO.

Since high frequency LSO cells are sensitive to both ITDs of the envelope as well as ILDs, a natural question arises regarding the relative contribution of each cue to the sensitivity of LSO cells in the free field, where

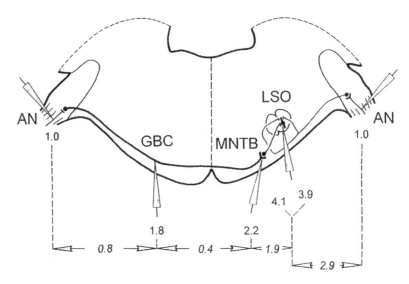

FIGURE 4.9. Summary diagram of delay measurements (in ms) on a tracing of a coronal section through the superior olivary complex. Italic numbers on *bottom* give differences in delay between cell classes. Long delay between AN and LSO to ipsilateral modulation (2.9 ms) probably derives from the small diameter and circuitous path of SBC axons, and their termination on distal LSO dendrites. Comparatively short delay between AN and LSO to contralateral modulation (3.1 ms in total) derives from fast axonal conduction and axosomatic placement of terminals along each relay point. (For abbreviations, see Table 4.1.) From Joris and Yin (1998).

both cues vary in a systematic fashion. Most dichotic studies (in the superior olive or elsewhere in the auditory system) have studied the sensitivity of cells to one cue in isolation, with the other cues held constant. There are virtually no dichotic studies in which the localization cues are covaried in any realistic manner. However, some guidance can be gleaned from the sensitivity of LSO cells to these two cues. By studying ITD sensitivity to AM stimuli at different ILDs, Joris and Yin (1995) showed that the sensitivity of LSO cells to ILDs of pure tones in the physiological range was on average about 4 times larger than their sensitivity to ITDs. A more direct assessment of the relative importance of each localization cue was obtained by using the virtual space technique. The relative salience of each of the three localization cues in shaping the azimuthal receptive field was determined by setting each of the three cues constant which letting the other two cues very naturally in azimuth, or by letting only one of the cues vary naturally (Tollin and Yin 1999). With all cues present, high frequency cells in the LSO had spatial receptive fields that were consonant with the IE binaural interaction. In other words, they responded to stimuli in the ipsilateral sound field where the signal level to the excitatory, ipsilateral ear is high, but were suppressed in the contralateral field where the signal level to the inhibitory, contralateral ear is high. Thus, LSO cells are unusual binaural cells in that they respond to stimuli in the ipsilateral, rather than the contralateral sound field. This ipsilateral bias in the LSO is converted to a contralateral bias by the unusual bilateral projections of the LSO to the IC; the excitatory crossed projection and the inhibitory uncrossed projection both produce EI responses in the IC.

Through selective manipulation of the localization cues, Tollin and Yin (1999) showed that the spatial receptive fields in the LSO were shaped primarily by the neuron's sensitivity to ILD (Tollin and Yin 1999), while spectral cues and ITDs contributed much less to the spatial receptive field. These results are in agreement with the psychophysical studies that show that envelope ITDs create a weak perceived laterality of the intracranial image despite low discrimination thresholds for detecting small ITD differences (Henning 1974a, 1974b; Bernstein and Trahiotis 1985; Trahiotis and Bernstein 1986), as well as a weak role of high frequency ITDs for localizing virtual free field sound sources (Wightman and Kistler 1992).

5.2.3 Low-Frequency Cells in Lateral Limb

Thus far, we have restricted our discussion to the physiology of high-frequency cells in the LSO. The earliest recordings from the LSO indicated that the lateral, low frequency limb was qualitatively different in that it consisted mostly of monaural, ipsilaterally excited cells (Boudreau and Tsuchitani 1968, Tsuchitani 1977; Guinan et al. 1972b). Anatomical studies seemed to support this difference in that the inhibitory projection to the lateral limb from the MNTB seemed less dense than inhibitory projections to the more medial limbs (Glendenning et al. 1985, 1991; Spangler et al. 1985; Sanes et al. 1987; Saint Marie et al. 1989b), although there is also anatomical evidence for input to the lateral limb from the MNTB (Glendenning et al. 1985; Smith et al. 1998). However, more recent physiological studies have reported mixed results on this issue: while some studies have shown that low-frequency cells in the lateral limb show the same IE binaural interaction that is seen elsewhere in the LSO (Finlayson and Caspary 1991; Joris and Yin 1995; Tollin et al. 2000), others report finding no low-frequency IE cells in the lateral limb of the LSO (rabbit, Batra et al. 1997a; gerbil, Spitzer and Semple 1995). On the other hand, Batra et al. (1997a) found a population of low-frequency cells near the LSO that had CPs near 0.5. These cells responded strongly to ipsilateral stimulation and weakly or not at all to contralateral stimulation, as expected from the classic IE interaction. They apparently did not test to see if a contralateral tone suppressed an ongoing ipsilateral tone, as they did for high-frequency neurons classified as IE. Tollin et al. (2000) found that the large majority of low-frequency cells in the LSO had the classic IE properties: they were driven by stimulation of the ipsilateral but not the contralateral ear, were sensitive to ILDs in that contralateral stimulation would suppress evoked ipsilateral responses, and showed characteristic delays with CPs near 0.5. It seems likely that the cells reported by Batra et al. (1997a) correspond to these IE cells. The low-frequency IE neurons studied by Finlayson and Caspary (1991) are unusual in that they responded better when the stimuli were in phase than when they were out of phase. In all other respects they resemble the IE neurons studied by Tollin et al. (2000).

5.2.4 Latency Model of ILD Coding: The Importance of Timing of Inputs

In his famous coincidence paper that proposed a model for encoding ITDs, Jeffress (1948) also briefly mentioned the possible importance of the effect of SPLs on response latency as another mechanism by which coincidence of inputs may determine ILD sensitivity, the so-called "latency hypothesis." The idea is attractive because it provides a natural mechanism for time/intensity trading, a well known psychophysical effect (see Hafter and Jeffress 1968). For example, in an LSO cell when the SPL to the contralateral inhibitory ear is low, the latency of inhibitory input is delayed relative to the excitatory input from the ipsilateral ear, and the cell will respond well (Irvine et al. 1998). As the SPL to the contralateral ear is raised, its latency shortens until its inhibitory effect arrives at the LSO at the same time as the excitation and thus suppresses the LSO cell. An important caveat to the latency hypothesis is that it would be expected to be effective only for the onset component of the response.

Evidence supporting the latency hypothesis was first provided by Yin et al. (1985) in the superior colliculus, where responses are generally transient. Since the LSO is the first point at which ILDs are encoded, it is natural to explore the relevance of the latency hypothesis there. Data supporting the hypothesis has been found, particularly in the bat where responses also tend to be transient (Pollak 1988; Park et al. 1996) or in response to transient stimuli in the cat (Joris and Yin 1995) and rat (Irvine et al. 1998). Irvine et al. (1998) made the most direct test of this hypothesis. These researchers measured the change in latency as a function of SPL for the monaural excitatory input of the ipsilateral ear and then tested binaurally under three conditions: (1) variable ILDs under the unusual condition of holding the inhibitory (contralateral) ear constant while varying the ipsilateral SPL so as to match the SPLs tested monaurally where latency measurements were available, (2) "equivalent ITDs" corresponding to the latency changes measured monaurally but with 0 ILDs and (3) "delay cancelled" ILDs in which ITDs opposite to those imposed in the "equivalent ITDs" condition were imposed along with the original ILDs. If a cell's ILD sensitivity were totally determined by the ITDs that are imposed by latency changes with level, then its response to the equivalent ITD condition should match the control. On the other hand, if the ILD sensitivity were dictated by the changing intensity independent of latency changes, then its response to the delay cancelled condition should match the control. The population of LSO cells divided roughly in equal thirds: one group that seemed to fit the latency hypothesis, a second group that seemed to be dominated by intensity rather than time and the last group, which fell in between (Irvine et al. 1998).

5.2.5 Studies of LSO in Brain Slices

The *in vitro* brain slice technique has been used to study several important features of LSO neurons that are not readily accessible in the whole animal. The current-voltage relationship of LSO cells is approximately linear and the cells fire trains of action potentials when depolarized (Sanes 1990; Wu and Kelly 1991; Adam et al. 1999). From a large sample of intracellular recordings in the rat, Adam et al. (1999) identified two classes of cells in the LSO: choppers that resembled those described by Wu and Kelly (1991) and Sanes (1990) and delay cells that display an initial transient depolarization followed by a slow depolarizing ramp of variable slope that caused a long delay before a late burst of spikes. Intracellular labeling with neurobiotin showed that the chopper cells were principal cells in the LSO while the delay cells had characteristics of olivocochlear cells, which in the rodent lie within the boundaries of the LSO.

The brain slice preparation also facilitates studies of the pharmacological features of the neurons. The ipsilateral excitatory input from the SBCs is glutamatergic acting through non-N-methyl-d-aspartate (non-NMDA) receptors since excitatory amino acid agonists of the nonNMDA type caused excitation at low concentrations while NMDA did not (Wu and Kelly 1992; Kandler and Friauf 1995). Likewise, nonNMDA antagonists blocked responses while NMDA antagonists had little effect. As expected, the inhibitory input from the contralateral side was found to be glycinergic, since it was blocked by strychnine. However, NMDA was found to block the inhibition as well, suggesting a role for NMDA receptors to modulate the contralateral inhibitory input (Wu and Kelly 1992). Based on area measurements of incremental PSPs, Sanes (1990) estimated that an LSO cell received an average of ten excitatory afferents and eight inhibitory ones and found that IPSPs were about twice as long as EPSPs.

The inhibitory neurotransmitters GABA and glycine both cause a concentration-dependent drop in the membrane resistance of LSO cells, suggesting that LSO cells could be sensitive to both inhibitory transmitters. During development, there appears to be a shift from GABAergic to glycinergic transmission in the LSO (Kotak et al. 1998).

5.2.6 Models of the LSO

Computer models of the LSO have addressed several different issues: Johnson et al. (1990) used function-based models to investigate whether an

IE binaural interaction is optimal for processing ILDs while Johnson et al. (1986) and Zacksenhouse et al. (1992) used point process modeling to try to reproduce the interval statistics of LSO cells. Reed and Blum (1990) used a computer model to show that a Jeffress-like model of spatially-graded variations in threshold could generate a spatial map of ILDs analogous to the spatial map of ITDs in the MSO. However, there is no experimental support for either the threshold gradient or the map of ILDs.

5.3 Development and Plasticity of the LSO Circuit

A thorough review of the developmental studies in the auditory system is beyond the scope of this review (see Sanes and Friauf 2000). Instead, I will touch upon interesting studies that take advantage of the binaural features of this circuit. The IE binaural interaction of the LSO circuit offers a significant advantage for studying various aspects of the plasticity and development of synapses in the central nervous system because the excitatory and inhibitory cells are physically separate and activated by stimulation of one ear or the other. Thus, the excitatory or inhibitory influences on LSO cells can be manipulated independently unlike most situations in the CNS where the cell bodies are intermingled. During development, the duration of EPSPs and IPSPs decline by nearly 2 orders of magnitude from postnatal day 1 to postnatal day 20 and there appears to be elimination of both excitatory and inhibitory synapses (Sanes 1993). During early development, the IPSPs are depolarizing until a few days after birth at which time they reverse and hyperpolarize, as would be expected for the glycine-mediated chloride channel (Kandler and Friauf 1995; Ehrlich et al. 1999). In parallel with these physiological changes during early development are considerable anatomical alterations: (1) there is a significant decrease in the complexity of dendritic arbors of LSO cells (Henkel and Brunso-Bechtold 1991; Rietzel and Friauf 1998), (2) during the first ten postnatal days, there is a spatial gradient of dendritic arbors such that low-frequency LSO cells have broader arbors in terms of total dendritic length and number of branch points than high-frequency cells (Sanes et al. 1990, 1992), (3) axonal arborizations of MNTB cells to the LSO show a corresponding reduction in complexity (Sanes and Siverls 1991), and (4) there is a parallel gradient of the density of glycine receptors along the tonotopic axis with higher concentrations in the high-frequency cells (Sanes and Wooten 1987).

Since the excitatory and inhibitory inputs to the LSO derive from different sides of the brainstem, the circuit offers a unique opportunity to manipulate one or the other of the inputs independently. Unilateral cochlear ablation denervates the ipsilateral LSO of excitatory input and the contralateral LSO of inhibitory input. Sanes et al. (1992) found considerable atrophy of the LSO on the ipsilateral side while the contralateral side exhibited a hypertrophic dendritic response with increased branching and spread. Interestingly, the loss of inhibitory input to the LSO contralateral to the cochlear ablation affected the strength of its excitatory input from the unmanipulated side (Kotak and Sanes 1996). Similar effects were found by silencing the inhibitory synapses by continuous perfusion of the glycine antagonist strychnine for 21 days. Thus, suppressing the glycinergic input to the LSO cell made the excitatory synapses stronger through the expression of NMDA receptors. Furthermore, it appears that normal inhibitory activity contributes to synapse formation or elimination since the inhibitory terminals are more spread out (though the IPSPs are smaller) when the glycinergic transmission is disrupted (Sanes and Takacs 1993; Kotak and Sanes 1996).

A unilateral cochlear ablation early in development can also result in abnormal anatomical projections in the superior olive. Using DiI injections following a unilateral cochlear ablation on postnatal day two of gerbils, Kitzes et al. (1995) found that the AVCN on the unoperated side projected to its normal targets as well as to areas of the superior olive normally innervated by the AVCN of the lesioned side. Thus, the AVCN was found to establish synaptic connections *bilaterally* to the MSO, MNTB and LSO. In the MSO, the projections were to both the medial and lateral facing dendrites while the MNTB projections made normal calyceal endings. EM analysis showed that the projections made functional synaptic endings and the projections were specific to auditory targets. These results suggest a remarkable plasticity in the anatomical and physiological processes in the auditory brainstem during early development.

6. Behavioral/Lesion Studies of Sound Localization in the Superior Olivary Complex

The importance of the SOC for sound localization and behavior has also been studied by examining the effects of lesions in a number of different animals. A difficulty in such experiments is specifying the precise neural elements ablated since there are so many fiber tracts coursing through the relevant nuclei. Thus, many of the early studies made lesions of the trapezoid body at the midline where there are no cell bodies. Such a lesion effectively deafferents much of the superior olive and results in significant deficits in a cat's ability to localize and move to one of two speakers (Moore et al. 1974; Casseday and Neff 1975). The nature of the deficit depended upon the size of the loss in the trapezoid body. Ablations restricted to other commissural pathways (corpus callosum or commissure of the IC) had no effect on this performance.

Jenkins and Masterton (1982) trained cats to choose one of seven speakers placed in the frontal hemifield and compared the effects of trapezoid body or superior olivary lesions with unilateral lesions of the output of the superior olive and higher levels of the auditory system. While lesions of the trapezoid body resulted in localization deficits in both hemifields, unilateral lesions severing the output of the superior olive in the lateral lemniscus or higher levels resulted in deficits in the contralateral hemifield only. Similar results were obtained in a head orientation task (Thompson and Masterton 1978). These results are in accord with the physiological responses that show a bias for sounds in the contralateral field above the level of the superior olive.

More recent studies of localization behavior have used kainic acid lesions, which destroy cell bodies while leaving fibers of passage intact. Kavanaugh and Kelly (1992) trained ferrets on a minimum audible angle task where the stimuli could be placed in the lateral hemifields, or centered on the midline. In general, unilateral lesions confined to the superior olivary complex resulted in severe localization deficits in both left and right sides and in more pronounced deficits in the lateral fields than on the midline. Bilateral lesions resulted in more severe behavioral deficits though there were no complete bilateral lesions. Unfortunately, the lesions could not be restricted to either the MSO or LSO. Similar results were found in rats (Van Adel and Kelly 1998). These studies provide evidence that the localization deficits result from damage to the cell bodies in the superior olive and not to damage of fibers of passage and suggest that accurate localization requires both SOCs. Ablation above the SOC on one side produces deficits on the side contralateral to the lesion.

7. Conclusions

I have reviewed the circuitry in the mammalian SOC that is thought to be important for encoding the two interaural localization cues, ITDs and ILDs. I have argued that these cues are important for localization along the horizontal meridian, and further, that sound localization is a critical task for the auditory system. In accordance with this imperative role, we can see that a considerable portion of the ascending auditory system is devoted to these tasks. In addition, there are interesting and extreme anatomical and physiological specializations that are unique to the nervous system in these circuits, strongly suggesting that they have evolved for specific reasons.

While the operation of the MSO and LSO circuits seems straightforward, there remain many unresolved problems. A principal puzzle is what happens as the information ascends to the IC and higher auditory centers. The convergence of many ascending systems at the level of the IC makes these circuits much more complicated and less stereotypical than the SOC. On the other hand, I am not suggesting that the MSO and LSO circuits are only engaged in computing sound localization cues. It is likely that they are important for some of the other functions that the auditory system performs so well.

Acknowledgments. For the problems reviewed in this chapter, I am grateful for the efforts of many talented individuals with whom I have had the pleasure of working: Shig Kuwada, Joseph Chan, Dexter Irvine, Laurel Carney, Phil Smith, Phillip Joris, B. Delgutte, and Don Tollin. Special thanks to P. Joris and D. Tollin for critically reading earlier drafts of this chapter. This work was supported by NIH grants DC00116 and DC02840.

References

Adam TJ, Schwarz DW, Finlayson PG (1999) Firing properties of chopper and delay neurons in the lateral superior olive of the rat. Exp Brain Res 124:489–502.

Adams JC (1979) Ascending projections to the inferior colliculus. J Comp Neurol 183:519–538.

Adams JC, Mugnaini E (1990) Immunocytochemical evidence for inhibitory and disinhibitory circuits in the superior olive. Hearing Res 49:281–298.

Agmon-Snir H, Carr CE, Rinzel J (1998) The role of dendrites in auditory coincidence detection. Nature 393:268–272.

Aitkin L, Schuck D (1985) Low frequency neurons in the lateral central nucleus of the cat inferior colliculus receive their input predominantly from the medial superior olive. Hear Res 17:87–93.

Anderson DJ, Rose JE, Hind JE, Brugge JF (1971) Temporal position of discharges in single auditory nerve fibers within the cycle of a sine-wave stimulus: frequency and intensity effects. J Acoust Soc Am 49:1131–1139.

Banks MI, Smith PH (1992) Intracellular recordings from neurobiotin-labeled cells in brain slices of the rat medial nucleus of the trapezoid body. J Neurosci 12: 2819–2837.

Barnes-Davies M, Forsythe ID (1995) Pre- and postsynaptic glutamate receptors at a giant excitatory synapse in rat auditory brainstem slices. J Physiol 488:387–406.

Batra R, Kuwada S, Stanford TR (1989) Temporal coding of envelopes and their interaural delays in the inferior colliculus of the unanesthetized rabbit. J Neurophysiol 61:257–268.

Batra R, Kuwada S, Fitzpatrick DC (1997a) Sensitivity to interaural temporal disparities of low- and high-frequency neurons in the superior olivary complex. I. Heterogeneity of responses. J Neurophysiol 78:1222–1236.

Batra R, Kuwada S, Fitzpatrick DC (1997b) Sensitivity to interaural temporal disparities of low- and high-frequency neurons in the superior olivary complex. II. Coincidence detection. J Neurophysiol 78:1237–1247.

Batteau DW (1966) The role of the pinna in human localization. Proc Roy Soc Lond B 168:158–180.

Beckius GE, Batra R, Oliver DL (1999) Axons from anteroventral cochlear nucleus that terminate in medial superior olive of cat: observations related to delay lines. J Neurosci 19:3146–3161.

Bernstein LR, Trahiotis C (1985) Lateralization of low-frequency, complex waveforms: the use of envelope-based temporal disparities. J Acoust Soc Am 77: 1868–1880.

Berrebi AS, Spirou GA (1998) PEP-19 immunoreactivity in the cochlear nucleus and superior olive of the cat. Neuroscience 83:535–554.

Blackburn CC, Sachs MB (1989) Classification of unit types in the anteroventral cochlear nucleus: PST histograms and regularity analysis. J Neurophysiol 62: 1303–1329.

Bledsoe SC Jr, Snead CR, Helfert RH, Prasad V, Wenthold RJ, Altschuler RA (1990) Immunocytochemical and lesion studies support the hypothesis that the projection from the medial nucleus of the trapezoid body to the lateral superior olive is glycinergic. Brain Res 517:189–194.

Borst JG, Helmchen F, Sakmann B (1995) Pre- and postsynaptic whole-cell recordings in the medial nucleus of the trapezoid body of the rat. J Physiol 489:825–840.

Boudreau JC, Tsuchitani C (1968) Binaural interaction in the cat superior olive S segment. J Neurophysiol 31:442–454.

Boudreau JC, Tsuchitani C (1970) Cat superior olive s-segment cell discharge to tonal stimulation. In: Neff WD (ed) Contributions to Sensory Physiology. New York: Academic. pp. 143–213.

Bourk TR (1976) Electrical Responses of Neural Units in the Anteroventral Cochlear Nucleus of the Cat. pp. 1–385. Cambridge, MA: Ph.D. Thesis, M.I.T.

Brawer JR, Morest DK (1975) Relations between auditory nerve endings and cell types in the cat's anteroventral cochlear nucleus seen with the Golgi method and Nomarski optics. J Comp Neurol 160:491–506.

Brownell WE, Manis PB, Ritz LA (1979) Ipsilateral inhibitory responses in the cat lateral superior olive. Brain Res 177:189–193.

Browner RH, Webster DB (1975) Projections of the trapezoid body and the superior olivary complex of the Kangaroo rat (*Dipodomys merriami*). Brain Behav Evol 11:322–354.

Brughera AR, Stutman ER, Carney LH, Colburn HS (1996) A model with excitation and inhibition for cells in the medial superior olive. Aud Neurosci 2:219–233.

Brunso-Bechtold JK, Linville MC, Henkel CK (1994) Terminal types on ipsilaterally and contralaterally projecting lateral superior olive cells. Hear Res 77:99–104.

Caird D, Klinke R (1983) Processing of binaural stimuli by cat superior olivary complex neurons. Exp Brain Res 52:385–399.

Campistron G, Buijs RM, Geffard M (1986) Glycine neurons in the brain and spinal cord. Antibody production and immunocytochemical localization. Brain Res 376: 400–405.

Cant NB (1984) The fine structure of the lateral superior olivary nucleus of the cat. J Comp Neurol 227:63–77.

Cant NB, Casseday JH (1986) Projections from the anteroventral cochlear nucleus to the lateral and medial superior olivary nuclei. J Comp Neurol 247:457–476.

Cant NB, Hyson RL (1992) Projections from the lateral nucleus of the trapezoid body to the medial superior olivary nucleus in the gerbil. Hear Res 58:26–34.

Cant NB, Morest DK (1984) The structural basis for stimulus coding in the cochlear nucleus of the cat. In: Berlin C (ed) Hearing Science. San Diego, CA: College-Hill Press. pp. 371–422.

Carney LH (1990) Sensitivities of cells in the anteroventral cochlear nucleus of cat to spatio-temporal discharge patterns across primary afferents. J Neurophysiol 64:437–456.

Carney LH (1993) A model for the responses of low-frequency auditory-nerve fibers in cat. J Acoust Soc Am 93:401–417.

Carney LH, Yin TCT (1989) Responses of low-frequency cells in the inferior colliculus to interaural time differences of clicks: excitatory and inhibitory components. J Neurophysiol 62:144–161.

Carr CE, Konishi M (1990) A circuit for detection of interaural time differences in the brainstem of the barn owl. J Neurosci 10:3227–3246.

Caspary DM, Backoff PM, Finlayson PG, Palombi PS (1994) Inhibitory inputs modulate discharge rate within frequency receptive fields of anteroventral cochlear nucleus neurons. J Neurophysiol 72:2124–2133.

Casseday JH, Neff WD (1973) Localization of pure tones. J Acoust Soc Am 54: 365–372.

Casseday JH, Neff WD (1975) Auditory localization: Role of auditory pathways in brainstem of the cat. J Neurophysiol 38:842–858.

Cherry C (1953) Some experiments on the recognition of speech, with one and two ears. J Acoust Soc Am 26:975–979.

Chuhma N, Ohmori H (1998) Postnatal development of phase-locked high-fidelity synaptic transmission in the medial nucleus of the trapezoid body of the rat. J Neurosci 18:512–520.

Clark GM (1969a) The ultrastructure of nerve endings in the medial superior olive of the cat. Brain Res 14:293–305.

Clark GM (1969b) Vesicle shape versus type of synapse in the nerve endings of the cat medial superior olive. Brain Res 15:548–551.

Colburn HS, Han YA, Cullota CP (1990) Coincidence model of MSO responses. Hearing Res 49:335–346.

Crow G, Rupert AL, Moushegian G (1978) Phase locking in monaural and binaural medullary neurons: implications for binaural phenomena. J Acoust Soc Am 64:493–501.

Ehrlich I, Lohrke S, Friauf E (1999) Shift from depolarizing to hyperpolarizing glycine action in rat auditory neurones is due to age-dependent Cl- regulation. J Physiol (Lond) 520:121–137.

Elverland HH (1978) Ascending and intrinsic projections of the superior olivary complex in the cat. Exp Brain Res 32:117–134.

Finlayson PG, Caspary DM (1991) Low-frequency neurons in the lateral superior olive exhibit phase-sensitive binaural inhibition. J Neurophysiol 65:598–605.

Forsythe ID (1994) Direct patch recording from identified presynaptic terminals mediating glutamatergic EPSCs in the rat CNS, in vitro. J Physiol 479:381–387.

Forsythe ID, Barnes-Davies M (1993) The binaural auditory pathway: excitatory amino acid receptors mediate dual timecourse excitatory postsynaptic currents in the rat medial nucleus of the trapezoid body. Proc R Soc Lond B Biol Sci 251: 151–157.

Friauf E, Ostwald J (1988) Divergent projections of physiologically characterized rat ventral cochlear nucleus neurons as shown by intra-axonal injection of horseradish peroxidase. Exp Brain Res 73:263–284.

Frisina RD, Smith RL, Chamberlain SC (1985) Differential encoding of rapid changes in sound amplitude by second-order auditory neurons. Exp Brain Res 60:417–422.

Furst M, Aharonson V, Levine RA, Fullerton BC, Tadmor R, Pratt H, Polyakov A, Korczyn AD (2000) Sound lateralization and interaural discrimination. Effects of brainstem infarcts and multiple sclerosis lesions. Hear Res 143:29–42.

Gardner MB, Gardner RS (1973) Problem of localization in the median plane: effect of pinnae cavity occlusion. J Acoust Soc Am 53:400–408.

Glendenning KK, Baker BN (1988) Neuroanatomical distribution of receptors for three potential neurotransmitters in the brainstem auditory nuclei of the cat. J Comp Neurol 275:288–308.

Glendenning KK, Masterton RB (1983) Acoustic chiasm: efferent projections of the lateral superior olive. J Neurosci 3:1521–1537.

Glendenning KK, Brunso-Bechtold JK, Thompson GC, Masterton RB (1981) Ascending auditory afferents to the nuclei of the lateral lemniscus. J Comp Neurol 197:673–703.

Glendenning KK, Hutson KA, Nudo RJ, Masterton RB (1985) Acoustic chiasm II: Anatomical basis of binaurality in lateral superior olive of cat. J Comp Neurol 232:261–285.

Glendenning KK, Masterton RB, Baker BN, Wenthold RJ (1991) Acoustic chiasm. III: Nature, distribution, and sources of afferents to the lateral superior olive in the cat. J Comp Neurol 310:377–400.

Glendenning KK, Baker BN, Hutson KA, Masterton RB (1992) Acoustic chiasm V: inhibition and excitation in the ipsilateral and contralateral projections of LSO. J Comp Neurol 319:100–122.

Goldberg JM, Brown PB (1968) Functional organization of the dog superior olivary complex: an anatomical and electrophysiological study. J Neurophysiol 31:639–656.

Goldberg JM, Brown PB (1969) Response of binaural neurons of dog superior olivary complex to dichotic tonal stimuli: some physiological mechanisms of sound localization. J Neurophysiol 32:613–636.

Goldberg JM, Brownell WE (1973) Discharge characteristics of neurons in the anteroventral and dorsal cochlear nuclei of cat. Brain Res 64:35–54.

Grothe B (1994) Interaction of excitation and inhibition in processing of pure tone and amplitude-modulated stimuli in the medial superior olive of the mustached bat. J Neurophysiol 71:706–721.

Grothe B (2000) The evolution of temporal processing in the medial superior olive, an auditory brainstem structure. Prog Neurobiol 61:581–610.

Grothe B, Neuweiler G (2000) The function of the medial superior olive in small mammals: temporal receptive fields in auditory analysis. J Comp Physiol [A] 186:413–423.

Grothe B, Sanes DH (1993) Bilateral inhibition by glycinergic afferents in the medial superior olive. J Neurophysiology 69:1192–1196.

Grothe B, Sanes DH (1994) Synaptic inhibition influences the temporal coding properties of medial superior olivary neurons: an in vitro study. J Neurosci 14:1701–1709.

Guinan JJ Jr, Li RYS (1990) Signal processing in brainstem auditory neurons which receive giant endings (calyces of Held) in the medial nucleus of the trapezoid body of the cat. Hearing Res 49:321–334.

Guinan JJ Jr, Guinan SS, Norris BE (1972a) Single auditory units in the superior olivary complex. I: Responses to sounds and classifications based on physiological properties. Int J Neurosci 4:101–120.

Guinan JJ Jr, Norris BE, Guinan SS (1972b) Single auditory units in the superior olivary complex. II: Location of unit categories and tonotopic organization. Int J Neurosci 4:147–166.

Hafter ER, Jeffress LA (1968) Two-image lateralization of tones and clicks. J Acoust Soc Am 44:563–569.

Held H (1893) Die centrale Gehörleittung. Arch Anat Physiol Abt 201–248.

Helfert RH, Schwartz IR (1986) Morphological evidence for the existence of multiple neuronal classes in the cat lateral superior olivary nucleus. J Comp Neurol 244:533–549.

Helfert RH, Schwartz IR (1987) Morphological features of five neuronal classes in the gerbil lateral superior olive. Am J Anat 179:55–69.

Helfert RH, Bonneau JM, Wenthold RJ, Altschuler RA (1989) GABA and glycine immunoreactivity in the guinea pig superior olivary complex. Brain Res 501: 269–286.

Helfert RH, Juiz JM, Bledsoe SC Jr, Bonneau JM, Wenthold RJ, Altschuler RA (1992) Patterns of glutamate, glycine, and GABA immunolabeling in four synaptic terminal classes in the lateral superior olive of the guinea pig. J Comp Neurol 323:305–325.

Henkel CK, Brunso-Bechtold JK (1991) Dendritic morphology and development in the ferret lateral superior olivary nucleus. J Comp Neurol 313:259–272.

Henkel CK, Brunso-Bechtold JK (1993) Laterality of superior olive projections to the inferior colliculus in adult and developing ferret. J Comp Neurol 331:458–468.

Henkel CK, Brunso-Bechtold JK (1995) Development of glycinergic cells and puncta in nuclei of the superior olivary complex of the postnatal ferret. J Comp Neurol 354:470–480.

Henkel CK, Gabriele ML (1999) Organization of the disynaptic pathway from the anteroventral cochlear nucleus to the lateral superior olivary nucleus in the ferret. Anat Embryol (Berl) 199:149–160.

Henkel CK, Spangler KM (1983) Organization of the efferent projections of the medial superior olivary nucleus in the cat as revealed by HRP and autoradiographic tracing methods. J Comp Neurol 221:416–428.

Henning GB (1974a) Detectability of interaural delay in high-frequency complex waveforms. J Acoust Soc Am 55:84–90.

Henning GB (1974b) Lateralization and the binaural masking-level difference. J Acoust Soc Am 55:1259–1262.

Hirsh IJ (1948) The influence of interaural phase on interaural summation and inhibition. J Acoust Soc Am 20:536–544.

Inbody SB, Feng AS (1981) Binaural response characteristics of single neurons in the medial superior olivary nucleus of the albino rat. Brain Res 210:361–366.

Irvine DRF (1986) The auditory brainstem: processing of spectral and spatial information. Berlin, Springer-Verlag. pp. 1–276.

Irvine DRF (1992) Physiology of the auditory brainstem. In: Popper AN and Fay RR (eds) The Mammalian Auditory Pathway: Neurophysiology. New York: Springer-Verlag. pp. 153–231.

Irvine DRF, Park VN, McCormick L (1998) Contributions of changes in the timing and amplitude of synaptic inputs to neural sensitivity to interaural intensity differences. In: Palmer AR, Rees A, Summerfield AQ, Meddis R (eds) Psychophysical and Physiological Advances in Hearing. London: Whurr Publishers, Ltd. pp. 359–367.

Javel E (1980) Coding of AM tones in the chinchilla auditory nerve: implications for the pitch of complex tones. J Acoust Soc Am 68:133–146.

Jean-Baptiste M, Morest DK (1975) Transneuronal changes of synaptic endings and nuclear chromatin in the trapezoid body following cochlear ablations in cats. J Comp Neurol 162:111–134.

Jeffress LA (1948) A place theory of sound localization. J Comp Physiol Psychol 41:35–39.

Jenkins WM, Masterton RB (1982) Sound localization: effects of unilateral lesions in central auditory system. J Neurophysiol 47:987–1016.

Johnson DH (1974) The response of single auditory nerve fibers in the cat to single tones: synchrony and average discharge rate. Cambridge, MA: Ph.D. Thesis, M.I.T.

Johnson D (1980) The relationship between spike rate and synchrony in responses of auditory-nerve fibers to single tones. J Acoust Soc Amer 68:1115–1122.

Johnson DH, Tsuchitani C, Linebarger DA, Johnson MJ (1986) Application of a point process model to responses of cat lateral superior olive units to ipsilateral tones. Hear Res 21:135–159.

Johnson DH, Dabak A, Tsuchitani C (1990) Function-based modeling of binaural processing: interaural level. Hear Res 49:301–319.

Joris PX (1996) Envelope coding in the lateral superior olive. II. Characteristic delays and comparison with responses in the medial superior olive. J Neurophysiol 76:2137–2156.

Joris PX, Yin TCT (1992) Responses to amplitude-modulated tones in the auditory nerve of the cat. J Acoust Soc Am 91:215–232.

Joris PX, Yin TCT (1995) Envelope coding in the lateral superior olive. I. Sensitivity to interaural time differences. J Neurophysiol 73:1043–1062.

Joris PX, Yin TCT (1998) Envelope coding in the lateral superior olive. III. Comparison with afferent pathways. J Neurophysiol 79:253–1269.

Joris PX, Carney LH, Smith PH, Yin TCT (1994a) Enhancement of neural synchronization in the anteroventral cochlear nucleus. I. Responses to tones at the characteristic frequency. J Neurophysiol 71:1022–1036.

Joris PX, Smith PH, Yin TCT (1994b) Enhancement of neural synchronization in the anteroventral cochlear nucleus. II. Responses in the tuning curve tail. J Neurophysiol 71:1037–1051.

Joris PX, Smith PH, Yin TCT (1998) Coincidence detection in the auditory system: 50 years after Jeffress. Neuron 21:1235–1238.

Kandler K, Friauf E (1995) Development of glycinergic and glutamatergic synaptic transmission in the auditory brainstem of perinatal rats. J Neurosci 15:6890–6904.

Kavanagh GL, Kelly JB (1992) Midline and lateral field sound localization in the ferret (*Mustela putorius*): contribution of the superior olivary complex. J Neurophysiol 67:1643–1658.

Kelly JB, Liscum A, van Adel B, Ito M (1998) Projections from the superior olive and lateral lemniscus to tonotopic regions of the rat's inferior colliculus. Hear Res 116:43–54.

Kettner RE, Feng J, Brugge JF (1985) Postnatal development of the phase-locked response to low frequency tones of auditory nerve fibers in the cat. J Neurosci 5:275–283.

Kil J, Kageyama GH, Semple MN, Kitzes LM (1995) Development of ventral cochlear nucleus projections to the superior olivary complex in gerbil. J Comp Neurol 353:317–340.

Kiss A, Majorossy K (1983) Neuron morphology and synaptic architecture in the medial superior olivary nucleus. Light- and electron microscope studies in the cat. Exp Brain Res 52:315–327.

Kitzes LM, Kageyama GH, Semple MN, Kil J (1995) Development of ectopic projections from the ventral cochlear nucleus to the superior olivary complex induced by neonatal ablation of the contralateral cochlea. J Comp Neurol 353: 341–363.

Klumpp R, Eady H (1956) Some measurements of interaural time difference thresholds. J Acoust Soc Am 28:859–860.

Kotak VC, Sanes DH (1996) Developmental influence of glycinergic transmission: regulation of NMDA receptor-mediated EPSPs. J Neurosci 16:1836–1843.

Kotak VC, Korada S, Schwartz IR, Sanes DH (1998) A developmental shift from GABAergic to glycinergic transmission in the central auditory system. J Neurosci 18:4646–4655.

Koyano K, Ohmori H (1996) Cellular approach to auditory signal transmission. Jpn J Physiol 46:289–310.

Kuwabara N, Zook JM (1991) Classification of the principal cells of the medial nucleus of the trapezoid body. J Comp Neurol 314:707–720.

Kuwabara N, Zook JM (1992) Projections to the medial superior olive from the medial and lateral nuclei of the trapezoid body in rodents and bats. J Comp Neurol 324:522–538.

Kuwabara N, DiCaprio RA, Zook JM (1991) Afferents to the medial nucleus of the trapezoid body and their collateral projections. J Comp Neurol 314:684–706.

Kuwada S, Batra R, Fitzpatrick D (1997) Neural processing of binaural temporal cues. In: Gilkey R, Anderson T (eds) Binaural and Spatial Hearing in Real and Virtual Environments. Mahwah, N.J.: Lawrence Erlbaum Assoc. pp. 399–426.

Kuwada S, Stanford TR, Batra R (1987) Interaural phase-sensitive units in the inferior colliculus of the unanesthetized rabbit: effects of changing frequency. J Neurophysiol 57:1338–1360.

Lavine RA (1971) Phase-locking in response of single neurons in cochlear nuclear complex of the cat to low-frequency tonal stimuli. J Neurophysiol 34:467–483.

Leakey D, Sayers B, Cherry C (1958) Binaural fusion of low- and high-frequency sounds. J Acoust Soc Am 30:222.

Lenn NJ, Reese TS (1966) The fine structure of nerve endings in the nucleus of the trapezoid body and the ventral cochlear nucleus. Am J Anat 118:375–390.

Levine RA, Gardner JC, Stufflebeam SM, Fullerton BC, Carlisle EW, Furst M, Rosen BR, Kiang NY (1993) Binaural auditory processing in multiple sclerosis subjects. Hear Res 68:59–72.

Liberman MC (1991) Central projections of auditory-nerve fibers of differing spontaneous rate. I. Anteroventral cochlear nucleus. J Comp Neurol 313:240–258.

Licklider JCR (1948) The influence of interaural phase relations upon the masking of speech by white noise. J Acoust Soc Am 20:150–159.

Licklider JC, Webster JC, Hedlun JM (1950) On the frequency limits of binaural beats. J Acoust Soc Am 22:468–473.

Lindsey BG (1975) Fine structure and distribution of axon terminals from the cochlear nucleus on neurons in the medial superior olivary nucleus of the cat. J Comp Neurol 160:81–103.

Majorossy K, Kiss A (1990) Types of neurons and synaptic relations in the lateral superior olive of the cat: normal structure and experimental observations. Acta Morphol Hung 38:207–215.

Masterton B, Thompson GC, Bechtold JK, RoBards MJ (1975) Neuroanatomical basis of binaural phase-difference analysis for sound localization: a comparative study. J Comp Physiolog Psychol 89:379–386.

Matsubara JA (1990) Calbindin D-28K immunoreactivity in the cat's superior olivary complex. Brain Res 508:353–357.

McAlpine D, Jiang D, Palmer AR (1996) Interaural delay sensitivity and the classification of low best-frequency binaural responses in the inferior colliculus of the guinea pig. Hear Res 97:136–152.

McAlpine D, Jiang D, Shackleton TM, Palmer AR (1998) Convergent input from brainstem coincidence detectors onto delay-sensitive neurons in the inferior colliculus. J Neurosci 18:6026–6039.

McFadden D, Pasanen EG (1976) Lateralization at high frequencies based on interaural time differences. J Acoust Soc Am 59:634–639.

Mills AW (1958) On the minimum audible angle. J Acoust Soc Am 30:237–246.

Mills A (1960) Lateralization of high-frequency tones. J. Acoust Soc Am 32:132–134.

Monsivais P, Yang L, Rubel EW (2000) GABAergic inhibition in nucleus magnocellularis: implications for phase locking in the avian auditory brainstem. J Neurosci 20:2954–2963.

Moore CN, Casseday JH, Neff WD (1974) Sound localization: The role of the commissural pathways of the auditory system of the cat. Brain Res 82:13–26.

Moore DR (1988) Auditory brainstem of the ferret: sources of projections to the inferior colliculus. J Comp Neurology 269:342–354.

Moore DR, Russell FA, Cathcart NC (1995) Lateral superior olive projections to the inferior colliculus in normal and unilaterally deafened ferrets. J Comp Neurol 357:204–216.

Moore MJ, Caspary DM (1983) Strychnine blocks binaural inhibition in lateral superior olivary neurons. J Neurosci 3:237–242.

Morest DK (1968) The collateral system of the medial nucleus of the trapezoid body of the cat, its neuronal architecture and relation to the olivo-cochlear bundle. Brain Res 9:288–311.

Morest DK (1973) Auditory neurons of the brainstem. Adv Otorhinolaryngol 20:337–356.

Moushegian G, Rupert A, Whitcomb MA (1964) Brain-stem neuronal response patterns to monaural and binaural tone. J Neurophysiol 27:1174–1191.

Moushegian G, Rupert AL, Gidda JS (1975) Functional characteristics of superior olivary neurons to binaural stimuli. J Neurophysiol 38:1037–1048.

Nuetzel J, Hafter E (1976) Lateralization of complex waveforms: effects of fine structure, amplitude, and duration. J Acoust Soc Am 60:1339–1346.

Oertel D (1983) Synaptic responses and electrical properties of cells in brain slices of the mouse anteroventral cochlear nucleus. J Neurosci 3:2043–2053.

Oliver DL, Beckius GE, Shneiderman A (1995) Axonal projections from the lateral and medial superior olive to the inferior colliculus of the cat: a study using electron microscopic autoradiography. J Comp Neurol 360:17–32.

Ostapoff EM, Benson CG, Saint Marie RL (1997) GABA- and glycine-immunoreactive projections from the superior olivary complex to the cochlear nucleus in guinea pig. J Comp Neurol 381:500–512.

Palmer AR (1982) Encoding of rapid amplitude fluctuations by cochlear-nerve fibres in the guinea-pig. Arch Otorhinolaryngol 236:197–202.

Palmer AR, Winter IM, Darwin CJ (1986) The representation of steady-state vowel sounds in the temporal discharge patterns of the guinea pig cochlear nerve and primarylike cochlear nucleus neurons. J Acoust Soc Am 79:100–113.

Park TJ, Grothe B, Pollak GD, Schuller G, Koch U (1996) Neural delays shape selectivity to interaural intensity differences in the lateral superior olive. J Neurosci 16:6554–6566.

Park TJ, Monsivais P, Pollak GD (1997) Processing of interaural intensity differences in the LSO: role of interaural threshold differences. J Neurophysiol 77:2863–2878.

Perkins RE (1973) An electron microscopic study of synaptic organization in the medial superior olive of normal and experimental chinchillas. J Comp Neurol 148:387–415.

Peyret D, Campistron G, Geffard M, Aran JM (1987) Glycine immunoreactivity in the brainstem auditory and vestibular nuclei of the guinea pig. Acta Otolaryngol 104:71–76.

Pfeiffer RR (1966) Anteroventral cochlear nucleus: wave forms of extracellularly recorded spike potentials. Science 154:667–668.

Pollak GD (1988) Time is traded for intensity in the bat's auditory system. Hear Res 36:107–124.

Ramon y Cajal S (1909) Histologie du Systeme Nerveux de l'Homme et des Vertebrates, Vol. I. pp. 754–838. Madrid: Instituto Ramon y Cajal.

Rayleigh LJS (1907) On our perception of sound direction. Philos Mag 6:214–232.

Reed MC, Blum JJ (1990) A model for the computation and encoding of azimuthal information by the lateral superior olive. J Acoust Soc Am 88:1442–1453.

Rhode WS, Kettner RE (1987) Physiological study of neurons in the dorsal and posteroventral cochlear nucleus of the unanesthetized cat. J Neurophysiol 57:414–442.

Rietzel HJ, Friauf E (1998) Neuron types in the rat lateral superior olive and developmental changes in the complexity of their dendritic arbors. J Comp Neurol 390:20–40.

Roberts RC, Ribak CE (1987) GABAergic neurons and axon terminals in the brainstem auditory nuclei of the gerbil. J Comp Neurol 258:267–280.

Rose JE, Gross NB, Geisler CD, Hind JE (1966) Some neural mechanisms in the inferior colliculus of the cat which may be relevant to localization of a sound source. J Neurophysiol 29:288–314.

Rose JE, Kitzes LM, Gibson MM, Hind JE (1974) Observations on phase-sensitive neurons of anteroventral cochlear nucleus of the cat: nonlinearity of cochlear output. J Neurophysiol 37:218–253.

Roth GL, Aitkin LM, Andersen RA, Merzenich MM (1978) Some features of the spatial organization of the central nucleus of the inferior colliculus of the cat. J Comp Neurol 182:661–680.

Rothman JS, Young ED (1996) Enhancement of neural synchronization in computational models of ventral cochlear nucleus bushy cells. Aud Neurosci 2:47–62.

Rothman JS, Young ED, Manis PB (1993) Convergence of auditory nerve fibers onto bushy cells in the ventral cochlear nucleus: implications of a computational model. J Neurophysiol 70:2562–2583.

Sachs M, Abbas P (1974) Rate versus level functions for auditory-nerve fibers in cats: tone-burst stimuli. J Acoust Soc Am 56:1835–1847.

Saint Marie RL, Morest DK, Brandon CJ (1989a) The form and distribution of GABAergic synapses on the principal cell types of the ventral cochlear nucleus of the cat. Hear Res 42:97–112.

Saint Marie RL, Ostapoff E-M, Morest DK, Wenthold RJ (1989b) Glycine immunoreactive projection of the cat lateral superior olive: possible role in midbrain ear dominance. J Comp Neurol 279:382–396.

Sanes DH (1990) An in vitro analysis of sound localization mechanisms in the gerbil lateral superior olive. J Neurosci 10:3494–3506.

Sanes DH (1993) The development of synaptic function and integration in the central auditory system. J Neurosci 13:2627–2637.

Sanes DH, Friauf E (2000) Development and influence of inhibition in the lateral superior olivary nucleus. Hear Res 147:46–58.

Sanes DH, Siverls V (1991) Development and specificity of inhibitory terminal arborizations in the central nervous system. J Neurobiol 22:837–854.

Sanes DH, Takacs C (1993) Activity-dependent refinement of inhibitory connections. Eur J Neurosci 5:570–574.

Sanes DH, Wooten GF (1987) Development of glycine receptor distribution in the lateral superior olive of the gerbil. J Neurosci 7:3803–3811.

Sanes DH, Geary WA, Wooten GF, Rubel EW (1987) Quantitative distribution of the glycine receptor in the auditory brainstem of the gerbil. J Neurosci 7:3793–3802.

Sanes DH, Goldstein NA, Ostad M, Hillman DE (1990) Dendritic morphology of central auditory neurons correlates with their tonotopic position. J Comp Neurol 294:443–454.

Sanes DH, Markowitz S, Bernstein J, Wardlow J (1992) The influence of inhibitory afferents on the development of postsynaptic dendritic arbors. J Comp Neurol 321:637–644.

Scheibel M, Scheibel A (1974) Neuropile organization in the superior olive of the cat. Exp Neurol 43:339–348.

Schwartz IR (1972) Axonal endings in the cat medial superior olive: coated vesicles and intercellular substance. Brain Res 46:187–202.

Schwartz IR (1977) Dendritic arrangements in the cat medial superior olive. Neurosci 2:81–101.

Schwartz IR (1980) The differential distribution of synaptic terminal on marginal and central cells in the cat medial superior olivary nucleus. Am J Anat 159:25–31.

Schwartz IR (1984) Axonal organization in the cat medial superior olivary nucleus. In: Neff WD (ed) Contributions to Sensory Physiology. New York: Academic Press. pp. 99–129.

Schwartz IR (1992) The superior olivary complex and lateral lemniscal nuclei. In: Webster DB, Popper AN, Fay RR (eds) The Mammalian Auditory Pathway: Neuroanatomy. New York: Springer-Verlag. pp. 117–167.

Shneiderman A, Henkel CK (1985) Evidence of collateral axonal projections to the superior olivary complex. Hear Res 19:199–205.

Smith DJ, Rubel EW (1979) Organization and development of brainstem auditory nuclei of the chicken: dendritic gradients in nucleus laminaris. J Comp Neurol 186:213–239.

Smith PH (1995) Structural and functional differences distinguish principal from nonprincipal cells in the guinea pig MSO slice. J Neurophysiology 73:1653–1667.

Smith PH, Rhode WS (1987) Characterization of HRP-labeled globular bushy cells in the cat anteroventral cochlear nucleus. J Comp Neurol 266:360–375.

Smith PH, Joris PX, Carney LH, Yin TCT (1991) Projections of physiologically characterized globular bushy cell axons from the cochlear nucleus of the cat. J Comp Neurol 304:387–407.

Smith PH, Joris PX, Yin TCT (1993) Projections of physiologically characterized spherical bushy cell axons from the cochlear nucleus of the cat: evidence for delay lines to the medial superior olive. J Comp Neurol 331:245–260.

Smith PH, Joris PX, Yin TCT (1998) Anatomy and physiology of principal cells of the medial nucleus of the trapezoid body (MNTB) of the cat. J Neurophysiol 79:3127–3142.

Smith RL, Brachman ML (1980) Response modulation of auditory-nerve fibers by AM stimuli: effects of average intensity. Hear Res 2:123–133.

Smolders JW, Klinke R (1986) Synchronized responses of primary auditory fibre-populations in *Caiman crocodilus* (L.) to single tones and clicks. Hear Res 24: 89–103.

Sommer I, Lingenhohl K, Friauf E (1993) Principal cells of the rat medial nucleus of the trapezoid body: an intracellular in vivo study of their physiology and morphology. Exp Brain Res 95:223–239.

Spangler KM, Warr WB, Henkel CK (1985) The projections of principal cells of the medial nucleus of the trapezoid body in the cat. J Comp Neurol 238:249–262.

Spirou GA, Berrebi AS (1997) Glycine immunoreactivity in the lateral nucleus of the trapezoid body of the cat. J Comp Neurol 383:473–488.

Spirou GA, Brownell WE, Zidanic M (1990) Recordings from cat trapezoid body and HRP labeling of globular bushy cell axons. J Neurophysiol 63:1169–1190.

Spitzer MW, Semple MN (1995) Neurons sensitive to interaural phase disparity in gerbil superior olive: diverse monaural and temporal response properties. J Neurophysiol 73:1668–1690.

Stevens S, Newman E (1936) The localization of actual sources of sound. Amer J Psychol 48:297–306.

Stotler WA (1953) An experimental study of the cells and connections of the superior olivary complex of the cat. J Comp Neurol 98:401–432.

Takahashi T, Konishi M (1986) Selectivity for interaural time difference in the owl's midbrain. J Neurosci 6:3413–3422.

Taschenberger H, von Gersdorff H (2000) Fine-tuning an auditory synapse for speed and fidelity: developmental changes in presynaptic waveform, EPSC kinetics, and synaptic plasticity. J Neurosci 20:9162–9173.

Thompson AM, Thompson GC (1991) Posteroventral cochlear nucleus projections to olivocochlear neurons. J Comp Neurol 303:267–285.

Thompson GC, Masterton RB (1978) Brainstem auditory pathways involved in reflexive head orientation to sound. J Neurophysiol 41:1183–1202.

Thompson GC, Cortez AM, Lam DM (1985) Localization of GABA immunoreactivity in the auditory brainstem of guinea pigs. Brain Res 339:119–122.

Thompson SP (1882) On the function of the two ears in the perception of space. Philos Mag 13:406–416.

Tolbert L, Morest D, Yurgelun-Todd D (1982) The neuronal architecture of the anteroventral cochlear nucleus of the cat in the region of the cochlear nerve root: Horseradish peroxidase labelling of identified cell types. Neurosci 7:3031–3052.

Tollin DJ, Joris PX, Yin TCT (2000) Coding of interaural phase differences in low-frequency cells in the lateral superior olive of the cat. Assoc Res Otolaryngol 23:113.

Tollin DJ, Yin TCT (1999) Spatial receptive fields of cells in the lateral superior olive of the cat. Soc Neurosci 29:267.13.

Trahiotis C, Bernstein LR (1986) Lateralization of bands of noise and sinusoidally amplitude-modulated tones: Effects of spectral locus and bandwidth. J Acoust Soc Am 79:1950–1957.

Tsuchitani C (1977) Functional organization of lateral cell groups of cat superior olivary complex. J Neurophysiol 40:296–318.

Tsuchitani C (1997) Input from the medial nucleus of trapezoid body to an inter-aural level detector. Hear Res 105:211–224.

Tsuchitani C, Boudreau JC (1966) Single unit analysis of cat superior olive S segment with tonal stimuli. J Neurophysiol 29:684–697.

Van Adel BA, Kelly JB (1998) Kainic acid lesions of the superior olivary complex: effects on sound localization by the albino rat. Behav Neurosci 112:432–446.

Van Gisbergen JA, Grashuis JL, Johannesma PI, Vendrik AJ (1975) Neurons in the cochlear nucleus investigated with tone and noise stimuli. Exp Brain Res 23:387–406.

Van Noort JV (1969) The structure and connections of the inferior colliculus. An investigation of the lower auditory system. pp. 1–118.

Wang LY, Gan L, Forsythe ID, Kaczmarek LK (1998) Contribution of the Kv3.1 potassium channel to high-frequency firing in mouse auditory neurones. J Physiol 509:183–194.

Wang X, Sachs MB (1994) Neural encoding of single-formant stimuli in the cat. II. Responses of anteroventral cochlear nucleus units. J Neurophysiol 71:59–78.

Warr WB (1966) Fiber degeneration following lesions in the anterior ventral cochlear nucleus of the cat. Exp Neurol 14:453–474.

Warr WB (1972) Fiber degeneration following lesions in the multipolar and globular cell areas in the ventral cochlear nucleus of the cat. Brain Res 40:247–270.

Warr WB (1992) Organization of olivocochlear efferent systems in mammals. In: Fay RR, Popper A (eds) The Mammalian Auditory Pathway: Neuroanatomy. New York: Springer Verlag. pp. 410–448.

Webster WR, Batini C, Buisseret-Delmas C, Compoint C, Guegan M, Thomasset M (1990) Colocalization of calbindin and GABA in medial nucleus of the trapezoid body of the rat. Neurosci Lett 111:252–257.

Wenthold RJ, Huie D, Altschuler RA, Reeks KA (1987) Glycine immunoreactivity localized in the cochlear nucleus and superior olivary complex. Neuroscience 22:897–912.

Wenthold RJ, Parakkal MH, Oberdorfer MD, Altschuler RA (1988) Glycine receptor immunoreactivity in the ventral cochlear nucleus of the guinea pig. J Comp Neurol 276:423–435.

Wightman FL, Kistler DJ (1992) The dominant role of low-frequency interaural time differences in sound localization. J Acoust Soc Am 91:1648–1661.

Winter IM, Palmer AR (1990) Responses of single units in the anteroventral cochlear nucleus of the guinea pig. Hear Res 44:161–178.

Wu SH, Kelly JB (1991) Physiological properties of neurons in the mouse superior olive: membrane characteristics and postsynaptic responses studied in vitro. J Neurophysiol 65:230–246.

Wu SH, Kelly JB (1992) Synaptic pharmacology of the superior olivary complex studied in mouse brain slice. J Neurosci 12:3084–3097.

Wu SH, Kelly JB (1994) Physiological evidence for ipsilateral inhibition in the lateral superior olive: synaptic responses in mouse brain slice. Hear Res 73:57–64.

Wu SH, Oertel D (1986) Inhibitory circuitry in the ventral cochlear nucleus is probably mediated by glycine. J Neurosci 6:2691–2706.

Yin TCT (1994) Physiological correlates of the precedence effect and summing localization in the inferior colliculus of the cat. J Neurosci 14:5170–5186.

Yin TCT, Chan JC (1990) Interaural time sensitivity in medial superior olive of cat. J Neurophysiol 64:465–488.

Yin TCT, Kuwada S (1983a) Binaural interaction in low-frequency neurons in inferior colliculus of the cat. II. Effects of changing rate and direction of interaural phase. J Neurophysiol 50:1000–1019.

Yin TCT, Kuwada S (1983b) Binaural interaction in low-frequency neurons in inferior colliculus of the cat. III. Effects of changing frequency. J Neurophysiol 50:1020–1042.

Yin TCT, Kuwada S, Sujaku Y (1984) Interaural time sensitivity of high-frequency neurons in the inferior colliculus. J Acoust Soc Am 76:1401–1410.

Yin TCT, Hirsch JA, Chan JC (1985) Responses of neurons in the cat's superior colliculus to acoustic stimuli. II. A model of interaural intensity sensitivity. J Neurophysiol 53:746–758.

Yin TCT, Chan JCK, Irvine DRF (1986) Effects of interaural time delays of noise stimuli on low-frequency cells in the cat's inferior colliculus. I. Responses to wideband noise. J Neurophysiol 55:280–300.

Young ED, Robert JM, Shofner WP (1988) Regularity and latency of units in ventral cochlear nucleus: implications for unit classification and generation of response properties. J Neurophysiol 60:1–29.

Young SR, Rubel EW (1983) Frequency-specific projections of individual neurons in chick brainstem auditory nuclei. J Neurosci 3:1373–1378.

Zacksenhouse M, Johnson DH, Tsuchitani C (1992) Excitatory/inhibitory interaction in the LSO revealed by point process modeling. Hear Res 62:105–123.

Zook JM, Leake PA (1989) Connections and frequency representation in the auditory brainstem of the mustache bat, *Pteronotus parnellii*. J Comp Neurol 290:243–261.

Zwislocki J, Feldman R (1956) Just noticeable differences in dichotic phase. J Acoust Soc Am 28:860–886.

5
Circuitry and Function of the Dorsal Cochlear Nucleus

Eric D. Young and Kevin A. Davis

1. Introduction

In Chapter 2 of this volume, Smith and Spirou describe the wonderful complexity of the brainstem auditory system. This system forms a collection of parallel pathways that diverge at the first auditory synapse in the brainstem, in the cochlear nucleus (CN), and then converge again, at least in a gross anatomical sense, in the inferior colliculus (for abbreviations, see Table 5.1). The CN is a well-studied collection of neural circuits that are diverse both in anatomical and physiological terms (reviewed by Cant 1992; Rhode and Greenberg 1992; Young 1998). These vary from the simplest system, the bushy cells of the ventral cochlear nucleus (VCN; see Yin, Chapter 4), to the most complex, in the dorsal cochlear nucleus (DCN). The DCN differs from other parts of the CN by having an extensive internal neuropil formed by groups of interneurons (Lorente de Nó 1981; Osen et al. 1990). As a result, the DCN makes significant changes in the auditory representation from its inputs to its outputs. In this chapter, the neural organization of the DCN is reviewed, paying most attention to data from the cat. The response properties of DCN neurons are discussed in the context of its neural organization and related to data on the functional role of the DCN in hearing.

2. Role of the Dorsal Cochlear Nucleus in Hearing

Understanding auditory processing in terms of the neural circuits in the brain depends on working out the roles of the multiple parallel pathways of the brainstem auditory system. An excellent example is provided by Chapter 4 of this book (see Yin, Chapter 4), in which the role of the system consisting of the bushy cells in the CN and the principal cells of the superior olivary complex is described. This circuit makes the precise comparisons of time of arrival and sound level in the two ears that are needed for binaural sound localization. In this chapter, the role of a second CN pathway, that formed by the principal cells of the DCN, is discussed.

TABLE 5.1. Abbreviations

ANF	auditory nerve fiber
BBN	broadband noise
BF	best frequency
CIA	central inhibitory area
CN	cochlear nucleus
CNIC	central nucleus of the inferior colliculus
DAS	dorsal acoustic stria
DCN	dorsal cochlear nucleus
EP	evoked potential
GABA	γ-aminobutyric acid
HRTF	head-related transfer function
ILD	interaural level difference
ITD	interaural time difference
MSN	medullary somatosensory nuclei
PST	post-stimulus time
STRF	spectro-temporal receptive field
VCN	ventral cochlear nucleus
WBI	wideband inhibitor

By the "role" of the DCN, we mean the computations done by the DCN for the other parts of the auditory system. This is an ill-defined concept in that there is no accepted method for defining the role of an arbitrary group of neurons in the brain. Several kinds of evidence can be offered. First, there is behavioral evidence, i.e., the appearance of deficits in certain kinds of auditory performance when the neurons are lesioned. Using an example from the previous chapter, deficits in sound localization performance are seen when lesions are made in the principal nuclei of the superior olivary complex (Kavanagh and Kelly 1992; Van Adel and Kelly 1998). Second, there is evidence from the projection sites of the principal cells of the nucleus. Again, using the example of the previous chapter, the bushy cells of the CN project essentially only to the superior olivary complex and are the primary source of inputs to the superior olive's principal nuclei. This fact means that CN bushy cells are the pathway through the CN for information about binaural cues for sound localization. Third, there is evidence from the response properties of the neurons. The responses may convey information about a particular aspect of the stimulus (e.g. Blackburn and Sachs 1990; Dabak and Johnson 1992) or respond to the stimulus in a way that corresponds well to some aspect of perception (e.g. Cariani and Delgutte 1996a,b; May et al. 1996). In either case, the responses are prima facie evidence that the neurons are participating in the corresponding aspect of auditory processing. Again, a clear example is provided by the bushy cell/superior olivary circuit. The superior olive is the first place in the auditory system where neurons respond sensitively to differences between the stimuli at the two ears, which argues that performing the basic calculations for binaural sound localization is the principal role of this structure.

Each of the three kinds of evidence has its own problems of interpretation. Strong conclusions can be drawn only when multiple lines of evidence converge to support the same set of hypotheses. In this chapter, we will focus mainly on the third kind of evidence for the case of the DCN, but the existing evidence of the first two kinds will be reviewed first.

The principal cells of the DCN project directly to the central nucleus of the inferior colliculus (CNIC; Osen 1972; Oliver 1984; Ryugo and Willard 1985). These cells are unique in the CN in that their activity reflects substantial interneuronal processing in the complex neuropil of the DCN (Lorente de Nó 1981; Berrebi and Mugnaini 1991; Weedman et al. 1996). They are also unique in that they serve as the convergence point for auditory and somatosensory information (Young et al. 1995). The main features of the interneuronal circuitry of the DCN have been worked out, so that the generation of DCN principal cell responses to sound and somatosensory stimuli is well understood. These results suggest that one role of the DCN is to process the complex spectral patterns produced by directionally-dependent filtering in the external ear (Young et al. 1992; Imig et al. 2000). These so-called head-related transfer functions (HRTFs) convey information about sound localization (e.g., in the cat, Musicant et al. 1990; Rice et al. 1992). Thus, one hypothesis for the role of the DCN in the auditory system is to process aspects of spectral sound localization cues. This chapter will review what is known about neural signal processing in the DCN as it relates to this hypothesis.

2.1 Evidence from Behavioral Studies of Sound Localization

Because the output axons of the DCN leave the nucleus through the dorsal acoustic stria (DAS; Fernandez and Karapas 1967; Adams and Warr 1976), which is separate from the other outputs of the CN, it is possible to interrupt the DCN's output with minimal effects on the projections from the rest of the CN. The intermediate acoustic stria, containing axons of multipolar cells in VCN (Adams and Warr 1976), is near the DAS and is probably damaged when the DAS is cut, but the largest output of the VCN, which passes through the trapezoid body, is unaffected. A surprising result from experiments in animals with DAS lesions is that the deficits produced are subtle. Masterton and colleagues showed that DAS lesions did not affect cats' absolute thresholds, masked thresholds, or ability to discriminate the location of a sound source (Masterton and Granger 1988; Masterton et al. 1994; Sutherland et al. 1998a).

The only reported positive finding from such experiments is that lesioned cats show a deficit in sound localization when tested in situations that require pointing to the actual location of the sound source. This is to be distinguished from situations in which the cat is asked to discriminate the location of two speakers, a task that can be done without necessarily knowing where either speaker is located. The deficit in sound localization has been shown in two ways. First, Sutherland and colleagues used untrained cats performing a reflex head orientation to a sound source (Sutherland et al. 1998b). The accuracy of the orientation was decreased after a DAS lesion. Second, May (2000) trained cats to point to the source of a sound by turning their heads. Although the results are complex, DAS lesions produced a clear deficit in performance, especially for the vertical component of localization. When the same cats were tested in speaker discrimination, however, they showed little or no deficit, consistent with results from Sutherland and colleagues mentioned above.

2.2 Sound Localization Cues in the Cat

Interpretation of the DCN's role in sound localization depends on understanding the nature of sound localization cues in the cat. Figure 5.1 shows a summary of the relevant aspects of these cues. Figures 5.1A and 5.1B show examples of HRTFs from a cat. These functions show the magnitude of the frequency-dependent gain from a speaker in free field to a microphone near the eardrum (Musicant et al. 1990; Rice et al. 1992). That is, they show the changes in a sound's spectrum that are produced by the acoustical effects of the head and external ear. If a speaker were to produce a broadband noise with a flat spectrum, i.e., equal energy at all frequencies, the signal at the cat's eardrum would have a spectral shape equal to the appropriate HRTF. The important point for the present discussion is that HRTFs are direction dependent. Figure 5.1A shows HRTFs with the sound source at three different azimuthal positions; that is, the source was moved in 30° steps along the horizon. The HRTFs at these locations differ in two important ways. First, the overall gain of the HRTF is higher at the larger azimuths. This is most clearly seen at frequencies below 8 kHz, where the HRTFs are smooth in shape. The change in gain means that the sound is louder as the source moves toward the ear; this effect produces a difference in interaural sound level, called an interaural level difference (ILD), as the sound source moves away from the midline. For example, if the speaker were located at 15° azimuth, the HRTF for 15° would apply to the right ear and the HRTF for –15° would apply to the left ear. The ILD would be the (frequency-dependent) difference between these two HRTFs.

The second difference in HRTFs with azimuth is in their detailed spectral shapes. The most obvious shape change is the movement in the center frequency of the prominent spectral notch located in the frequency region between 5 and 18 kHz (marked "FN region" in Fig. 5.1A). The center frequency of the notch increases as the azimuth increases. This notch, called the first notch, is a potent spectral cue to sound source location (Rice et al.

1992). At higher frequencies, there are more complex and variable changes in the spectral cues.

Figure 5.1B shows HRTFs from three different elevations at the same azimuth. Similar changes in spectral shape are seen, with the first notch frequency increasing with elevation, Note, however, that the overall gain does

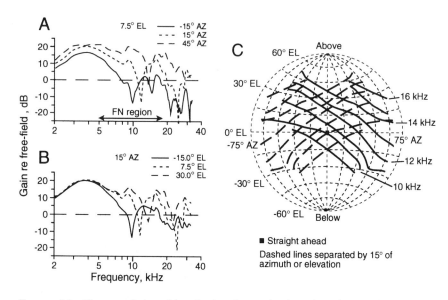

FIGURE 5.1. Characteristics of head related transfer functions (HRTFs) in the cat. These functions show the ratio between the amplitude of a sound at a microphone near the cat's eardrum to the amplitude of the sound at the same point in space in the absence of the cat. They thus measure the effect of the cat's head and external ear in perturbing the sound field. A: HRTFs for three different azimuths at fixed elevation, i.e., moving parallel to the horizon in a left-right direction. Azimuth and elevation are defined below. B: HRTFs for three elevations at fixed azimuth, i.e., moving in a down to up direction. In both A and B, notice the prominent notches in the transfer functions at frequencies between 8 and 17 kHz (first notch or FN). C: Contours of constant first-notch frequency for locations in space in front of a cat. Light dashed lines show contours of constant azimuth and elevation. Each heavy line shows all the speaker locations for which there would be a first notch of a certain frequency in one ear. Solid lines are for the right ear and dashed lines are for the left ear; left ear contours are actually just mirror images of the right ear contours, assuming symmetric ears. *Azimuth* refers to positions along the horizon: 0° is straight ahead and positive azimuths are for positions on the right, where the measuring microphone was located. For example, a speaker at 20° azimuth is centered on a line which makes a 20° angle with respect to straight ahead. *Elevation* refers to vertical positions and is measured with respect to the horizon; 0° elevation is on the horizon, positive elevations are above the horizon. (Reprinted from Hearing Research, Vol. 58, Rice et al., Pinna-based spectral cues for sound localization in cat, pp. 132–152, copyright 1992, with permission from Elsevier Science.)

not change at low frequencies. Apparently, the principal cue to elevation is spectral shape.

Accompanying the differences in interaural level and spectral shape is a difference in the time of arrival of the stimuli at the two ears, an interaural time difference (ITD), which is not shown in Fig. 5.1 (Roth et al. 1980; Kuhn 1987). Psychophysical analysis of humans and cats have shown that all three cues to sound localization are used (reviewed by Middlebrooks and Green 1991; Populin and Yin 1998). Spectral cues are most important, in human observers, for elevation and for discriminating front from back.

In cat, the first-notch frequency is a potential spectral cue for both azimuth and elevation. Figure 5.1C shows a plot of the distribution of first-notch frequencies for locations in front of a cat. The lines show contours of constant first-notch frequency, i.e., locations in space having the same first-notch frequencies. The solid lines show contours for the right ear and the dashed lines show contours for the left ear (assumed to mirror the right ear data). This plot shows that a cat can localize a sound over a significant region of space based on a knowledge only of the first-notch frequencies in the two ears.

If cats do use the first-notch cue to localize sound, then their sound localization performance should be sensitive to manipulations that interfere with the first-notch information. Huang and May (1996a) trained cats to localize sounds from speakers in the frontal field, i.e., roughly over the range of locations shown in Fig. 5.1C. The cats were trained with broadband noise so that all the spectral cues were present along with the interaural cues (ILD and ITD). The cats were required to respond by making a head orientation to the sound source, to guarantee that the task involved actually localizing the sound. Cats were tested with tones and with filtered bands of noise. Their performance was best when the stimulus contained noise in the frequency band between 5 and 18 kHz. Because this is the frequency region where the first notch cue is located, this result suggests that cats depend on the first notch cue for sound localization.

In an additional experiment, cats were trained to discriminate the sounds coming from two different speakers (Huang and May 1996b). Again, the cats were trained with broadband noise and tested with noise filtered to remove certain frequency bands. In this case, removing the signal energy above 18 kHz had the largest effect on performance, suggesting that the cats depended on changes in the stimulus spectrum at high frequencies to discriminate speakers. Apparently, cats can hear the spectral cues at higher frequencies and discriminate changes in the spectrum, but the variation of spectral shape with sound source direction is so complex at high frequencies that it cannot easily be used as a cue for sound location.

The apparent difference between the cues that cats use for sound localization versus direction discrimination is consistent with the behavioral results, discussed above, suggesting that the DCN is involved in sound localization but not in direction discrimination. The shape of the signal spectrum at the eardrum is represented multiple times in the output pathways of the CN. For example, the chopper neurons of the VCN provide a robust representation of spectral shape for vowel-like stimuli (Blackburn and Sachs 1990) and so probably also provide a good representation of the spectral shapes created by HRTFs. Evidence will be summarized below to show that DCN principal cells give a strong inhibitory response to spectral notches; this response is a specialized mechanism in the DCN that is not present in VCN. Thus, it is possible that the outputs of the VCN are sufficient to support discrimination behavior (so that this behavior is not affected by DAS lesions) whereas the outputs of the DCN are necessary for reflex orientation and sound localization. The existence of the notch response in DCN principal cells provides evidence that the DCN is involved in processing spectral sound localization cues and ties the behavioral and lesion data together.

2.3 Evidence from the Targets of Efferent Projections from the DCN

The axons of DCN principal cells project mainly to auditory targets in the CNIC, bypassing the superior olive and the nuclei of the lateral lemniscus (Fernandez and Karapas 1967; Osen 1972; Ryugo and Willard 1985). The effects of DCN activity on cells in CNIC is beginning to be understood, and this topic will be discussed in the last section of this chapter. However, knowledge about the targets within CNIC of DCN axons is in a preliminary state, and it is not possible to draw strong conclusions about the DCN's role from its projection to CNIC.

One intriguing hypothesis is that there are DCN-like or DCN-related subsystems of the auditory system which exist in the inferior colliculus, medial geniculate, and auditory cortex. Imig and colleagues have argued for the existence of such subsystems based on the presence of cells in thalamus and cortex that are sensitive to monaural (and therefore spectral) sound localization cues as opposed to binaural cues (Samson et al. 1993; Imig et al. 1997). Further evidence for a segregation of DCN outputs is provided by the finding that evoked potentials produced by electrical stimulation of the DAS are larger and shorter in latency in the anterior auditory field than in the primary auditory cortex of the cat (Imig and Samson 2000). The existence of a separate DCN-related module of cells in CNIC is suggested by the response characteristics of CNIC neurons (Ramachandran et al. 1999) and by the effects on CNIC neurons of blocking the DAS with lidocaine (K.A. Davis, unpublished); this evidence will be reviewed in the last section of this chapter.

The DCN projects to nonauditory structures, among them the caudal pontine reticular nucleus (Lingenhöhl and Friauf 1994). This nucleus is important for acoustic startle reflexes, and the DCN has been shown to contribute a component of startle (Meloni and Davis 1998). These data do not provide much insight into the auditory role of the DCN, however, because the nature of auditory processing in the startle reflex has not been analyzed. For example, it would be useful to know if there are late components of startle that require auditory processing, such as a knowledge of the location of the sound source that induced the startle.

3. Structural Organization of Neural Circuits in the DCN

Figure 5.2A is a sketch of a frontal section of the cat DCN, showing the main principal cell types and some of the interneurons of the nucleus. The DCN consists of three layers which are cut in cross section in this figure. The layers are organized around the pyramidal cells, principal cells whose cell bodies form the second layer (Blackstad et al. 1984), along with the cell bodies of granule cells (Mugnaini et al. 1980b). The principal cells' somata are shown as black shapes in Fig. 5.2A; the prominent layer of pyramidal cell bodies running parallel to the surface of the nucleus is evident. A full drawing of one pyramidal cell is shown (P). These cells are bipolar, with an apical dendritic tree extending into the molecular layer, i.e., toward the free surface of the nucleus, and a basal dendrite extending into the deep layer.

The superficial (molecular) layer lies above the pyramidal cell somata. It contains the apical dendritic trees of pyramidal cells as well as several kinds of interneurons, including the inhibitory cartwheel (C), Golgi, and stellate cells, and their associated neuropil. The cartwheel cell somata are scattered in a thick layer extending from the pyramidal cell layer into the molecular layer (Berrebi and Mugnaini 1991). Stellate cells are not shown in Fig. 5.2A, but are located in the superficial layer (Wouterlood et al. 1984), and Golgi cells, also not shown, are located in the granule cell regions (Mugnaini et al. 1980a). The inputs to the molecular layer are parallel fibers (pf), which are the axons of granule cells. Parallel fibers run roughly parallel to the plane of the frontal section shown in Fig. 5.2A and make synapses on the interneurons and principal cell dendrites located in the molecular layer. Granule cell somata are concentrated in the pyramidal cell layer of the DCN and in six other areas mostly around the surface of the DCN and VCN (Mugnaini et al. 1980b). Inputs to the granule cells are made by mossy fiber terminals of axons originating from a variety of auditory and nonauditory sources (reviewed in Weedman and Ryugo 1996 and Wright and Ryugo 1996). Among the best studied of these are inputs from the somatosensory dorsal column and spinal trigeminal nuclei (here called MSN for medullary somatosensory nuclei; Itoh et al. 1987; Weinberg and Rustioni 1987), but additional inputs come from the vestibular system (Burian and Gestoettner 1988; Kevetter and Perachio 1989), from pontine

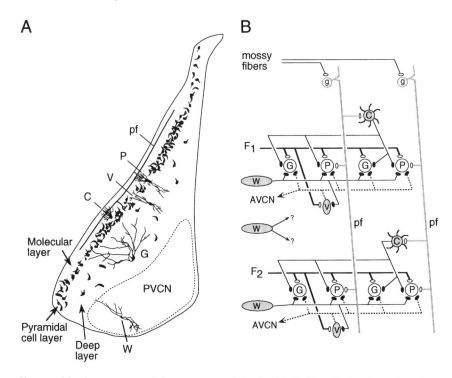

FIGURE 5.2. A summary of the anatomy of the DCN. A: Sketch of a frontal section of the DCN showing the positions of the cell types that will be discussed in this chapter. The section has roughly the shape and layout of the cat DCN, although the drawing is a combination of descriptions from both cat and rodent. The dark shapes are somata of DCN principal cells (pyramidal and giant cells) antidromically filled from an injection of HRP in the CNIC (based on drawings in Ryugo and Willard 1985). Drawings of individual examples of six cell types are shown to illustrate the typical positions of their cell bodies and dendritic trees in the DCN and VCN. The layers are identified at bottom left. The axon of one granule cell is shown forming a parallel fiber (pf) in the molecular layer; the axon originates in a granule cell in the pyramidal cell layer. B: Schematic circuit diagram of the DCN, drawn in a plane parallel to the layers of the nucleus. F_1 and F_2 represent excitatory inputs from auditory nerve fibers and perhaps VCN T-multipolar neurons with best frequencies F_1 and F_2, forming two isofrequency sheets. Excitatory connections are unfilled while inhibitory neurons are shaded and have filled terminals. Abbreviations for cell types: g—granule; C—cartwheel; G—giant; P—pyramidal; pf—parallel fiber; V—vertical; W—wideband inhibitor (VCN D-multipolar or radiate neuron). (A is drawn from figures in Osen 1969; Osen 1983; Rhode et al. 1983; Ryugo and Willard 1985; Smith and Rhode 1989; Osen et al. 1990; and Berrebi and Mugnaini 1991.)

nuclei (Ohlrogge et al. 2001), and probably also from other sources not yet described.

The deep layer of the DCN is a heterogeneous region located deep relative to the pyramidal cell bodies. It contains the basal dendritic trees of pyramidal cells along with the somata and most of the dendritic trees of the second DCN principal cell type, the giant cell (G). Giant cell somata can be seen as scattered, large profiles in the deep layer in Fig. 5.2A. The deep layer also contains the vertical cells (V), also called tuberculoventral cells, whose somata are located deep to the pyramidal cells (Lorente de Nó 1981; Saint Marie et al. 1991; Zhang and Oertel 1993b). The main inputs to the deep layer are auditory axons from the auditory nerve (Osen 1970; Ryugo and May 1993) and from VCN multipolar cells (Adams 1983; Smith and Rhode 1989; Oertel et al. 1990). Auditory nerve fibers travel roughly perpendicular to the plane of section in Fig. 5.2A. They innervate neurons along their path, forming isofrequency sheets of cells. The directions of travel of the parallel fibers in the superficial layer and the auditory nerve axons in the deep layer are roughly orthogonal. The dendritic trees of pyramidal cells, cartwheel cells, and vertical cells are flattened in the direction parallel to the auditory nerve fibers so that these cells tend to receive inputs within a narrow range of frequencies (Osen 1983; Blackstad et al. 1984; Berrebi and Mugnaini 1991).

The arrangement of synaptic connections among the main neuron types of DCN is shown schematically in Fig. 5.2B. This figure shows a view of the DCN looking down on the surface, in the plane of the paper; the isofrequency sheets run from left to right and the parallel fibers run from top to bottom. Two isofrequency sheets are shown, for frequencies F1 and F2. The sheets are bound together by the axons of auditory nerve fibers and possibly also VCN multipolar neurons. Each sheet contains a complement of pyramidal (P) and giant (G) cells, along with a vertical cell (V) representing the vertical cells in the sheet. The vertical cell axons distribute in parallel to isofrequency sheets and terminate on both principal cell types.

There are two cell types in the VCN which send axon collaterals to the DCN (Smith and Rhode 1989; Oertel et al. 1990; Doucet and Ryugo 1997). The planar, or T-multipolar cells are VCN principal cells whose axons travel to the CNIC in the trapezoid body. These neurons have dendritic trees oriented in the direction of the auditory nerve fibers and make terminals with excitatory morphology (asymmetric with round vesicles) in DCN. The radiate, or D-multipolar cells are VCN principal cells whose axons probably leave the CN in the intermediate acoustic stria. Their dendrites are oriented across the array of auditory nerve fibers so that they receive inputs from a broad range of frequencies. Radiate neurons are glycinergic (Doucet et al. 1999) and make terminals with inhibitory morphology in DCN. They correspond to the element called the wideband inhibitor (W) in Fig. 5.2. Evidence to associate the wideband inhibitor and the radiate neuron is discussed in a later section. Whether these neurons make tonotopic connections within an isofrequency sheet, as shown in Fig. 5.2B, is not settled; this uncertainty is indicated by the "W" neuron with projections ending in question marks.

The circuitry of the molecular layer of the DCN is represented in Fig. 5.2B by the mossy fiber inputs, granule cells, and cartwheel cells. Like other

DCN circuits, cartwheel cell axons distribute predominantly in parallel with the isofrequency sheets, as they are drawn (Berrebi and Mugnaini 1991). Although no source of auditory inputs to cartwheel cells is shown in Fig. 5.2B, these cells do respond to sound (Parham and Kim 1995; Davis and Young 1997), so either some mossy fibers are auditory or there are additional connections that are not shown. Auditory nerve fibers do not terminate in the molecular layer (Merchán et al. 1985; Ryugo and May 1993).

The circuitry of the molecular layer is substantially more complex than is shown in Fig. 5.2B. Four additional cell types (Golgi cells, stellate cells, unipolar brush cells, and chestnut cells; Mugnaini et al. 1980a; Wouterlood et al. 1984; Weedman et al. 1996) have been described in the molecular layer. However, the details of their participation in DCN circuits have not been worked out from either the anatomical or physiological perspective.

4. Response Properties of DCN Neurons

The circuit diagram of Fig. 5.2B shows that the DCN is well endowed with inhibitory interneurons. In fact, the responses of DCN neurons to sound display substantial inhibitory influences, and, in the work discussed in this chapter, it has proven most useful to characterize responses in the DCN on the basis of the nature and extent of inhibitory inputs. Figure 5.3 shows typical response maps of elements of the DCN circuit. These maps are based on responses to tones of various frequencies and sound levels; they show discharge rate versus frequency at a series of fixed sound levels (actually attenuations, as explained in the figure caption). The horizontal line in each plot is the spontaneous discharge rate. Increases in rate above spontaneous are colored black to show excitatory response regions; decreases in rate, inhibitory regions, are shaded.

Figure 5.3A shows a type IV response map, which is characteristic of most principal cells in the unanesthetized cat DCN (Young and Brownell 1976; Young 1980; Rhode and Kettner 1987). Type IV response maps vary considerably from neuron to neuron, but the one shown in Fig. 5.3A displays the characteristics that are most typical of this response type in decerebrate cat (Spirou and Young 1991). There is a small excitatory area near threshold, centered on the neuron's best frequency (BF, 17.5 kHz). At higher sound levels, the response near BF is inhibitory over a V-shaped area called the central inhibitory area (CIA, centered on 16–17 kHz in this case). The upper frequency edge of the CIA is bounded by a small excitatory ridge (near 22 kHz here). Usually there is another inhibitory area at higher frequencies, above 22 kHz in this example, and a large excitatory area at high levels and low frequencies, below 10 kHz in this example. Several examples of type IV response maps are given in subsequent figures in this chapter; these illustrate the variability in type IV maps. The only required

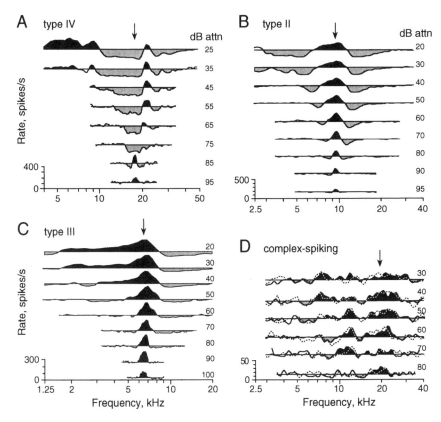

FIGURE 5.3. Response maps for DCN neurons. The maps are labeled at upper left with the corresponding response type. Each map is a collection of plots of discharge rate versus frequency obtained from the presentation of 100 tone bursts of duration 200 ms; the tone frequencies are interpolated logarithmically over a range varying from 1–4 octaves, as shown on the abscissa. For each plot, the attenuation was held constant at the value given by the number at the right of the curve. The actual sound level varies with the acoustic calibration, but 0 dB attenuation corresponds to roughly 100 dB SPL. The acoustic calibration is reasonably flat (e.g., Rice et al. 1995) and typically varies less than ±10 dB over the frequency range of a response map. The horizontal line in each plot is the spontaneous rate. The rate scale at lower left applies to the plot at the lowest sound level; other plots are shifted vertically to prevent overlap, but use the same rate scale. Arrows at the top point to the BFs. Type III and IV responses (A and C) are recorded from principal cells (pyramidal and giant cells), and type II (B) responses are recorded from vertical cells. The type II response map was constructed in the presence of a weak exciter tone of fixed level and frequency (9.35 kHz, 90 dB attn.) for the reasons described in the text. Complex-spiking units (D) are recorded from cartwheel cells, which show widely varying response maps (Davis and Young 1997). For the example shown here, two repetitions of the map are superimposed (solid and dashed lines). (B redrawn from Spirou et al. 1999 and D redrawn from Davis and Young 1997 with permission. D: Copyright 1997 by the Society for Neuroscience.)

features of a type IV map are the low-level excitatory area at BF and the CIA. A model for the type IV response map is presented below. Quantitative variations in this model have been shown to be capable of producing the range of type IV response maps observed in the DCN (Reed and Blum 1995).

Type II responses (Fig. 5.3B) have a narrow V-shaped excitatory area centered on BF and a significant inhibitory surround. These units are characterized by two features in addition to their response maps: (1) type II units do not have spontaneous activity; and (2) type II units give weak or no response to broadband noise (Spirou et al. 1999). Because of the lack of spontaneous activity, inhibitory responses are demonstrated in response maps like Fig. 5.3B by presenting a low-level BF tone of fixed attenuation and frequency; this tone produces a low rate of background activity against which both excitatory and inhibitory responses can be seen. Type II responses are recorded from vertical cells in DCN, based on the fact that they can be antidromically stimulated from the VCN where vertical cells project an axon collateral (Young 1980) and based on identification of type II neurons by dye filling (Rhode 1999).

Type III neurons (Fig. 5.3C) have response maps like those of type II units, with a central V-shaped excitatory area centered on BF and inhibitory sidebands. They are recorded from principal cells and perhaps also other cell types (Young 1980). Type III responses differ from type II in that type III units have at least some spontaneous activity and respond to broadband noise about as strongly as they do to tones (Young and Voigt 1982). To discriminate type II and type III units, a spontaneous rate of 2.5 per second and a relative noise responsiveness (maximum driven rate to noise divided by maximum driven rate to BF tones) of 0.35 are typically used. Type III units are relatively rare in the decerebrate cat DCN (Shofner and Young 1985) but are much more common in anesthetized cats or in rodents like the gerbil, regardless of anesthetic state (Evans and Nelson 1973; Davis et al. 1996a; Joris 1998). Pentobarbital anesthesia converts type IV units to type III (Evans and Nelson 1973; Young and Brownell 1976), presumably by reducing the potency of the vertical cell inhibitory circuit. This issue is discussed in more detail below.

Two additional classes of DCN response maps are not shown in Fig. 5.3: (1) type IV-T units which appear to be an intermediate between types III and IV. They are like type IV units, except with weaker inhibition in the CIA (Spirou and Young 1991); and (2) type I/III units which have no spontaneous activity and have response maps like those of type II units. They differ from type II units in the quantitative details of their BF-tone responses and in that they respond strongly to broadband noise (Spirou et al. 1999). The criteria used to distinguish type II and type I/III units are somewhat arbitrary, and it is clear that they show overlapping properties. However, there is an important functional difference between the two unit types, in that good evidence exists to associate type II responses with the vertical cell and with inhibitory inputs to principal cells (see below; Young 1980; Voigt and Young 1990; Rhode 1999) whereas there is no evidence that type I/III units serve an inhibitory function. There is also no evidence as to the identity of type I/III units.

The remaining response type in Fig. 5.3 is the complex-spiking neuron, whose response map is shown in Fig. 5.3D. Complex-spiking neurons are those whose action potentials, in extracellular recording, show short bursts of spikes. These bursts resemble the mixed calcium-sodium spike bursts recorded intracellularly in cartwheel cells (Zhang and Oertel 1993a; Manis et al. 1994; Golding and Oertel 1997). Such bursts are seen in extracellular recordings only in the superficial DCN, at depths that correspond to the location of the cartwheel cell bodies (Parham and Kim 1995; Davis and Young 1997). Based on this evidence, we assume that complex-spiking responses are recorded from cartwheel cells. Almost all complex-spiking neurons respond to sound, but their responses are highly variable, as illustrated by the response map shown in Fig. 5.3D. Two repetitions of the map are shown (solid and dashed lines); the variability between these two is typical of complex-spiking neurons. These neurons also typically show fluctuations in their spontaneous rates, and it is often hard to assign them a BF. Their response maps show a mixture of excitatory and inhibitory areas, but these are not organized in any typical patterns.

The characterization of type IV units provided by tone response maps suggests that these units respond to stimulus energy in a mainly inhibitory fashion. However, when broadband noise is used as the stimulus, the conclusion is different. Figure 5.4 shows responses to broadband noise and to noise notches for two type IV units. The response maps of the units are shown in Figs. 5.4A and 5.4B. The unit in Fig. 5.4A has a response map similar to the one in Fig. 5.3A, with the same general features. The unit in Fig. 5.4B has an almost entirely inhibitory response map, a pattern of response that is also commonly seen. Figures 5.4C and 5.4D show the responses of the same two units to BF tones and to broadband noise (BBN), plotted as discharge rate versus sound level. The BF-tone rate-level functions show the strongly nonmonotonic shape that is typical of type IV units. Of course, the nonmonotonicity occurs because the tone rate-level function corresponds to moving vertically through the response map at BF, first through the excitatory area near BF threshold and then through the CIA. More important, however, is that the responses to BBN are excitatory at all sound levels despite the predominantly inhibitory responses to tones. The contrast is clear for the unit in the right column. From the response map, one would predict that the unit's responses to a broadband stimulus should be inhibitory because inhibitory responses are observed at all frequencies in the response map over a 60 dB range of sound levels. Nevertheless, the net response of the unit to BBN is excitatory. There are clear signs of inhibitory effects in the BBN rate-level functions, in that they are nonmonotonic; nevertheless, the noise responses

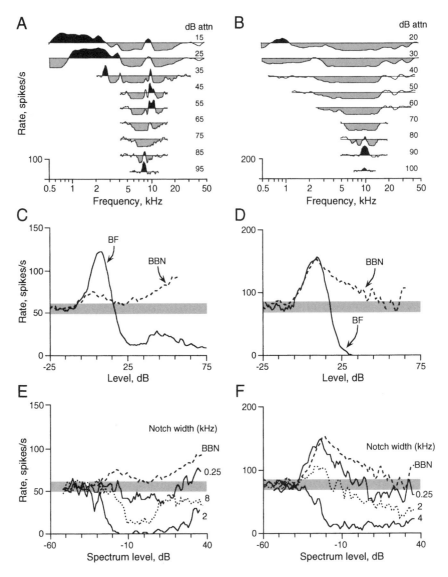

FIGURE 5.4. Response maps and responses to broadband noise for two type IV units. Each column shows data from one unit. Response maps are shown in the top row (A and B). Plots of discharge rate versus sound level are shown in the middle row (C and D) for 200 ms BF-tone bursts and noise bursts. The shaded bars show the range of spontaneous rates. For both units, the response to noise is excitatory at all levels, despite the predominantly inhibitory responses to tones. The bottom plots (E and F) show rate-level functions for notch noise, which is broadband noise with a narrow notch or bandstop region centered arithmetically on BF; a sketch of the spectrum is shown in Fig. 5.7A. The bandwidth of the notch is shown at the right of each curve. Note that the notch responses are inhibitory, which is also not expected from the response maps. Sound levels are given as dB SPL for the BF tones in C and D and as passband dB spectrum level (dB re $20\,\mu\mathrm{Pa/Hz^{1/2}}$) in E and F. The BBN rate functions in C and D are aligned at threshold with the BF-tone rate functions, so the abscissa scale is meaningless for them. (Redrawn from Nelken and Young 1994 with permission.)

remain excitatory. This is the typical behavior of type IV units for noise responses.

Figures 5.4E and 5.4F show another unexpected characteristic of type IV unit responses to broadband stimuli (Spirou and Young 1991; Nelken and Young 1994). In this case, the stimulus is BBN with a notch, or bandstop region, of varying width centered on the unit's BF. The BBN rate-level functions are repeated in Figs. 5.4E and 5.4F along with responses to three notch widths. As the notch is widened, the response becomes strongly inhibitory (e.g., the 2 kHz notch width in Fig. 5.4E and 4 kHz notch width in Fig. 5.4F). At the widest notch widths, the response becomes less inhibitory, as in the 8 kHz notch width in Fig. 5.4E, and ultimately becomes excitatory again (not shown; Spirou and Young 1991). The inhibitory response to notches is not expected from the tone response maps by the following argument. The energy that is removed from the noise to make the notch is centered on unit BF; because responses to tone energy near BF are inhibitory in the response map, removing this energy from a BBN should produce an excitatory effect instead of the inhibitory one that is actually observed. This argument is quite clear for the unit in the right column; it must be made quantitatively for units like the one in the left column because of the small excitatory areas that are present at most levels (Spirou and Young 1991).

The responses of type IV units to broadband stimuli show that these units integrate energy within their response areas in a nonlinear fashion. This point has been demonstrated in several different ways (Nelken and Young 1997; Nelken et al. 1997; Yu and Young 2000). One direct demonstration is to compare tone response maps, like Figs. 5.3A, 5.4A, and 5.4B, with spectro-temporal receptive fields (STRFs), which show the equivalent of a response map constructed from responses to BBN. Whereas tone response maps for type IV units are predominantly inhibitory, the STRFs are predominantly excitatory. In the next section, the nature of the DCN's auditory circuits is analyzed in a way that shows how the nonlinearity of type IV responses arises.

5. The Interactions in DCN Neural Circuits That Generate Its Responses to Sound

The nonlinear response characteristics of DCN principal cells can be accounted for by the basic DCN circuit of Fig. 5.2B. Two inhibitory interneurons are particularly important for responses to sound: the vertical cell, which serves the role of a narrowband inhibitor, and the D-multipolar or radiate neuron of the VCN, which serves the role of a wideband inhibitor. The characteristics of these two inhibitory neurons are shown in Fig. 5.5, which shows plots of discharge rate versus sound level for BF tones and BBN for a type II unit (Fig. 5.5A) and an onset-C unit (Fig. 5.5B). Type II

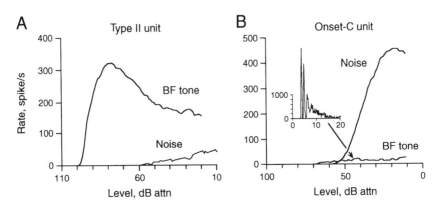

FIGURE 5.5. Properties of narrowband and wideband inhibitors in DCN. Discharge rate versus sound level for responses of a type II unit (A) and an onset-C unit (B) to BF tones and broadband noise, as labeled. Rates are calculated from responses to 200 ms tone or noise bursts presented once per second. Sound level is given as dB attenuation. The inset in B shows a PST histogram of this unit's responses to the first 20 ms of 50 ms BF-tone bursts at 45 dB attn.

units are recorded from vertical cells in DCN, as discussed above. Onset-C responses are recorded in the VCN and have been shown to come from multipolar neurons (Smith and Rhode 1989) whose anatomical characteristics are the same as those of the D-multipolar (Oertel et al. 1990) or radiate neurons (Doucet and Ryugo 1997). Most importantly, these neurons project an axon collateral to the DCN and are glycinergic and therefore inhibitory interneurons (Doucet et al. 1999). These are the cells marked W in Fig. 5.2.

Type II neurons give a strong excitatory response to BF tones (Fig. 5.5A) but a weak response to BBN. In this sense, they are a narrowband inhibitor which is active in DCN for stimuli like tones or narrow bands of noise. Natural stimuli, like speech, which have relatively narrow peaks of energy at certain frequencies (the formants in the case of speech), are also likely to activate these units. That type II units inhibit type IV units has been shown using cross-correlation analysis, in experiments where simultaneous recordings were made from a type II and a type IV unit (Voigt and Young 1980, 1990). In appropriate pairs, there is a dip in the discharge probability of the type IV unit immediately after spikes in the type II unit; a so-called inhibitory trough. This is the feature expected of a monosynaptic inhibitory synapse. Inhibitory troughs are seen when the BF of the type II unit is near that of the type IV unit, actually at or just below the type IV BF. This result is consistent with the tonotopic distribution of vertical-cell axons. When a type II and a type IV unit show an inhibitory trough, the excitatory portion of the type II response map usually corresponds well to the CIA of the type

IV unit (Young and Voigt 1981), suggesting that the CIA is produced by type II inputs.

By contrast, onset-C neurons give weak responses to BF tones and other narrowband stimuli, but strong responses to BBN (Winter and Palmer 1995). This behavior is illustrated by the rate-level functions in Fig. 5.5B. Palmer and colleagues have characterized the responses of onset-C neurons as resulting from wideband facilitation, meaning that inputs from different frequency ranges interact in a strongly facilitatory way (Winter and Palmer 1995; Palmer et al. 1996). Thus the responses of onset-C neurons increase as bandwidth is widened to a bandwidth well beyond the integrating bandwidth of other CN neurons. The inset in Fig. 5.5B shows the post-stimulus time (PST) histogram of an onset-C response to BF tones at the sound level indicated by the arrow. The strong onset character of the response and the relatively weak steady state response are evident.

The evidence to associate onset-C neurons with wideband inhibition in DCN is indirect. The primary evidence, discussed above, is that onset-C responses are recorded from a neuron in the VCN that is glycinergic and projects axon collaterals to the DCN (Smith and Rhode 1989; Oertel et al. 1990; Doucet et al. 1999). Onset-C neurons are hypothesized to provide the inhibition to type II neurons that prevents them from responding to BBN (Winter and Palmer 1995), and the inhibition to DCN type IV units that produces inhibitory responses to notch noise (see below; Nelken and Young 1994). In both cases, onset-C neurons have exactly the characteristics needed to produce the inhibition observed in DCN. This is clearest for wideband inhibition of type II units where the bandwidth of inhibition, measured by broadening a band of noise, is consistent with the excitatory bandwidth of onset-C neurons but is wider than the bandwidth of other neurons in the CN (Palmer et al. 1996; Spirou et al. 1999).

The tone response map of DCN type IV neurons can be reconstructed qualitatively from the properties of the two inhibitory sources discussed above. The model at left in Fig. 5.6 shows the interconnections of a type II, a type IV, and a wideband inhibitor (simplified from Fig. 5.2B). The inset at top center of the figure shows the overlapping excitatory receptive fields of the model's three inputs to the type IV unit. The receptive fields are drawn as if they were plotted on standard tuning curve axes of frequency (abscissa) and sound level (ordinate) although the axes are not shown. The white tuning curve marked ANF represents the excitatory input to the cell from auditory nerve fibers and perhaps also from T-multipolars of the VCN. The two black tuning curves represent the excitatory areas of type II neurons. These are aligned with the excitatory tuning curve in a way that is consistent with the data from previous studies. First, the thresholds of the type II units are elevated with respect to the ANF threshold, which is consistent with the finding that type II units have higher thresholds than either low-threshold (high spontaneous rate) auditory nerve fibers or type IV units (Young and Brownell 1976; Davis and Young 2000). Second, the type

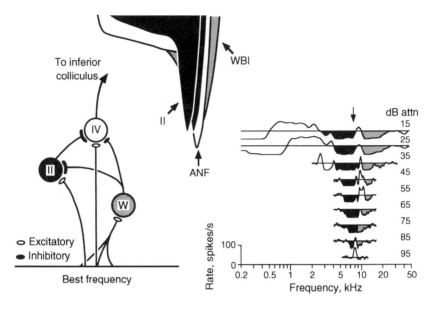

FIGURE 5.6. Qualitative model to account for type IV tone response maps. At left is a circuit showing the postulated interconnections of type II and type IV units and the wideband inhibitor (W). Excitatory and inhibitory connections are indicated by unfilled and filled ovals, respectively, as shown in the legend. The size of each terminal is an indication of its postulated strength. The horizontal line at bottom represents the tonotopic array of excitatory inputs to the circuit, with BF varying from left to right. The BF of the type IV unit is determined by the BF of its excitatory input; the BF of the type II unit is slightly below that of the type IV. The wideband facilitation of the wideband inhibitor is represented by the broad range of BFs of its inputs. At top center, the outlines of the excitatory areas are shown for the three sources of input to the type IV unit: (1) the excitatory input from auditory nerve fibers (*ANF*) or T-multipolar cells is shown unfilled; (2) the excitatory areas of two type II units are shown in black; and (3) the excitatory tuning curve of a wideband inhibitor is shown shaded. There is probably another inhibitory input with tuning similar to the wideband inhibitor, as is discussed in the next section. At right is a type IV response map with the inhibitory areas shaded, to show the responses contributed by inhibitory inputs in the model.

II BFs are shifted somewhat toward frequencies below the ANF BF. This is based on evidence that, in type II-type IV pairs with an inhibitory trough, the BF of the type II unit tends to be below the BF of the type IV unit (Voigt and Young 1990). The type II input is assumed to be strong enough to produce the type IV unit's CIA in the areas of the type II excitatory response. The type IV unit's low-level excitation near BF results from the high thresholds of the type II input; the small excitatory region at the high-frequency edge of the CIA is produced by the slight downward shift of the type II BFs. The type IV's excitatory area at high levels and low frequencies results from the high tail thresholds of the type II units, relative to ANFs (Young and Voigt 1982). The remaining element of the circuit is the wideband inhibitor. The tuning curves of onset-C units tend to have high thresholds and wide bandwidths, which is consistent with the wideband facilitation model for these cells (Winter and Palmer 1995; Jiang et al. 1996; Palmer et al. 1996). The gray tuning curve shows the hypothesized contribution of the wideband inhibitor to the type IV response map, which is mainly the inhibitory area at frequencies above BF. However, for tones, the wideband inhibitor's input is weak because these units are weakly activated by tones (Fig. 5.5B), and it is likely that another inhibitory source also contributes to the upper inhibitory sideband (see below; Davis and Young 2000).

The response map at right in Fig. 5.6 is repeated from Fig. 5.4A; the inhibitory areas are colored black and gray to show the portions of the map that are hypothesized in the model to derive from the two inhibitory sources. A similar picture could be drawn starting with the response map in Fig. 5.3A. The response map in Fig. 5.4B can be produced from the same elements by increasing the bandwidth and strength of the type II inhibitory inputs (Reed and Blum 1995).

Figure 5.7 shows how the model of Fig. 5.6 can be used to account for the responses to broadband stimuli shown in Fig. 5.4. The plots in Figs. 5.7B, C, and D are sketches of the rate versus sound level functions predicted by the model for narrowband stimuli (tones or narrow noise bands, Fig. 5.7B), notch noise (Fig. 5.7C) and broadband noise (Fig. 5.7D). Figs. 5.7E and 5.7F show actual rate-level functions from a type IV unit, for reference. In the model plots, the rate-level functions for the type IV unit and its excitatory input (ANF) are plotted in the positive direction and the rate-level functions of the inhibitory inputs are plotted in the negative direction. The assumption used in constructing these plots is that the inputs add to produce the type IV output after weighting by synaptic strengths. For narrowband stimuli (Fig. 5.7B), the excitatory input (ANF) has a monotonic rate-level function typical of both auditory nerve fibers and T-multipolar cells of VCN (Sachs and Abbas 1974; Shofner and Young 1985). The type II unit has a strong response with the characteristically nonmonotonic shape typical of these units (e.g., Fig. 5.5A). The wideband inhibitor (WBI) gives only a weak response, as illustrated in Fig. 5.5B and contributes little. Because the type II unit has a higher threshold than the excitatory input, the type IV unit is excited at low sound levels but then is inhibited when the type II unit begins to fire. The correspondence of the type II excitatory threshold and the type IV inhibitory threshold, meaning the sound level at which the type IV rate reaches a peak and begins to decline, has been demonstrated for type II-type IV pairs that show an inhibitory trough (Young and Voigt 1981) and also in the population of type II and type IV units (Davis and Young 2000). The type II inhibitory input is strong enough to inhibit the discharge of the type IV unit at high sound levels, resulting

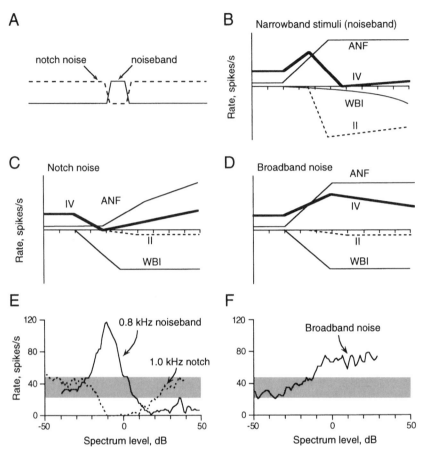

FIGURE 5.7. Qualitative explanation of the rate-level functions of type IV neurons and their derivation from inputs, using the model of Fig. 5.6. A: Spectra of a noise-band and a notch noise. In all cases, the noise band or the notch are centered on the type IV BF. B, C, and D: Sketches of rate-level functions for a type IV unit (IV, heavy line), its excitatory input (ANF), and two sources of inhibitory input (type II, dashed line, and the wideband inhibitor, WBI). The excitatory input is shown as positive and the inhibitory inputs are negative. The rate-level functions have approximately correct relative amplitudes and thresholds, according to data. The spontaneous rate of the type IV unit is assumed to be partly intrinsic and partly due to spontaneous activity in the excitatory inputs. The synapse strengths are fixed in B, C, and D and are constrained in the following ways: (1) the type II synapse must be strong enough to overcome the excitatory input in B and produce partial or complete inhibition of the type IV unit for narrowband stimuli; (2) the WBI synapse on the type IV must be weak enough to produce only partial inhibition of the type IV unit for broadband noise in D, but still strong enough to produce inhibition with the weakened excitatory input in C; and (3) the WBI synapse on the type II unit must be strong enough to produce near-zero rate in the type II for both broadband noise and notch noise. E and F show examples of noiseband, notch noise, and broadband noise responses from a type IV unit. (E and F redrawn from Nelken and Young 1994 with permission.)

in the CIA and the characteristic nonmonotonicity of type IV units for narrowband stimuli centered near BF (Figs. 5.4C, 5.4D, and 5.7E).

For broadband noise (Fig. 5.7D), the excitatory input behaves approximately the same as for tones. The wideband inhibitor gives a strong response in this case. Because of the inhibition from the wideband inhibitor, the type II unit gives a weak response and contributes little to the type IV response. The type IV response is the result of the excitatory input minus the inhibition from the wideband inhibitor. The strength of the inhibitory synapse from the wideband inhibitor is presumed to be relatively weak, so that it produces only partial inhibition of the type IV unit and gives the weak excitatory response typical of type IV units for broadband noise (Figs. 5.4C, 5.4D, and 5.7F).

Finally, with notch noise, the situation is similar to the case of broadband noise, in that the wideband inhibitor is strongly activated and the type II unit is not (Fig. 5.7C). Activation of the wideband inhibitor occurs because these neurons receive and integrate excitatory inputs across a relatively broad array of BFs; the broadband facilitation mentioned above. For notch widths that are narrower than the bandwidth of the wideband inhibitor's excitatory inputs, it will still receive enough input to be activated. By contrast, the excitatory input to the type IV unit is narrowband; the energy removed to make the notch reduces substantially the excitatory input to the ANF. The effect is a shift in threshold and a decrease in the slope of the ANF rate function (Schalk and Sachs 1980; Poon and Brugge 1993). The type II response is also weak because the wideband inhibitor is activated and because there is reduced excitatory input near BF. As a result, the wideband inhibitor is the strongest input to the type IV unit for notch noise, and the net response of the type IV unit is inhibitory (Figs. 5.4E, 5.4F, and 5.7E). Type IV units show a range of sensitivities to notch width (Nelken and Young 1994). The net effect of a particular notch width on a type IV unit presumably depends on the relative integrating bandwidths of the excitatory and wideband inhibitor inputs. As long as more energy is removed from the excitatory than from the inhibitory inputs, the type IV response will be inhibitory. The qualitative model shown in Figs. 5.6 and 5.7 has been quantified to show that it can successfully account for most properties of type IV neurons in DCN (Blum et al. 1995; Hancock et al. 1997; Blum and Reed 1998).

According to the model, the nonlinearity of DCN principal cell responses results from the fact that the DCN circuit switches from being dominated by type II units for narrowband stimuli to a cirouit that is dominated by the wideband inhibitor for broadband stimuli. Because neither inhibitor is spontaneously active, the switch behaves like a rectifier, producing nonlinear responses. For example, the lack of spontaneous activity in type II units means that a broadband stimulus produces no effect through the type II circuit; if type II units were spontaneously active, the inhibition of type II units by broadband stimuli through the wideband inhibitor would appear as disinhibition in type IV neurons. The result would be a smooth transition from inhibitory to disinhibitory effects in type IV units as the bandwidth widens instead of a sudden disappearance of the type II unit at zero rate. In fact, many of the nonlinearities in DCN principal cell responses occur at the point where type II units reach threshold (Nelken and Young 1997).

6. Inhibitory and Excitatory Inputs to Principal Cells from the Superficial DCN and the Effects of Somatosensory Inputs

The model presented in the previous sections does not include the effects of the neuropil in the superficial DCN. As described in Fig. 5.2, DCN pyramidal cells receive excitatory inputs from the parallel fibers of the molecular layer and inhibitory inputs from interneurons located there. Giant cells in deep DCN do not have extensive dendritic trees in the molecular layer; however, on physiological grounds, they receive inputs from at least a part of the circuitry of the superficial layer (Davis et al. 1996b; Golding and Oertel 1997). Figure 5.8 shows typical effects of electrical stimulation of the somatosensory inputs to the DCN from the MSN; data are shown for type IV (Figs. 5.8A,B), type II (Fig. 5.8C), and complex-spiking units (Fig. 5.8D). For each plot, the top trace shows the extracellular evoked potential (EP) at the DCN recording site. The stimulus was a sequence of four shocks delivered to the MSN at the times of the arrows at the top of the figure. The evoked potentials serve as a marker of the synaptic currents produced in the molecular layer by activation of the parallel fibers (Young et al. 1995). The bottom traces are PST histograms of the responses to the stimulation for DCN units that were otherwise firing spontaneously, except for the type II unit in Fig. 5.8C, which was activated by a BF tone 10 dB above threshold. The response of the type IV unit in Fig. 5.8A shows three components: (1) a short-latency inhibitory component that precedes the onset of the EP (marked by the dashed lines); (2) a transient excitatory component (bold) just after the onset of the EP; and (3) a long-latency inhibitory component which follows the excitatory component. Approximately half of all type IV units show this response pattern to MSN stimulation while most of the remainder show only the third long-latency inhibitory component, as for the unit in Fig. 5.8B. Note that the long-latency inhibitory component begins after the onset of the evoked potential in Fig. 5.8B. Although the source of the short-latency inhibition is unknown, the excitatory component likely reflects direct excitation of principal cells by parallel fibers (Manis 1989; Waller et al. 1996), and the long-latency inhibition is probably produced by input from cartwheel cells (see below; Davis and Young 1997). Type II inhibitory inputs to DCN principal cells are themselves

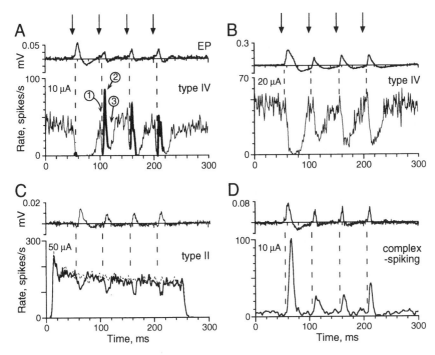

FIGURE 5.8. Responses of two DCN type IV units (A and B), a type II unit (C), and a complex-spiking unit (D) to electrical stimulation in the MSN. Arrows at the top show the times of the 4 electrical pulses, which were spaced 50 ms apart and were presented once per second; the current level is given above the histograms. In C, a 250 ms tone at BF, 10 dB above threshold, was presented to produce background activity (dashed line); the solid line shows the responses to the electrical stimulus in the presence of the tone. For each panel, the top plot shows the evoked potential (EP) at the recording site and the bottom plot shows a PST histogram of the responses. Vertical dashed lines are aligned with the onsets of the EPs at the DCN recording site. The three components of the type IV response in A are numbered to correspond to the discussion in the text. The type IV unit in B shows only the third long-latency component. The type II unit in C shows only weak long-latency inhibition. The complex-spiking unit in D shows an excitatory response that corresponds in latency and adaptation properties to the long latency components in A and B. Histograms were constructed from 400 repetitions of the stimulus using a binwidth of 1 ms. In all cases, the electrical stimulus was applied at a site in the MSN somatotopic map where the pinna was represented. (A and B redrawn from Davis et al. 1996b and D redrawn from Davis and Young 2000 with permission.)

weakly inhibited by MSN stimulation (Fig. 5.8C; Young et al. 1995); onset C neurons are weakly excited by MSN stimulation, but only in the presence of an acoustic stimulus (K.A. Davis unpublished). Thus the long latency inhibition of type IV units cannot be produced by excitation of either type II or onset-C inhibitory circuits.

Evidence that cartwheel cells are the source of the long-latency inhibitory component is shown in Fig. 5.8D. In contrast to type IV units, complex-spiking units are strongly excited by MSN stimulation. This excitatory response coincides in time with the onset of the EP (dashed lines) and shows a characteristic adaptation pattern: strongest at the first pulse, weakest at the second pulse, and then increasing in strength at subsequent pulses. Comparison of Figs. 5.8B and 5.8D shows that the latency and four-pulse amplitudes of the complex-spiking unit responses correspond exactly to the long-latency inhibitory responses of type IV units, suggesting that cartwheel cells are the source of this component. These comparisons are made in more detail elsewhere (Davis and Young 1997).

Somatosensory inputs to DCN can be activated by stimulating many parts of the body (Saadé et al. 1989; Young et al. 1995), but by far the strongest effects are from the pinna. The EPs in DCN are largest when electrical stimulation is applied to the pinna representation of the somatosensory map in the MSN (Young et al. 1995) and are also largest when spinal nerve C2 is stimulated (Kanold and Young 2001). C2 is the nerve that carries sensory fibers from both the skin and muscles associated with the pinna (Hekmatpanah 1961; Abrahams et al. 1984a,b). The modality of the somatosensory effects in DCN is clearly muscle sense, not skin sense (Kanold and Young 2001). That is, touching the skin or hairs on the pinna and surrounding skin rarely causes any effect in DCN. Instead, the effective stimuli are actions that stretch the muscles that move the pinna. Both lateral extension of the pinna and vibration applied to the pinna muscles are strong stimuli; both actions should primarily stimulate muscle stretch receptors and tendon organs. Thus the somatosensory inputs to the DCN seem to convey information about the contractile state of the muscles that move the pinna, and, therefore, about the orientation of the pinna. The circuitry in the superficial DCN conveys this information to the DCN principal cells where it is integrated with auditory information from the circuitry in the deeper layers.

7. Evidence for DCN Circuits from Pharmacological Manipulations

The three inhibitory circuits described above, (i.e., the vertical, radiate, and cartwheel cells) are all glycinergic (Oertel and Wickesberg 1993; Golding and Oertel 1997; Doucet et al. 1999). A test of the model described in the previous section is to apply pharmacological antagonists to glycine and γ-aminobutyric acid (GABA), the other common inhibitory neurotransmitter, and examine whether responses change in the expected way. Figure 5.9 shows the effects on DCN responses of iontophoretic application of strychnine, a glycinergic antagonist, and bicuculline, a GABAergic antagonist. The antagonists were applied to the recording site of the neurons from a pipette

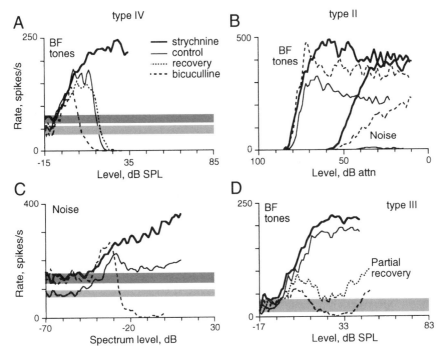

FIGURE 5.9. Comparison of the effects of strychnine and bicuculline on the BF-tone and noise rate-level functions of DCN type IV, type II, and type III units. The line weight identifies the conditions, as described in the legend. Thin solid lines show responses before application of any agent, and thin dotted lines show recovery data. The recovery curves in A and D are after bicuculline but before strychnine application. The shaded bars show the range of spontaneous discharge rates before (light shading) and during (heavy shading) application of a drug (strychnine in A; strychnine and bicuculline in C). The stimulus (BF-tone or noise) is given at left in each panel, except in B where stimulus type is identified next to the curves. Note that strychnine abolished the inhibitory responses of the type IV unit at BF (A) or in response to noise (C) whereas bicuculline enhanced inhibitory responses (i.e., units were inhibited at lower sound levels and more strongly). In contrast, both drugs increased the discharge rate of the type II unit (B) and lowered the excitatory threshold slightly. Bicuculline converted the type III unit into a type IV unit (D). (A, C, and D redrawn from Davis and Young 2000 and B redrawn from Spirou et al. 1999 with permission.)

glued to the recording electrode (Davis and Young 2000). This figure shows that strychnine has the effects expected from the model (see also Caspary et al. 1987; Evans and Zhao 1993). Figure 5.9A shows BF rate-level functions for a type IV unit before, during, and after iontophoretic application of strychnine and bicuculline. Under control conditions (thin solid line), the type IV unit had a nonmonotonic rate-level function typical of type IV units. In the presence of strychnine (heavy line), the unit lost entirely its inhibitory responses to high-level tones, with the result that the function became monotonic; there was also an increase in the spontaneous discharge rate (dark gray band). Figure 5.9C shows that, similarly, strychnine releases type IV unit responses to broadband noise. Here, the unit showed a weak, excitatory, nonmonotonic rate-level function to noise in control conditions (thin solid line) and a strong, monotonic response during strychnine application (heavy line); again, there was an increase in spontaneous activity.

The strychnine data can be interpreted as follows: Fig. 5.9A shows release of inhibition produced primarily by vertical cells, which are the most active inhibitory interneuron during presentation of tones; Fig. 5.9C shows release of inhibition produced primarily by the wideband inhibitor, which is most active during presentation of noise. Increases in spontaneous rate were observed in about half the cases (Davis and Young 2000). Because neither vertical cells nor wideband inhibitors are spontaneously active, changes in spontaneous rate must reflect inhibition from other, as yet unidentified, sources. Not shown in Fig. 5.9 is that strychnine also blocks both the short- and long-latency inhibition produced by MSN stimulation in type IV units (Davis and Young 2000). In the case of the long-latency inhibition, the strychnine effect presumably acts by blocking inhibition from cartwheel cells.

Although GABAergic neurons are not part of the model described above, there are many GABAergic cells in the CN; especially in the superficial DCN (Mugnaini 1985; Roberts and Ribak 1987; Osen et al. 1990). Golgi cells in the granule cell areas, including those in the DCN, and small cells scattered throughout the CN are GABAergic. Cartwheel cells colocalize GABA and glycine although they seem to be functionally glycinergic (Golding and Oertel 1997; Davis and Young 2000). In addition, there are GABAergic efferent projections to the CN from other parts of the auditory system (Ostapoff et al. 1990, 1997). Although the responses to sound of these GABAergic systems are not known, some properties of a GABAergic inhibitory system can be inferred from the effects of GABA antagonists on DCN responses. Figure 5.9 shows that application of the GABA-A antagonist bicuculline has unexpected effects on DCN principal cells. The heavy dashed line in Fig. 5.9A shows that bicuculline did not abolish the inhibition produced by BF tones; instead, the inhibition was stronger in the presence of bicuculline, in that the inhibitory threshold was lower. A similar result is shown for noise-driven responses in Fig. 5.9C. The noise rate-level function became strongly nonmonotonic with bicuculline, demonstrating stronger inhibition (heavy dashed line) in the presence of this antagonist.

The effects of both strychnine and bicuculline on type II units are what one expects for inhibitory antagonists. Figure 5.9B shows BF-tone and noise rate-level functions for a type II unit under control conditions and during iontophoretic application of strychnine and bicuculline. Under control conditions (thin solid lines), the type II unit had no spontaneous activity, gave excitatory responses to BF tones at all sound levels, and responded with a rate near zero to broadband noise. While neither drug alone (or in combination) endowed the unit with spontaneous activity, both strychnine (heavy solid line) and bicuculline (heavy dashed line) produced substantial increases in response to both tones and noise. Both drugs also reduced the threshold slightly. The effects of strychnine on type II units are consistent with the hypothesized strong inhibitory input to these neurons from the wideband inhibitor. The effects of bicuculline imply that there is an additional strong GABAergic input to type II neurons. The source of the GABAergic input is not known.

The most dramatic bicuculline effects occur in type III units, illustrated in Fig. 5.9D. In control conditions, type III units show monotonic or near-monotonic rate-level functions to BF tones (thin solid line). Bicuculline converted this type III rate function into a nonmonotonic curve resembling a type IV unit (heavy dashed line). While the unit was still recovering from the bicuculline (dotted line marked partial recovery), strychnine was applied, resulting in abolition of the inhibition and a monotonic rate-level curve (heavy solid line). This result suggests that the bicuculline-enhanced inhibition in type III units is mediated by a glycinergic interneuron. Given that the stimulus was a BF tone, the glycinergic interneuron was most likely a type II unit.

The effects of bicuculline on type III and IV units in Fig. 5.9 can be explained by assuming that the iontophoresed drug is spreading from the recording site of the principal cell to nearby type II neurons. Type II neurons often can be recorded on the same microelectrode with type III and type IV neurons (Voigt and Young 1980), which implies that their cell bodies are close together. Moreover, the strongest inhibitory cross-correlation is observed for type II/type IV pairs that are close together (Voigt and Young 1990). Thus it is reasonable to suppose that the substantial release from inhibition observed in type II units when bicuculline is applied (Fig. 5.9B) also occurs when bicuculline is iontophoresed near a type III or type IV unit. The changes in type III and IV response seen in Fig. 5.9 are then explained as a secondary consequence of release from inhibition in type II neurons, with the additional assumption that there are few GABAergic synapses on principal cells, or that those synapses are weak. Consistent with this suggestion, the bicuculline-induced changes of the tone thresholds in type II units are in the same direction as and similar in amplitude to the changes in inhibitory thresholds in type IV units (−4 dB versus −3 dB; Davis and Young 2000).

Bicuculline has little or no effect on the inhibition of type IV units by electrical stimulation of the MSN, (either short- or long-latency; Davis and Young 2000), supporting the conclusion that the cartwheel-cell inputs to principal cells are glycinergic and suggesting that the unknown short-latency inhibitor is also glycinergic. However, the responses of cartwheel cells to MSN stimulation are increased following bicuculline application at their recording sites. Thus the cartwheel cells must receive direct or indirect GABAergic inhibition activated by the mossy fiber inputs to DCN. A likely substrate for this inhibition is the Golgi cell, which could inhibit both granule cells and cartwheel cells.

The discussion above is based on the effects of antagonists at BF only; off-BF responses show new features which demonstrate the presence of an additional inhibitory input to type IV neurons. Figure 5.10 shows the effects of strychnine and bicuculline on the response maps of two type IV units. Neither drug alone can eliminate all the inhibition present in type IV response maps. The top half of Fig. 5.10 shows a type IV response map in control conditions (Fig. 5.10A) and during strychnine application (Fig. 5.10B). The vertical line in Fig. 5.10A is aligned with the excitatory edge of the CIA and separates this unit's inhibitory response into its CIA and an upper inhibitory sideband. In the presence of strychnine (Fig. 5.10B), the CIA is eliminated and replaced with excitation, consistent with the model and previous results (Fig. 9A). The inhibition above BF, however, is reduced but not eliminated by strychnine. The residual inhibition in these response maps is likely to be GABAergic in origin, as is illustrated by the second example in Fig. 5.10.

The bottom half of Fig. 5.10 compares the response map of another type IV unit in control conditions (Fig. 5.10C) and with bicuculline (Fig. 5.10D). The bicuculline enhanced the CIA, which is most clearly seen at 70 dB where the near-BF excitatory response was abolished. However, the opposite effect is observed for frequencies away from BF; the upper inhibitory sideband was converted to excitation and a weak excitatory response below BF was enhanced by the bicuculline. These results suggest that there is an additional inhibitory input to principal cells which is GABAergic. It is apparently weaker than the type II glycinergic inhibition that produces the CIA, so it is only clearly seen at frequencies outside the CIA. The upper inhibitory sideband in type IV units is apparently due to two effects: this GABAergic inhibitor and the glycinergic wideband inhibitor as in Fig. 5.6. The GABAergic effect appears to be stronger for tones away from BF (compare Fig. 5.10D with Fig. 5.10B), but the glycinergic effect dominates in broadband noise (Fig. 5.9C).

8. Circuit Model of the DCN

The results of this and the previous section imply that the circuit model of Fig. 5.6 is incomplete with regard to the synaptic inputs of DCN neurons. At least two additional elements must be added: the cartwheel cells of the superficial DCN and at least one GABAergic source, of unknown origin. Figure 5.11 shows a modified circuit model that includes these two elements as well as a granule cell of the superficial layer. The cartwheel cells synapse

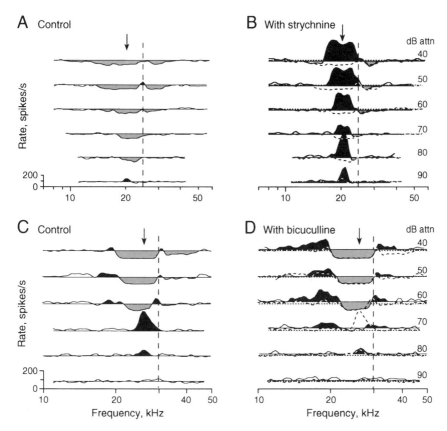

FIGURE 5.10. Frequency response maps of DCN type IV units before (left) and during (right) iontophoretic application of strychnine (top) or bicuculline (bottom). The control response maps are repeated in the right column with dashed lines to allow comparison. The vertical lines separate the CIA to the left of the line from the upper inhibitory sideband to the right. The arrows at top point to BF. A and B show the effects of strychnine on one unit. In B, strychnine abolished most of the CIA and reduced, but did not eliminate, the upper inhibitory sideband. C and D show the effects of bicuculline on a different neuron. Note the enhancement of the CIA near BF (e.g., the response at 70 dB attenuation), but the weakening of inhibition or enhancement of excitation away from BF (at levels from 40–70 dB attenuation). (Redrawn from Davis and Young 2000 with permission.)

on type IV neurons and weakly or not at all on the type II and WBI interneurons (e.g., those neurons give little or no inhibitory response to MSN stimulation; Fig. 5.8C and Young et al. 1995). The GABAergic source must be activated by both somatosensory and auditory inputs in order to explain the results summarized in the previous section and must terminate on type II, type IV, and cartwheel cells. The effects of the GABAergic source are strongest on the cartwheel and type II neurons, and are relatively weak on principal neurons. It is not clear that there is only one GABAergic source, but only one is shown in Fig. 5.11 for parsimony. Evidence that there is more than one is provided by the fact that bicuculline iontophoresis at type IV recording sites affects these cells' responses to off-BF tones but not to MSN stimulation. This result seems to require two GABAergic sources: one more strongly driven by sound and one more strongly driven by the granule cell circuits.

One apparent role of the GABAergic neuron is to regulate the strength of type II inhibition in DCN. Type III and type IV neurons can be understood as forming a continuum of principal cell responses in which type IV neurons are seen when the GABAergic inhibition of type II neurons is weak and type III neurons are seen with strong GABAergic inhibition. This hypothesis is consistent with the effects of pentobarbital anesthesia on the DCN. In anesthetized animals, type IV units are rare, and, in decerebrate preparations, type IV units are converted to type III when pentobarbital is given intravenously (Evans and Nelson 1973; Young and Brownell 1976). Because one effect of pentobarbital is to potentiate GABAergic inhibition (e.g. Richter and Holtman 1982), type II units would be more strongly inhibited by their GABAergic inputs in anesthetized animals, which should bias the principal cells toward type III. Another possibility is that the level of GABAergic tone in an animal could account for the difference in prevalence of type III versus type IV neurons between species. In decerebrate cat, type IV units predominate (57% of principal cells; Shofner and Young

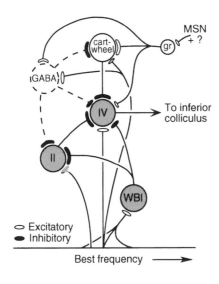

FIGURE 5.11. Circuit model of the DCN that includes all the effects described above. Shaded neurons are the primarily auditory neurons described in Fig. 5.6. The granule cell and cartwheel cell of the superficial layer and a GABAergic neuron of unknown identity have been added. The relative size of synaptic terminals in this figure corresponds roughly to the relative synaptic strength needed to account for the effects described above. (Redrawn from Davis and Young 2000 with permission.)

1985), whereas in decerebrate gerbil, type III units predominate (85%, Davis et al. 1996a).

9. Responses of DCN Neurons to Acoustic Cues for Sound Localization

The evidence discussed at the beginning of this chapter suggests that the DCN is involved in integrating information about sound localization. Both behavioral results (Sutherland et al. 1998b; May 2000) and the fact that the DCN receives somatosensory inputs that carry information about the position of the pinna (Kanold and Young 1996) support this idea. Given this hypothesis, it is interesting to ask how DCN principal cells respond to the acoustic cues that convey information about sound location. Sound localization cues are both binaural and monaural. The DCN is primarily a monaural nucleus, although it does receive inputs from the contralateral ear (Mast 1970; Young and Brownell 1976). However, the contralateral inputs are weak and do not produce sensitive responses to either interaural time differences or interaural level differences. Certainly the binaural responses in DCN are insensitive in comparison to the binaural responses of cells in the superior olive and inferior colliculus. In contrast, DCN type IV units are particularly sensitive to monaural cues, that is, to spectral sound localization cues of the kind produced by the cat external ear (Young et al. 1992, 1997). This sensitivity has been shown to give DCN principal cells a sensitivity to the direction of a sound source (Imig et al. 2000).

Figure 5.12 shows an example of the responses of a type IV unit to spectral cues. The stimulus used in this experiment was broadband noise filtered by the HRTF (Fig. 5.1) from one direction in space; thus, the power spectrum of the noise was equal to the magnitude spectrum of a HRTF filter. This situation simulates the presentation of a broadband noise in free field from a speaker located in the direction of the HRTF. The strongest response produced by DCN type IV units to HRTF-filtered stimuli is an inhibitory response when a spectral notch is centered on BF. Fig. 5.12B shows the response map of a DCN type IV unit and Fig. 5.12A shows part of the power spectrum of a HRTF-filtered noise on the same frequency scale. The power spectrum is repeated three times (a, c, and e). These three spectra are from one stimulus that was presented at three different sampling rates of the D/A converter used to generate the stimuli; this manipulation has the effect of shifting the spectrum to the left or right along the (logarithmic) frequency scale. The unit's sensitivity to particular features in the stimulus spectrum can be investigated by studying the responses with those features located at various frequencies relative to BF. Figures 5.12C and 5.12D show discharge rate versus sound level for the HRTF stimulus with its sampling rate adjusted to place the notch at five different frequencies, labeled a through e in Fig. 5.12B. When the notch was centered on BF

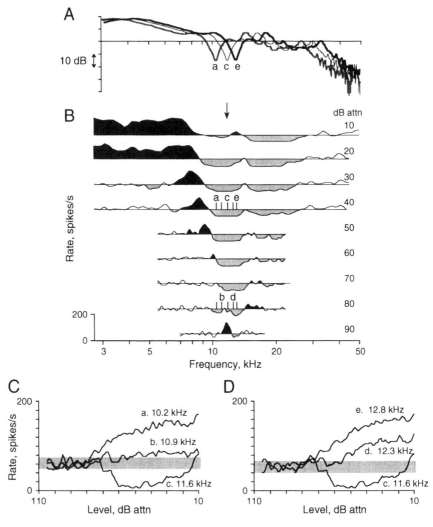

FIGURE 5.12. A: Power spectra of broadband noise filtered with a HRTF spectrum and presented as a stimulus to the type IV unit shown below. The sampling rate of the D/A converter used to generate the stimuli was varied in order to shift the stimulus spectrum along the frequency axis. The positions of the center of the prominent spectral notch for five sampling rates are labeled a through e in this figure. Three of the spectra are plotted in A. B: Response map of a type IV unit. The positions of the centers of the notches are marked with vertical ticks on the 40 dB and 80 dB lines of the map. The notch was at BF for stimulus c. C and D: Plots of discharge rate versus sound level for the HRTF-filtered stimuli at the five sampling rates; rates are for responses to 200 ms bursts of noise presented once per second. The labels on each curve identify the notch center frequencies with the actual frequencies and the labels a through e. Notice the strong inhibitory response when the notch is at BF (c) which is replaced by an excitatory response as the notch moves away from BF in either direction. (Redrawn from Young et al. 1997 with permission.)

(c), the response was strongly inhibitory across almost the entire range of sound levels. As the notch was moved away from BF, the inhibitory response disappeared and was replaced by an excitatory response reminiscent of this unit's response to broadband noise. At positions a and e, the notch center was only −0.19 and 0.14 octaves away from BF; nevertheless, the response to the notch had disappeared, showing that this unit had a very sensitive response to the position of a notch relative to BF. As discussed above, this response cannot be understood from the response map, but is accounted for by the circuit model of Figs. 5.6 and 5.7.

The inhibitory response to spectral notches shown in Figs. 5.12C and 5.12D is a property only of DCN principal cells. Neurons in VCN show no sign of the inhibitory response, and even DCN cells vary considerably in their sensitivity to the notch. This behavior is illustrated by the data in Fig. 5.13. Figures 5.13A through C show plots of rate versus sound level for HRTF-filtered stimuli. In each case the sampling rate was adjusted to place a notch at BF (labeled B), just below BF (U), or just above BF (L). When the notch was below or above BF, the sampling rate was chosen to place the BF just at the shoulder of the notch as in cases a and e in Fig. 5.12 (U means the BF is at the upper shoulder of the notch and L refers to the lower shoulder). In most units, when the notch is below or above BF, the response resembles the unit's response to broadband noise (as in Fig. 5.12). It is the response to the notch at BF that varies most among units. The case in Fig. 5.13A is like the case in Fig. 5.12 in that the response is strongly inhibitory at BF. This inhibitory response is seen only in type III and type IV units in DCN. The unit in Fig. 5.13C shows responses typical of VCN neurons and auditory nerve fibers (Poon and Brugge 1993) and some DCN neurons. There is only a shift in the rate function to higher sound levels when the notch is at BF. In the VCN and the auditory nerve, the shift in the rate function is presumably due to the reduction in stimulus power within the unit's tuning curve when the notch is located at BF. In addition, there is sometimes a reduction in the saturation rate with the notch at BF; this reduction reflects inhibition in the VCN or cochlear two-tone suppression. The case in Fig. 5.13B is intermediate between the other two examples; this type IV unit shows a reduction of rate to near zero when the notch is located at BF. From the model of type IV units and the explanation in Fig. 5.7C, it seems likely that the differences among the type IV units in Figs. 5.13A and 5.13B can be explained by differences in the excitatory and inhibitory integrating bandwidths of the neurons' inputs, as discussed previously, or by differences in the synaptic strength of the wideband inhibitor.

Figure 5.13D shows a summary of the notch responses of DCN and VCN neurons. The ordinate of this plot is an index of notch sensitivity equal to the ratio of the saturation rate with the notch at BF (rate minus spontaneous rate), divided by the average saturation rate with the notch above and below BF (also rate minus spontaneous rate). Units like Fig. 5.13A have negative notch sensitivity values, units like Fig. 5.13B have notch sen-

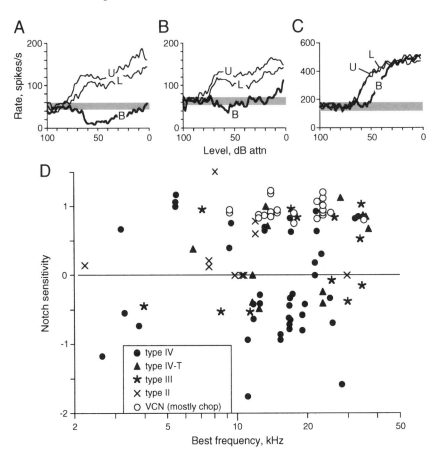

FIGURE 5.13. Sensitivity of cochlean nucleus neurons to spectral notches. A, B, C: Discharge rate versus sound level for HRTF-filtered noise stimuli with spectra like those in Figs. 5.1A, 5.1B, and 5.12A. The HRTFs used have been shown elsewhere (Fig. 5.1 in Rice et al. 1995). The sampling rate of the D/A converter used to produce the stimulus was adjusted to place a notch at BF (B) or so that the BF was at the upper (U) or lower (L) shoulder of the notch, as illustrated by positions c, a, and e respectively in Fig. 5.12B. Sound level is given as attenuation, but the spectrum level of the stimulus was the same for the three curves in each plot. Rates are computed from 200 ms stimuli presented once per second; shaded bars show spontaneous discharge rates. Units in A and B are DCN type IV, unit in C is from VCN. D: Scatter plot of notch sensitivity versus BF for a population of DCN and VCN neurons. Notch sensitivity is the ratio of the saturation rate (re spontaneous rate), of responses with the notch at BF (B in parts A–C) to the average saturation rate (re spontaneous), of the responses with the notch above or below BF (L and U). The saturation rates were taken as the maximum or minimum rate near the first inflection point in the rate plot (e.g., at −60 dB, −68 dB, and −56 dB for B, U, and L respectively in A). The notch sensitivity is −0.6 for the data in A, −0.3 for the data in B, and +1 for the data in C. Neuron types are identified by symbols, defined in the legend. VCN neurons were mostly choppers, with a few primarylike units included.

sitivities near 0, and units like Fig. 5.13C have notch sensitivities near +1. Thus the notch sensitivity is an index of the presence of the inhibitory response with the notch at BF. In Fig. 5.13D, notch sensitivity is plotted versus the units' BFs. Symbols identify the different unit types, as defined in the legend. Units from VCN (unfilled circles) all show notch sensitivities near +1 (median of 0.92). These units are not inhibited or only weakly inhibited by the notch at BF. DCN type II units (Xs) have notch sensitivities which range between +1 and 0 (median of 0.13); of course, these units cannot have negative notch sensitivities because their spontaneous rates are zero. DCN principal cells (types III, IV-T, and IV, filled symbols) have notch sensitivities between +1 and −1 (medians of 0.20, 0.53, and −0.33 respectively). The most sensitive notch responders, with sensitivities near −1, are all type IV units. The distribution of notch sensitivities in DCN principal cell types is actually bimodal (not shown), with one mode near 0.75 and another near −0.5. The DCN neurons with notch sensitivities near +1 are probably neurons whose notch responses are inhibitory only for wider notches. There is a range in behaviors in type IV units in terms of the notch width at which the inhibitory response appears (Spirou and Young 1991; Nelken and Young 1994); the range is not determined by BF since neurons with the same or similar BFs can have substantially different responses at the same notch width. This diversity is illustrated by the two examples in Fig. 5.4.

An important point about Fig. 5.13D is that there is no trend with BF toward good or poor notch sensitivity. Although there are few low BF (<10 kHz) neurons in the sample, notch sensitivity appears to be as prevalent in low BF neurons as it is in high BF neurons. This result is important because it suggests that the notch sensitivity is a part of a general DCN response property and not something developed specifically for processing notches produced by the pinna, which are seen in cat at frequencies above about 8 kHz only (Musicant et al. 1990; Rice et al. 1992).

10. Targets of DCN Axons in the Inferior Colliculus

Axons of the principal cells leave the DCN through the DAS and project to the contralateral CNIC (Osen 1972; Oliver 1984; Ryugo and Willard 1985). There they terminate in broad swaths that extend through the entire isofrequency sheet. In the decerebrate cat, neurons in the CNIC display four main response types (Ramachandran et al. 1999). One of these, called type O, appears to be associated particularly with the DCN. Type O units have tone response maps that are essentially identical to DCN type IV response maps (e.g., Fig. 5.14A). When inhibitory antagonists (strychnine or bicuculline) are iontophoresed at a type O recording site, the response map is unchanged for most units (Davis 1999). This result suggests that the inhibitory areas of these type O units are not created by direct inhibition within the CNIC. Evidence that these response maps are, in fact, inherited from DCN type IV units is provided by the results of blocking the contralateral DAS with lidocaine (K.A. Davis, unpublished). Blocking the output of the DCN in this way abolishes all activity in most CNIC type O

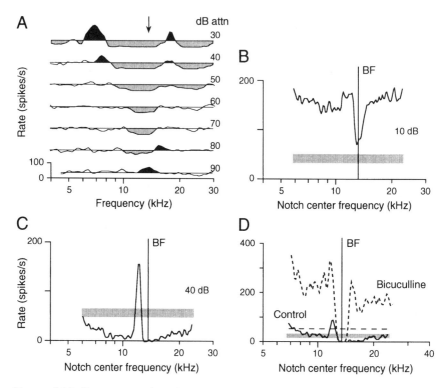

FIGURE 5.14. Some properties of a type O unit in the inferior colliculus. Neurons of this type seem to receive their primary input from DCN type IV units. A: Response map of the neuron. The arrow points to the BF. Note the features that are similar to those of type IV response maps, including the central inhibitory area, excitatory areas at threshold and at the upper frequency edge of the CIA, and the upper inhibitory sideband. B and C: Responses of the same unit to notch noise; broadband noise was filtered to have a 3.2 kHz stopband centered at various frequencies; the center frequency of the notch is given on the abscissa. The shaded bar shows the spontaneous rate and the vertical line shows BF. Data in B were obtained at about 10 dB re threshold, and data in C were obtained at about 40 dB re threshold. D: Effects of the GABA antagonist bicuculline on the responses to notches of a different type O unit. Rate versus center frequency for notch noise in control conditions (solid line) and during bicuculline iontophoresis (dashed line). The shaded bar is the spontaneous rate under control conditions and the horizontal dashed line is the spontaneous rate during bicuculline.

units. It appears that about 70–80% of CNIC type O units inherit their tone responses from the DCN. In the remainder, iontophoresis of inhibitory antagonists abolishes most of the inhibition in the response map, showing that some type O units are created at the level of the CNIC. The two subtypes of type O units can be distinguished by their rate responses; the discussion of this section applies only to the low-rate unit type that is associated with the DCN.

When broadband stimuli are presented, type O units differ from DCN type IV units in several ways. Most interesting are the effects of noise notches, shown in Fig. 5.14B and 5.14C. These figures show responses to broadband noise that has been filtered to have a notch (stopband) of 3.2 kHz width centered at a range of frequencies, shown on the abscissa. The data in Fig. 5.14B are at a low sound level (10 dB re threshold). The type O unit gives a predominantly excitatory response to the broadband noise, with a dip in rate when the notch is centered on the unit's BF. This response is like that seen in type IV units (e.g., the example in Fig. 5.13B), except that the type O unit does not show an inhibitory response with the notch centered on BF. Thus the type O unit appears to receive an excitatory response to broadband stimuli, which shifts up the discharge rate from the type IV input without eliminating the notch response. At higher sound levels, the response changes to that seen in Fig. 5.14C. At these levels, broadband noise inhibits type O units, including when the notch is centered at BF (vertical line). However, when the notch is just below BF, there is a strong excitatory response that is very specific to the location of the notch. The shift of noise responses from excitatory to inhibitory as noise level is raised is a characteristic of type O units. These units produce nonmonotonic rate functions to noise that are similar to the BF-tone rate functions of both type O units and DCN type IV units but are more strongly nonmonotonic than type IV noise responses.

The relationship of the excitatory response with the notch just below BF to the response of the DCN type IV input is shown in Fig. 5.14D. The solid line is the response of a CNIC type O neuron to a notch, plotted against the center frequency of the notch. The same inhibitory/excitatory effect is seen, with the excitatory response occurring when the notch is just below BF. When bicuculline was iontophoresed, the same stimulus set produced the response rate shown by the dashed line. There was a large release from inhibition, and the notch response became like that expected of a DCN type IV neuron; that is, there was a large excitatory response when the notch was away from BF which became inhibitory when the notch was centered on BF. Notice that the inhibitory response to the notch near BF forms the upper frequency edge of the excitatory response to the notch in the absence of bicuculline. Apparently there are strong inhibitory inputs to type O units which change the encoding of notches from an inhibitory dip at BF to an excitatory peak just below BF.

The data on the CNIC suggest two conclusions about the neural processing of sounds in the DCN. First, type IV responses clearly serve a role in the auditory system distinct from that of the more standard auditory response types in the VCN. Not only are type IV-like responses preserved in the type O population of the CNIC, but new units of this type are created by inhibitory inputs within the CNIC. Thus, neurons with strong inhibitory inputs and specific non-linear responses to certain spectral shapes apparently provide a useful stimulus representation to the auditory system. This representation is not just a useful intermediate that is generated at one level and then combined into another response type at a higher level. Instead, type IV responses are themselves preserved as a parallel system at the higher level. The second conclusion that can be drawn from data on the CNIC is that, the representation of spectral notches, which is a special feature of DCN type IV units, is reencoded in the CNIC in a different, but no less specific, format. This reencoding depends on both excitatory and inhibitory processes, some of which are quite strong (e.g., the data in Fig. 5.14D). The means by which the outputs of the DCN are reencoded in the CNIC remains as an essential piece of the DCN puzzle and is necessary to the ultimate interpretation of signal representation in DCN.

11. Summary: What Is the Role of the DCN?

The notch responses of DCN type IV neurons demonstrate a specific sensitivity to sharp spectral features. These neurons are inhibited by spectral notches with exquisite sensitivity to the location of the notch with respect to BF. They are also inhibited by the opposite spectral feature, which is a peak in the spectrum such as might be produced by a formant frequency in a speech sound or a resonant energy peak in the sound produced by some vibrating object. Thus, DCN principal cells show a particular response (inhibition) at BFs where sharp spectral features, peaks, or notches, are located in the stimulus spectrum. Sharp features of this kind often convey useful information about the environment, including the location of sound sources or the identity or quality of the sound source. Thus, it has been suggested that the DCN serves the role of identifying for the rest of the auditory system the frequencies at which interesting acoustic information is located (Nelken and Young 1996).

DCN principal cells are also predominantly inhibited by somatosensory stimuli. The most effective somatosensory stimulus seems to be stretching the muscles or tendons connected to the pinna as would happen when the pinna is moved away from its rest state by muscle contraction (Kanold and Young 2001). At present, it is hard to relate the somatosensory input to a useful computational mode for the animal because the correspondence between the nature of the pinna muscle activation and the effect in DCN is not known. It is not yet clear, for example, what pinna movements would be effective in producing DCN inhibition, or whether there is a mapping of specific patterns of muscle activity or pinna movement onto the array of parallel fibers. However, it is clear that movement and the position of the pinna are important sensory variables for a cat. As the cat's external ear is moved, its directional gain characteristics change dramatically (Young et al. 1996). The mapping of notch frequency into position in space, for example, moves approximately with the pinna. Thus, the map in Fig. 5.1C applies only for the pinna in its relaxed state. In addition, movement of the pinna causes the directional transfer functions of the pinna to sweep across a stationary sound source, producing substantial time varying changes in the spectrum of the sound reaching the eardrum. Poon and Brugge (1993) showed, in an approximate simulation of this situation, that auditory nerve fibers respond strongly to the resulting spectral changes. DCN neurons would be expected to be even more sensitive to situations like transient notches in the spectrum of a sound source. In addition, a cat can substantially change its listening conditions for a particular sound source by changing the orientation of its pinnae (Young et al. 1996). The ear can be optimized for location or movement sensitivity by placing notches in the HRTF near the source, or it can be optimized for identification of the source by moving the notches away (thereby eliminating the spectral distortions of the HRTF). Proper use of the information provided by the external ear requires coordination of the auditory processing of signals from the ear with the position of the pinna. The convergence of auditory and somatosensory information in the DCN thus could represent a form of sensory-motor coordination for optimizing auditory processing. This hypothesis is similar to the hypothesized role of the cerebellum for sensory-motor coordination (Gao et al. 1996).

At the beginning of this chapter, three types of evidence for the role of a neural structure were discussed. Evidence of each of the three types has been put forward to support a role for the DCN in processing of spectral sound localization cues, and perhaps integrating them with somatosensory information is important for interpreting the spectral cues. Although the evidence concerning the projection sites of the DCN's axons cannot be easily interpreted at present, it is clear that the information about spectral notches is preserved and reencoded in the CNIC in a specifically DCN-related system. Thus, we can conclude that the DCN is probably involved in some aspects of sound source localization. However, there are reasons to believe that the DCN's role extends beyond spectral sound localization cues. Among these are the fact that the specializations for notch detection are not confined to the frequency range in which notches are found in HRTFs (Fig. 5.13D), and the fact that there is enormous computational power in the circuitry of the molecular layer. Unfortunately, good hypotheses for the computations being done there have not yet emerged. Thus, we have hints as to the role of the DCN, but much more to discover.

Acknowledgments. This work was supported by grants DC00023, DC00979, and DC00115 from the National Institute for Deafness and Other Communications Disorders. Thanks are due to Phyllis Taylor for preparation of figures and to our colleagues in the Johns Hopkins Center for Hearing Sciences for useful discussions of the ideas presented here. The research summarized in this chapter was done in collaboration with H.F. Voigt, W.P. Shofner, B.J. May, G.A. Spirou, I. Nelken, J.J. Rice, and P.O. Kanold.

References

Abrahams VC, Lynn B, Richmond FJ (1984a) Organization and sensory properties of small myelinated fibres in the dorsal cervical rami of the cat. J Physiol (Lond) 347:177–187.

Abrahams VC, Richmond FJ, Keane J (1984b) Projections from C2 and C3 nerves supplying muscles and skin of the cat neck: a study using transganglionic transport of horseradish peroxidase. J Comp Neurol 230:142–154.

Adams JC (1983) Multipolar cells in the ventral cochlear nucleus project to the dorsal cochlear nucleus and the inferior colliculus. Neurosci Lett 37:205–208.

Adams JC, Warr WB (1976) Origins of axons in the cat's acoustic striae determined by injection of horseradish peroxidase into severed tracts. J Comp Neurol 170: 107–122.

Berrebi AS, Mugnaini E (1991) Distribution and targets of the cartwheel cell axon in the dorsal cochlear nucleus of the guinea pig. Anat Embryol 183:427–454.

Blackburn CC, Sachs MB (1990) The representation of the steady-state vowel sound /ɛ/ in the discharge patterns of cat anteroventral cochlear nucleus neurons. J Neurophysiol 63:1191–1212.

Blackstad TW, Osen KK, Mugnaini E (1984) Pyramidal neurones of the dorsal cochlear nucleus: A Golgi and computer reconstruction study in cat. Neuroscience 13:827–854.

Blum JJ, Reed MC (1998) Effects of wide band inhibitors in the dorsal cochlear nucleus. II. Model calculations of the responses to complex tones. J Acoust Soc Am 103:2000–2009.

Blum JJ, Reed MC, Davies JM (1995) A computational model for signal processing by the dorsal cochlear nucleus. II. Responses to broadband and notch noise. J Acoust Soc Am 98:181–191.

Burian M, Gestoettner W (1988) Projection of primary vestibular afferent fibres to the cochlear nucleus in the guinea pig. Neurosci Lett 84:13–17.

Cant NB (1992) The cochlear nucleus: Neuronal types and their synaptic organization. In: Webster DB, Popper AN, Fay RR (eds) The Mammalian Auditory Pathway: Neuroanatomy. Berlin: Springer-Verlag, pp. 66–116.

Cariani PA, Delgutte B (1996a) Neural correlates of the pitch of complex tones. I. Pitch and pitch salience. J Neurophysiol 76:1698–1716.

Cariani PA, Delgutte B (1996b) Neural correlates of the pitch of complex tones. II. Pitch shift, pitch ambiguity, phase invariance, pitch circularity, rate pitch, and the dominance region for pitch. J Neurophysiol 76:1717–1734.

Caspary DM, Pazara KE, Kossl M, Faingold CL (1987) Strychnine alters the fusiform cell output from the dorsal cochlear nucleus. Brain Res 417:273–282.

Dabak AG, Johnson DH (1992) Function-based modeling of binaural processing: interaural phase. Hear Res 58:200–212.

Davis KA (1999) The basic receptive field properties of neurons in the inferior colliculus of decerebrate cats are rarely created by local inhibitory mechanisms. Soc Neurosci Abstr 25:667.

Davis KA, Young ED (1997) Granule cell activation of complex-spiking neurons in dorsal cochlear nucleus. J Neurosci 17:6798–6806.

Davis KA, Young ED (2000) Pharmacological evidence of inhibitory and disinhibitory neural circuits in dorsal cochlear nucleus. J Neurophysiol 83:926–940.

Davis KA, Ding J, Benson TE, Voigt HF (1996a) Response properties of units in the dorsal cochlear nucleus of unanesthetized decerebrate gerbil. J Neurophysiol 75:1411–1431.

Davis KA, Miller RL, Young ED (1996b) Effects of somatosensory and parallel-fiber stimulation on neurons in dorsal cochlear nucleus. J Neurophysiol 76:3012–3024.

Doucet JR, Ryugo DK (1997) Projections from the ventral cochlear nucleus to the dorsal cochlear nucleus in rats. J Comp Neurol 385:245–264.

Doucet JR, Ross AT, Gillespie MB, Ryugo DK (1999) Glycine immunoreactivity of multipolar neurons in the ventral cochlear nucleus which project to the dorsal cochlear nucleus. J Comp Neurol 408:515–531.

Evans EF, Nelson PG (1973) The responses of single neurons in the cochlear nucleus of the cat as a function of their location and the anaesthetic state. Exp Brain Res 17:402–427.

Evans EF, Zhao W (1993) Varieties of inhibition in the processing and control of processing in the mammalian cochlear nucleus. Prog Brain Res 97:117–126.

Fernandez C, Karapas F (1967) The course and termination of the striae of Monakow and Held in the cat. J Comp Neurol 131:371–386.

Gao J-H, Parsons LM, Bower JM, Xiong J, Li J, Fox PT (1996) Cerebellum implicated in sensory acquisition and discrimination rather than motor control. Science 272:545–547.

Golding NL, Oertel D (1997) Physiological identification of the targets of cartwheel cells in the dorsal cochlear nucleus. J Neurophysiol 78:248–260.

Hancock KE, Davis KA, Voigt HF (1997) Modeling inhibition of type II units in dorsal cochlear nucleus. Biol Cybern 76:419–428.

Hekmatpanah J (1961) Organization of tactile dermatomes, C_1 through L_4, in cat. J Neurophysiol 24:129–140.

Huang AY, May BJ (1996a) Sound orientation behavior in cats. II. Mid-frequency spectral cues for sound localization. J Acoust Soc Am 100:1070–1080.

Huang AY, May BJ (1996b) Spectral cues for sound localization in cats: Effects of frequency domain on minimum audible angles in the median and horizontal planes. J Acoust Soc Am 100:2341–2348.

Imig TJ, Samson F (2000) Differential projection of dorsal and intermediate acoustic striae upon fields AAF and AI in cat auditory cortex. Assoc Res Otolaryngol 23:11.

Imig TJ, Poirier P, Irons WA, Samson FK (1997) Monaural spectral contrast mechanism for neural sensitivity to sound direction in the medial geniculate body of the cat. J Neurophysiol 78:2754–2771.

Imig TJ, Bibikov NG, Poirier P, Samson FK (2000) Directionality derived from pinna-cue spectral notches in cat dorsal cochlear nucleus. J Neurophysiol 83: 907–925.

Itoh K, Kamiya H, Mitani A, Yasui Y, Takada M, Mizuno N (1987) Direct projection from the dorsal column nuclei and the spinal trigeminal nuclei to the cochlear nuclei in the cat. Brain Res 400:145–150.

Jiang D, Palmer AR, Winter IM (1996) Frequency extent of two-tone facilitation in onset units in the ventral cochlear nucleus. J Neurophysiol 75:380–395.

Joris PX (1998) Response classes in the dorsal cochlear nucleus and its output tract in the chloralose-anesthetized cat. J Neurosci 18:3955–3966.

Kanold PO, Young ED (2001) Proprioceptive information from the pinna provides somatosensory input to cat dorsal cochlear nucleus. J Neurosci 21:7848–7858.

Kavanagh GL, Kelly JB (1992) Midline and lateral field sound localization in the ferret (Mustela putorius): Contribution of the superior olivary complex. J Neurophysiol 67:1643–1658.

Kevetter GA, Perachio AA (1989) Projections from the sacculus to the cochlear nuclei in the Mongolian Gerbil. Brain Behav Evol 34:193–200.

Kuhn GF (1987) Physical acoustics and measurements pertaining to directional hearing. In: Yost WA, Gourevitch G (eds) Directional Hearing. Berlin: Springer-Verlag, pp. 3–25.

Lingenhöhl K, Friauf E (1994) Giant neurons in the rat reticular formation: A sensorimotor interface in the elementary acoustic startle circuit? J Neurosci 14:1176–1194.

Lorente de Nó R (1981) The Primary Acoustic Nuclei. New York: Raven Press.

Manis PB (1989) Responses to parallel fiber stimulation in the guinea pig dorsal cochlear nucleus in vitro. J Neurophysiol 61:149–161.

Manis PB, Spirou GA, Wright DD, Paydar S, Ryugo DK (1994) Physiology and morphology of complex spiking neurons in the guinea pig dorsal cochlear nucleus. J Comp Neurol 348:261–276.

Mast TE (1970) Binaural interaction and contralateral inhibition in dorsal cochlear nucleus of the chinchilla. J Neurophysiol 33:108–115.

Masterton RB, Granger EM (1988) Role of the acoustic striae in hearing: contribution of dorsal and intermediate striae to detection of noises and tones. J Neurophysiol 60:1841–1860.

Masterton RB, Granger EM, Glendenning KK (1994) Role of acoustic striae in hearing—mechanism for enhancement of sound detection in cats. Hear Res 73:209–222.

May BJ (2000) Role of the dorsal cochlear nucleus in the sound localization behavior of cats. Hear Res 148:74–87.

May BJ, Huang A, LePrell G, Hienz RD (1996) Vowel formant frequency discrimination in cats: Comparison of auditory nerve representations and psychophysical thresholds. Aud Neurosci 3:135–162.

Meloni EG, Davis M (1998) The dorsal cochlear nucleus contributes to a high intensity component of the acoustic startle reflex in rats. Hear Res 119:69–80.

Merchán MA, Collia FP, Merchán JA, Saldaña E (1985) Distribution of primary afferent fibres in the cochlear nuclei. A silver and horseradish peroxidase (HRP) study. J Anat 141:121–130.

Middlebrooks JC, Green DM (1991) Sound localization by human listeners. Ann Rev Psychol 42:135–159.

Mugnaini E (1985) GABA neurons in the superficial layers of rat dorsal cochlear nucleus: light and electron microscopic immunocytochemistry. J Comp Neurol 235:537–570.

Mugnaini E, Osen KK, Dahl AL, Friedrich Jr. VL, Korte G (1980a) Fine structure of granule cells and related interneurons (termed Golgi cells) in the cochlear nuclear complex of cat, rat, and mouse. J Neurocytol 9:537–570.

Mugnaini E, Warr WB, Osen KK (1980b) Distribution and light microscopic features of granule cells in the cochlear nuclei of cat, rat, and mouse. J Comp Neurol 191:581–606.

Musicant AD, Chan JCK, Hind JE (1990) Direction-dependent spectral properties of cat external ear: New data and cross-species comparisons. J Acoust Soc Am 87:757–781.

Nelken I, Young ED (1994) Two separate inhibitory mechanisms shape the responses of dorsal cochlear nucleus type IV units to narrowband and wideband stimuli. J Neurophysiol 71:2446–2462.

Nelken I, Young ED (1996) Why do cats need a dorsal cochlear nucleus? Rev Clin Basic Pharm 7:199–220.

Nelken I, Young ED (1997) Linear and non-linear spectral integration in type IV neurons of the dorsal cochlear nucleus: I. Regions of linear interaction. J Neurophysiol 78:790–799.

Nelken I, Kim PJ, Young ED (1997) Linear and non-linear spectral integration in type IV neurons of the dorsal cochlear nucleus: II. Predicting responses using non-linear methods. J Neurophysiol 78:800–811.

Oertel D, Wickesberg RE (1993) Glycinergic inhibition in the cochlear nuclei: evidence for tuberculoventral neurons being glycinergic. In: Merchán MA, Juiz JM, Godfrey DA, Mugnaini E (eds) The Mammalian Cochlear Nuclei: Organization and Function. New York: Plenum, pp. 225–237.

Oertel D, Wu SH, Garb MW, Dizack C (1990) Morphology and physiology of cells in slice preparations of the posteroventral cochlear nucleus of mice. J Comp Neurol 295:136–154.

Ohlrogge M, Doucet JR, Ryugo DK (2001) Projections from the pontine nuclei to the cochlear nucleus in rats. J Comp Neurol 436:290–303.

Oliver DL (1984) Dorsal cochlear nucleus projections to the inferior colliculus in the cat: A light and electron microscopic study. J Comp Neurol 224:155–172.

Osen KK (1969) Cytoarchitecture of the cochlear nuclei in the cat. J Comp Neurol 136:453–482.

Osen KK (1970) Course and termination of the primary afferents in the cochlear nuclei of the cat. Arch Ital Biol 108:21–51.

Osen KK (1972) Projection of the cochlear nuclei on the inferior colliculus in the cat. J Comp Neurol 144:355–372.

Osen KK (1983) Orientation of dendritic arbors studied in Golgi sections of the cat dorsal cochlear nucleus. In: Webster WR, Aitkin LM (eds) Mechanisms of Hearing. Clayton: Monash Univ Press, pp. 83–89.

Osen KK, Ottersen OP, Storm-Mathisen J (1990) Colocalization of glycine-like and GABA-like immunoreactivities. A semiquantitative study of individual neurons in the dorsal cochlear nucleus of cat. In: Ottersen OP, Storm-Mathisen

J (eds) Glycine Neurotransmission. New York: John Wiley and Sons, pp. 417–451.

Ostapoff EM, Morest DK, Potashner SJ (1990) Uptake and retrograde transport of [3H]GABA from the cochlear nucleus to the superior olive in the guinea pig. J Chem Neuroanat 3:285–289.

Ostapoff EM, Benson CG, Saint Marie RL (1997) GABA- and glycine-immunore-active projections from the superior olivary complex to the cochlear nucleus in guinea pig. J Comp Neurol 381:500–511.

Palmer AR, Jiang D, Marshall DH (1996) Responses of ventral cochlear nucleus onset and chopper units as a function of signal bandwidth. J Neurophysiol 75:780–794.

Parham K, Kim DO (1995) Spontaneous and sound-evoked discharge characteristics of complex-spiking neurons in the dorsal cochlear nucleus of the unanesthetized decerebrate cat. J Neurophysiol 73:550–561.

Poon PWF, Brugge JF (1993) Sensitivity of auditory nerve fibers to spectral notches. J Neurophysiol 70:655–666.

Populin LC, Yin TCT (1998) Behavioral studies of sound localization in the cat. J Neurosci 18:2147–2160.

Ramachandran R, Davis KA, May BJ (1999) Single-unit responses in the inferior colliculus of decerebrate cats I. Classification based on frequency response maps. J Neurophysiol 82:152–163.

Reed MC, Blum JJ (1995) A computational model for signal processing by the dorsal cochlear nucleus, I: responses to pure tones. J Acoust Soc Am 97:425–438.

Rhode WS (1999) Vertical cell responses to sound in cat dorsal cochlear nucleus. J Neurophysiol 82:1019–1032.

Rhode WS, Greenberg S (1992) Physiology of the cochlear nucleus. In: Popper AN, Fay RR (eds) The Mammalian Auditory Pathway: Neurophysiology. Berlin: Springer-Verlag, pp. 94–152.

Rhode WS, Kettner RE (1987) Physiological studies of neurons in the dorsal and posteroventral cochlear nucleus of the unanesthetized cat. J Neurophysiol 57: 414–442.

Rhode WS, Smith PH, Oertel D (1983) Physiological response properties of cells labeled intracellularly with horseradish peroxidase in cat dorsal cochlear nucleus. J Comp Neurol 213:426–447.

Rice JJ, May BJ, Spirou GA, Young ED (1992) Pinna-based spectral cues for sound localization in cat. Hear Res 58:132–152.

Rice JJ, Young ED, Spirou GA (1995) Auditory-nerve encoding of pinna-based spectral cues: Rate representation of high-frequency stimuli. J Acoust Soc Am 97:1764–1776.

Richter JA, Holtman JR (1982) Barbiturates: their in vivo effects and potential biochemical mechanisms. Neurobiol 18:275–319.

Roberts RC, Ribak CE (1987) GABAergic neurons and axon terminals in the brainstem auditory nuclei of the gerbil. J Comp Neurol 258:267–280.

Roth GL, Kochhar RK, Hind JE (1980) Interaural time differences: Implications regarding the neurophysiology of sound localization. J Acoust Soc Am 68: 1643–1651.

Ryugo DK, May SK (1993) The projections of intracellularly labeled auditory nerve fibers to the dorsal cochlear nucleus of cats. J Comp Neurol 329:20–35.

Ryugo DK, Willard FH (1985) The dorsal cochlear nucleus of the mouse: A light microscopic analysis of neurons that project to the inferior colliculus. J Comp Neurol 242:381–396.

Saadé NE, Frangieh AS, Atweh SF, Jabbur SJ (1989) Dorsal column input to cochlear neurons in decerebrate-decerebellate cats. Brain Res 486:399–402.

Sachs MB, Abbas PJ (1974) Rate versus level functions for auditory-nerve fibers in cats: tone-burst stimuli. J Acoust Soc Am 56:1835–1847.

Saint Marie RL, Benson CG, Ostapoff EM, Morest DK (1991) Glycine immunoreactive projections from the dorsal to the anteroventral cochlear nucleus. Hear Res 51:11–28.

Samson FK, Clarey JC, Barone P, Imig TJ (1993) Effects of ear plugging on single-unit azimuth sensitivity in cat primary auditory cortex. I. Evidence for monaural directional cues. J Neurophysiol 70:492–511.

Schalk T, Sachs MB (1980) Nonlinearities in auditory-nerve fiber response to bandlimited noise. J Acoust Soc Am 67:903–913.

Shofner WP, Young ED (1985) Excitatory/inhibitory response types in the cochlear nucleus: Relationships to discharge patterns and responses to electrical stimulation of the auditory nerve. J Neurophysiol 54:917–939.

Smith PH, Rhode WS (1989) Structural and functional properties distinguish two types of multipolar cells in the ventral cochlear nucleus. J Comp Neurol 282: 595–616.

Spirou GA, Young ED (1991) Organization of dorsal cochlear nucleus type IV unit response maps and their relationship to activation by bandlimited noise. J Neurophysiol 65:1750–1768.

Spirou GA, Davis KA, Nelken I, Young ED (1999) Spectral integration by type II interneurons in dorsal cochlear nucleus. J Neurophysiol 82:648–663.

Sutherland DP, Glendenning KK, Masterton RB (1998a) Role of acoustic striae in hearing: Discrimination of sound-source elevation. Hear Res 120:86–108.

Sutherland DP, Masterton RB, Glendenning KK (1998b) Role of acoustic striae in hearing: reflexive responses to elevated sound-sources. Behav Brain Res 97:1–12.

Van Adel BA, Kelly JB (1998) Kainic acid lesions of the superior olivary complex: Effects on sound localization by the albino rat. Behav Neurosci 112:432–446.

Voigt HF, Young ED (1980) Evidence of inhibitory interactions between neurons in the dorsal cochlear nucleus. J Neurophysiol 44:76–96.

Voigt HF, Young ED (1990) Cross-correlation analysis of inhibitory interactions in dorsal cochlear nucleus. J Neurophysiol 64:1590–1610.

Waller HJ, Godfrey DA, Chen K (1996) Effects of parallel fiber stimulation on neurons of rat dorsal cochlear nucleus. Hear Res 98:169–179.

Weedman DL, Ryugo DK (1996) Projections from auditory cortex to the cochlear nucleus in rats: Synapses on granule cell dendrites. J Comp Neurol 371:311–324.

Weedman DL, Pongstaporn T, Ryugo DK (1996) Ultrastructural study of the granule cell domain of the cochlear nucleus in rats: Mossy fiber endings and their targets. J Comp Neurol 369:345–360.

Weinberg RJ, Rustioni A (1987) A cuneocochlear pathway in the rat. Neurosci 20: 209–219.

Winter IM, Palmer AR (1995) Level dependence of cochlear nucleus onset unit responses and facilitation by second tones or broadband noise. J Neurophysiol 73:141–159.

Wouterlood FG, Mugnaini E, Osen KK, Dahl A-L (1984) Stellate neurons in rat dorsal cochlear nucleus studied with combined Golgi impregnation and electron microscopy: synaptic connections and mutual coupling by gap junctions. J Neurocytol 13:639–664.

Wright DD, Ryugo DK (1996) Mossy fiber projections from the cuneate nucleus to the dorsal cochlear nucleus of rat. J Comp Neurol 365:159–172.

Young ED (1980) Identification of response properties of ascending axons from dorsal cochlear nucleus. Brain Res 200:23–38.

Young ED (1998) The cochlear nucleus. In: Shepherd GM (ed) Synaptic Organization of the Brain. New York: Oxford Press, pp. 121–157.

Young ED, Brownell WE (1976) Responses to tones and noise of single cells in dorsal cochlear nucleus of unanesthetized cats. J Neurophysiol 39:282–300.

Young ED, Voigt HF (1981) The internal organization of the dorsal cochlear nucleus. In: Syka J, Aitkin L (eds) Neuronal Mechanisms of Hearing. New York: Plenum, pp. 127–133.

Young ED, Voigt HF (1982) Response properties of type II and type III units in dorsal cochlear nucleus. Hear Res 6:153–169.

Young ED, Spirou GA, Rice JJ, Voigt HF (1992) Neural organization and responses to complex stimuli in the dorsal cochlear nucleus. Phil Trans R Soc Lond B Biol Sci 336:407–413.

Young ED, Nelken I, Conley RA (1995) Somatosensory effects on neurons in dorsal cochlear nucleus. J Neurophysiol 73:743–765.

Young ED, Rice JJ, Tong SC (1996) Effects of pinna position on head-related transfer functions in the cat. J Acoust Soc Am 99:3064–3076.

Young ED, Rice JJ, Spirou GA, Nelken I, Conley RA (1997) Head-related transfer functions in cat: neural representation and the effects of pinna movement. In: Gilkey RH, Anderson TR (eds) Binaural and Spatial Hearing in Real and Virtual Environments. Mahwah, NJ: Lawrence Erlbaum Assoc, pp. 475–498.

Yu JJ, Young ED (2000) Linear and nonlinear pathways of spectral information transmission in the cochlear nucleus. Proc Nat Acad Sci 97:11780–11786.

Zhang S, Oertel D (1993a) Cartwheel and superficial stellate cells of the dorsal cochlear nucleus of mice: Intracellular recordings in slices. J Neurophysiol 69:1384–1397.

Zhang S, Oertel D (1993b) Tuberculoventral cells of the dorsal cochlear nucleus of mice: Intracellular recordings in slices. J Neurophysiol 69:1409–1421.

6
Ascending Pathways Through Ventral Nuclei of the Lateral Lemniscus and Their Possible Role in Pattern Recognition in Natural Sounds

Donata Oertel and Robert E. Wickesberg

1. Introduction

In all higher vertebrates, the nuclei of the lateral lemniscus lie interposed between the cochlear nuclei and the inferior colliculi. The integrative roles of these neurons in the largely monaural, ventral nuclei of the lateral lemniscus are not well understood, but they are intriguing. The suggestion has been made that neurons in the ventral nuclei of the lateral lemniscus are involved in pattern recognition. It is, however, difficult to pinpoint the role of a specific small nucleus in such a complex function. The goal of this chapter is to examine what is known about the ventral lemniscal nuclei to yield clues about their possible role in the identification of sounds.

The description of the structure and function of the ventral lemniscal nuclei will be placed in the context of the necessity of the auditory system to detect and recognize patterns in natural sound stimuli. Presumably, at least part of the task of extracting the details of the temporal and spectral fine structure in a sound or determining the timing of transients, which are important for the recognition of natural sounds such as speech, occurs in the brainstem auditory nuclei where that information is still encoded directly. It seems likely that the ventral nuclei of the lateral lemniscus contribute to this integrative task with inhibitory input to the inferior colliculi, but that notion is at present supported only by fragmentary, circumstantial evidence.

This chapter will review what is known about the structure and function of the ventral nuclei of the lateral lemniscus, together with what is known about the functional properties of their inputs and targets, in an attempt to understand what the integrative role is of these nuclei in the mammalian auditory pathway. This review, which is based on studies of many different species conducted and reported over a long period, is made with the goal of providing a framework for future studies. While this chapter will emphasize the consistent features of the ventral nuclei of the lateral lemniscus, there are notable variations between species; variations which themselves could reflect the differences in the acoustical environments in which animals live and in the sounds they recognize.

2. Structure of the Nuclei of the Lateral Lemniscus

The nuclei of the lateral lemniscus are clusters of cell bodies that lie among the fibers that form the lateral lemniscus: a fiber bundle that feeds into the inferior colliculus (see Chapter 1, this volume, and Schwartz 1992). These nuclei are located ventral and posterior to the inferior colliculus. In these clusters, several anatomically and functionally distinct groups have been identified. These groups are represented in differing proportions and are intermingled to varying degrees in different species. In some species, four such clusters can be identified while in others fewer areas can be clearly distinguished. A ventral cluster of neurons or group of clusters, called the ventral (VNLL) and the intermediate nucleus of the lateral lemniscus (INLL) by some authors, lumped and together called the ventral nucleus of the lateral lemniscus (VNLL) by other authors, and termed the ventral complex of the lateral lemniscus (VCLL) by yet others, is distinct in all mammals from the dorsal cluster of neurons that forms the dorsal nucleus of the lateral lemniscus (DNLL). The ventral nuclei receive input mainly from the contralateral ventral cochlear nucleus and are largely monaural in most species. In contrast, the DNLL receives input from the ipsilateral medial superior olivary nucleus (MSO), from both the ipsilateral and contralateral lateral superior olivary nucleus (LSO), and from the contralateral DNLL and is binaural (Schwartz 1992). The present discussion concerns the ventral, largely monaural, lemniscal nuclei.

The distinction between the ventral and intermediate nuclei of the lateral lemniscus is more clear in some mammals and some preparations than in others, raising the question whether these should be considered to be separate nuclei or subdivisions of a single nuclear complex. Figure 6.1 shows a schematic diagram of the general mammalian pattern, using the terminology that has been applied to species in which the cellular groups can be separated (bats, cats, and guinea pigs; Glendenning et al. 1981; Covey and Casseday 1986; Adams 1997; Schofield and Cant 1997). In some species, including cats and guinea pigs, the fascicles of lemniscal fibers that pass laterally and medially around a ventral cluster come together and pass through the dorsal cluster, separating the multipolar neurons into dorsoventral bands. This difference in clustering, together with the differential innervation of the dorsal cluster by the medial nucleus of the trapezoid body (MNTB), prompted the naming of the dorsal group the intermediate nucleus of the lateral lemniscus (INLL) and the naming of a ventral cluster

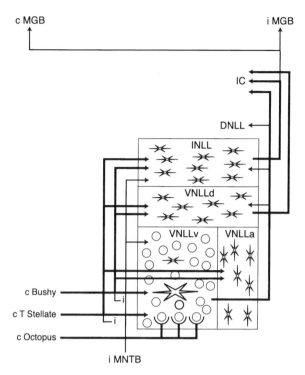

FIGURE 6.1. Schematic representation of the mammalian organization of the ventral nuclei of the lateral lemniscus (VNLL). In many species a distinct intermediate nucleus of the lateral lemniscus (INLL) lies dorsal to the VNLL. The lemniscal fibers that course around the VNLL come together and course through the INLL. The multipolar neurons of the INLL tend to be lined up in dorsoventral columns with dendrites generally spread perpendicularly across the lemniscal fibers. Three distinct regions of the VNLL have been identified. The VNLLv is identified by the presence of spherical neurons (circles) which are innervated by endbulbs (arcs) from octopus cells in the contralateral (c) posteroventral cochlear nucleus. In some species multipolar cells (stars) like those in VNLLd and occasional giant cells (large stars) are intermingled among the spherical cells. The VNLLd and VNLLa are devoid of spherical cells. VNLLa is distinct from VNLLd in that it contains multipolar neurons whose dendrites lie parallel to the lemniscal fibers. This diagram was made using the terminology defined by Schofield and Cant (1997) for guinea pigs. The major excitatory inputs come from the contralateral cochlear nucleus and are indicated by heavy lines. A few bushy and stellate cells from the ipsilateral side (i) have also been shown to project to the ventral lemniscal nuclei. The INLL receives strong input from the ipsilateral medial nucleus of the trapezoid body (MNTB). The major projections are to the inferior colliculus (IC) and are indicated by heavy lines. The minor projections are indicated with light lines and include the dorsal nucleus of the lateral lemniscus (DNLL), and the medial geniculate body (MGB).

the ventral nucleus of the lateral lemniscus (VNLL) (Glendenning et al. 1981, Schofield and Cant 1997). However, these nuclei cannot be distinguished on the basis of clustering in some species, e.g., rats and oppossums (Willard and Martin 1983; Merchán and Berbel 1996), nor can the projection from the MNTB be used as an unambiguous criterion for identifying the INLL. Neurons of the MNTB innervate most or all of the ventral nuclei of the lateral lemniscus but in differing proportions (Spangler et al. 1985; Sommer et al. 1993; Huffman and Covey 1995; Smith et al. 1998). Even when they can be distinguished, many anatomists have recognized that similar multipolar cells occupy both subdivisions and receive similar inputs, though possibly in differing proportions, and that the major projection of all regions is to the inferior colliculus. The similarities in morphology, inputs, and projection patterns between the VNLL and INLL suggest that these groups of neurons should perhaps be considered subdivisions of a single nucleus rather than as separate nuclei. To convey the notion of a single nucleus with subnuclei while using terminology that is consistent with the existing body of literature, we will refer to the "ventral nuclei of the lateral lemniscus." In many mammals, three distinct regions, the INLL, a ventral division of the VNLL (VNLLv), and a dorsal division of VNLL (VNLLd) can be distinguished in the ventral nuclei of the lateral lemniscus. These will be considered in turn.

2.1 The Intermediate Nucleus of the Lateral Lemniscus

The INLL lies in the most dorsal position within the ventral nuclei of the lateral lemniscus and forms a distinct cluster of neurons in a diverse group of animals including cats, bats, mice and moles. This group of neurons receives a heavier projection from the MNTB than neurons located more ventrally in the nuclei of the ventral lemniscus (Glendenning et al. 1981; Spangler et al. 1985). While the major projection of the INLL is to the ipsilateral inferior colliculus in all mammals, some neurons in the INLL (but not the VNLL) of cats and its presumed counterpart in the dorsal VNLL of rats have been reported to project to the medial geniculate body of the thalamus (Henkel 1983; Whitley and Henkel 1984; Hutson et al. 1991).

2.2 The Ventral Division of the Ventral Nucleus of the Lateral Lemniscus

A distinctive cluster of cells that comprises part of the VNLL (VNLLv in Figure 6.1) has been identified in humans, cats, bats, mice, guinea pigs and moles. The VNLLv is characterized by the presence of neurons that resemble cochlear nuclear spherical bushy cells not only in their general shape (Willard and Ryugo 1983; Kudo et al. 1990; Vater and Feng 1990; Adams 1997; Schofield and Cant 1997; Vater et al. 1997) but also in receiving input through endbulbs (Vater and Feng 1990; Schofield 1995; Adams 1997; Vater et al. 1997) and perhaps in their biophysical specializations (Wu 1999). In guinea pigs, these neurons are intermingled with multipolar cells and occasional giant cells in the VNLLv (Schofield and Cant 1997). The relative size of VNLLv varies considerably between species, consituting only 4% of the volume of the VNLL in cats but 38% of the volume of the VNLL in humans (Adams 1997). In studies of rats and oppossums, the VNLLv has not been distinguished (Willard and Martin 1983; Merchán and Berbel 1996). In rats, round, bushy neurons such as those in VNLLv are scattered in the VNLL (M.S. Malmierca, per com). In bats of the genus *Eptesicus*, a prominent group of round, bushy neurons that resemble neurons in VNLLv of other species form the columnar area (Covey and Casseday 1986, 1991; Vater et al. 1997).

2.3 The Ventral Nucleus of the Lateral Lemniscus (Dorsal)

The dorsal portion of the VNLL and more ventral regions that surround the VNLLv (VNLLd) are characterized by the absence of spherical bushy cells. In the VNLLd, the majority of cells are multipolar with dendrites oriented mainly perpendicular to the lemniscal fibers (Schofield and Cant 1997). These regions appear as a homogeneous stack of flattened multipolar cells (Merchán and Berbel 1996; Malmierca et al. 1998).

2.4 The Ventral Nucleus of the Lateral Lemniscus (Anterior)

An anterior cluster of neurons within ventral nuclei of the lateral lemniscus, the VNLLa, also contains only multipolar cells. These multipolar cells differ from those in the VNLLv and VNLLd in that they are oriented parallel to the lemniscal fibers (Schofield and Cant 1997). This region was considered to be part of the ventral nuclei of the lateral lemniscus on the basis of its receiving input from the same group of cells that provide input to VNLLd, bushy, and stellate cells of the ventral cochlear nucleus (VCN) (Schofield and Cant 1997).

3. All Groups of Principal Cells of the VCN Project to the Ventral Nuclei of the Lateral Lemniscus

The major excitatory ascending pathways to the ventral nuclei of the lateral lemniscus arise in the principal cells of the contralateral VCN (Cant 1992; Schwartz 1992). All groups of VCN cells have been reported specifically to project to the ventral lemniscal nuclei: small spherical bushy cells (Huffman and Covey 1995), large spherical bushy cells (Smith et al. 1993b), globular bushy cells (Friauf and Ostwald 1988; Smith et al. 1991), stellate cells of the subtype that also projects to the inferior colliculus (Smith et al. 1993a; Schofield and Cant 1997), and octopus cells (Friauf and Ostwald 1988; Vater and Feng 1990; Smith et al. 1993a; Schofield 1995; Adams 1997; Schofield and Cant 1997; Vater et al. 1997). The subnuclei of the VNLL vary, however, in the proportion of input they get from the various groups of principal cells in the VCN. Bushy and stellate cells of the cochlear nucleus project to all parts of the INLL and VNLL whereas octopus cells innervate only the VNLLv. While the input is largely from the contralateral side, a few bushy and stellate cells from the ipsilateral side have been shown to innervate the VNLL of guinea pigs (Schofield and Cant 1997). The neotropical fruit bat is unusual among mammals in that the ipsilateral projection is heavy (Hutson 2000). Octopus cells have not been observed to innervate the ipsilateral side in any species.

The projection from the ventral cochlear nucleus to the spherical bushy cells of the VNLLv is through endbulbs. Most studies have concluded that the endbulbs arise from octopus cells. A projection of octopus cells to the VNLL through endbulbs was described briefly in early studies (Warr 1969; Adams 1986) but has recently been documented in detail. Endbulbs identified as originating from octopus cells in the contralateral, posteroventral cochlear nucleus have been seen in bats (Vater and Feng 1990), guinea pigs (Schofield 1995), cats, and humans (Adams 1997). Multiple endbulbs can converge on individual spherical bushy cells (Adams 1997). Inputs to the columnar area of the VNLL of *Eptesicus* bats have been reported to arise also from spherical bushy cells of the anteroventral cochlear nucleus (AVCN) (Covey and Casseday, 1986; Huffman and Covey, 1995). Examples of the morphology of endbulbs in bats and cats are shown in Figure 6.2. Endbulbs in humans, labeled by calretinin, are shown in Figure 6.3.

In addition to the direct projections from the ventral cochlear nucleus, the ventral lemniscal nuclei receive a significant indirect, inhibitory projection from contralateral globular bushy cells through the ipsilateral MNTB. In cats and bats, the INLL has been shown to receive more input from the MNTB than the VNLL (Glendenning et al. 1981; Spangler et al. 1985; Sommer et al. 1993; Huffman and Covey 1995). Individual neurons of the MNTB have, however, been shown to terminate widely in the VNLL (Sommer et al. 1993; Smith et al. 1998). In bats, inputs to the spherical bushy cells of the VNLLv include terminals that probably mediate glycinergic inhibition from the MNTB (Vater et al. 1997).

3.1 Functional Characteristics of Input to the VNLL from Octopus Cells

Octopus cells have anatomical and biophysical specializations which allow them to detect coincident firing of groups of auditory nerve fibers and to convey the timing of that firing with exceptional temporal precision (Oertel

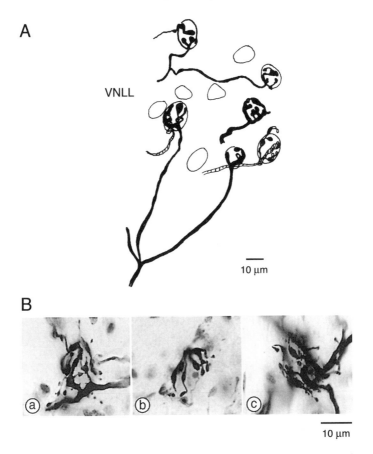

FIGURE 6.2. Endbulb endings in the VNLLv visualized by injections of label into the contralateral, posteroventral cochlear nucleus. A: Endbulbs in VNLL of the horseshoe bat. (From Vater and Feng, Functional organization of ascending and descending connections of the cochlear nucleus of horseshoe bats, Journal of Comparative Neurology 292: 373–395, copyright 1990, by permission of Wiley-Liss, Inc., a subsidiary of John Wiley & Sons, Inc.). B: a, b, c Endbulbs in the VNLL of cats. (Reprinted from Adams, 1997, copyright OPA (Overseas Publishers Association) N.V., by permission of Gordon & Breach Publishers.)

et al. 2000; Chapter 3, this volume). The dendrites of octopus cells cross the bundle of auditory nerve fibers, enabling them to receive convergent synaptic input from many fibers. The summed synaptic currents from auditory nerve inputs are shaped by the biophysical properties of octopus cells to produce synaptic potentials that are unusually brief, and whose peaks are consistent to within fractions of a millisecond over a wide range of amplitudes (Oertel et al. 2000).

Recordings have been made from octopus cells in vitro. In tissue slices from mice that contain most of the nucleus, the responses of octopus cells to brief shocks that activate a single action potential in auditory nerve fibers

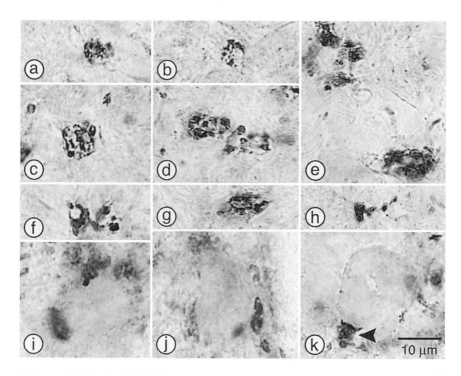

FIGURE 6.3a–k. Endbulbs in the human VNLLv visualized by immunostaining for calretinin, a calcium-binding protein. (Reprinted from Adams, 1997, copyright OPA (Overseas Publishers Association) N.V., by permission of Gordon & Breach Publishers.) Octopus cells also stain heavily for calretinin.

have been measured. In such experiments, weak shocks activate only a small number of auditory nerve fibers and more and more auditory nerve fibers are activated as the strength of shocks is increased (Golding et al. 1995). Figure 6.4 shows how two octopus cells responded as the strength of shocks to the cut end of the auditory nerve increased in fine increments. Weak shocks produced a small depolarization; with increasing shock strength the depolarization produced a small action potential, recognizable by the inflection (marked with a dot) and shown in other experiments to be blocked by tetrodotoxin (Golding et al. 1999). With further increases in the strength of shocks to the auditory nerve, the response continued to grow, but the timing of the action potential changed by only tens of microseconds. These results show that octopus cells require the summation of roughly between 10 and 25% of inputs over a one-millisecond time period to fire. A rough estimate can be made of the numbers of auditory nerve fibers that impinge on a single octopus cell. In mice, the 200 octopus cells in each cochlear nucleus (Willott and Bross 1990) receive input from 12,000 auditory nerve fibers (Ehret 1979); if each auditory nerve fiber innervates only a single octopus cell, each octopus cell would be expected to receive input from, on average, at least 60 auditory nerve fibers.

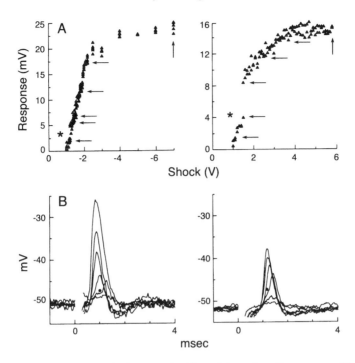

FIGURE 6.4. Two octopus cells detect coincident firing of auditory nerve fibers. A: The plots in the top panels show that the amplitude of synaptic potentials in responses to shocks to the auditory nerve in two octopus cells vary as a function of shock strength in two cells (left and right). While auditory nerve fiber inputs are recruited stepwise as each is brought to threshold, responses grow as smooth, graded functions, indicating that each auditory nerve fiber contributes only a small increment to the response and that octopus cells have many auditory nerve inputs. In each cell, there is a single jump in amplitude that is associated with the presence of an action potential (marked with a *). The finding that amplitude varies with shock strength below the threshold indicates that summation of multiple inputs is required to bring an octopus cell to threshold. B: Traces in the lower panels show examples of synaptic responses to shocks (0.1 ms duration) recorded with intracellular, sharp microelectrodes in the same two octopus cells. Responses shown in traces are marked with arrows in the plots above. Breaks in the traces mark the shock artifacts. Synaptic potentials rose from the resting potential after a synaptic delay to potentials that varied as a function of shock strength and were followed by an afterhyperpolarization. In the traces associated with the jump in amplitude, a small action potential appeared on the synaptic response (marked by a dot). (Reprinted with permission from Golding et al. 1995.)

The small size, consistent latency of peaks, and brevity of synaptic potentials results from the action of synaptic currents on postsynaptic cells. Octopus cells have very low input resistances (about $6\,M\Omega$; Bal and Oertel 2000). Low input resistances endow octopus cells with short time constants but also require octopus cells to be activated by large synaptic currents (see Chapter 3, this volume). The low input resistances result from the presence of high densities of two types of voltage-sensitive ion channels and large resting currents: hyperpolarization-activated mixed-cation channels (which mediate I_h) and low-threshold potassium channels (which mediate I_{KL}) (Golding et al. 1995, 1999; Bal and Oertel 2000, 2001). The strong, low-threshold, voltage-sensitive potassium conductance contributes to the short duration and constant latency of the peaks of synaptic responses (reviewed by Oertel et al. 2000).

As expected from the observation that octopus cells detect synchronous firing in groups of auditory nerve fibers, octopus cells are broadly tuned, and their responses are sharply timed (Godfrey et al. 1975; Rhode and Smith 1986; Oertel et al. 2000). Response areas are generally derived from the measurement of the number of action potentials in response to tones at various frequencies and intensities (Rhode and Greenberg 1992). Except at low frequencies, octopus cells respond to tones with only a single, sharply-timed action potential at the onset of the tone because generally action potentials in responses to tones are coincident only at the onset of tones (see Rhode and Greenberg 1992). Response areas from two presumed octopus cells are shown in Figure 6.5. These cells have relatively high thresholds to tones presumably because only relatively loud tones recruit a large enough number of auditory nerve fibers to produce sufficient

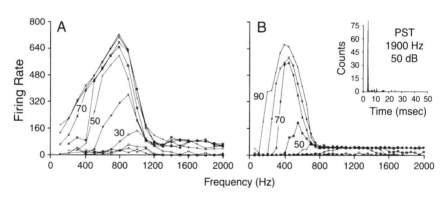

FIGURE 6.5. Responses of two presumed octopus cells (A and B) as a function of frequency and intensity of tones (25 ms duration). For each cell, plots are shown of the spike rate (action potentials per second) as a function of frequency. Intensity is indicated for each of the isointensity curves in dB SPL. For each cell, tuning extends over the entire frequency range tested. Each cell had a threshold of 30 dB SPL. Characteristic frequencies were: A, 950 Hz; B, 2,200 Hz. Inset in B shows a histogram of responses to 250 repetitions of a tone. It is typical of octopus cells that at high frequencies they fire once at the onset of the tone. It is also typical that, at frequencies below 800 Hz and at high intensities, octopus cells can fire on every cycle, resulting in firing rates of nearly 800 spikes/s. (Reprinted with permission from Rhode and Smith 1986.)

summation to bring octopus cells to threshold. It is not obvious what the characteristic frequency (the frequency at which the cells have the lowest threshold) is of these cells. The single action potential at the onset of a tone over a wide range of frequencies makes the response areas flat above about 2 kHz. At low frequencies, auditory nerve fibers fire in phase with the sounds, producing synchrony in firing not only at the onset but with every cycle. As a consequence, the firing rate of octopus cells is equal to the frequency of tones up to about 800 Hz, giving the response areas a bulge at low frequencies (Fig. 6.5) (Rhode and Smith 1986). These responses show how well octopus cells respond to periodic stimuli and also show that octopus cells can sustain unusually high firing rates (maximal firing rates for auditory nerve fibers are about 300 Hz.).

Octopus cells respond to broadband and periodic sounds as well as to frequency sweeps. Trains of clicks are periodic, broad-band sounds to which octopus cells respond robustly. Recordings by P. Joris and P.H. Smith (Fig. 6.6) document the precision in the timing of firing of octopus cells (Oertel et al. 2000). The histogram and dot raster of the timing of action potentials in response to 10 repetitions of the train illustrate the clock-like precision in the timing of firing. The timing of responses as a function of the 2 ms period of the sound shows that individual action potentials generally fall into a 200 μs window. Some octopus cells are sensitive to the direction of frequency sweeps (Godfrey et al. 1975; Rhode and Smith 1986; Rhode and Greenberg 1992). Responses to more complex, natural sounds have not been recorded directly from octopus cells. From what is known of the convergence of auditory nerve input and their biophysical characteristics, one would predict that octopus cells would respond strongly to periodicity in harmonic complexes (Cariani and Delgutte 1996a,b) and to the peaks in the responses to normal and whispered consonants in populations of auditory nerve fibers (Stevens and Wickesberg 1999).

The input to the VNLL from octopus cells would be expected to have short latencies because they are propagated through thick axons. The conduction velocity of myelinated axons is a function of the axonal diameter. The exceptionally large axons of octopus cells would be expected to conduct faster than bushy and stellate cells (Schofield 1995; Schofield and Cant 1997).

3.2 Functional Characteristics of Input to the VNLL from Bushy Cells

Bushy cells in the ventral cochlear nucleus have anatomical (see Cant 1992) and biophysical (Oertel, 1997) specializations which enable them to convey the fine structure in the firing patterns of auditory nerve fibers (see Chapter 3, this volume). While they generally have a short, profusely branching dendrite, bushy cells receive most synaptic input at the cell body. They receive

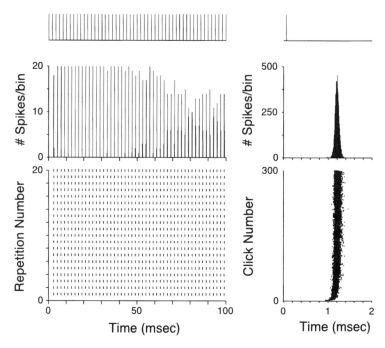

FIGURE 6.6. Intraaxonal recording from a cell in a cat whose responses to tones identified it as an octopus cell shows that octopus cells fire with exquisite temporal precision. The cell had a characteristic frequency at 13.4 kHz. Panels at the left show the responses of this cell to 20 trains of clicks presented 20 dB above the click treshold and at 500/s. At top left, the panel illustrates the timing of the clicks in a train. The middle left panel illustrates responses in histogram form. At bottom left the panel illustrates the same responses as dot rasters of the spike times. The precision in the timing of firing is revealed more clearly when responses are examined as a function of the 2-ms period of the sound stimulus on the right. The top right panel shows the timing of the click in the period while the middle right panel shows responses in histogram form. Finally the bottom right panel shows those similar responses as dot rasters. After prolonged firing, the timing shifted slightly. These unpublished data were generously provided by Philip X. Joris and Philip H. Smith.

input from a small number of auditory nerve fibers: spherical bushy cells in mammals and their avian homologues receive input from about three auditory nerve fibers through somatic, calyceal endings whereas globular bushy cells receive input from between about four and ten auditory nerve fibers in mice (Oertel 1985) and about 40 auditory nerve fibers in cats (Liberman 1993). As predicted from the small number of auditory nerve inputs, bushy cells are sharply tuned and have firing patterns that resemble those of auditory nerve fibers ("primary-like" and "primary-like-with-notch"; Rhode and Greenberg 1992). Responses of auditory nerve fibers are discussed by Ruggero (1992).

In several ways, the responses of cochlear nuclear bushy cells show neuronal processing. While the responses of bushy cells are similar to those of auditory nerve fibers in encoding the fine structure of sounds within their response area, the convergence of inputs removes temporal jitter and makes the encoding of timing of onset transients and periodicity sharper than in their auditory nerve inputs, especially for globular bushy cells (Joris et al. 1994a, b). Bushy cells also receive inhibitory input. Some of the glycinergic inhibition arises from neurons in the dorsal cochlear nucleus and is closely matched tonotopically to the excitatory inputs of bushy cells (Wickesberg and Oertel 1988, 1990; Winter and Palmer, 1990). In a comparison of how vowels are encoded by neurons of the ventral cochlear nucleus, bushy cells (primary-like units) were found to encode the spectrum of vowels in the timing of their discharges even when those vowels are loud (Recio and Rhode, 2000). They encode the spectrum of vowels less well in their firing rates.

3.3 Functional Characteristics of Input to the VNLL from Stellate Cells

The multiple dendrites of stellate cells give them their name (Cant 1992). Of two types of stellate cells in the ventral cochlear nucleus, one type provides input to the ventral nuclei of the lateral lemniscus. Two types of stellate cells have been distinguished on the basis of the prevalence of somatic inputs (Cant 1981), on their responses to sound (Smith and Rhode 1989), on their dendritic morphology (Oertel et al. 1990; Doucet and Ryugo 1997), on their projections (Cant and Gaston 1982; Oliver 1987; Oertel et al. 1990), and on whether they are likely to be inhibitory and glycinergic (Wenthold 1987). The ventral nuclei of the lateral lemniscus are innervated by those stellate cells that (in cats) lack somatic input, those that respond to tones with sustained regular firing (chopper) patterns, and those that are excitatory and whose axons exit the cochlear nuclei through the trapezoid body and project to the contralateral inferior colliculus but not to the contralateral cochlear nuclei.

In some species, the dendrites of stellate cells in the ventral cochlear nucleus are obviously oriented parallel to auditory nerve fibers (Oertel et al. 1990; Doucet and Ryugo 1997). Consistent with the anatomical arrangement, functional counts of inputs show that stellate cells in mice receive input from only about five auditory nerve fibers (Ferragamo et al. 1998). In all species in which it has been measured, stellate cells in the ventral cochlear nucleus that respond to tones with sustained, regular firing are correspondingly narrowly tuned (see Rhode and Greenberg 1992). In response to tones, stellate cells fire tonically and, in many cases, with so much regularity from one trial to the next that histograms of responses show modes for each of the consecutive action potentials. This series of modes inspired the term "chopping" to describe these cells' responses to tones (Rhode and Greenburg 1992).

In considering the contribution of these inputs to pattern recognition, several features of the responses of stellate cells that affect responses to complex sounds are noteworthy. First, the narrowness of the first mode chopper responses shows that stellate cells encode the onset of a tone with considerable temporal precision (Blackburn and Sachs 1989; Rhode and Greenberg 1992). Second, except at very low frequencies, the fine structure and details of the firing of auditory nerve fibers are lost in the steady firing of stellate cells. The conversion of the adapting response pattern found in the auditory nerve to steady, tonic firing suggests that an additional, delayed, "feed-forward" source of excitation supplements the initial excitation by auditory nerve fibers. In mice, such excitation comes through collateral inputs from other stellate cells (Ferragamo et al. 1998). The third noteworthy feature of stellate cell responses is that stellate cells receive glycinergic inhibition from at least two sources. They receive glycinergic inhibition that is closely matched tonotopically to their excitation through tuberculoventral cells of the dorsal cochlear nucleus (Wickesberg and Oertel 1988, 1990). A second source of inhibition is through local glycinergic neurons whose widespread dendrites suggest that they are broadly tuned (Ferragamo et al. 1998). This inhibition is presumably the source of the sideband inhibition that is so commonly observed in stellate cells (choppers) and that sharpens their frequency selectivity (Rhode and Greenberg 1992). This inhibition presumably also contributes to the termination of responses at the offset of tones. Finally, the responses of stellate cells are modulated through slow inputs. Such modulation is difficult to document in vivo because the circuits that mediate the responses of stellate cells are likely to be affected by anesthetics. Although GABAergic inhibitory synaptic potentials and synaptic currents have not been directly observed, subtle inhibition mediated through the $GABA_A$ sub-family of receptors that seemed to be generated far from the cell body was detected (Ferragamo et al. 1998). Cholinergic modulation of firing through neurons in the ventral nucleus of the trapezoid body that include olivocochlear efferents (Warr 1992) is mediated through both nicotinic and muscarinic receptors (Fujino and Oertel 2001).

The combination of excitatory and inhibitory inputs endows stellate cells with unexpected properties. In stellate cells, for example, the encoding of amplitude modulation is enhanced over some frequencies in comparison with their auditory nerve inputs (Frisina et al. 1990; Rhode and Greenberg 1992). The cholinergic efferent innervation may enhance signaling in noisy environments (Fujino and Oertel 2001). The combination of the narrowly tuned and broadly tuned inhibition has unexpected consequences on responses to clicks, too (Wickesberg 1996).

Stellate cells stand alone among the classes of neurons in the ventral cochlear nucleus in encoding the spectra of sounds in their firing rates. In responses to speech sounds, stellate cells encode only the very low frequencies in the spectra of vowels in their temporal firing patterns whereas they encode the entire spectrum in their firing rates (Recio and Rhode, 2000).

4. Characteristics of Responses to Sound in Neurons of the Ventral Nuclei of the Lateral Lemniscus

Two patterns of responses to tones have been observed in recordings from the ventral lemniscal nuclei. In cats as in bats, some neurons respond to tones with regular, sustained firing and are sharply tuned; others respond with a sharply timed action potential at the onset of a tone and are broadly tuned (Aitkin et al. 1970; Guinan et al. 1972a, b; Covey and Casseday 1991; Adams 1997). In the unanesthetized rabbit, the majority of neurons displayed a sustained or transient response to contralateral stimuli, but a variety of response patterns were observed (Batra and Fitzpatrick 1997, 1999). While many authors consider the ventral lemniscal nuclei to be monaural, many studies shown that neurons in this area are indeed consistently driven through the contralateral ear. However, some cells, especially those neurons near the medial boundary of the ventral lemniscal nuclei, are in addition driven through the ipsilateral ear (Aitkin et al. 1970; Guinan et al. 1972b; Batra and Fitzpatrick 1997, 1999). What distinguished neurons in the ventral lemniscal nuclei from those in the other binaural nuclei was that they were excited by both ears, rather than being excited by one and inhibited by the other.

Several lines of evidence indicate that the spherical bushy cells in VNLLv are broadly tuned and respond at the onset of sounds. First, prepotentials, which have been recorded in neurons with calyceal inputs in the anteroventral cochlear nucleus (Pfeiffer, 1966; Rhode and Greenberg 1992) and the MNTB (Guinan and Li, 1990; Smith et al. 1998), are recorded from neurons that respond with a sharply-timed action potential at the onset of a tone in the VNLLv (Fig. 6.7) (Adams 1997). In spherical bushy neurons of the VNLLv in cats, the prepotentials presumably reflect the firing of the calyceal input from octopus cells (Adams 1997). A second line of evidence indicates that, in bats where the spherical bushy cells are clustered in the columnar region, neurons fire at the onset of tones with short latencies, little temporal jitter, and constant latency (Fig. 6.8A) (Covey and Casseday 1991). Both in the bat and in the cat, these cells did not fire spontaneously. Covey and Casseday (1991) noted that the latencies of many of the onset units was particularly short and that temporal variability could be as low as

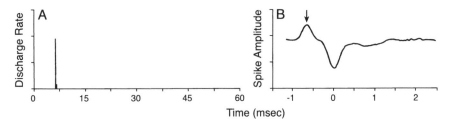

FIGURE 6.7. Extracellular recording from a neuron in VNLLv whose action potentials were preceded by a prepotential. A: Peristimulus time histogram of responses to repeated, 50 ms tone bursts. The unit responded only at the onset of the tone. B: The unit had a prepotential (arrow). (Reprinted from Adams, 1997, copyright OPA (Overseas Publishers Association) N.V., by permission of Gordon & Breach Publishers.)

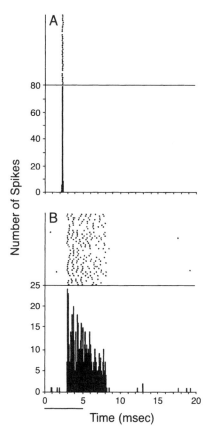

FIGURE 6.8. Extracellular recordings from two neurons in the ventral lemniscal nuclei of bats. A: One neuron in the columnar area (VNLLv) responded with precisely timed firing at the onset of the tone. Upper panel: A subset of 40 of the responses are shown as dot rasters. Lower panel: Peristimulus time histogram of 100 responses of the same unit. B: Sustained responses of a unit in the multipolar cell area. Upper panel: A subset of responses are shown as dot rasters. Lower panel: Peristimulus time histogram of 100 responses. (Reprinted with permission from Covey and Casseday 1991.)

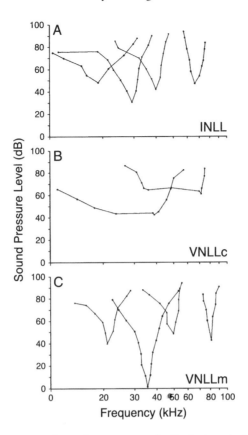

FIGURE 6.9. Tuning curves from three neurons in the bat ventral lemniscal nuclei. Neurons from the INLL and from the multipolar cell region of the VNLL are sharply tuned. Those from the columnar area are broadly tuned. (Reprinted with permission from Covey and Casseday 1991.)

30 µsec. Neurons in the columnar area in bats responded preferentially to frequency sweeps from high to low (Huffman et al. 1998). Those neurons are broadly tuned (Covey and Casseday 1991) (Fig. 6.9B). The responses to sound of neurons in the columnar region of the bat and the VNLLv of the cat resemble responses to sound of octopus cells in many ways. Like neurons in the columnar area of the VNLL of bats, octopus cells responded to the directionality of sweeps (Godfrey et al. 1975; Rhode and Smith 1986; Covey and Casseday 1991). In bats, however, those responses may have arisen from the convergence of primary-like inputs from the AVCN rather than from octopus cells (Covey and Casseday 1986; Huffman and Covey 1995).

Other neurons in the VNLL and INLL (presumably the multipolar neurons) respond to tones with regular, sustained firing or "chopping" (Aitkin et al. 1970; Guinan et al. 1972b; Covey and Casseday 1991)

(Fig. 6.8E). These neurons were sharply tuned (Covey and Casseday 1991) (Fig. 6.9A and C). Their nonmonotonic firing rates as a function of intensity indicated that inhibition contributed to responses to sounds (Covey and Casseday 1991).

Responses to tones in ventral lemniscal neurons are consistent with what is known of their inputs. Bushy and stellate cells that provide the major excitatory input to the more sharply tuned and tonically firing multipolar cells of the ventral lemniscal nuclei are sharply tuned and respond to tones with more sustained firing (Rhode and Greenberg 1992). These comparisons are necessarily somewhat crude in that these conclusions are drawn from measurements in various species.

At the cellular level, neurons in the ventral nuclei of the lateral lemniscus display two sets of intrinsic electrical characteristics (Wu 1999). One group of neurons responds to steady, depolarizing current pulses with sustained firing. The pattern of the sustained firing varied somewhat among these cells, indicating that perhaps they fall into several functional classes. A second group of neurons responded to suprathreshold depolarizing current pulses with only a single action potential and a rectification; a pattern that is also evident in bushy and octopus cells (see Chapter 3 in this volume; Oertel 1997).

5. Tonotopic Organization of the Ventral Lemniscal Nuclei

Studies to date have generally not revealed a very clear tonotopic organization within the VNLL. In recordings from cats, occasional electrode penetrations revealed a tonotopic progression, but such findings are not consistent (Aitkin et al. 1970; Guinan et al. 1972a). A tonotopic organization was also not detected when the activity of neurons was monitored with 2-deoxyglucose, a nonmetabolizable analog of glucose that is thought to label metabolically active neurons (Glendenning and Hutson 1998). In recordings from bats whose ventral lemniscal nuclei are hypertrophied, it has been reported that, except in the columnar area in which neurons are broadly tuned, high frequencies seem to be represented in the center and that lower frequencies are represented around the core (Covey and Casseday 1991). The conclusion that the ventral nuclei of the lateral lemniscus have a general tonotopic organization that is concentric is intriguing in light of anatomical experiments in rats, which show that the organization of the projections is roughly concentric or polarized with respect to the tonotopic map of the inferior colliculus, as discussed further below (Merchan and Berbel 1996; Malmierca et al. 1998). In interpreting these anatomical experiments, however, it is important to keep in mind that the frequency representation of the inferior collicular targets need not necessarily match the best frequency of impinging inhibitory neurons.

6. Projections of the Ventral Nuclei of the Lateral Lemniscus

6.1 Projection of the VNLL to the Ipsilateral Inferior Colliculus

The major projection of most or all of the neurons in the ventral lemniscal nuclei is to the ipsilateral inferior colliculus. Injection of label into the inferior colliculus fills many cell bodies in the ipsilateral INLL and VNLL retrogradely in all mammals in which the projection has been studied. The projection is stronger to the rostral and ventrolateral than to the caudal and dorsomedial portions of the inferior colliculus in cats (Brunso-Bechtold et al. 1981; Kudo 1981; Whitley and Henkel 1984). Anterograde tracing methods show that while the projection of INLL (called the dorsal zone of VNLL) and the VNLLd (the middle zone) are generally similar, they are not identical (Whitley and Henkel 1984). Occasionally, neurons of the INLL and VNLL have been seen to project to the contralateral inferior colliculus in cats and gerbils but bilaterally projecting neurons are rare (Adams 1979; Brunso-Bechtold et al. 1981; Nordeen et al. 1983).

Unlike the connections among many of the brainstem auditory nuclei with the inferior colliculus, which follow the tonotopical organization, the projection of the ventral lemniscal nuclei to the inferior colliculus is broad and widespread and strikingly uneven. Restricted injections of retrograde label into the inferior colliculus or anterograde label into the ventral nuclei of the lateral lemniscus suggest that the projections are patchy in all species in which they have been investigated (Roth et al. 1978; Adams 1979; Zook and Casseday 1982; Whitley and Henkel 1984; Covey and Casseday 1986; Merchán and Berbel 1996; Glendenning and Hutson 1998; Malmierca et al. 1998). While the pattern of projection appears at first glance to be disorderly, the pattern of patches varies systematically. In rats, the projection appears to be concentrically organized with respect to the tonotopic organization of the target (Merchan and Berbel 1996). In cats, there is a mediolateral gradient; patches that project to high-frequency areas of the inferior colliculus are more common laterally and patches projecting to low-frequency areas more common medially (Malmierca et al. 1998) (Fig. 6.10). A consistent feature of these patches is that they interdigitate in a three dimensional mosaic (Glendenning and Hutson 1998; Malmierca et al. 1998). Such an organization promotes the juxtaposition in the ventral nuclei of the lateral lemniscus of neurons with targets to different frequency regions of the inferior colliculus. Such a juxtaposition promotes neuronal interactions across frequencies. Whether or not the frequency tuning of the ventral lemniscal neurons matches that of their targets, the mosaic organization promotes cross-frequency interactions. Those interactions occur either in the ventral nuclei of the lateral lemniscus or in the inferior colliculus. Cross-

High freq.   Middle freq.   Low freq.

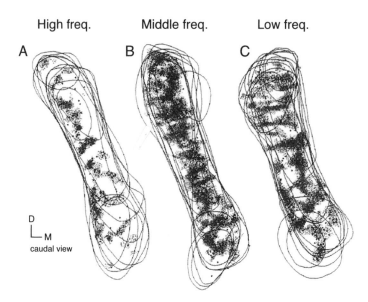

D
└ M
caudal view

FIGURE 6.10. Projection pattern of neurons in the ventral nuclei of the lateral lemniscus of a cat to the inferior colliculus is patchy. Neurons that project to the high frequency region of the inferior colliculus are more likely to be located laterally and those that project to low frequency areas to be located medially. Each of the panels shows the location of labeled cells (black dots) and labeled fibers (smaller gray dots) after deposition of biotinylated dextran in the inferior colliculus. A: Cells and fibers labeled when an injection was made in the ventromedial, high frequency region of the inferior colliculus. B: Labeling after an injection into the mid-frequency region of the inferior colliculus. C: Labeling after an injection into the dorsolateral, low frequency region of the inferior colliculus. (Reprinted with permission Malmierca et al., 1998.)

frequency interactions have been documented. Portfors and Wenstrup (2000) have demonstrated that some neurons in the ventral nuclei of the lateral lemniscus and more in the inferior colliculus show facilitated responses to combinations of tones, suggesting that there is a convergence of neurons with quite distinct frequency tuning in these regions of the auditory pathway.

An interesting exception to the apparently disorderly projection from the ventral lemniscal nuclei to the inferior colliculus is the projection of the hypertrophied area of spherical bushy cells, (columnar area) in bats that corresponds to VNLLv. Covey and Casseday (1986) show that sheets of neurons projected topographically to the inferior colliculus; the ventral bands of the columnar area of the VNLL project ventrally in the inferior colliculus and dorsal bands project dorsally. It is striking that this systematic projection is detected in those cells in which tuning is difficult to measure. Neurons in the columnar area are broadly tuned with poorly defined tips to their tuning curves (Fig. 6.9, middle panel). These results show that the tuning properties of VNLLv neurons and their targets are not similarly tuned; neurons of the VNLLv are presumably more broadly tuned than their targets in the inferior colliculus.

Most neurons in the ventral lemniscal nuclei are probably inhibitory. Many, including the spherical bushy cells of the VNLLv, are probably glycinergic while some multipolar cells may be GABAergic. In cats, over 90% of cell bodies in the ventral lemniscal nuclei are immunoreactive for glycine and/or GABA; 81% were immunoreactive for glycine alone (Saint Marie et al. 1997). In mustached bats, immunocytochemical staining with antibodies against glycine conjugates labels all neurons of VNLLv, the columnar area, and many neurons in VNLLd and INLL (Winer et al. 1995). Neurons of the ventral nuclei of the lateral lemniscus also take up glycine (Saint Marie and Baker 1990). In rats, cell bodies and fibers were labeled by antibodies against glutamic acid decarboxylase (GAD), the enzyme that catalyzes the synthesis of GABA (Moore and Moore 1987). Both in cats and bats, some neurons in VNLLd and INLL were double labeled for glycine and GABA.

How is somatic labeling correlated with neurotransmitter release? Studies have not yet been done in the lemniscal nuclei, but in other parts of the auditory pathway, glycine-positive and glycine- and GABA-double labeled neurons have been shown functionally to be glycinergic (Wickesberg et al. 1994; Golding and Oertel 1997). Glycine-positive puncta, thought to be glycine-releasing terminals, are prominent in the ventral inferior colliculus where ventral lemniscal neurons terminate most abundandly; GABA-positive puncta are most prominent dorsally (Winer et al. 1995). Glycine receptors, too, are densest in the ventral inferior colliculus (Fubara et al. 1996; Friauf et al. 1997). The prevalence of glycine-positive punta and the presence of glycine receptors thus matches the pattern of projections from the ventral lemniscal nuclei determined by anterograde and retrograde labeling. Thus, we tentatively conclude that many ventral lemniscal neurons and all of the spherical neurons in the VNLLv, are glycinergic and inhibitory. Some multipolar neurons are likely to be GABAergic. Whether excitatory neurons are intermingled among the inhibitory neurons is unclear. If they exist, excitatory neurons are likely to represent only a small proportion of neurons in the ventral nuclei of the lateral lemniscus.

6.2 Minor Projections of the VNLL

In addition to the major projection to the ipsilateral inferior colliculus, ventral lemniscal neurons have minor projections to the thalamus, the DNLL, the VNLL, the dorsomedial periolivary nucleus (DMPO), and the ventral nucleus of the trapezoid body (VNTB) (Whitley and Henkel 1984; Hutson et al. 1991). While these studies agree that some neurons in the INLL project to the medial geniculate body of the thalamus bilaterally, only

the earlier study reports a projection of the VNLLv to the medial geniculate body. Neurons in the VNLLv also project to more dorsal regions of the
VNLLd, INLL, and DNLL, indicative of a role as interneurons in the
ventral lemniscal nuclei.

7. Possible Integrative Role of the VNLL in Pattern Recognition

The ventral lemniscal nuclei, which comprise neuronal circuits that are
interposed between the cochlear nuclei and the midbrain, are ubiquitous
not only in mammals but also birds and reptiles (Carr and Code 2000), but
their roles are not well understood in any species. A consideration of their
functional role is therefore necessarily speculative and should be regarded
as a working hypothesis. It is useful to formulate such a hypothesis because
it provides a framework into which to place the anatomical and electrophysiological studies.

The ubiquitous presence of the ventral lemniscal nuclei is one indication
that they are involved in fundamental integrative processes of the auditory
system for localizing and/or understanding the meaning of sounds. Ventral
lemniscal nuclei are predominantly monaural in most, but not all, species.
This leads to the question of which characteristics of a sound can be distinguished using predominantly monaural information?

Abundant psychophysical, behavioral, anatomical, and electrophysiological evidence has led to the conclusion that localization of sound in the horizontal plane depends on interaural time and level cues, which involve the
binaural neuronal circuits of the medial and lateral superior olivary nuclei
(see Chapter 4). As these neuronal circuits bypass the ventral lemniscal
nuclei, it is unlikely that the ventral lemniscal nuclei play an integrative role
in sound localization in the horizontal plane.

Can the ventral lemniscal nuclei be involved in localization of sound in
the vertical plane? While a role in localization in elevation cannot be ruled
out, there is evidence that at least some of the neuronal circuits that are
involved bypass the ventral nuclei of the lateral lemniscus. Spectral cues are
used to localize sound in the vertical plane and can be used to localize sound
in the horizontal plane (see Chapter 5, this volume). The use of spectral
cues does not necessarily require binaural interactions. Exactly which neuronal pathways are required for making use of spectral cues is not completely understood in any species. In birds, it has been shown that nucleus
angularis, which contains the avian homologues of stellate cells (but not
bushy cells), is involved in the localization of sound in elevation (Takahashi
et al. 1984). Whether this function in birds involves the lemniscal nuclei or
the direct projection of stellate cells to the midbrain was not tested in those
experiments; however, the presence of a strong projection of stellate cells
directly to the midbrain makes the pathway through the lemniscal nuclei

unnecessary in accounting for those experimental observations. In mammals, the dorsal cochlear nucleus has been implicated in contributing to the detection of spectral cues for localization of sounds (see Chapter 5, this volume). The principal neurons of the dorsal cochlear nucleus also project directly to the inferior colliculus, bypassing the ventral nuclei of the lateral lemniscus.

It is by process of elimination that we arrive at the working hypothesis that the ventral lemnical nuclei are involved in the recognition and interpretation of the meaning of sounds. This hypothesis makes intuitive sense. Understanding speech, for example, is a monaural function that relies heavily on temporal pattern recognition (Shannon et al. 1995); people or other animals who lose hearing in one ear do not lose the ability to understand speech or to recognize other natural sounds. The recognition of sounds involves the distinction of temporal features of sound that are well encoded by the bushy, stellate, and octopus cells of the cochlear nuclei that provide the input to the ventral lemniscal nuclei. Spectral features, on the other hand, could be transmitted by stellate cells directly to the inferior colliculus. The conclusion that the ventral lemniscal nuclei participate in the temporal processing of acoustic information and pattern recognition was initially suggested by Covey and Casseday (1986, 1991) on the basis of their studies of the VNLL in bats. The hypothesis that the neuronal circuits through the ventral nuclei of the lateral lemniscus are involved in the integrative processes that underlie recognition of temporal patterns does not exclude the possibility that other pathways contribute to the task.

One common temporal pattern that is detected by the mammalian auditory system is the interval between acoustic transients. Detecting that pattern or enabling its detection may be a function of the monaural auditory system, and one that can be used in several different ways. Acoustic transients produce coincident firing of auditory nerve fibers across the tonotopic array that is detected by octopus cells and transmitted to the lateral lemniscus (Oertel et al. 1990; Golding et al. 1995, 1999). Periodic sounds, including low-frequency tones, harmonic complexes, and click trains behave like repeated acoustic transients and activate octopus cells effectively (as shown in Fig. 6.6). Cariani and Delgutte (1996a, b) demonstrated that pitch estimates of complex signals were correlated with pooled interspike interval distributions. These investigators postulated the existence of a central processor that could analyze interval patterns across the population of auditory nerve fibers. The octopus cells are anatomically and biophysically specialized to detect and convey exactly such patterns of activation in groups of auditory nerve fibers. Phase-locked responses to components of harmonic complexes in auditory nerve fibers would be expected to sum in octopus cells and maximal coincidence would be expected to recur at the pitch period, conclusions that have been explored by Kim et al. (1986).

Acoustic transients often occur at the onset and offset of a sound. Covey and colleagues proposed that the information from the lateral lemniscus arrives at the target neurons in the IC as a series of signals that arrive at different times (Covey and Casseday 1991; Casseday et al. 1994). The variety of latencies enables the target neurons to encode specific time delays or temporal patterns such as stimulus duration (Chapter 7).

The detection of the temporal intervals between acoustic transients is useful for detecting the voice onset time (VOT) associated with the differentiation of stop consonants (Sinex and MacDonald 1988, 1989; Stevens and Wickesberg 1999). Even in whispered stop consonants, the duration of the interval distinguishes /d/, which has a short gap, from /t/, which has a longer gap. A significant contribution of high-frequency-encoding auditory nerve fibers was detected in the signaling the occurrence of an acoustic transient, especially for whispered syllables (Stevens and Wickesberg 1999). This contribution is consistent with the improved gap discrimination observed with broadband and high frequency signals. The pattern of responses to the con-

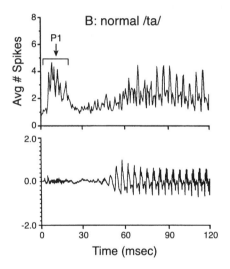

FIGURE 6.11. Global average peristimulus time histogram of responses of a representative population of auditory nerve fibers to /ta/ spoken by a female voice. Responses to 50 repetitions of the stimulus were recorded from each of 71 auditory nerve fibers with low, medium and high spontaneous firing whose characteristic frequencies ranged from 100 to 17,000 Hz. Stimuli were at a constant intensity with peak amplitudes matched to the amplitude of a 60 dB, 1,000 Hz tone. Peristimulus time histograms of the population of fibers were then averaged. In response to the spoken syllable, /ta/, auditory nerve fibers first fired in a broad peak (P1) over about 15 ms. This peak was followed by period asynchronous firing at just over the spontaneous firing rate that lasted about 30 ms. Voicing of the vowel was accompanied by firing with synchronous peaks at the fundamental period, the period corresponding to the pitch of the voice. The lower panel shows the normalized pressure waveform of the stimulus. (Reprinted from Hearing Research, Vol. 131, Stevens and Wickesberg, Ensemble responses of the auditory nerve to normal and whispered stop consonants, pp 47–62, copyright 1999, with permission from Elsevier Science.)

sonants was analyzed experimentally by Stevens and Wickesberg (1999) using ensemble or global peristimulus time histograms from a sample auditory nerve fibers (Fig. 6.11). Octopus cells perform a similar analysis cellularly.

One temporal attribute of a sound that is detected primarily monaurally is the presence of a gap or silent period. Phillips et al. (1997) demonstrated that the monaural detection of a gap is extremely good, while a gap that occurs between presentations to opposite ears is much more difficult to detect. Gap detection is a function of the bandwidth of the stimulus, with the best detection occurring for broadband stimuli (Grose et al. 1989; Grose 1991; Eddins et al. 1992; Hall et al. 1996). A gap within a broadband stimulus is essentially the interval between the two acoustic transients that occur as the stimulus is turned off and then back on again. The better detectability of a gap with wideband or high-frequency stimuli suggests that it is the temporal sharpness of the acoustic transients that is important. While gap detection may be predominantly monaural, binaural inputs such as those described by Batra and Fitzpatrick (1997, 1999) and Hutson (2000) are not excluded and may improve performance. Patients with auditory neuropathy sometimes have abnormally low synchrony in auditory nerve discharge, which is detected as small auditory brainstem responses. Such patients have deficits in speech recognition that are disproportionate to their hearing losses and show dramatically increased thresholds for gap detection (Zeng et al. 1999).

References

Adams JC (1979) Ascending projections to the inferior colliculus. J Comp Neurol 183:519–538.

Adams JC (1986) Neuronal morphology in the human cochlear nucleus. Arch Otolaryngol Head Neck Surg 112:1253–1261.

Adams JC (1997) Projections from octopus cells of the posteroventral cochlear nucleus to the ventral nucleus of the lateral lemniscus in cat and human. Aud Neurosci 3:335–350.

Aitkin LM, Anderson DJ, Brugge JF (1970) Tonotopic organization and discharge characteristics of single neurons in nuclei of the lateral lemniscus of the cat. J Neurophysiol 33:421–440.

Bal R, Oertel D (2000) Hyperpolarization-activated, mixed-cation current (Ih) in octopus cells of the mammalian cochlear nucleus. J Neurophysiol 84:806–817.

Bal R, Oertel D (2001) Potassium currents in octopus cells of the mammalian cochlear nucleus. J Neurophysiol (in press).

Batra R, Fitzpatrick DC (1997) Neurons sensitive to interaural temporal disparities in the medial part of the ventral nucleus of the lateral lemniscus. J Neurophysiol 78:511–515.

Batra R, Fitzpatrick DC (1999) Discharge patterns of neurons in the ventral nucleus of the lateral lemniscus of the unanesthetized rabbit. J Neurophysiol 82:1097–1113.

Blackburn CC, Sachs MB (1989) Classification of unit types in the anteroventral cochlear nucleus: PST histograms and regularity analysis. J Neurophysiol 62: 1303–1329.

Brunso-Bechtold JK, Thompson GC, Masterton RB (1981) HRP study of the organization of auditory afferents ascending to central nucleus of inferior colliculus in cat. J Comp Neurol 197:705–722.

Cant NB (1981) The fine structure of two types of stellate cells in the anterior division of the anteroventral cochlear nucleus of the cat. Neurosci 6:2643–2655.

Cant NB (1992) The Cochlear Nucleus: Neuronal Types and Their Synaptic Organization. In: Webster DB, Popper AN, Fay RR (eds) The Mammalian Auditory Pathway: Neuroanatomy. pp. 66–116. New York: Springer-Verlag.

Cant NB, Gaston KC (1982) Pathways connecting the right and left cochlear nuclei. J Comp Neurol 212:313–326.

Cariani PA, Delgutte B (1996a) Neural correlates of the pitch of complex tones. II. Pitch shift, pitch ambiguity, phase invariance, pitch circularity, rate pitch, and the dominance region for pitch. J Neurophysiol 76:1717–1734.

Cariani PA, Delgutte B (1996b) Neural correlates of the pitch of complex tones. I. Pitch and pitch salience. J Neurophysiol 76:1698–1716.

Carr CE, Code RA (2000) The central auditory system of reptiles and birds. In: Dooling R, Popper AN, Fay RR (eds) The Auditory System of Birds and Reptiles. pp. 197–248. New York: Springer-Verlag.

Casseday JH, Ehrlich D, Covey E (1994) Neural tuning for sound duration: role of inhibitory mechanisms in the inferior colliculus. Science 264:847–850.

Covey E, Casseday JH (1986) Connectional basis for frequency representation in the nuclei of the lateral lemniscus of the bat *Eptesicus fuscus*. J Neurosci 6: 2926–2940.

Covey E, Casseday JH (1991) The monaural nuclei of the lateral lemniscus in an echolocating bat: parallel pathways for analyzing temporal features of sound. J Neurosci 11:3456–3470.

Doucet JR, Ryugo DK (1997) Projections from the ventral cochlear nucleus to the dorsal cochlear nucleus in rats. J Comp Neurol 385:245–264.

Eddins DA, Hall JW, III, Grose JH (1992) The detection of temporal gaps as a function of frequency region and absolute noise bandwidth. J Acoust Soc Am 91: 1069–1077.

Ehret G (1979) Quantitative analysis of nerve fibre densities in the cochlea of the house mouse (*Mus musculus*). J Comp Neurol 183:73–88.

Ferragamo MJ, Golding NL, Oertel D (1998) Synaptic inputs to stellate cells in the ventral cochlear nucleus. J Neurophysiol 79:51–63.

Friauf E, Ostwald J (1988) Divergent projections of physiologically characterized rat ventral cochlear nucleus neurons as shown by intra-axonal injection of horseradish peroxidase. Exp Brain Res 73:263–284.

Friauf E, Hammerschmidt B, Kirsch J (1997) Development of adult-type inhibitory glycine receptors in the central auditory system of rats. J Comp Neurol 385: 117–134.

Frisina RD, Smith RL, Chamberlain SC (1990) Encoding of amplitude modulation in the gerbil cochlear nucleus: I. A hierarchy of enhancement. Hear Res 44: 99–122.

Fubara BM, Casseday JH, Covey E, Schwartz-Bloom RD (1996) Distribution of $GABA_A$, $GABA_B$, and glycine receptors in the central auditory system of the big brown bat, *Eptesicus fuscus*. J Comp Neurol 369:83–92.

Fujino K, Oertel D (2001) Cholinergic modulation of stellate cells in the mammalian ventral cochlear nucleus. J Neurosci 21:7372–7383.

Glendenning KK, Hutson KA (1998) Lack of topography in the ventral nucleus of the lateral lemniscus. Microsc Res Tech 41:298–312.

Glendenning KK, Brunso-Bechtold JK, Thompson GC, Masterton RB (1981) Ascending auditory afferents to the nuclei of the lateral lemniscus. J Comp Neurol 197:673–703.

Godfrey DA, Kiang NYS, Norris BE (1975) Single unit activity in the posteroventral cochlear nucleus of the cat. J Comp Neurol 162:247–268.

Golding NL, Oertel D (1997) Physiological identification of the targets of cartwheel cells in the dorsal cochlear nucleus. J Neurophysiol 78:248–260.

Golding NL, Ferragamo M, Oertel D (1999) Role of intrinsic conductances underlying transient responses of octopus cells of the cochlear nucleus. J Neurosci 19:2897–2905.

Golding NL, Robertson D, Oertel D (1995) Recordings from slices indicate that octopus cells of the cochlear nucleus detect coincident firing of auditory nerve fibers with temporal precision. J Neurosci 15:3138–3153.

Grose JH (1991) Gap detection in multiple narrow bands of noise as a function of spectral configuration. J Acoust Soc Am 90:3061–3068.

Grose JH, Eddins DA, Hall JW, 3d (1989) Gap detection as a function of stimulus bandwidth with fixed high-frequency cutoff in normal-hearing and hearing-impaired listeners. J Acoust Soc Am 86:1747–1755.

Guinan JJ, Jr., Li RY (1990) Signal processing in brainstem auditory neurons which receive giant endings (calyces of Held) in the medial nucleus of the trapezoid body of the cat. Hear Res 49:321–334.

Guinan JJ, Jr., Norris BE, Guinan S (1972a) Single auditory units in the superior olivary complex II: Locations of unit categories and tonotopic organization. Intern J Neurosci 4:147–166.

Guinan JJ, Jr., Guinan SS, Norris BE (1972b) Single auditory units in the superior olivary complex I: Responses to sounds and classifications based on physiological properties. Intern J Neurosci 4:101–120.

Hall JW, 3rd, Grose JH, Joy S (1996) Gap detection for pairs of noise bands: effects of stimulus level and frequency separation. J Acoust Soc Am 99:1091–1095.

Henkel CK (1983) Evidence of sub-collicular auditory projections to the medial geniculate nucleus in the cat: an autoradiographic and horseradish peroxidase study. Brain Res 259:21–30.

Huffman RF, Covey E (1995) Origin of ascending projections to the nuclei of the lateral lemniscus in the big brown bat, *Eptesicus fuscus*. J Comp Neurol 357:532–545.

Huffman RF, Argeles PC, Covey E (1998) Processing of sinusoidally amplitude modulated signals in the nuclei of the lateral lemniscus of the big brown bat, *Eptesicus fuscus*. Hear Res 126:181–200.

Hutson KA (2000) Bilateral projections from cochlear nucleus to nuclei of the lateral lemniscus in a neotropical fruit bat, *Artibeus jamaicensis*. Assoc Res Otolaryngol Abstr 23:36.

Hutson KA, Glendenning KK, Masterton RB (1991) Acoustic chiasm. IV: Eight midbrain decussations of the auditory system in the cat. J Comp Neurol 312:105–131.

Joris PX, Carney LH, Smith PH, Yin TC (1994a) Enhancement of neural synchronization in the anteroventral cochlear nucleus. I. Responses to tones at the characteristic frequency. J Neurophysiol 71:1022–1036.

Joris PX, Smith PH, Yin TC (1994b) Enhancement of neural synchronization in the anteroventral cochlear nucleus. II. Responses in the tuning curve tail. J Neurophysiol 71:1037–1051.

Kim DO, Rhode WS, Greenberg SR (1986) Responses of cochlear nucleus neurons to speech signals: neural encoding of pitch, intensity and other parameters. In: Moore BCJ, Patterson RD (eds) Auditory Frequency Selectivity. pp. 281–288. Plenum Publishing Corp.

Kudo M (1981) Projections of the nuclei of the lateral lemniscus in the cat: an autoradiographic study. Brain Res 221:57–69.

Kudo M, Nakamura Y, Tokuno H, Kitao Y (1990) Auditory brainstem in the mole (*Mogera*): nuclear configurations and the projections to the inferior colliculus. J Comp Neurol 298:400–412.

Liberman MC (1993) Central projections of auditory nerve fibers of differing spontaneous rate, II: Posteroventral and dorsal cochlear nuclei. J Comp Neurol 327: 17–36.

Malmierca MS, Leergaard TB, Bajo VM, Bjaalie JG, Merchan MA (1998) Anatomic evidence of a three-dimensional mosaic pattern of tonotopic organization in the ventral complex of the lateral lemniscus in cat. J Neurosci 18:10603–10618.

Merchan MA, Berbel P (1996) Anatomy of the ventral nucleus of the lateral lemniscus in rats: a nucleus with a concentric laminar organization. J Comp Neurol 372:245–263.

Moore JK, Moore RY (1987) Glutamic acid decarboxylase-like immunoreactivity in brainstem auditory nuclei of the rat. J Comp Neurol 260:157–174.

Nordeen KW, Killackey HP, Kitzes LM (1983) Ascending projections to the inferior colliculus following unilateral cochlear ablation in the neonatal gerbil, Meriones unguiculatus. J Comp Neurol 214:144–153.

Oertel D (1985) Use of brain slices in the study of the auditory system: spatial and temporal summation of synaptic inputs in cells in the anteroventral cochlear nucleus of the mouse. J Acoust Soc Amer 78:328–333.

Oertel D (1997) Encoding of timing in the brainstem auditory nuclei of vertebrates. [Review] [14 refs]. Neuron 19:959–962.

Oertel D, Wu SH, Garb MW, Dizack C (1990) Morphology and physiology of cells in slice preparations of the posteroventral cochlear nucleus of mice. J Comp Neurol 295:136–154.

Oertel D, Bal R, Gardner SM, Smith PH, Joris PX (2000) Detection of synchrony in the activity of auditory nerve fibers by octopus cells of the mammalian cochlear nucleus. Proc Nat Acad Sci 97:11773–11779.

Oliver DL (1987) Projections to the inferior colliculus from the anteroventral cochlear nucleus in the cat: possible substrates for binaural interaction. J Comp Neurol 264:24–46.

Pfeiffer RR (1966) Anteroventral cochlear nucleus: wave forms of extracellularly recorded spike potentials. Science 154:667–668.

Phillips DP, Taylor TL, Hall SE, Carr MM, Mossop JE (1997) Detection of silent intervals between noises activating different perceptual channels: some properties of "central" auditory gap detection. J Acoust Soc Am 101:3694–3705.

Portfors CV, Wenstrup JJ (2000) Responses to complex sounds in the nuclei of the lateral lemniscus. Assoc Res Otolargyngol Abstr 23:287.

Recio A, Rhode WS (2000) Representation of vowel stimuli in the ventral cochlear nucleus of the chinchilla. Hearing Res 146:167–184.

Rhode WS, Smith PH (1986) Encoding timing and intensity in the ventral cochlear nucleus of the cat. J Neurophysiol 56:261–286.

Rhode WS, Greenberg S (1992) Physiology of the cochlear nuclei. In: Fay RR, Popper AN (eds) Auditory Research, Volume 2: The Physiology of the Mammalian Auditory Central Nervous System. pp. 94–152.

Roth GL, Aitkin LM, Andersen RA, Merzenich MM (1978) Some features of the spatial organization of the central nucleus of the inferior colliculus of the cat. J Comp Neurol 182:661–680.

Ruggero MA (1992) Physiology and Coding of Sound in the Auditory Nerve. In: Popper AN, Fay RR (eds) The Mammalian Auditory Pathway: Neurophysiology. pp. 34–93. New York: Springer-Verlag.

Saint Marie RL, Baker RA (1990) Neurotransmitter-specific uptake and retrograde transport of [³H] glycine from the inferior colliculus by ipsilateral projections of the superior olivary complex and nuclei of the lateral lemniscus. Brain Res 524: 244–253.

Saint Marie RL, Shneiderman A, Stanforth DA (1997) Patterns of gamma-aminobutyric acid and glycine immunoreactivities reflect structural and functional differences of the cat lateral lemniscal nuclei. J Comp Neurol 389:264–276.

Schofield BR (1995) Projections from the cochlear nucleus to the superior paraolivary nucleus in guinea pigs. J Comp Neurol 360:135–149.

Schofield BR, Cant NB (1997) Ventral nucleus of the lateral lemniscus in guinea pigs: cytoarchitecture and inputs from the cochlear nucleus. J Comp Neurol 379: 363–385.

Schwartz IR (1992) The superior olivary complex and lateral lemniscal nuclei. In: Webster DB, Popper AN, Fay RR (eds) The Mammalian Auditory Pathway: Neuroanatomy. pp. 117–167. New York: Springer-Verlag.

Shannon RV, Zeng FG, Kamath V, Wygonski J, Ekelid M (1995) Speech recognition with primarily temporal cues. Science 270:303–304.

Sinex DG, McDonald LP (1988) Average discharge rate representation of voice onset time in the chinchilla auditory nerve. J Acoust Soc Am 83:1817–1827.

Sinex DG, McDonald LP (1989) Synchronized discharge rate representation of voice-onset time in the chinchilla auditory nerve. J Acoust Soc Am 85:1995–2004.

Smith PH, Rhode WS (1989) Structural and functional properties distinguish two types of multipolar cells in the ventral cochlear nucleus. J Comp Neurol 282: 595–616.

Smith PH, Joris PX, Carney LH, Yin TCT (1991) Projections of physiologically characterized globular bushy cell axons from the cochlear nucleus of the cat. J Comp Neurol 304:387–407.

Smith PH, Joris PX, Banks MI, Yin TCT (1993a) Responses of cochlear nucleus cells and projections of their axons. In: Merchan MA, Juiz JM, Godfrey DA, Mugnaini E (eds) The Mammalian Cochlear Nuclei, Organization and Function. pp. 349–360. New York and London: Plenum Press.

Smith PH, Joris PX, Yin TC (1993b) Projections of physiologically characterized spherical bushy cell axons from the cochlear nucleus of the cat: evidence for delay lines to the medial superior olive. J Comp Neurol 331:245–260.

Smith PH, Joris PX, Yin TCT (1998) Anatomy and physiology of principal cells of the medial nucleus of the trapezoid body. J Neurophysiol 79:3127–3142.

Sommer I, Lingenhohl K, Friauf E (1993) Principal cells of the rat medial nucleus of the trapezoid body: an intracellular in vivo study of their physiology and morphology. Exp Brain Res 95:223–239.

Spangler KM, Warr WB, Henkel CK (1985) The projections of principal cells of the medial nucleus of the trapezoid body in the cat. J Comp Neurol 238:249–262.

Stevens HE, Wickesberg RE (1999) Ensemble responses of the auditory nerve to normal and whispered stop consonants. Hear Res 131:47–62.

Takahashi T, Moiseff A, Konishi M (1984) Time and intensity cues are processed independently in the auditory system of the owl. J Neurosci 4:1781–1786.

Vater M, Feng AS (1990) Functional organization of ascending and descending connections of the cochlear nucleus of horseshoe bats. J Comp Neurol 292:373–395.

Vater M, Covey E, Casseday JH (1997) The columnar region of the ventral nucleus of the lateral lemniscus in the big brown bat (Eptesicus fuscus): synaptic arrangements and structural correlates of feedforward inhibitory function. Cell Tiss Res 289:223–233.

Warr WB (1969) Fiber degeneration following lesions in the posteroventral cochlear nucleus of the cat. Exp Neurol 23:140–155.

Warr WB (1992) Organization of olivocochlear efferent systems in mammals. In: Webster DB, Popper AN, Fay RR (eds) The Mammalian Auditory Pathway: Neuroanatomy. pp. 410–448. New York: Springer.

Wenthold RJ (1987) Evidence for a glycinergic pathway connecting the two cochlear nuclei: an immunocytochemical and retrograde transport study. Brain Res 415:183–187.

Whitley JM, Henkel CK (1984) Topographical organization of the inferior collicular projection and other connections of the ventral nucleus of the lateral lemniscus in the cat. J Comp Neurol 229:257–270.

Wickesberg RE (1996) Rapid inhibition in the cochlear nuclear complex of the chinchilla. J Acoust Soc Am 100:1691–1702.

Wickesberg RE, Oertel D (1988) Tonotopic projection from the dorsal to the anteroventral cochlear nucleus of mice. J Comp Neurol 268:389–399.

Wickesberg RE, Oertel D (1990) Delayed, frequency-specific inhibition in the cochlear nuclei of mice: A mechanism for monaural echo suppression. J Neurosci 10:1762–1768.

Wickesberg RE, Whitlon DS, Oertel D (1994) In vitro modulation of somatic glycine-like immunoreactivity. J Comp Neurol 339:311–327.

Willard FH, Martin GF (1983) The auditory brainstem nuclei and some of their projections to the inferior colliculus in the North American opossum. Neurosci 10:1203–1232.

Willard FH, Ryugo DK (1983) Anatomy of the central auditory system. In: Willott JF (ed) The Auditory Psychobiology of the Mouse. pp. 201–304. Springfield: Charles C Thomas.

Willott JF, Bross LS (1990) Morphology of the octopus cell area of the cochlear nucleus in young and aging C57BL/6J and CBA/J mice. J Comp Neurol 300:61–81.

Winer JA, Larue DT, Pollak GD (1995) GABA and glycine in the central auditory system of the mustache bat: structural substrates for inhibitory neuronal organization. J Comp Neurol 355:317–353.

Winter IM, Palmer AR (1990) Responses of single units in the anteroventral cochlear nucleus of the guinea pig. Hear Res 44:161–178.

Wu SH (1999) Physiological properties of neurons in the ventral nucleus of the lateral lemniscus of the rat: Intrinsic membrane properties and synaptic responses. J Neurophysiol 81:2862–2874.

Zeng FG, Oba S, Garde S, Sininger Y, Starr A (1999) Temporal and speech processing deficits in auditory neuropathy. Neuroreport 10:3429–3435.

Zook JM, Casseday JH (1982) Origin of ascending projections to inferior colliculus in the mustache bat, *Pteronotus parnellii*. J Comp Neurol 207:14–28.

7
The Inferior Colliculus: A Hub for the Central Auditory System

John H. Casseday, Thane Fremouw, and Ellen Covey

1. Introduction

The inferior colliculus (IC) (Fig. 7.1) occupies a strategic position in the central auditory system. Evidence reviewed in this chapter indicates that it is an interface between lower brainstem auditory pathways, the auditory cortex, and motor systems (For abbreviations see Table 7.1). The IC receives ascending input, via separate pathways, from a number of auditory nuclei in the lower brainstem. Moreover, it receives crossed input from the opposite IC and descending input from auditory cortex. These connections suggest that (1) the IC integrates information from various auditory sources and (2) at least some of the integration utilizes cortical feedback. The IC also receives input from ascending somatosensory pathways, suggesting that auditory information is integrated with somatosensory information at the midbrain. Motor-related input to the IC arises from the substantia nigra and globus pallidus. These connections raise the possibility that sensory processing in the IC is modulated by motor action. The major output of the IC is to the auditory thalamocortical system. However, it also transmits information to motor systems such as the deep superior colliculus, and the cerebellum, via the pontine gray. These connections suggest that processing in the IC not only prepares information for transmission to higher auditory centers but also modulates motor action in a direct fashion. In short, the IC is ideally suited to process auditory information based on behavioral context and to direct information for guiding action in response to this information (Aitkin 1986; Casseday and Covey 1996).

Here we review information about the structure and function of the IC in order to develop an overview of its organization and a working theory of what the IC does. It would be impossible in this space to review the entire literature on the IC. Instead, we have focused on structural and functional properties that seem to emerge from interactions within the IC. It is our view that the emergent and specialized forms of processing provide the most revealing clues about what the IC does.

FIGURE 7.1. Three-dimensional overview showing the position of the left inferior colliculus (IC) in the brainstem. Fibers ascend from lower brainstem nuclei in the lateral lemniscus (LL). Within the IC, one fibrodendritic lamina is shown, to provide an idea of how the laminae are tilted from medial to lateral and rostral to caudal. Just rostral to the IC is the superior colliculus (SC). (From Fitzpatrick, Cellular architecture and topographic organization of the inferior colliculus of the squirrel monkey, Journal of Comparative Neurology 164: 185–207, Copyright 1975, Wiley-Liss, Inc. Reprinted by permission of Wiley-Liss, Inc., a subsidiary of John Wiley & Sons, Inc.)

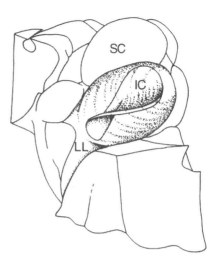

In order to understand the function of the IC, it is particularly important to view its neuroanatomy and physiology within the context of behavior and evolution. The IC, like any other structure of the brain, acquired its connections and functions through evolution and became what it is because it plays some role in helping the species survive. During the course of evolution, different species have come to use hearing for very different purposes. We should not be surprised, therefore, to find species differences in the specific processing that occurs in the IC. However, it is reasonable to propose that these processing differences have been superimposed on a fundamental processing plan and that in all species the IC performs processing tasks that allow the animal to make appropriate responses to sounds in the environment. That is, processing in the IC helps select sounds for a behavioral context. Such processing might include facilitation of relevant sounds and screening or inhibition of nonrelevant sounds, in order to help select information that is needed to initiate action. Therefore, it is important to examine both the sensory and motor connections of the IC.

We begin by reviewing evidence for the well-established structure-function relationship between tonotopy and fibrodendritic laminae and ask the obvious question of what lies orthogonal to the tonotopic organization. Although the answer to this question is far from complete, there are a number of clues. For example, we will review evidence that there are both connectional and biochemical gradients orthogonal to isofrequency contours, and that these gradients are important for shaping the specialized response properties of IC neurons.

TABLE 7.1. Abbreviations

5-HT	5-hydroxytryptamine (serotonin)
AMPA	∝-amino-3-hydroxy-5-methyl-4-isoxazole-4-propionate
AVCN	anteroventral cochlear nucleus
BF	best frequency
BMF	best modulation frequency
DCN	dorsal cochlear nucleus
DNLL	dorsal nucleus of the lateral lemniscus
EE	excitatory/excitatory
EI	excitatory/inhibitory
EPSC	excitatory post synaptic current
EPSP	excitatory post synaptic potential
FM	frequency modulation
GABA	gamma-amino-butyric acid
GP	globus pallidus
IC	inferior colliculus
ICc	central nucleus of the inferior colliculus
ICdc	dorsal cortex of the inferior colliculus
ICx	external cortex of the inferior colliculus
IID	interaural intensity difference
INLL	intermediate nucleus of the lateral lemniscus
IPSC	inhibitory post synaptic current
IPSP	inhibitory post synaptic potential
ITD	interaural time difference
LSO	lateral superior olive
MSO	medial superior olive
NMDA	N-methyl-D-aspartate
PVCN	posteroventral cochlear nucleus
SAM	sinusoidal amplitude modulation
SC	superior colliculus
SCd	deep layer of the superior colliculus
SFM	sinusoidal frequency modulation
SN	substantia nigra
VCN	ventral cochlear nucleus
VNLL	ventral nucleus of the lateral lemniscus
VNLLc	columnar cell division of VNLL
VNLLm	multipolar cell division of VNLL

2. The Structure of the Inferior Colliculus

The internal structure of the IC is characterized by the distribution of different cell types, the patterns of dendritic branching of these cells, and the patterns of termination of afferent, intrinsic, and commissural fibers. Further characterization of the internal organization of the IC includes the biochemical makeup of cells as seen by immunocytochemistry and the cellular membrane properties as seen by electrophysiology. Although some of these areas have only recently begun to be studied, they are important new directions for understanding the function of the IC.

Over the years, there have been many different attempts to subdivide the IC based on anatomical criteria such as Golgi staining, Nissl staining, and

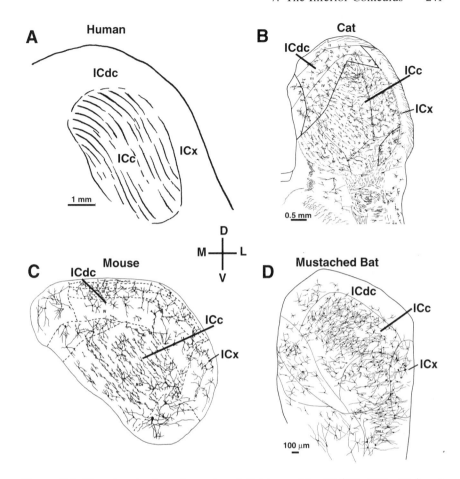

FIGURE 7.2. Drawings to show the main subdivisions of the IC (ICc, ICdc, ICx) and the orientation of disk-shaped cells and fibrodendritic laminae in the IC, as seen in frontal sections through Golgi-stained tissue. A: Orientation of laminae in the IC of the human (from Geniec and Morest 1971); B: cat (from Morest and Oliver, The neuronal architecture of the inferior colliculus in the cat: defining the functional anatomy of the auditory midbrain, Journal of Comparative Neurology 222: 209–236, Copyright 1984, Wiley-Liss, Inc. Reprinted with permission from Wiley-Liss, Inc., a subsidiary of John Wiley & Sons, Inc.); C: mouse (reprinted from Neuroscience, Vol. 17, No. 4, Meininger, Pol and Derer. The inferior colliculus of the mouse, p. 1173, Copyright 1986, with permission from Elsevier Science); D: mustached bat (from Zook et al., Topology of the central nucleus of the mustache bat's inferior colliculus: correlation of single unit properties and neuronal architecture, Journal of Comparative Neurology, Vol. 231, No. 4, Copyright 1985, Wiley-Liss, Inc. Reprinted with permission from Wiley-Liss, Inc., a subsidiary of John Wiley & Sons, Inc.)

immunocytochemistry. Here we will describe only the most basic and simple anatomical subdivisions that are easily related to sensory neurophysiology and leave it to the reader to explore the literature on more complicated cytoarchitectural views.

Based on the criteria of Golgi and Nissl staining, the IC has been sub-divided into three main areas (Fig. 7.2): the central nucleus (ICc), the dorsal cortex (ICdc), and the external cortex (ICx) (e.g., human, Geniec and Morest 1971; cat, Morest and Oliver 1984; bat, Zook et al. 1985; rat, Faye-Lund and Osen 1985; mouse, Meininger et al. 1986). These subdivisions can be related to the patterns of ascending and descending projections, and together, they serve as a framework for discussing the structure and function of the IC.

2.1 Cell Morphology, Fibrodendritic Laminae, and Tonotopy in the IC

Although it has neither the rigid segregation of cell types seen in neocortex nor the connectional interdigitation of the lateral geniculate nucleus, the organization of the ICc is, nevertheless, referred to as laminar. The laminar structure of the central nucleus is defined by the orientation of the "disc-shaped" cells and by the course of input fibers and terminals. The laminae have approximately the same orientation as the isofrequency contours.

The disc-shaped cells, shown in Fig. 7.3, have elongated cell bodies and a disc-like arrangement of their dendrites. The dendrites and cell bodies of the disc-shaped cells that make up a lamina are oriented parallel to one another within a thin, two-dimensional plane that extends dorsomedial to ventrolateral and anterior to posterior (Fig. 7.1). These dendrites are oriented parallel to the incoming terminal fields of ascending and descending auditory fibers and parallel to the local axons of the cells. The local axons of the disc-shaped cells follow the same orientation. Thus, there is a congruent organization of input fibers, dendritic laminae, and intrinsic connections (e.g., human, Geniec and Morest 1971; cat, Rockel and Jones 1973a,b; Morest and Oliver 1984; Oliver 1984; Oliver and Morest 1984; mouse, Willard and Ryugo 1983; rat, Faye-Lund and Osen 1985; Saldaña and Merchán 1992; Saldaña et al. 1996; bat, Zook et al. 1985; marmoset, Garey and Webster 1989; guinea pig, Malmierca et al. 1995; see also Oliver and Heurta 1992 and Ehret 1997).

Figure 7.3 shows the orientation of disc-shaped cells relative to incoming axons from ascending pathways. The fact that most of the incoming fibers are oriented parallel to the orientation of the disc-shaped cells has implications for signal processing in the IC. Most of the afferent connnections form a system in which terminals are organized into separate sheets that are roughly congruent with tonotopy (e.g., bat, Zook and Casseday 1982a, 1987; cat, Oliver 1987; Shneiderman and Henkel 1987; Shneiderman et al. 1988; rat, Merchán et al. 1994). These sheets or laminae have a ventrolateral to dorsomedial orientation and extend through most of the length of the ICc from anterior to posterior (Fig. 7.1, Fig. 7.4). However, terminals from different sources are distributed along slightly different gradients that peak at different points along the lamina (e.g., cat, Oliver et al. 1997; see Oliver

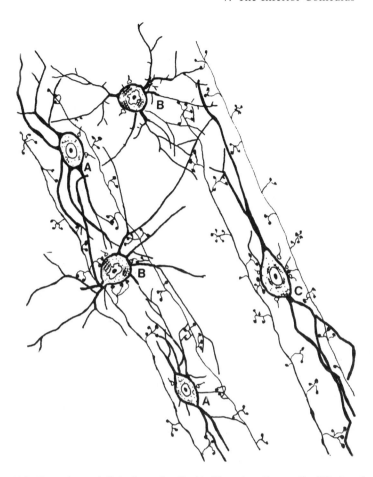

FIGURE 7.3. Drawings of disk-shaped cells (A,C) and stellate cells (B) showing the relation of their dendrites to the terminal fields of ascending fibers to the IC. (Reprinted from Neuroscience, Vol. 11, No. 2, Oliver, Neuron types in the central nucleus of the inferior colliculus that project to the medial geniculate body, Copyright 1984, with permission from Elsevier Science).

and Huerta 1992 for review), a point that will be important later. In the cat (Shneiderman and Henkel 1987) and rat (Saldaña and Merchán 1992; Saldaña et al. 1996), the lateral part of the IC contains shorter projection laminae, oriented in a ventromedial to dorsolateral direction. In frontal sections, the central part of the fiber lamina appears longest and extends throughout the ICc and into the ICdc. The large, central part of each lamina is aligned with the long axis of the disk-shaped cells (and their dendritic fields), and both lie along the axis of isofrequency contours. This organization is no doubt the connectional basis for the single tonotopy found throughout the ICc. The organization and function of the lateral part of the IC, including the ICx, is not well understood, and we shall have much less to say about it than about the ICc.

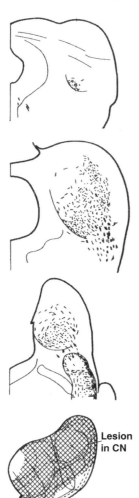

FIGURE 7.4. Drawings to show the pattern of ascending fibers to the IC (top three sections) following a large lesion in the cochlear nucleus of the cat (bottom section) The IC is shown in frontal sections. The cochlear nucleus is shown in a parasagittal view. (From Osen, Projection of the cochlear nuclei on the inferior colliculus in the cat, Journal of Comparative Neurology, Vol. 144, No. 3, Copyright 1972, Wiley-Liss, Inc. Reprinted with permission from Wiley-Liss, Inc., a subsidiary of John Wiley & Sons, Inc.)

Lesion
in CN

2.1.1 The Relationship of Laminae to Tonotopy

The fact that the projections to the IC laminae result in a single tonotopic organization is remarkable when one considers the following facts. First, most, if not all auditory structures in the lower brainstem have their own complete tonotopy, and each appears to have a different function (e.g., rat, Saint Marie et al. 1999; see Irvine 1986, and Covey and Casseday 1995 for reviews). Second, evolutionary pressures to adapt to different sensory environments have produced clear species variations in the structural features of the cochlea and the lower brainstem nuclei and in their frequency maps (e.g., bat, Suga et al. 1975; Kössl and Vater 1985; Covey and Casseday 1986; Casseday et al. 1988; Neuweiler 1990; Covey et al. 1991).

Electrophysiology, connectional tracing, 2-DG labeling, and c-fos mapping all show essentially the same basic tonotopic organization with respect to the main fibrodendritic laminae (e.g., monkey, FitzPatrick 1975; ferret, Moore et al. 1983; cat Servière et al. 1984; mouse, Stiebler and Ehret 1985; bat, Pollak and Casseday 1989; Casseday and Covey 1992; rat, Saldaña and Merchán 1992; Saldaña et al. 1996; see Caird 1991, and Ehret 1997 for reviews). For example, electrophysiological studies of the responses of single units or multiunits to pure tones reveal that the best frequencies (BFs) of neurons are organized in an orderly fashion within the IC. As illustrated in Fig. 7.5, the typical result is that the low BFs are dorsal, lateral and anterior in the ICc and ICdc. Progressively higher BFs are successively located in the ventral, medial and posterior parts of the ICc. Therefore, the isofrequency contours follow the three-dimensional structure of fibrodrendritic laminae in the ICc and extend into the ICdc. An important physiological finding that may help relate structural laminae to isofrequency contours, described in more detail later, is that electrophysiological measures show abrupt frequency transitions orthogonal to the laminae (cat, Schreiner and Langner, 1997). The precise frequency organization of the lateral bands and the ICx is not clear.

The general pattern of tonotopic organization is quite similar across the representative species shown in Fig. 7.5. Nevertheless, there are, of course, species differences in the range of BFs and the relative magnification factor for different frequency ranges. Perhaps the most pronounced example is the mustached bat (*Pteronotus parnellii*). The tonotopic organization of the mustached bat's IC is of special interest because it is the only known species in which tonotopic arrangement deviates from the general mammalian plan. Although the frequencies represented in the IC of this bat range from 10 kHz to 120 kHz, the frequencies representing the dominant harmonic (~60 kHz) of the constant frequency portion of the echolocation call occupy most of the posterior and dorsal part of the IC and are so greatly expanded that the tonotopy appears distorted. Zook et al. (1985) hypothesized that during evolution the 60 kHz region expanded and widened dorsally so that it eventually displaced the areas representing higher and lower frequencies. The expansion is not uniform along the isofrequency contour. Instead, the contour progressively increases in size from the inner, ventromedial IC, to the dorsoposterior surface.

In the mustached bat and horseshoe bat (*Rhinolophus rouxi*), the expanded frequency representations in the IC largely reflect expansion along the highly specialized basilar membrane of the cochlea of frequencies contained in the dominant constant-frequency harmonic of the call. An interesting question is whether the expanded cochlear representation undergoes additional neural magnification in the IC (cat, Merzenich and Reid 1974). The possibility of neural magnification in the IC is also raised by electrophysiological results on the big brown bat (*Eptesicus fuscus*). This species has no cochlear specialization nor does it have an expansion of any

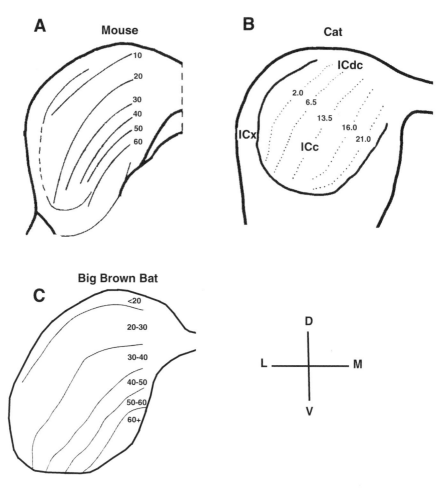

FIGURE 7.5. Drawings to show tonotopic organization in the IC as seen in a frontal section approximately midway through the IC. A: Mouse redrawn from Stiebler and Ehret 1985. B: Cat redrawn from Ehret and Romand 1997, and C: the big brown bat redrawn from Casseday and Covey 1992, copyright from Wiley-Liss, Inc. Note that drawings are not on same scale. Numbers indicate frequency in kHz.

particular frequency range at levels below the IC. Thus, it would not be expected to have any expanded representation in the IC. Nevertheless, the frequencies from 23 kHz to 29 kHz do appear to have an expanded representation in the IC relative to that in lower brainstem nuclei (Poon et al. 1990; Casseday and Covey 1992). These frequencies correspond to the frequencies contained in the quasi-constant frequency signal that the bat uses when searching for prey. If there is neural magnification of specific frequency ranges in the IC of bats or other mammals, it will be further evidence for the evolution of species-specific adaptations in the IC.

Regardless of how it arises, the expanded representation of a frequency range in the IC forms an expanded substrate for other types of processing. For example, the 60kHz area of the mustached bat's IC has provided a useful system for studies of neural processing within one specific frequency band, including binaural processing (e.g., Wenstrup et al. 1986; Park and Pollak 1993b, 1994; Klug et al. 1995; Pollak 1997) or investigations of temporal relationships such as the time interval between pulse and echo (e.g., Mittmann and Wenstrup 1995; Yan and Suga 1996a; Portfors and Wenstrup 2001; Wenstrup and Leroy 2001). These topics will be dealt with in Section 3 of this chapter.

2.1.2 Organization Against the Frequency Grain

So far, we have discussed the connectional basis for integration parallel to the laminae. Not all cells in the IC have dendrites that are oriented parallel to the laminae. Stellate cells, found in both the ICc and the ICdc (Fig. 7.3), have dendrites that are unoriented or that have a mediolateral orientation. In contrast to the disc-shaped cells, the dendrites of the stellate cells typically cross several laminae (cat, Oliver 1984; Oliver and Morest 1984; Oliver et al. 1991). Stellate cells are thus positioned to sample across a wide range of inputs in three dimensions, whereas the disc-shaped cells are positioned to sample within a range of inputs that is restricted in at least one dimension. In fact, there is evidence to support the idea that stellate cells integrate across frequencies (rat, Poon et al. 1992). In the rat IC, some cells respond to frequency modulated signals (FM) but not to pure tones ("FM-specialized cells"). Other cells respond to both FM sweeps and pure tones. Poon et al. (1992) used HRP to fill a population of cells that included both FM-specialized cells and unspecialized cells and found that the specialized cells, but not the unspecialized cells, were characterized by dendrites with large three-dimensional fields, extensive branching, and many spines (Fig. 7.6). Moreover, the best FM range of the cells was highly correlated with how far their dendrites extended across laminae.

The orientation of dendrites is not the only structural basis for cross-talk among laminae. Some intrinsic connections cross laminae (e.g., cat, Oliver and Morest 1984; Oliver et al. 1991). All of these nonparallel connections are potential anatomical substrates for processing that involves frequency integration, lateral inhibition, frequency modulation sensitivity, or frequency combination sensitivity. Moreover, there appears to be even broader integration of frequencies than would be suggested by the pattern of afferent termination. Recent evidence indicates that some IC neurons are sensitive to widely separated combinations of frequencies, such as harmonics. In fact, some of the most interesting questions concern interactions among frequencies and the relationships among different kinds of processing within or among isofrequency contours (Section 3).

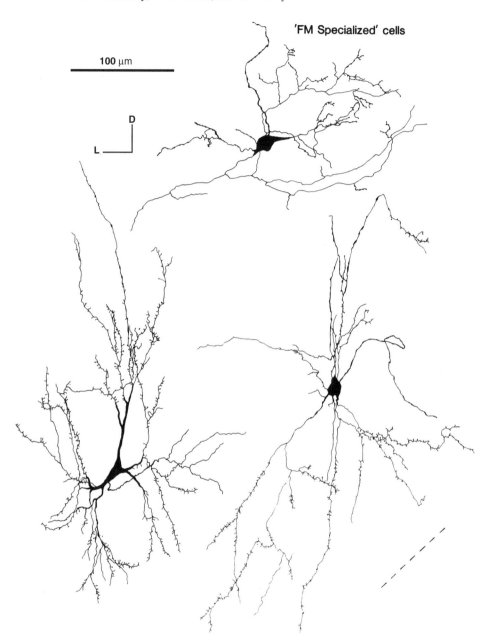

FIGURE 7.6. FM-selective neurons in the IC of the rat labeled intracellularly with HRP to show the extensive branching pattern of their dendrites. The dashed line at bottom right indicates the approximate orientation of the fibrodendritic laminae. (From Poon, Chen and Cheung 1992.)

2.2 Ascending Auditory Pathways to the IC

The lower auditory brainstem consists of more than 10 auditory centers that are distinct in structure and function. These separate structures all converge at the IC in ways that are still poorly understood. The principal pathways arise from the cochlear nuclei, the superior olivary complex, and the nuclei of the lateral lemniscus. Figure 7.7 shows a highly schematic overview of these pathways. Important points that will come up repeatedly in this chapter arise from evidence that (1) the different pathways to the IC can create delay lines as they converge and (2) that some of them invert the incoming neural signal from excitation to inhibition. In the lower brainstem, the pathways that integrate binaural information are anatomically delimited from those that do not. Thus, there are monaural and binaural pathways to the IC (Fig. 7.7). The extent to which the binaural pathways converge with the monaural pathways at the single cell level in the IC is an important but unresolved issue.

The anteroventral cochlear nucleus (AVCN), posteroventral cochlear nucleus (PVCN), and dorsal cochlear nucleus (DCN) all project directly to the contralateral IC and thus provide the IC with direct "monaural" input

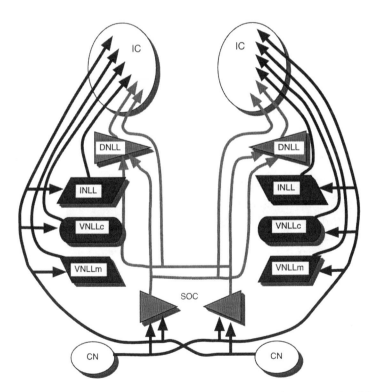

FIGURE 7.7. Schematic overview of major ascending pathways to the IC. The binaural pathways (SOC and DNLL) are shown in more detail in Figure 14. (For abbreviations, see Table 7.1.)

(e.g., cat, Brunso-Bechtold et al. 1981; Cant 1982; Oliver 1987; Thompson 1998; mouse, Ryugo and Willard 1985; bat, Zook and Casseday 1982a, 1985; Vater and Feng 1990; rat, González-Hernández et al. 1996; see also Chapter 2 and Rouiller 1997). Within the IC, the AVCN and PVCN projections are most dense ventrally and the DCN projections are most dense dorsally (cat, Osen 1972; bat, Zook and Casseday 1985; rat, González-Hernández et al. 1996). The AVCN and PVCN project contralaterally to the nuclei of the lateral lemniscus (LL), in a position just below the dorsal nucleus (DNLL) (Chapter 6). These nuclei are especially large and well differentiated in echolocating bats (Schweizer 1981; Zook and Casseday 1982b; Covey and Casseday 1986). They consist of the intermediate (INLL) and ventral nuclei (VNLL) of the lateral lemniscus. In bats, the VNLL has two prominent subdivisions, the columnar (VNLLc) and multipolar cell (VNLLm) nuclei. Because these nuclei receive input almost exclusively from the contralateral side (Chapter 6), they send predominantly monaural information to the IC. Thus, the IC receives monaural information directly from AVCN, PVCN, and DCN and indirectly via INLL, VNLLc, and VNLLm, with the latter three nuclei providing additional processing circuits and synaptic delays (see Covey and Casseday 1995). In some bats and perhaps all mammals, these nuclei transmit special timing information to the IC (Section 3).

The AVCN also projects to the principal binaural structures, the medial superior olive (MSO) and lateral superior olive (LSO). In turn, the MSO and LSO project directly to the IC as well as to the dorsal nucleus of the lateral lemniscus (DNLL), which projects bilaterally to the IC. Thus, the IC receives binaural input directly, from LSO and MSO, as well as indirectly, via a synaptic delay and further processing at DNLL (see Schwartz 1992: Covey and Casseday 1995).

Many of these inputs to the IC are inhibitory. The input from the ipsilateral LSO is mainly inhibitory (e.g., cat, Saint Marie et al. 1989; Glendenning et al. 1992), as is the input from DNLL of both sides (e.g., rat, González-Hernández et al. 1996; Zhang et al. 1998). Many of the cells in the INLL and VNLL stain for γ-amino butyric acid (GABA), glycine, or both and thus provide inhibitory input to the IC (chinchilla and gerbil, Roberts and Ribak 1987; Saint Marie and Baker 1990; bat, Winer et al. 1995; Vater et al. 1997; rat, González-Hernández et al. 1996; Ueyama et al. 1999; Riquelme et al. 2001; cat, Saint Marie et al. 1997).

2.3 Commissural and Intrinsic Connections

The IC on one side is connected to the opposite IC via a large pathway, the commissure of the IC (e.g., cat, Aitkin and Phillips 1984; Hutson et al. 1991; ferret, Moore 1988; guinea pig, Malmicerca et al. 1995; mouse, González-Hernández et al. 1986; rat, Saldaña and Merchán 1992). As shown in Fig. 7.8, crossed IC connections follow the same laminar organization as the ascending connections (rat, Saldaña and Merchán 1992). The ICc projects to the ICx of both sides but most densely to the ipsilateral side. The crossed pro-

1 mm

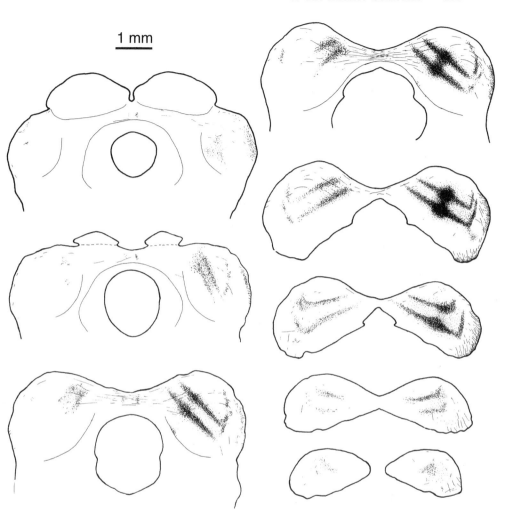

FIGURE 7.8. Intrinsic and crossed IC connections as seen by anterograde labeling after two injections of PHA-L in the right IC of the rat. Frontal sections are arranged from most rostral at upper left to most caudal at lower right. (From Saldana and Merchan, Intrinsic and commissural connections of the rat inferior colliculus, Journal of Comparative Neurology, Vol. 319, No. 3, Copyright 1992, Wiley-Liss, Inc. Reprinted with permission form Wiley-Liss, Inc., a subsidiary of John Wiley & Sons, Inc.)

jections tend to have the most dense termination in the ICdc and dorsal ICc and thus seem to form a dorsoventral gradient (e.g., cat, Brunso-Bechtold et al. 1981; rat, Saldaña and Merchán 1992; González-Hernández et al. 1996). The projections intrinsic to the ICc follow the laminar arrangement, and projections to the ICx are in parallel to the laminar arrangement of the ICc. Similar patterns of intrinsic projections within the ICc and to the ICx have been seen in several other species, such as mouse, opossum (*Didelphis virginiana*), cat, bat (*Rhinolophus rouxi*), and bush baby (*Galago crassicaudatus*) (see Saldaña and Merchán 1992 for review). The projections from one IC to the other appear to be both inhibitory (gerbil, Moore et al. 1998; rat, Smith 1992; González-Hernández et al. 1996) and excitatory (chinchilla, Saint Marie 1996; gerbil, Moore et al. 1998; rat, Smith 1992).

2.4 Descending Pathways from Auditory Cortex

Projections to the IC arise from pyramidal cells in layer V of the auditory cortex. Axons from the pyramidal cells terminate mainly in the ICdc, the ICx, and, to a lesser extent, in the ICc (monkey, Fitzpatrick and Imig 1978; guinea pig, Feliciano and Potashner 1995; Budinger et al. 2000; rat, Saldaña et al. 1996; Druga et al. 1997; cat, Winer et al. 1998; see also Huffman and Henson 1990 and Chapter 2).

As shown in Fig. 7.9 the descending input from the auditory cortex is related to the laminae and subdivisions of the IC (rat, Saldaña et al. 1996). Saldaña et al. (1996) used the highly sensitive tracer *Phaseolus vulgaris*-leucoagglutinin (PHA-L) to observe anterograde projections from the auditory cortex. Projections from the auditory cortex form two bands, one in the lateral part of the IC and one located centrally. The lateral band is confined to the ICx; the medial band extends through the ICdc and into the ICc.

These results allow us to bring together a number of observations on how the laminar structure of the IC relates to the ascending and descending auditory pathways. The bands of labeled terminals in the ICc are parallel to the fibrodendritic laminae and to the orientation of the ascending projections and thus, are parallel to isofrequency contours. However, the ascending projections differ in an important aspect from the descending and crossed projections. Terminals of the ascending projections are most dense ventrally whereas terminals of the descending projections are most dense dorsally and least dense ventrally, as are the crossed projections from one IC to another (cat, Brunso-Bechtold et al. 1981; rat, Saldaña and Merchán 1992; Saldaña et al. 1996).

Electron microscopy shows that corticocollicular fiber terminals have round synaptic vesicles that are densely and evenly packed, which are characteristic of excitatory synapses (rat, Saldaña et al. 1996). This and other evidence (guinea pig, Feliciano and Potashner 1995) strongly suggest that the descending cortical projections to the IC are excitatory.

2.5 Connections Outside the Auditory Pathway

2.5.1 Somatosensory Input

The nucleus gracilis and nucleus cuneatus have been shown to project to the ICx in a number of species (cat, Hand and Van Winkle 1977; Aitkin et al. 1978;

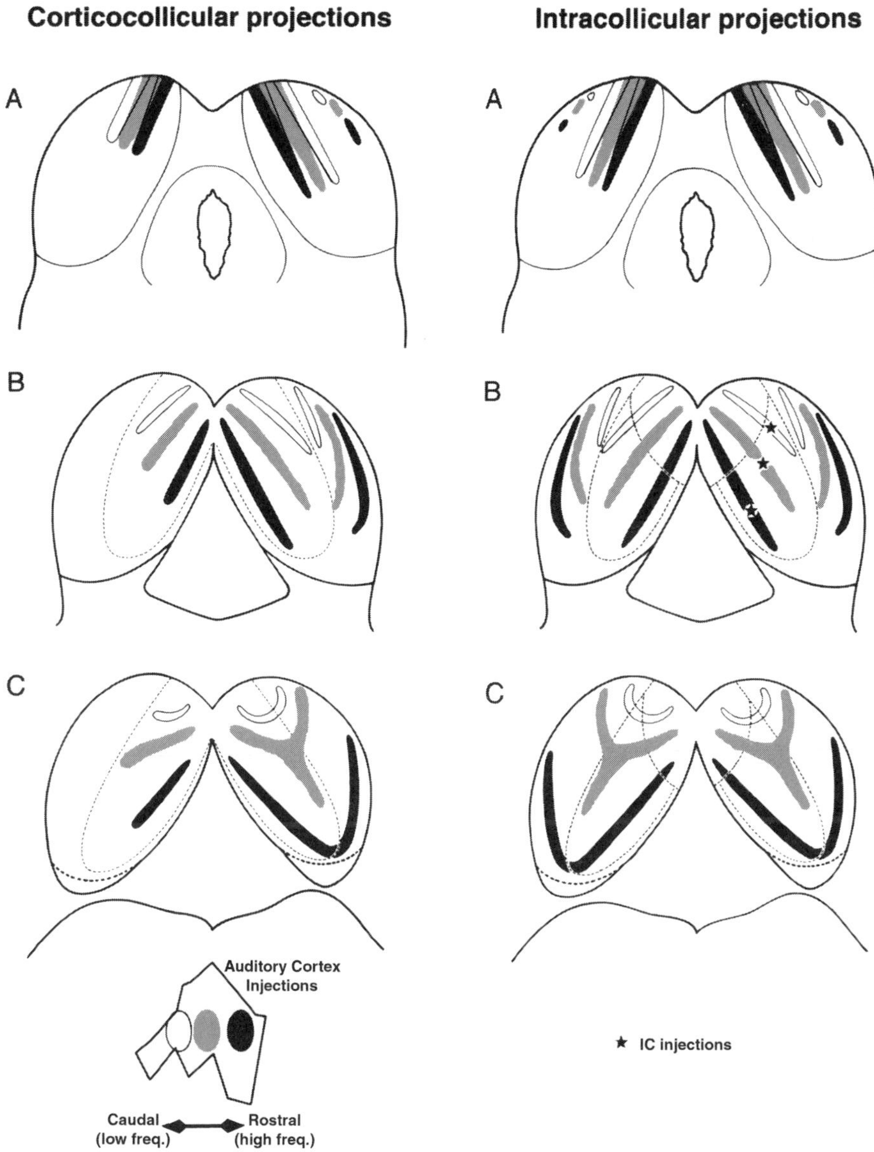

Corticocollicular projections   **Intracollicular projections**

FIGURE 7.9. Summary diagrams to compare the pattern of termination of descending fibers from auditory cortex (left column) with intrinsic and crossed projections (right column) in the IC of the rat. (From Saldana, Feliciano and Mugnaini, Distribution of descending projections from primary auditory neocortex to inferior colliculus mimics the topography of intracollicular projections, Journal of Comparative Neurology, Vol. 371, No. 1, Copyright 1996, Wiley-Liss, Inc. Reprinted with permission from Wiley-Liss, Inc., a subsidiary of John Wiley & Sons, Inc.)

Aitkin et al. 1981; Bjorkel and and Boivie 1984a,b; Wiberg and Blomqvist 1984; Cooper and Dostrovsky 1985; Paloff and Usunoff, 1992; rat, Massopust et al. 1985; Coleman and Clerici 1987; Li and Mizuno 1997a,b; opossum, RoBards et al. 1976; Robards 1979; Willard and Martin 1983; hedgehog (*Echinops telfairi*), Kunzle 1993). In the monkey, there are dense projections from dorsal column nuclei to both ICdc and ICx (Wiberg et al. 1987).

In the ICx, bimodal neurons have been found that respond to both cutaneous and auditory stimulation (cat, Aitkin et al. 1981). Furthermore, the projection area of the DCN can indirectly receive somatosensory information because the DCN receives somatosensory input (see Chapter 5), and some neurons in the dorsal column project both to the DCN and to the ICx (rat, Li and Mizuno 1997b).

In the rat, Tokunaga et al. (1984) showed by retrograde transport that the dorsal IC, the "cortical zone" and the ICx receive more nonauditory input (sensory and motor) than does the ICc. They suggested that the areas external to the ICc act to integrate multisensory and motor information.

2.5.2 Inputs from Motor Systems

The principal nonauditory inputs to the IC are from forebrain areas involved in the initiation of motor actions: the globus pallidus (GP) and the substantia nigra (SN). It has been known for some time that the SN projects to the IC of several species (cat, Adams 1980; Olazábal and Moore 1989; rat, Moriizumi and Hattori 1991a; and bat, Olazábal and Moore 1989). The projection cells are located in the dorsolateral portion of SN and appear to utilize the inhibitory neurotransmitter GABA (rat, Yasui et al. 1991; Moriizumi et al. 1992). In addition, the GP projects to the ICx (rat, Moriizumi and Hattori 1991a; Shammah-Lagnado et al. 1996). The projections from SN are mainly to ICdc and ICx (rat, Yasui et al. 1991). The fact that the GP also projects to the SN further implicates both structures as candidates for extraauditory control of processing in the IC. The fact that the GP has connections with cortical and thalamic areas as well as other striatopallidal areas suggests that it may be part of one or more feedback loops having to do with sensory-motor processing (rat, Shammah-Lagnado et al. 1996) which would include audiomotor processing in the IC.

In summary, these projections are significant for function of the IC because the GP and SN are motor structures associated with the initiation of intentional motor actions. Therefore, an important hypothesis is that some processing in the IC may be modulated by the animal's actions. Exactly what sort of processing is modulated is unknown. It is tempting to speculate that an operation similar to saccadic suppression, where eye movements momentarily "suppress" perception, might be at work (see Casseday and Covey 1996). Further definition of the IC targets of SN and GP will be important for defining functional areas or functional gradients in the IC.

2.5.3 Outputs to Motor Systems

2.5.3.1 The Pathway to the Superior Colliculus

The IC projects to the deep layers of the superior colliculus (SC), which are premotor areas for movements of the eyes, pinnae, and head. Retrograde tracers injected into the SC show that projections arise from the ICx (cat, Edwards et al. 1979; Appell and Behan 1990; ferret: King et al. 1998) and the rostral pole of the IC (cat, Harting and Van Lieshout 2000). In the mustached bat, the projections arise from both ICx and ICc, with the ICc projections terminating mainly in the central part of the deep layers of SC (Covey et al. 1987). Thus the IC (especially the ICx, which also has somatosensory input) can influence motor outputs for orientation of the head, eyes, and pinnae via the deep SC (e.g., see Jay and Sparks 1987a,b).

The pathway from ICx to SC may be the means by which information about the body is integrated with auditory information, with the resulting integrated information being sent to pathways in the SC that initiate orientation reflexes. Although there is extensive evidence for such processing in the barn owl's midbrain (e.g., Knudsen 1982), the possibility that similar operations may occur in mammalian midbrain has received little attention.

2.5.3.2 The Pathway to the Cerebellum Via the Pontine Gray

Projections from the nuclei of the pontine gray provide the cerebellum with sensory information from a number of sources. In cats (Aitkin and Boyd 1978) and bats (Schuller et al. 1991; Wu et al. 1995), cells in the pontine gray respond to sound. One source of auditory input to the pontine gray appears to be the IC although not all studies of this projection in cats and rats agree on this point (see Schuller et al. 1991). In two species of echolocating bats, the evidence for this pathway is quite clear (mustached bat, *Pteronotus parnellii*, Frisina et al. 1989; Wenstrup et al. 1994; horseshoe bat, *Rhinolophus rouxi*, Schuller et al. 1991). Schuller et al. (1991) injected retrograde tracers into areas in the pontine gray that contain neurons responsive to sound. The results showed a massive input from ICc and ICx to the pontine gray. The pontine gray also receives input from auditory cortex (e.g., rat, Wiesendanger and Wiesendanger 1982; rabbit, Knowlton et al. 1993), raising the possibility that auditory inputs from cortex and midbrain converge at the pontine gray.

2.6 Connectional Gradients

Connectional evidence suggests that there are two opposing gradients which extend along the fibrodendritic and isofrequency laminae and include both the ICc and ICdc (Fig. 7.10). The ascending fibers from the various auditory nuclei in the lower brainstem (with the exception of DCN) terminate most densely in the ventral portion of each lamina and least densely

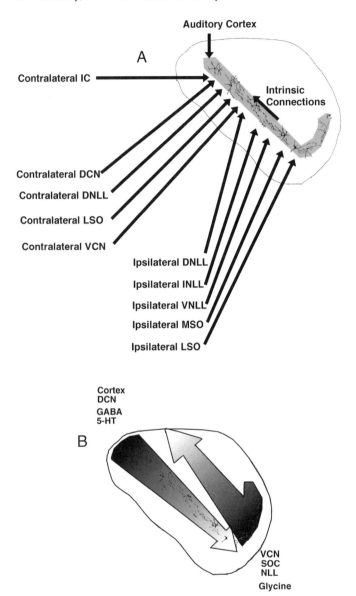

FIGURE 7.10. A: Summary diagram showing the inputs to a fibrodendritic lamina in the IC. B: summary diagram showing the opposing density gradients for groups of inputs and immunocytochemical markers. (For abbreviations, see Table 7.1.)

in the dorsal portion. The descending fibers also have the same orientation, but these terminate most densely in the dorsal part of each lamina and least densely ventrally (Fig. 7.10). This observation raises the hypothesis that descending and ascending gradients oppose one another, such that the ventral part of the ICc is mostly under the influence of ascending pathways that originate in VCN whereas the dorsal ICc and the ICdc are mostly under the influence of pathways from the auditory cortex, contralateral IC, and the DCN. Within the gradient, there is a large area of overlap, suggesting that the responses of many cells could be influenced by the VCN system, the DCN system, the IC commissural projection, and the auditory cortex.

In summary, there are multiple inputs to each lamina: from local cells, from the contralateral IC, from multiple sources in the lower brainstem, and from auditory cortex, with the relative contributions from each source changing in a graded manner throughout a lamina.

2.7 The Distribution of Receptors and Terminals for Neurotransmitters

2.7.1 γ-Amino Butyric Acid and Glycine

Receptor binding studies show that the IC contains receptors to the two major inhibitory neurotransmitters, GABA and glycine. The density of the $GABA_A$ subfamily of receptors in parts of the IC is as great if not greater than in any other part of the brain (rat, Edgar and Schwartz 1990; bat, Fubara et al. 1996). The distribution of $GABA_A$ receptors is not uniform (Fig. 7.11). They are most dense in the dorsal medial parts of the IC and least dense in a crescent-shaped area in the lateral and ventral part of the IC. In contrast, the density of glycine receptors is approximately reciprocal to the $GABA_A$ receptors, being greatest in the crescent-shaped area described above (Fig. 7.11). The density of GABA and glycine-containing terminals matches the receptor densities for these two neurotransmitters (bat, Winer et al. 1995). However, in spite of the contrasting distribution of terminals and receptors for $GABA_A$ and glycine, physiological evidence (Section 3) suggests that most IC cells are inhibited by both GABA and glycine. A large component of the GABAergic input arises from the DNLL (e.g., cat, Adams and Mugnaini 1984; Shneiderman and Oliver 1989; rat, González-Hernández et al. 1996; Zhang et al. 1998). Other GABAergic inputs arise from the LSO (cat, Saint Marie et al. 1989; Glendenning et al. 1992), the INLL, and the VNLL (rat, González-Hernández et al. 1996; Zhang et al. 1998; bat, Winer et al. 1995; Vater et al. 1997).

The IC may recieve GABAergic input from nonsensory sources as well. For example, projections from the substantia nigra appear to be GABAergic (rat, Yasui et al. 1991; Moriizumi et al. 1992). However, glycinergic input, as far as is known, is exclusively sensory and arises mainly from cells in the LSO (Saint Marie et al. 1989; Glendenning et al. 1992), the VNLL and the INLL (chincilla and gerbil, Saint Marie and Baker 1990; bats, Winer et al. 1995; Vater et al. 1997). In bats, the VNLLc, a structure that provides the IC with precise timing about stimulus onset, is an important source of glycinergic input (bats, Vater et al. 1997).

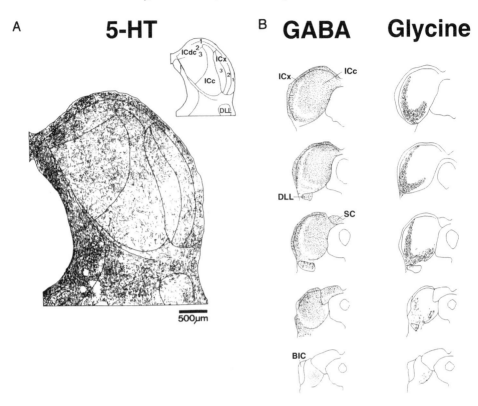

FIGURE 7.11. Distribution of immunoreactivity for 5-HT in the IC of the rat (A; Reprinted from Brain Research, Vol. 557, Iss. 1–2, Klepper and Herbert, Distribution and origin of noradrenergic and serotonergic fibers in the cochlear nucleus and inferior colliculus of the rat, p. 194, Copyright 1991, with permission from Elsevier Science) and GABA and glycine in the IC of the big brown bat. (B; From Johnson 1993.)

2.7.2 Other Neurotransmitters

N-methyl-D-aspartate (NMDA) receptors are clearly present in the IC (e.g., rat, Ishii et al. 1993; mouse, Watanabe et al. 1994) and appear to be differentially distributed by region. In the rat, Monaghan and Cotman (1985) reported higher levels of NMDA receptors in the dorsal medial region than in the ventral lateral region. Petralia et al. (1994) found high levels of NMDA receptors in the ICx, lower levels in the ICdc, and few receptors in the ICc in the rat. Thus, NMDA receptors appear to be distributed along a dorsomedial to ventrolateral gradient. The functional implications of such NMDA receptor patterning are unclear. However, given that the pattern seems to roughly match those of the descending projections from the auditory cortex, and that in bats the descending projections from the auditory cortex appear to cause long-lasting (>2 hrs) changes in the response properties of IC neurons (e.g., Yan and Suga 1996a, 1999; Zhang et al. 1997; see Suga et al. 2000 for review), a logical hypothesis is that the NMDA receptors in the IC play a role in neuronal plasticity. This idea is reinforced by an in vitro study showing that IC neurons exhibit long-term potentiation (rat: Zhang and Wu 2000).

In addition to NMDA glutamate receptors, AMPA (α-amino-3-hydroxy-5-methyl-4-isoxazole proprionic acid) glutamate receptors are also present in the IC (review: Parks 2000). The AMPA receptor subunits (GluR1–4) also appear to be differentially expressed in the IC. Overall, there appear to be higher levels of GluR2 and 3 than GluR1 and 4 (rat, Petralia and Wenthold 1992; Sato et al. 1993; Gaza and Ribak 1997; Caicedo and Eybalin 1999). In addition, there may be regional and cell type differences (rat, Gaza and Ribak 1997; Caicedo and Eybalin 1999). The functional implications of AMPA patterning are unclear.

Like NMDA receptors, serotonin (5-HT) terminals and receptors appear to be distributed along a dorsoventral gradient with the highest levels seen in the dorsocaudal region and weakest in the ventrolateral/rostral region (rat, Klepper and Herbert 1991; Thompson et al. 1994; bat, Kaiser and Covey 1997) (Fig. 7.11). A similar distribution pattern appears to be present for muscarinic and nicotinic acetylcholine receptors (rat, Regenold et al. 1989; Wada et al. 1989) and for noradrenergic fibers (rat, Klepper and Herbert 1991) although in this last case, the weakest density appears to be in the most lateral portion of the ICx.

2.8 Summary: Laminae and Gradients

A large proportion of the ascending, descending, commissural, and intrinsic connections follow a laminar-like arrangement that parallels the tonotopy of the ICc, thus preserving a single tonotopy in the face of converging input. This pattern of connections indicates that each isofrequency sheet provides a channel or module within which considerable integration of information centered upon a restricted bandwidth can take place. Evidence for discrete modules comes from recordings that measure the BF of neurons as a function of depth in the IC (e.g., Schreiner and Langner 1997). Figure 7.12 shows discontinuities and plateaus that may indicate that the electrode passed from one isofrequency contour to another.

From an anatomical perspective, a relevant issue regarding the multiple inputs to the IC is whether or not there is a nucleotopic organization, in which separate auditory centers in the lower brainstem have distinct and separate targets in the IC. The answer has not been forthcoming, and it appears that there is no obvious distinct and separate relation between the lower auditory centers and specific regions within the IC. Instead, the rule seems to be one of graded convergence in which different inputs have their peak density at different points within a lamina.

The evidence on the organization of projections from various sources to the IC reveals two important organizational principles. First, although the ascending and descending inputs converge, the density of termination from

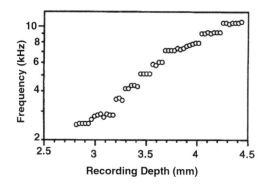

FIGURE 7.12. Characteristic frequencies of units recorded at different depths in the IC of the cat show plateaus and discontinuities that may correspond to boundaries of fibrodendritic laminae or modules. (Schreiner and Langner 1997, reprinted with permission from Nature.)

most of the ascending sources has a gradient that is opposite the gradient of the descending sources. The descending projections decrease in density from dorsal to ventral, and from the ICdc to the ICc. Thus, while the ascending and descending inputs are arranged in bands that overlap one another, the cortex makes a proportionately greater contribution dorsally and the lower brainstem makes a proportionately greater contribution ventrally within each band. The DCN projections approximately match the cortical projections. The second organizational principle in the IC is that the density of receptors or terminals of many neurotransmitters follow similar, though certainly not identical, gradients. For example, the density of $GABA_A$ receptors is greatest dorsally and least ventrally, roughly matching the cortical projection gradient. The density of glycine receptors and terminals is greatest ventrally and least dorsally in the ICc, roughly matching the gradient of ascending projections. Taken together, these findings have implications for the function of the IC. Within an isofrequency contour, the dorsal part receives more input from the auditory cortex than does the ventral part. This input is probably excitatory; GABAergic inhibition probably has the greatest influence in the dorsal part, but the source of the GABAergic input is from ascending auditory pathways or from nonauditory pathways, or both. If the auditory cortex and nonauditory inputs have greater input to the dorsal ICc and ICdc than to the ventral part, then we might expect more complex processing dorsally than ventrally.

An obvious question from the foregoing description of the IC, especially the graded convergence, is whether there is any kind of organization other than the tonotopic one. A quest for a second organization has been explicit or implicit in many electrophysiological studies of the IC, and, as we shall see, there is some evidence to support the idea that other organizations exist.

3. Functional Properties of the IC

3.1 Processing Frequency Information

In all mammals, fibers of the auditory nerve have V-shaped tuning curves (see Ruggero 1992). In the central auditory system, this basic pattern is modified through the convergence of multiple excitatory inputs, or through the convergence of excitatory and inhibitory inputs (see Section 2.2). Although changes in frequency tuning occur at levels below the IC (most notably in the dorsal cochlear nucleus, see Chapter 5), posteroventral cochlear nucleus (see Rhode and Greenberg, 1992 and Chapter 6), and in the ventral nuclei of the lateral lemniscus (see Chapter 6), the convergent nature of inputs to the IC also produces alterations in frequency tuning (Fig. 7.13).

3.1.1 Convergence of Excitatory Inputs and Broadband Frequency Response

Some IC neurons have extremely broad frequency tuning to pure tones while other fail to respond to pure tones but may respond best or exclusively to broad-band noise (e.g., bat, Grinnell 1963; Suga 1969; Schmidt et al. 1991; Casseday and Covey 1992). Neurons that are selective for broadband sounds are more common in the ICdc and ICx than in ICc (rabbit, Aitkin et al. 1972; cat, Aitkin et al. 1975; rat, Syka et al. 2000). Some units in the IC are not only broadly tuned, but tend to habituate after two or three presentations of the same stimulus (rabbit, Aitkin et al. 1972; cat, Aitkin et al. 1975). Frequently, they will respond again for a few trials if the stimulus frequency is changed. Many units in the ICx have this "novelty" response (rabbit and cat, Aitkin et al. 1972, 1975, 1994). These response patterns suggest that some IC neurons receive input across a broad frequency range. As will be described later, other IC neurons respond best or exclusively to temporal sequences of specific frequency combinations, or to frequency-modulated (FM) stimuli. Many of these cell types are seen in the ICc.

3.1.2 Convergence of Excitatory and Inhibitory Inputs and the Shaping of Frequency Response

The fact that many IC neurons receive convergent excitatory and inhibitory inputs raises the possibility that tuning of IC neurons is sharpened by lateral inhibition within the IC and the possibility that inhibition within the IC shapes frequency tuning in other ways. For example, inhibition may adjust thresholds, producing upper thresholds or creating nonmonotonic rate-level functions. Several lines of evidence indicate that, for some IC neurons, neural inhibition suppresses responses to frequencies at the upper and lower flanks of the tuning curve thereby narrowing the frequency response area

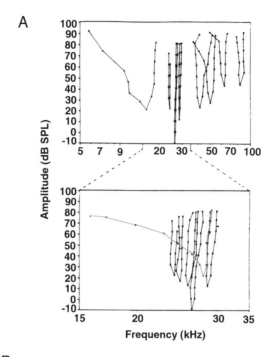

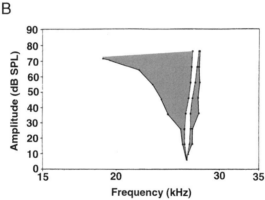

FIGURE 7.13. Representative examples of different types of frequency response areas in the IC of the big brown bat. A: V-shaped and narrow, level-tolerant response areas. B: Excitatory and inhibitory response areas for one neuron with a narrow, level-tolerant excitatory frequency response area. The unit was excited by tones within the clear area; its response to an excitatory test tone was inhibited by a conditioning tone that preceded the test tone by 1 ms when the conditioning tone was in the shaded area. C: Closed response areas. D: A broad, double-peaked response area. (From Casseday and Covey, Frequency tuning properties of neurons in the inferior colliculus of an FM bat, Journal of Comparative Neurology, Vol. 319, No. 1, Copyright 1992, Wiley-Liss, Inc. Reprinted with permission from Wiley-Liss, Inc., a subsidiary of John Wiley & Sons, Inc.)

C

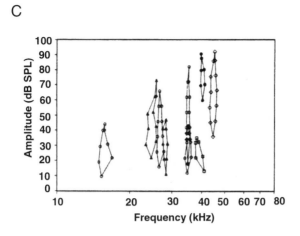

D

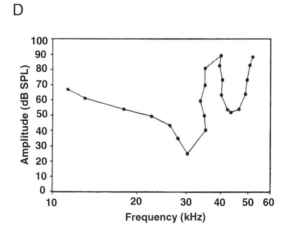

FIGURE 7.13. *Continued*

and increasing the Q_{10}. For example, extracellular recordings of responses to pairs of sounds show that, for some IC neurons, a tone at a frequency outside the borders of the excitatory response area inhibits the response to a tone that is normally excitatory (e.g., cat, Ehret and Merzenich 1988; bat, Casseday and Covey 1992). When $GABA_A$ receptors are blocked by application of bicuculline, the frequency tuning curves of many IC neurons widen at suprathreshold sound levels (e.g., bat, Vater et al. 1992; Yang et al. 1992; Pollak and Park 1993, guinea pig, Le Beau et al. 1996; chinchilla, Palombi and Caspary 1996). These results suggest that inhibition around the flanks of the excitatory tuning curve sharpens frequency tuning. This idea is supported by direct evidence from intracellular recordings which show that some IC neurons exhibit inhibitory events in response to sounds at frequencies above or below the frequency range that elicits an excitatory event (cat, Nelson and Erulkar 1963; Kuwada et al. 1997; bat, Covey et al. 1996). Thus, in at least some IC neurons, frequency tuning may be maintained or sharpened by neural inhibition at the flanks of the tuning curve.

However, for most IC neurons, inhibition appears to shape their response areas through mechanisms other than lateral inhibition. In the chincilla, Palombi and Caspary (1996) found that blocking GABA$_A$ receptors increased the response rate throughout the excitatory response area for 85% of IC neurons. In only a few cases was the increase greater at one or both flanks of the tuning curve than in the center of the excitatory response area. The authors argued that the predominant pattern of effects throughout the frequency response area is inconsistent with the lateral inhibition hypothesis.

It is interesting to note that, even after the constriction of frequency tuning produced by either lateral or across-the-board inhibition, many tuning curves of IC neurons are no narrower than those of neurons in lower auditory centers that project to the IC. All of these findings raise the question of what roles neural inhibition plays other than frequency sharpening. When neural inhibition is blocked, thresholds at BF decrease for many IC neurons (e.g., chincilla, Palombi and Caspary 1996; rat, Faingold et al. 1991). Furthermore, nonmonotonic rate-level functions become monotonic, and closed tuning curves open up to a V-shape (bat, Vater et al 1992; Yang et al. 1992; Pollak and Park 1993). In addition, intracellular recordings show that IPSPs occur in some IC neurons at sound levels below the excitatory threshold and/or at levels above the upper threshold or both (bat, Covey et al. 1996). Thus, inhibition clearly modifies frequency-intensity response area in some IC neurons, creating band-pass filters for sound level as well as for frequency.

The interpretation of how frequency tuning is shaped by inhibition in the IC is further complicated by the fact that it certainly could have some other role, such as complex temporal processing (Palombi and Caspary 1996). In Section 3.3.4, we discuss how temporally offset sequences of excitatory and inhibitory input may be important for producing selectivity for specific sequences of sounds.

3.2 Binaural Processing

3.2.1 The Anatomical Substrate

Binaural processing in the lower brainstem is covered in detail in Chapter 4. However, to provide a background for understanding binaural interactions at the IC, we summarize the inputs from the binaural centers in the lower brainstem to the IC (Fig. 7.14). The ascending projections of the LSO in all mammals are bilateral, terminating in the DNLL as well as the IC. The ipsilateral projections from the LSO arise mainly from inhibitory

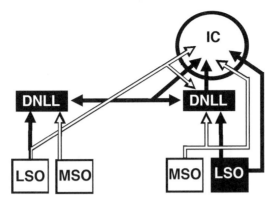

FIGURE 7.14. Summary diagram showing the main binaural pathways to the IC. Predominantly excitatory pathways are white, predominantly inhibitory pathways are black.

neurons whereas the contralateral projections arise from both excitatory and inhibitory neurons. Thus, in terms of excitatory input to the IC, the LSO pathway is mainly crossed; in terms of inhibitory input, it is bilateral (cat, Saint Marie et al. 1989; Glendenning et al. 1992; rat, González-Hernández et al. 1996). The ascending projections of the MSO are presumed to be mainly excitatory and, in most mammals, terminate almost exclusively in the ipsilateral DNLL and IC (e.g., cat, Adams 1979; Brunso-Bechtold et al. 1981; bat, (*Pteronotus parnellii*) Zook and Casseday 1982a, 1987; Vater et al. 1995). However, this pattern is not universal. For example, in the mole (*Mogera robusta*) (Kudo et al. 1988), opossum (*Didelphis virginiana*) (Willard and Martin 1983), and Australian native cat (*Dasyurus hallucatus*) (Aitkin et al. 1986) the MSO projects bilaterally.

The LSO and MSO are not the only sources of binaural input to the IC. Another binaural pathway arises from collaterals of the fibers that ascend from the LSO and MSO to the IC, but which also terminate in the DNLL. The DNLL, in turn, projects bilaterally to the IC. Cells in the DNLL also project via the commissure of Probst to the contralateral DNLL, so that the DNLLs of both sides are linked via reciprocal connections. Because the DNLL appears to consist mainly of GABAergic neurons, it is generally thought that it provides inhibitory input to IC neurons as well as to the contralateral DNLL (rat, González-Hernández et al. 1996; Zhang et al. 1998; cat, Saint Marie et al. 1997). The terminal fields of axons from MSO, LSO and DNLL axons largely overlap (Fig. 7.15), suggesting that there is convergence of direct and indirect, ipsilateral and contralateral, and excitatory and inhibitory projections on IC cells (bat, Vater et al. 1995; cat, Oliver et al. 1995, 1997). Thus, neurons in the IC could potentially receive input that has undergone one initial stage of binaural processing in the superior olive as well as input that has undergone at least one additional stage of processing in the DNLL.

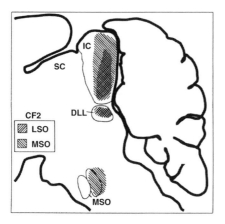

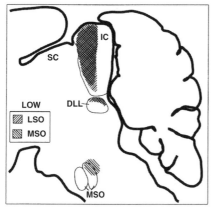

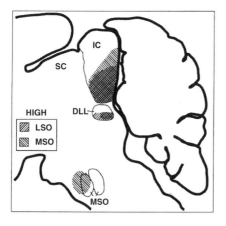

FIGURE 7.15. Summary diagrams of parasagittal sections to show convergence of inputs from the MSO and LSO on the DNLL and IC of the mustached bat from injections in three different frequency regions. In both DNLL and IC, the projections are largely overlapping. *Top panel* shows projections from injections in the frequency region of the main harmonic (61 kHz) or constant frequency (CF2) of the echolocation call. The middle panel shows projections from low frequency (<61 kHz) regions. The *lower panel* shows projections from high frequency regions. (From Vater, Casseday and Cover, Convergence and divergence of ascending binaural and monaural pathways from the superior olives of the mustached bat, Journal of Comparative Neurology 351: 632–646, Copyright 1995, Wiley-Liss, Inc. Reprinted by permission of Wiley-Liss, Inc., a subsidiary of John Wiley & Sons, Inc.)

It should be kept in mind that input from binaural pathways in the lower brainstem almost certainly converges with input from many other sources, including the ascending monaural pathways and inputs from nonauditory sources. In addition, there is convergence with other, potentially binaural inputs via intrinsic and crossed IC projections and descending projections from the auditory cortex (Sections 2.3–5). It is outside the scope of this chapter to give a detailed account of the binaural processing in the IC. Rather, we will focus on some ways in which binaural processing in the IC differs from that in the LSO, MSO or DNLL.

3.2.2 Evidence for Direct Binaural Interaction at IC Cells

A number of lines of evidence indicate that at least some IC neurons perform binaural operations that depend on interactions between excitatory and inhibitory inputs, and that it is the direct interaction of these two classes of input at the IC neuron that determines its binaural response characteristics. Iontophoretic application of antagonists of GABA and glycine has shown that neural inhibition actively contributes to the shaping of binaural response properties of many IC neurons and that at least some of this information comes from the DNLL. Studies in rats (e.g., Faingold et al. 1989, 1991) and bats (Vater et al. 1992; Park and Pollak 1992, 1993; Klug et al. 1995; Pollak 1997) have shown that blocking GABAergic or glycinergic inhibition at binaural IC neurons changes the pattern of their responses to binaural stimuli. The changes caused by eliminating inhibition can take various forms. For example, in cells that are excited by contralateral sounds and inhibited by ipsilateral sounds, one commonly seen effect of blocking inhibition is a general enhancement of spike counts evoked by binaural stimulation and a concomittant shifting of interaural intensity differences (IID) functions. Although highly suggestive of binaural interactions at IC neurons, this response enhancement is not in itself conclusive evidence for binaural interaction of excitation and inhibition at the IC. Although the enhanced response to binaural stimuli could be due to the blocking of ipsilaterally-evoked inhibition that normally cancels contralaterally-evoked excitation at specific IIDs, it could also simply be the result of blocking contralaterally-evoked inhibition that, under normal conditions, monaurally suppresses the response to contralaterally-evoked excitation (e.g., bat, Vater et al. 1992). Typically, blocking inhibition at IC neurons, especially GABAergic inhibition, causes a many-fold increase in the spike counts evoked by monaural contralateral stimuli as well as binaural stimuli (bat, Vater et al. 1992; Park & Pollak, 1993a, 1994). The altered binaural functions seen when inhibition is blocked may simply be the result of increasing the strength of an excitatory input from a source with predetermined binaural properties. This may be one way in which monaural inhibitory input contributes to the shaping of binaural properties. Li and Kelly (1992) have shown that reversible inactivation of the DNLL alters the IID functions of IC neurons in rats, suggesting that binaural interaction is shaped by inhibitory input from the DNLL.

Although monaural inhibition clearly plays a role in determining the responses of IC neurons to sound, there is direct evidence for binaural interaction between excitation and inhibition in IC neurons from intracellular recordings. Nelson and Erulkar (1963) were the first to report that some neurons in the IC of cats exhibit an IPSP when sound is presented at one ear and an EPSP when it is presented at the other ear. More recently, similar results have been reported in the IC of bats (Covey et al. 1996), guinea pigs (Torterolo et al. 1995) and cats (Kuwada et al. 1997). Examples of excitatory and inhibitory synaptic events that are evoked by binaural stimulation are shown in Figs. 7.16 and 7.17. In addition to providing direct evidence of the existence of binaural inhibition at IC neurons, intracellular recordings provide important information about the time course of the inhibition and the latency of the inhibition relative to the excitation. There is some evidence that ipsilaterally-evoked inhibition may last longer than contralaterally-evoked excitation in owls (Moiseff 1985) and mammals (Covey et al. 1996). Moreover, the pattern of inputs from either side may be more complex than simple inhibition or excitation. For example, an IC neuron may receive inhibition followed by excitation when the sound is at one ear and excitation followed by inhibition when the sound is at the other ear (Fig. 7.16) (cat, Kuwada et al. 1997). Binaural neurons in the IC may

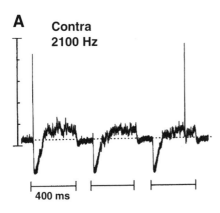

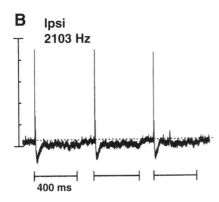

FIGURE 7.16. Postsynaptic potentials evoked by contralateral and ipsilateral stimuli recorded from a neuron in the IC of the cat using sharp electrode intracellular recording. This neuron exhibited both excitation and inhibition in response to sound at either ear, but the sequence and time course of IPSPs and EPSPs was different. A: In response to a contralateral sound, the first event was an IPSP, followed by a sustained EPSP. B: in response to an ipsilateral sound, the first event was a very short EPSP and action potential followed by an IPSP. (From Kuwada et al. Copyright 1997 by the Society for Neuroscience.)

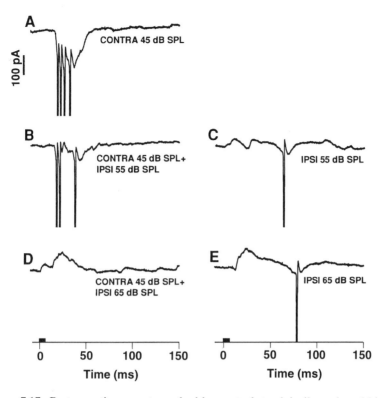

FIGURE 7.17. Postsynaptic currents evoked by contralateral, ipsilateral, and binaural stimuli recorded from an IC neuron in the big brown bat (whole-cell patch clamp technique). This neuron was excited by contralateral sound and inhibited by ipsilateral sound. Note the excitatory events following the IPSCs evoked by ipsilateral stimulation, suggesting an excitatory rebound. While these traces were recorded in the voltage clamp mode, the finding that action potentials were driven synaptically indicates that the voltage could not be clamped in the cell's processes. However, the lack of space clamping presumably does not alter the sign of synaptic responses and, presumably, changes the relative amplitudes little. (From Covey et al. Copyright 1996 by the Society for Neuroscience.)

receive inputs activated by the ipsilateral and contralateral sound with very different latencies (cat, Kuwada et al. 1997; bat, Klug et al. 1999).

Even though intracellular evidence shows that binaural processing occurs at IC neurons, this does not mean that binaural processing has not already occurred in the pathways that provide the ipsilateral and contralateral inputs to IC neurons. In fact, it is highly probable that the inputs to binaural neurons reflect binaural comparisons that have already occurred at a lower level, and that this process of "comparing the comparisons" may be an important component of binaural processing in at least some IC neurons. Convergent processing of binaural inputs in the IC may be important for processing information about dynamic binaural stimuli, including sequences of binaural sounds, multiple sound sources, and auditory motion.

3.2.3 What Does Binaural Processing in the IC Accomplish?

Given the amount of binaural processing that takes place in the pathways to the IC, it might be supposed that all binaural comparisons have already taken place in the lower brainstem and that no further interaction of ipsilateral and contralateral input need occur at the level of the IC. In fact, some studies of the IC have operated under this assumption, with responses of IC neurons being regarded as directly reflecting the responses of neurons in the LSO or MSO (e.g., rabbit, Stanford et al. 1992). To some extent this may be true, since the binaural response functions of at least some IC neurons appear to be unaffected by blocking neural inhibition (e.g., bat, Vater et al. 1992; Park and Pollak 1993b). However, many IC neurons do appear to perform further comparisons and integration of bilateral binaural information from multiple sources. Superficially, the interaction of contralateral excitatory input with ipsilateral inhibitory input is highly reminiscent of the binaural interactions that first occur in the LSO to create IID functions, simply with the laterality reversed. It is not completely understood why such processing should occur a second time in the IC, apparently producing IID functions de novo based on inputs that originate, at least in part, in cells with their own binaural IID or ITD response functions.

Given that IC neurons receive binaural inputs from different sources and that these inputs may be excitatory or inhibitory, have different discharge patterns, and arrive with different latencies, the obvious question is whether the convergence of projections from structures that have already performed basic binaural comparisons at levels below the IC actually produces some emergent form of binaural processing. It is tempting to speculate that the purpose of the convergence of binaural pathways at the mammalian IC is to combine IID, ITD, and spectral information to produce a fixed two-dimensional topographic representation of auditory space. This is the way that IID and ITD information is combined at the barn owl's midbrain to produce a two-dimensional topographic map of auditory space (see Konishi 1993 for review). It may seem puzzling that a similar auditory space map has not been seen in the mammalian IC. However, binaural processing may be more complicated in mammals. One reason for the complexity of the mammalian binaural system may simply be that most mammals, unlike the barn owl, have pinnae and eyes that move. Recent evidence from monkeys shows that the responses of IC neurons to sounds presented from loudspeakers at different azimuthal locations vary as a function of eye position. This indicates that input from the oculomotor system can modify spatial representation in the IC (Groh et al. 2001). It should be kept in mind that binaural processing does not necessarily have to be synonymous with the creation of a rigid auditory space representation. Binaural processing can have other roles such as enhancing the detection of signals in noise, separating complex sounds into multiple streams, and minimizing the effects of echoes. Finally, it should be kept in mind that monaural cues also contribute to auditory space processing in mammals (Musicant and Butler 1984; Wightman and Kistler 1997).

Although the much sought-after mammalian auditory space map has been elusive, there is nevertheless some evidence that binaural properties of IC cells vary in topographical ways. For example, neurons that are excited by one ear and inhibited by the other tend to be grouped together at specific points along the extent of an isofrequency contour and are somewhat separate from those that are excited by both ears and those with predominantly monaural responses (bat, Wenstrup et al. 1986; gerbil, Brückner and Rübsamen 1995). There is also some evidence for progressive shifts in the balance of excitatory and inhibitory inputs from the two ears. The shift is seen when the 50% points (or cutoffs) of IID functions are measured across an isofrequency contour. The 50% points create a gradient of cells with IID functions that extend progressively further into ipsilateral space (bat, Wenstrup et al. 1986). Figure 7.18 shows grouping of neurons with similar binaural properties within the expanded 60 kHz isofrequency contour of the mustached bat. This figure illustrates the effect of stimulus location on patterns of activity in a population of cells with systematically varied IID cutoffs. However, these types of measures are complicated by other response properties of IC neurons such as nonmonotonic rate-intensity functions for the excitatory (usually contralateral) ear (e.g., bat, Grothe et al. 1996). Spatial receptive field properties of IC cells are highly interactive with other parameters such as sound amplitude, frequency spectrum, modulation pattern, movement of the sound source, etc. (e.g., gerbil, Semple and Kitzes 1987; bat, Grothe et al. 1996; guinea pig, Ingham et al. 2001), so it is unlikely that individual IC neurons convey unambiguous information about the location of a sound source.

In nature, sound sources seldom are completely stationary with respect to the ears of a listener. The sound source itself may move, the listener's head and/or body may move, and many animals move their pinnae, either together or independently. Instead of creating a static map of auditory space, it may be that binaural processing in the mammalian IC is concerned with the analysis of dynamic properties of sound source location. Examples of dynamic binaural processing might include the location of a sound relative to head, eye, and/or pinna position (e.g., monkey, Jay and Sparks 1987a,b; Groh et al. 2001), the location of a sound source relative to other sound sources, the grouping of spectral components with different binaural profiles in order to separate multiple sound sources, the processing of information about the direction in which a sound source is moving relative to the head, or all of the above.

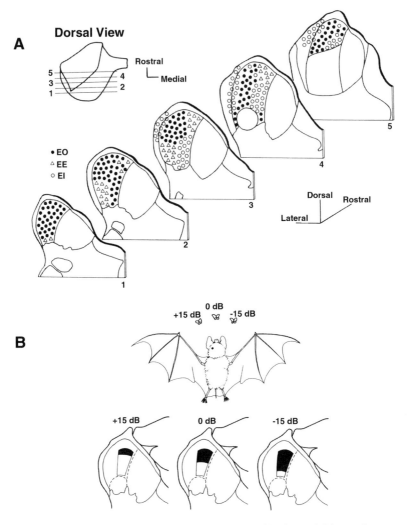

FIGURE 7.18. A: Schematic illustration of the distribution of binaural response classes in the expanded 60 kHz isofrequency contour of the mustached bat's IC. EO, monaural, excited by contralateral sound (solid circle); EE, excited by sound at either ear (opentriangle); EI, excited by contralateral sound, inhibited by ipsilateral sound (open circle). B: Diagram showing the hypothetical spread of activity (black area) within the population of EI neurons in the 60 kHz isofrequency contour of the left IC when sound originates from the ipsilateral field (+15), from straight ahead (0), or from the contralateral field (−15). (From Wenstrup et al. Copyright 1986 by the Society for Neuroscience.)

3.2.4 Delay Lines in Binaural Pathways

A key difference between binaural processing in the superior olive versus the IC is that synaptic delays of several milliseconds are introduced as the activity evoked by a stimulus is transmitted through the various pathways that lead from the lower brainstem auditory nuclei to the IC (see Covey and Casseday 1999). This means that the neural activity evoked by a single, instantaneous, binaural stimulus may arrive at a given IC cell at slightly different times via different pathways. For example, it is likely that input from the LSO or MSO would precede inhibitory input from the DNLL. The arrival of binaural input via delay lines could have consequences for tasks such as processing of echoes in reverberant environments, processing information about moving sound sources, and the precedence effect.

3.2.5 Binaural Processing and the Precedence Effect

One psychophysical manifestation of binaural processing that may have its origin at least partly in the IC is the "precedence effect." This term refers to the tendency of human listeners to perceive two sounds presented in rapid succession from two different points in space as originating from the location of the first sound (Wallach et al. 1949). It has been suggested that the precedence effect is behaviorally adaptive because it facilitates the localization of sound sources in reverberant environments (e.g., Hartmann 1983). Recordings from neurons in the IC of the cat (Yin 1994; Litovsky and Yin 1998a,b) have shown that when sounds are presented from two different locations, one after the other, either in the free-field or dichotically via earphones with different IIDs, the response to the second of the two sounds is suppressed for periods up to about 100 ms, but generally less than 50 ms. This period of suppression includes the range of interstimulus intervals over which humans experience the precedence effect (2–8 ms for clicks and 30 ms for speech: Blauert 1983). Experiments using monaural stimulation to measure recovery time have shown that IC neurons exhibit the same sort of paired-pulse suppression under these conditions as they do under free-field or dichotic stimulation (cat, Yin 1994; Litovsky and Yin 1998b). Therefore, it is not clear what role, if any, binaural circuitry plays in suppressing responses to echo-like stimuli or whether monaural circuits that create long recovery times in IC neurons operate in the same neurons that perform binaural comparisons. Nevertheless, there is little question that binaurally responsive neurons in the IC are sensitive to temporal and spatial context. The effects of spatiotemporal context are especially evident in studies of IC neurons' responses to moving sound sources or simulations of auditory motion.

3.2.6 Binaural Processing and Perception of Auditory Motion

It has been demonstrated repeatedly that IC neurons' responses to sounds originating from specific locations are determined in part by whether sounds have originated from other parts of the auditory field. It is also well known that spatial receptive fields that were measured using "stationary" binaural sounds differ from those measured using a real or simulated moving sound source (e.g., guinea pig, Ingham et al. 2001). Here we will consider several issues: (1) In what ways do IC neurons' responses to real or simulated motion of a sound source differ from those of neurons at lower levels, i.e., what are the emergent properties of binaural processing at the IC as they relate to a moving sound source? (2) In what ways does the interaction of excitatory and inhibitory synaptic inputs contribute to processing of auditory motion? and (3) In what ways do intrinsic properties such as the adaptation of excitation and the rebound from inhibition contribute?

3.2.7 Comparison of Auditory Motion Processing in the IC and Lower Brainstem Levels

Most studies of LSO and MSO neurons' responses to binaural sound have used static IIDs or ITDs as the test stimuli. However, a few studies have compared responses to dynamically changing stimuli at the level of the superior olive and IC. For example, when ITDs of low-frequency sounds are varied to simulate the changes that would occur as a sound source moves from one side of the head to the other (e.g., cat and gerbil, Spitzer and Semple 1993, 1998; guinea pig, McAlpine et al. 2000), the responses of IC neurons to a particular ITD are determined by the preceeding portions of the stimulus, the response history of the neuron, or both. In contrast, the responses of phaselocking neurons in the superior olive to a particular ITD are typically independent of other portions of the stimulus or the neurons' response histories. Thus, one function of binaural processing in the IC seems to be to produce context effects that cause neurons' spatial receptive fields to vary, depending on the motion of the sound source. Receptive fields are typically skewed toward the direction from which the stimulus is moving regardless of the direction of motion (e.g., bat, Kleiser and Schuller 1995; Wilson and O'Neill 1998; guinea pig, Ingham et al. 2001). An example of the skewing of the receptive field is shown in Fig. 7.19. In experiments where multiple speakers in a horizontal array were activated, the stimulus repetition rate (equivalent to speed of motion) seems to be highly interactive with the degree to which receptive fields are skewed. Higher repetition rates are correlated with lower spike counts, narrowing of receptive fields, and larger azimuthal skew of receptive fields toward the direction from which the sound originates. Thus, IC neurons typically fire most vigorously when a moving sound source approaches or first enters their spatial receptive field, and reduce or cease their firing as the stimulus passes through the remainder of the receptive field as defined using stationary stimuli.

IC neurons potentially receive a variety of excitatory and inhibitory synaptic inputs from binaural pathways, and some of the motion sensitivity

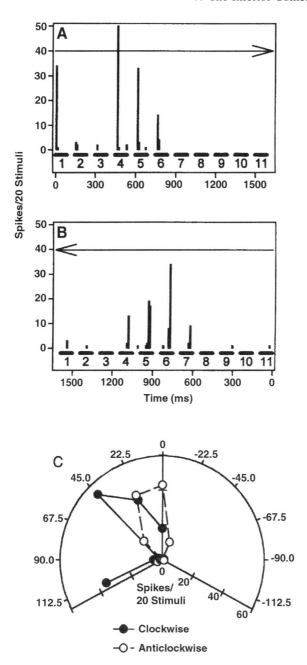

FIGURE 7.19. Comparison of the responses evoked by each of 11 azimuthal loud-speaker positions when the loudspeakers were sequentially activated in the clock-wise direction (from 1 to 11, panel A) versus when they were activated in the anticlockwise direction (from 11 to 1, panel B). Recording is from a neuron in the IC of the guinea pig (from Ingham et al. 2001). The azimuthal receptive fields measured using clockwise and anticlockwise sequences of loudspeaker activation are shown in C.

described above may be due to interaction of these inputs. However, many IC neurons also have spatial receptive fields that are clearly delineated in response to monaural stimuli, so monaural interaction of excitation and inhibition may also play a role in auditory motion processing, especially the effects of repetition rate.

3.2.8 Synaptic Interactions

Studies of IC neurons' responses to stationary binaural sounds have shown that the latencies and time courses of synaptic potentials evoked by ipsilateral sound often differ from those of the synaptic potentials evoked by contralateral sound. Intracellular recordings have shown that the inhibitory post-synaptic currents IPSCs or IPSPs evoked by an ipsilateral sound can last longer than the duration of the sound itself (owl, Moiseff 1985; bat, Covey et al. 1996). Similarly, when IC neurons are driven by iontophoresis of glutamate, an ipsilateral sound can inhibit the glutamate-driven action potentials for a prolonged period of time. Blocking GABA with bicuculline shortens the period of inhibition, and blocking both GABA and glycine eliminates it (see Pollak 1997 for review). A likely source of the short-latency ipsilaterally-evoked glycinergic inhibition is the ipsilateral LSO. A possible source of the delayed and longer-lasting ipsilaterally-evoked GABAergic inhibition is the DNLL (rat, Li and Kelly 1992; see also Pollak 1997). This prolonged, GABAergic inhibition may be responsible for suppressing responses to sounds that originate from locations within the neuron's spatial receptive field subsequent to the location that produced the initial burst of spikes.

Because some of the inhibitory, glycinergic, LSO neurons project to the ipsilateral DNLL as well as the IC, an ipsilateral sound could, at some point following a stimulus, cause DNLL neurons to be inhibited, thus releasing IC neurons from the GABAergic inhibition that would otherwise be provided by the DNLL after a short delay. Such a mechanism might be responsible for the skewing of spatial receptive fields into the ipsilateral part of auditory space that is observed when a stimulus moves from ipsi- to contralateral but not in the opposite direction.

3.2.9 Influence of Monaural Processing and Intrinsic Properties on Binaural Responses

The responses of IC neurons to moving stimuli are influenced by stimulus repetition rate (guinea pig, Ingham et al. 2001). There are several factors that may act together to produce this finding. The first factor is that ipsilateral inhibition may outlast contralateral excitation. The second factor is the monaural phenomenon of long recovery times following excitation of IC neurons, which Ingham et al. (2001) have termed "adaptation of excitation". More evidence for monaural and/or intrinsic effects on auditory motion processing come from experiments performed by Sanes and colleagues (e.g., gerbil, Sanes et al. 1998), in which they showed that IC neurons' responses to a given IID were profoundly influenced by the preceding IID. Their analysis was based on modulating the level of an ipsilateral sound around a level equal to that of a contralateral sound that was held constant. They observed that the IC neuron's response to an IID of Ø (both sides equal) was greatly enhanced if the prior ipsilateral level had been higher than the contralateral level; however, the response was suppressed if the prior level of the ipsilateral sound had been lower than the contralateral level (Fig. 7.20). Application of GABA or glycine to the IC neuron resulted in a suppression of its firing, followed by essentially the same conditioned enhancement effect observed with modulation of the ipsilateral sound. This result suggests that the conditioned enhancement caused by high ipsilateral sound levels was due to the intrinsic properties of the neuron (e.g., a prolonged period of rebound from inhibition). Evidence for inhibitory rebound has been observed in IC neurons in vitro in response to hyperpolarization (gerbil, Moore et al. 1998; Peruzzi et al. 2000) and in vivo following the inhibition produced by an ipsilateral sound (bat, Covey et al. 1996; cat, Kuwada et al. 1997).

3.2.10 Summary

What does IC binaural processing accomplish? There have been numerous studies of IC neurons' responses to real or virtual changes in the location of a sound source, some of which have operated under explicit or implicit assumptions derived from what is known about motion processing in the visual system. One such assumption is that there should be neurons that act as "motion detectors": cells that respond preferentially or exclusively to a particular direction and speed of motion. Another assumption is that binaural processing in the IC should result in a clearly defined map of auditory space within which motion detector neurons could operate. It seems likely that neither of these models is completely valid for the mammalian IC. Rather, IC neurons' responses to moving stimuli are shaped by the context of stimulus history, response history, and other properties of the stimulus such as average binaural intensity, spectral and temporal properties, as well as input from nonauditory sources that may include the visual system.

3.3 Temporal Processing

In the following section, we describe how ascending inputs from various sources, especially VNLL, produce delay lines and other temporal patterns that may be the basis for the first stage of temporal selectivity in the IC. In a later section, we will show how the cortex also contributes to temporal selectivity.

3.3.1 Anatomical Substrates for Temporal Pattern Detection

At the beginning of this chapter we portrayed the complexity of the ascending auditory pathways. The array of ascending pathways produces, if

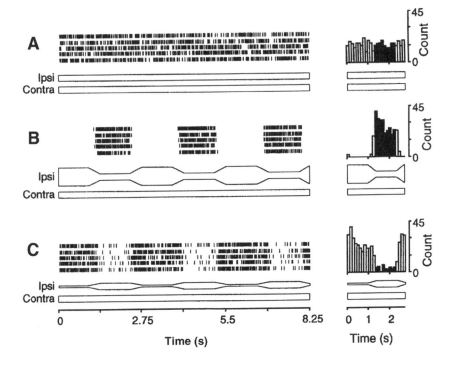

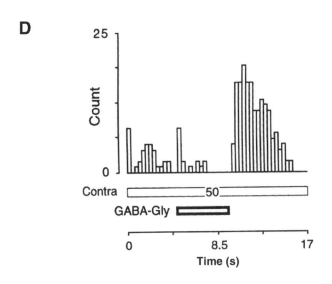

nothing else, an array of delay lines and temporally reshaped response patterns that alter the ways in which the neural representation of any given stimulus is distributed in time as it impacts an IC cell. What we mean by this is that in the lower brainstem pathways the responses of different neurons to the same stimulus provide a range of latencies and different response patterns, such as onset, sustained, and offset discharges. In addition, some of these IC inputs are excitatory while others are inhibitory. Even if only some of these inputs converge on a single IC neuron, the synaptic input following a simple sound can be a complex sequence of IPSPs and EPSPs distributed over a time interval far longer than the duration of the original sound. As described below, a number of studies demonstrate that most IC units receive both kinds of inputs and that they are indeed distributed over time following the occurrence of a sound.

We suggested earlier that a useful simplification is to divide the pathways into the basic functional components: binaural and monaural. We now briefly review why some of the monaural systems, particularly those in the nuclei of the lateral lemniscus, are primarily concerned with transforming various aspects of temporal features of sound and transmitting this information to the IC. We will consider how these temporally modified inputs from multiple pathways, combined with excitation and inhibition, may be the basis for creating selectivity to temporal features of sounds.

3.3.2 The Temporal Patterns of Inputs from the Lower Brainstem: A Potential Basis for Temporal Selectivity in the IC

If the IC deals with analyzing and selecting temporal patterns of sounds, it does so by utilizing the information it receives in the form of temporal patterns of EPSPs and IPSPs contributed by the auditory neurons that project to it. Insofar as new response properties emerge from afferent projections to the IC, they emerge from the combination of these different

---

FIGURE 7.20. Dot rasters and histograms illustrating the effects of a conditioning interaural intensity difference (IID) on the response of a neuron in the IC of the gerbil to sounds at equal level at the two ears (A). The neuron responds at a moderately high rate in the control condition in which the ipsilateral and contralateral levels are equal (IID = 0). B: When the ipsilateral stimulus starts at a higher level than the contralateral and is subsequently reduced, the neuron's response to the control condition of IID = 0 is approximately doubled. C: When the ipsilateral stimulus starts at a level lower than the contralateral and is subsequently raised, the response to the control condition of IID = 0 is greatly reduced. Thus, the response to an IID of 0 is highly context dependent. D: iontophoresis of a mixture of GABA and glycine onto the neuron suppressed its response to the contralateral stimulus presented alone. Once drug application stopped, the neuron's response greatly increased, suggesting a long-lasting rebound from inhibition. (From Sanes et al. Copyright 1998 by the Society for Neuroscience.)

input patterns and act upon a neuron whose state may have been altered by a prior input from the same or even a prior sensory stimulus. Therefore, it is important to know the response patterns and neurotransmitters of the inputs.

The idea that neural delay lines might be used to analyze auditory temporal patterns is one that has been around for at least 50 years. Jeffress (1949) utilized delay lines and coincidence detection in a model to encode binaural time differences. Licklider (1959) proposed a model for frequency or pitch discrimination based on similar mechanisms. Both models use very short delays generated by fiber length, and the Licklider model also incorporated synaptic delay. In principle, the same sort of mechanism can be used to encode information about time-varying events on any scale (see Casseday and Covey 1995).

However, many biological sounds have a slow cadence relative to fiber conductance and synaptic delay times. For example, consider the duration and interval between individual components of animal calls, echolocation sounds, and speech. For all of these sounds, intervals range from somewhat less than 1 ms to tens or hundreds of milliseconds. Likewise, the duration of these sounds is within the same range. If the system is selective for sounds on this time scale, it needs temporal filters or windows of excitation and inhibition that are temporally related to the sounds. A necessary component of this kind of temporal filter is a mechanism that produces temporal delays longer than any that can reasonably be produced by fiber length or even by a small number of synaptic delays. This is where inhibition becomes useful. To slow the pace to encode patterned sounds, such as wingbeats, footfalls, or syllables, we need two additions to the Jeffress-Licklider type of model. The first is to place inhibitory neurons in the chain of delay lines, and the other is to use neurons with different temporal response properties, such as those just mentioned: onset, sustained, or offset.

Offset responses are particularly important in generating stimulus-related delays. Obviously, the latency of the offset response, relative to sound onset, will vary according to sound duration. The auditory nerve contributes only sustained responses. Onset and offset responses are generated at various levels in the auditory brainstem below the IC (e.g., bat, Covey and Casseday 1991; Covey et al. 1991; Grothe 1994; Grothe et al. 1992, 1997). Thus, in bats, the IC receives signals that are correlated with both the offset and onset of sound (see also Covey et al. 1996). In other mammals, the IC receives transient onset information (cat, Kuwada et al. 1997; guinea pig, Syka 2000; see Ehret 1997 for review), but it is not clear if it also receives stable transient offset information (see Ehret 1997 for review). Later in this chapter we describe how offset information might be used in the construction of neurons that respond to restricted ranges of sound duration.

Regarding the response to sound onset, the nuclei of the LL may be especially important. These nuclei have been studied mainly in echolocating bats because of their hypertrophy and because of the elegance of their organization (e.g., Zook and Casseday 1982a,b, 1987; Covey and Casseday 1986, 1991). A type of onset responder in one part of the lateral lemniscus of bats, the VNLLc, is of special interest as a source of input to the IC for two reasons. First, these cells respond with remarkable precision to the onset of sound. They fire one spike, which has a variability from trial to trial of a few tens of microseconds. Moreover, the spike latency remains constant over a wide range of intensities and frequencies (Covey and Casseday 1991; Chapter 6). Thus, these input neurons to the IC seem to be an evolutionary development to signal the onset of sound very precisely. The second reason the VNLLc is of special interest is that these neurons use the inhibitory neurotransmitter glycine (Vater et al. 1997). This very precise onset information is apparently transmitted to the IC in the form of inhibition. Because the projections of VNLL are widespread throughout the IC, this rapid burst of inhibition may provide a "reset" signal that simultaneously brings IC neurons to a uniform state and prepares them for processing subsequent information. Another possible use for this inhibition will be proposed later in discussing the neural calculation of sound duration.

Other response patterns seen in the VNLL are like those seen in the cochlear nuclei; they are more or less sustained for the duration of the sound (Chapter 6). The chopping pattern provides a very regular oscillatory pattern, which contrasts with the constant input of sustained responders and with the decaying response of primary-like responders. The pauser type provides sharp onset information followed by a temporal gap that precedes a sustained burst of responses. If two or more of these patterns converged on an IC neuron, either simultaneously or separated by a delay, it is obvious that a complex pattern of synaptic potentials could ensue. The variety and complexity of the postsynaptic potentials is even greater when we consider that some of these inputs are excitatory and others inhibitory. For example, suppose that a chopper and a pauser were inhibitory. Convergence of inputs from these two cells could produce a pattern of postsynaptic inhibition that mirrors their temporal response patterns. The gaps in the response patterns would provide a window through which an excitatory input, if it arrived at the right time, could fire the IC cell. In this way, a very simple filter for temporal patterns is created. While this is of course speculative, it points to the need for additional intracellular studies to monitor subthreshold, postsynaptic potentials in response to sound.

In summary, the ascending input to the IC provides different temporal transformations of the sound envelope. These include onset responses, offset responses, and sustained responses. The same stimulus can elicit different responses across a range of latencies, even among neurons with the same frequency tuning characteristics. Finally, any of these inputs may be excitatory or inhibitory. As these inputs converge on an IC cell, they could generate temporal filters for the sound envelope or for temporal patterns of frequency change. We turn to this sort of temporal filtering next.

3.3.3 Response Patterns in the IC

The consequence of these ascending inputs, together with intrinsic properties of the neuron, is a complex temporal sequence of excitation and inhibition in response to sound. Evidence for this is apparent in intracellular
recordings in response to sound or in recordings made while blocking
inhibitory neurotransmitters. Intracellular recordings show that that for
some IC neurons, (1) inhibition is the first response to sound, (2) inhibition
can be followed by excitation, and (3) excitation may in turn be followed
by inhibition (Nelson and Erulkar 1963; bat, Covey et al. 1996; cat, Kuwada
et al. 1997; guinea pig, Pedmonte et al. 1997). Different IC neurons may
have a different sequence of excitatory and inhibitory responses, and the
sequence may change as a function of parametric changes in the stimulus.
Likewise, when GABA or glycine is blocked at IC neurons, response latency
may decrease, the duration of the response may increase (Fig. 7.21), or both
may occur (rat, Faingold 1989, 1991; bat, Park and Pollak 1993a; Casseday
et al. 2000). However, a decrease in latency has not been found in all experiments (guinea pig, Le Beau et al. 1996). The fact that GABA and glycine

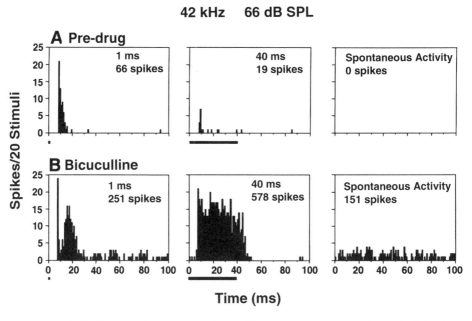

FIGURE 7.21. Increased response of an IC neuron to monaural, contralateral
stimulation after blocking GABAa receptors with bicuculline. Top row: evoked
responses and spontaneous activity under normal conditions; Bottom row: evoked
responses to the same stimuli and spontaneous activity during application of bicuculline. (From Casseday et al. 2000.)

can modulate latency suggests that GABA and glycine may play a role in creating the dorsomedial to ventrolateral gradient in latency that has been found within an isofrequency contour (cat, Schreiner and Lagner 1988; bat, Park and Pollak 1993a). Taken together, all of these results suggest that auditory stimulation sets up specific patterns of excitatory-inhibitory sequences in IC neurons that endow such neurons with specific types of selectivity for temporal features of sounds.

3.3.4 Temporal Processing in the IC

Examples of the sorts of temporal selectivity that appear to be generated by interaction of synaptic inputs in the IC include: tuning to sound duration, tuning to the delay between two sounds, tuning to the direction of frequency-modulated (FM) sweeps, and tuning to the rate of modulation of repetitive FM sounds. It is important to emphasize again that this kind of temporal processing involves analysis on a time scale of ~1 ms to hundreds of milliseconds, as opposed to the tens of microseconds range in binaural time coding. We will show how the initial steps in selectivity for temporal patterns appear to be set up below the IC and how this selectivity is further refined in the IC to create, for example, units that respond exclusively to repetitive FM sounds, units that are tuned to the rate of periodic frequency modulations, and units that respond only to certain durations of sound.

3.3.4.1 Duration Selectivity

Some neurons in the auditory midbrain respond to a specific range of sound durations, but fail to respond to shorter or longer sounds (e.g., frog, Narins and Capranica 1980; bat, Casseday et al. 1994; Fuzessery and Hall 1999; Zhou and Jen 2001). In other words, they are "bandpass" filters for sound duration. Here, bandpass, duration-tuned neurons will be referred to simply as "duration selective" neurons. It appears that duration selective neurons are located in the caudal parts of the IC (Ehrlich et al. 1997). An example of bandpass tuning to duration is shown in Fig. 7.22. As in the case of binaural time differences, duration tuning is achieved in the central nervous system by neural calculation through "tuning" of neurons to specific values of the parameter. In the case of duration, each neuron responds to a different range of durations and each has a best duration. Therefore, duration is a temporal sound parameter that is converted from the original temporal representation (firing throughout the length of a sound for all sound durations) to a place-like representation. That is, a specific cell responds to a particular duration, but not others. Duration selectivity may seem rather counterintuitive from the point of view of sensory integration time; in psychophysics, the detectability of short sounds increases as sound duration increases. So why should some neurons selectively respond to short sounds and fail completely to respond to long sounds? The answer may be forthcoming if we ask a different psychophysical question. If we ask how we can

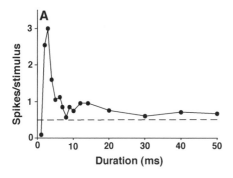

FIGURE 7.22. Examples of two band-pass-type duration-tuned neurons in the IC of the big brown bat. (Redrawn from Ehrlich et al. 1997.)

discriminate between sounds of slightly different duration (Wright and Dai 1994), then the answer seems clear: somewhere in the brain there must be a neural representation of different durations. From a biological point of view it makes sense to have a filter for sound duration. Sensory filtering is generally adapted to the needs of a species, especially to its needs in predation, predator avoidance or mate identification. Short sounds are likely to be especially biologically relevant, so duration tuning may simply be a biological filter that is used in parsing the important sounds from the unimportant sounds.

Is selectivity for short duration sounds a specialization for echo location? Fuzessery (e.g., 1994) has used the pallid bat (*Antrozous pallidus*) as a model to compare processing for echolocation sounds versus processing for sounds not containing the frequencies in the echolocation signal. The pallid bat is a gleaner, that is, it does not use echolocation for prey capture but rather locates prey on the ground by passive listening to the sounds made by small animals and insects. The frequencies contained in the bat's echolocation call (30 kHz to 80 kHz) are represented in the ventromedial part of the IC, and the remaining frequencies to which this bat is sensitive (3 kHz to 30 kHz) are represented in the dorsal and lateral parts of the IC. The lateral part of the IC is further distinguished by the fact that many of its neurons are selective to brief noise transients and respond poorly or not at all to pure tones. Fuzessery and Hall (1999) found neurons that are selective for short sound durations (< 7 ms) in the lateral IC as well as in the ventral IC. Duration selective neurons were also found in the dorsal IC, and most of these were tuned to long sound durations ("long-pass" tuning). Therefore, duration selective neurons were found in areas involved in both echolocation and passive hearing.

Finally, duration selective neurons have been found in the IC of chinchillas (Chen 1998) and mice (Brand et al. 2000). In the mouse IC, more than 50% of neurons were duration sensitive: either short pass, long pass, or band-pass (Brand et al. 2000). However, for most of these neurons a change in stimulus parameter such as frequency or binaurality eliminated or altered duration selectivity. A few of the duration sensitive neurons, mainly the band-pass neurons, retained their tuning in the face of changes in stimulus parameters and were therefore classified as duration selective. This issue of stability across stimulus paramaters is clearly very important for determining what role duration selectivity plays in IC sensory processing.

3.3.4.2 Mechanisms for Duration Tuning

The mechanisms underlying duration tuning are beginning to be understood. Duration selective neurons have been studied with extracellular recordings when antagonists to the inhibitory transmitters GABA or glycine have been applied, as well as with intracellular methods that show the time course of synaptic events. When antagonists to GABA or glycine are applied to duration selective neurons, most lose their bandpass tuning to duration and, like other neurons, respond to sounds of any duration beyond a minimum of 1 ms or so (bat, Casseday et al. 1994; 2000). This observation indicates that inhibitory input plays a major role in creating duration selectivity and that duration selectivity is created in the IC. Furthermore, in many duration selective neurons, blocking inhibition decreases the latency of the first spikes of the response, and changes the timing of the response from offset to onset. Intracellular recordings from a duration selective neuron, shown in Fig. 7.23, show that inhibition dominates the early part of the response for sounds of all durations, and inhibition lasts for the duration of the sound. Thus, for the example shown in Fig. 7.23, outward current precedes the spikes for a 10 to 20 ms sound, but the entire response is dominated by outward current when the sound is more than 20 ms long. These observations suggest that there is a temporal sequence of synaptic events in which inhibition is the first input, excitation arrives later, and further inhibition may follow the excitatory input. At the end of the sound, there is an offset excitation, possibly due to rebound from inhibition. It appears that "leading" and "lagging" inhibition allows an excitatory input to generate spikes if it occurs within a narrow time window. It is not clear whether lagging inhibition is present or needed for all duration-tuned neurons.

Based on these results, we suggested a model for duration tuning (Casseday et al. 1994; Fig. 7.24). The model is similar to that proposed by Narins

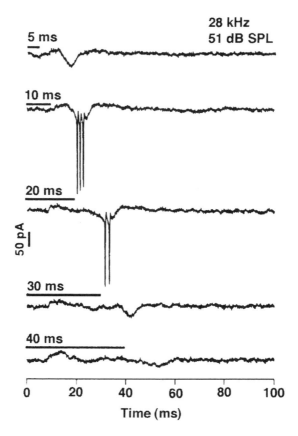

FIGURE 7.23. Intracellular recordings from a duration-tuned neuron in the IC of the big brown bat, obtained using whole-cell patch clamp recording. In response to a 5 ms stimulus (top trace), there was an IPSC followed by a subthreshold EPSC. The 10 ms stimulus evoked a large EPSC and action potentials, and the 20 ms stimulus evoked a smaller EPSC clearly associated with the offset of the sound. 20 ms was the longest duration at which the neuron produced spikes. Stimuli 30 and 40 ms in duration evoked an IPSC associated with sound onset and a small subthreshold EPSC asociated with offset. While these traces were recorded in the voltage clamp mode, the finding that action potentials were driven synaptically indicates that the voltage could not be clamped in the cell's processes. However, the lack of space clamping presumably does not alter the sign of synaptic responses and presumably changes the relative amplitudes little. (From Covey et al. Copyright 1996 by the Society for Neuroscience.)

and Capranica (1980), in that there are two excitatory events: one locked to the offset of sound and one locked to the onset. However, in the present model, inhibition also plays an inportant role. Models using similar mechanisms have been proposed for delay tuning (Suga et al. 1995) and for rate tuning (Grothe 1994).

   In the model shown in Fig. 7.24, there are three events that differ in their timing: (1) inhibition arrives first and lasts for the duration of the sound,

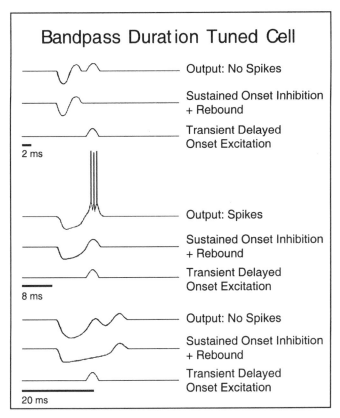

FIGURE 7.24. A model for the creation of bandpass duration tuning in an IC neuron. Each set of three traces represents a long-latency, transient, onset-evoked EPSP (lowest of the three traces), a short-latency, sustained, onset-evoked IPSP followed by rebound (middle of the three traces) and the resulting change in membrane potential and output of the IC neuron (topmost of the three traces). When the duration of the sound is short (2 ms, top three traces), the IPSP lasts for only 2 ms, followed by a short rebound. The latency of the EPSP (trace) is such that it occurs after the rebound from inhibition. Because there is no coincidence between the subthreshold EPSP and the rebound from inhibition, the neuron does not reach threshold (top trace). When the duration of the sound is 8 ms (middle three traces), the IPSP lasts long enough so that the rebound at offset coincides with the onset-evoked EPSP, bringing the neuron to threshold. When the duration of the sound is longer (20 ms, bottom three traces), the rebound at the end of the IPSP occurs after the EPSP is over. There is no coincidence between the EPSP and rebound, so the neuron does not fire.

(2) excitation, either from an extrinsic input or as rebound from inhibition, is locked to sound offset, and (3) excitation is locked to sound onset but is delayed relative to inhibition and is rendered subthreshold by inhibition. The cell fires when the two excitatory components coincide in time.

The source of offset excitation is not yet clear. In our initial model, we favored the idea that the offset excitation was an intrinsic property of the cell membrane. That is, a rebound from inhibition occurs when the sustained hyperpolarization ends. This idea was attractive because it is the simplest mechanism for duration tuning; it requires only two inputs, sustained inhibition, and delayed excitation. Furthermore, IC neurons appear to exhibit rebound from inhibition (rat, Peruzzi et al. 2000). However, the MSO contains a population of neurons that respond to the offset of sound (e.g., bat, Grothe et al. 1997). Because these neurons project to the IC and are presumably excitatory, they are a potential source of offset excitation to duration selective neurons.

The fact that the response latency of a few duration-tuned neurons is not locked to the offset of the sound (bat, Ehrlich et al. 1997; Fuzessery and Hall 1999) indicates that the model above does not account for the duration selectivity of these few neurons. That is, the rebound mechanism does not explain the behavior of some duration-tuned neurons. To account for this behavior, Fuzessary and Hall (1999) proposed a variation of the components of the model above, which they termed an "anti-coincidence" model. In their model, inhibition and excitation are both locked to the onset of sound. The inhibitory input arrives first and is sustained for the duration of the sound. The excitatory input is delayed and transient. Therefore, when the two inputs do not coincide, i.e., at short durations, the neuron fires. However, as the sound increases in duration, so does the inhibitory input so that it eventually overlaps with the excitatory input, suppressing it.

3.3.4.3 Delay Tuning

Delay tuning is the second example of tuning to biologically relevant sound patterns in the IC. Delay tuning was first found in the midbrain of the big brown bat (Feng et al. 1978). Delay tuning is biologically important in that it is the putative mechanism for determining the distance of an echo-producing object. In sonar, the distance to an object is measured by the elapsed time between an outgoing sound pulse and the echo that returns from the object. Echolocating bats have neural mechanisms, consisting of delay lines and coincidence detectors, that perform a similar measurement. The most well-studied example is the mustached bat. This bat uses an echolocation pulse that has multiple harmonics. Suga and his colleagues (e.g., Suga et al. 1978, 1983; O'Neil and Suga 1979, 1982) discovered neurons in the auditory cortex that responded poorly or not at all to individual FM components of the echolocation pulse. However, they responded preferentially to combinations of the fundamental FM (FM1) component and one of the higher harmonic FM (FMn) components, but only provided there was a specific time delay between FM1 and FMn. Neurons of this type are called "combination sensitive, delay-tuned" neurons. Delay-tuned neurons have subsequently been found in the IC of the mustached bat (Mittman and Wenstrup 1995; Yan and Suga 1996b) and in intercollicular areas in the big brown bat (Dear and Suga 1995). These neurons have best delays between 0 and 9 ms. One model for creating delay tuning is based on a circuit in which FM1 evokes an early inhibition that is followed by long latency excitation. When the excitation from FM1 coincides with the short-latency excitatory response to the later occurring FMn (echo), the neuron responds (Suga et al. 1995). Note the similarity to the duration tuning model, which was based in part on the same principles as the delay tuning model.

Why is delay tuning created at the IC and again at the thalamus and cortex? There may be two reasons for this apparent redundancy. First, the representation at the IC provides access to the motor systems with which the IC is connected. Second, the representation at the auditory cortex provides a mechanism for modulating the IC's access to motor systems. Indeed, as will be seen later, the auditory cortex plays a role in sharpening the delay tuning in the medial geniculate and IC (Yan and Suga 1996a, 1999).

3.3.4.4 Selectivity to Frequency Modulated (FM) Sounds

Suga (e.g., Suga 1965, 1968, 1969; reviewed in Casseday and Covey 1996) showed that the ICc of bats contains neurons that respond best or exclusively to FM sweeps and are selective for the direction of the FM. The existence of FM specialization in bats supports a common assumption that FM selectivity is a specialization for echolocation. In support of this notion, in the pallid bat (*Antrozous pallidous*), the ventral ICc represents the frequencies in the echolocation call, and in this part of the IC selectivity for FM sounds is enhanced or expanded. About half of all the ventral ICc neurons respond selectively to downward FM sweeps, and respond weakly or not at all to upward FM sweeps, pure tones, or noise. The dorsal IC contains very few of these "FM specialists" (Fuzessery 1994). Thus, the FM specialists are well suited for the perceptual requirements of this bat.

However, the idea that FM selectivity is a specialization just for echolocating bats does not appear to be justified. Other mammals, such as rats, use FM components in communication calls, and in fact, the ICc of rats also contains neurons that respond preferentially or exclusively to FM sweeps. As noted earlier (Section 2.1.2), there is evidence that these cells have a more extensive dendritic arbor than do cells that are less specialized (rat, Poon et al. 1992).

Is FM selectivity or FM direction selectivity created in the IC? Fuzessery and Hall (1996) used bicuculline to block GABAergic inhibition in neurons in the pallid bat that were specialized for FM direction. Under the blocked condition, most of the neurons lost their directional selectivity and responded equally well to both upward and downward sweeps. Interestingly, in some neurons, the directional FM selectivity was eliminated without eliminating the neuron's selective response to FM sounds. In other neurons, both directional selectivity and FM specialization were eliminated so that the neuron responded not only to FM, but also to other types of sound such as pure tones and noise.

A number of models have been proposed to account for selectivity to FM direction (for review see Fuzessery and Hall 1996). Most of these models involve some sort of coincidence mechanism in which high-frequency inputs are delayed with respect to low-frequency inputs or vice versa. As in the models for duration tuning, it is not clear where and how the delays come about. The delays could be established at inputs to the IC, by interneurons in the IC, or by where the inputs terminate on the cell (e.g., soma or dendrite) (Fuzessery and Hall 1996).

Finally, there is evidence that nonauditory inputs can influence FM sweep processing. In the Mexican free-tailed bat (*Tadarida brasiliensis*), Hurley and Pollak (1999, 2001) showed that the predominant effect in the IC of iontophoretically applied 5-HT is suppression for pure-tone and FM sweep-evoked responses. 5-HT also has different effects on FM sweeps, depending on sweep depth and/or rate. It appears, therefore, that IC neurons' responses to frequency sweeps may be adjusted by levels of neuromodulators that vary according to the general state of the animal.

In summary, specialization for downward FM sounds is created and refined in the IC, and this is probably a general feature of the mammalian IC. However, there is species-specific variation in the quantity of specialized neurons and in the frequency range to which they are tuned.

3.3.4.5 Rate Tuning

Most natural sounds contain amplitude and frequency modulations. However, the difficulties of interpreting data obtained from natural sounds have led most researchers to simplify these sounds and construct sounds that contain only amplitude modulations or frequency modulations. Commonly used repetitive stimuli are sinusoidal modulations of amplitude (SAM) and sinusoidal modulations of frequency (SFM).

A characteristic of many IC neurons is that they do not seem to be able to respond reliably to repetitive stimulation. For example, when tested with SAM sounds, there is often only a response to the onset of the sound rather than to every cycle (e.g., bat, Burger and Pollak 1998). For those neurons that do phaselock to SAM, the response rate at modulation rates greater than 100 and 300 Hz typically falls to near spontaneous levels or is no longer synchronized to the phase of the modulation frequency. The review to follow will emphasize several points concerning IC neurons' specializations for processing the rate of repetitive stimuli. First, for AM of pure tones, some aspects of the processing mechanism are either set up in centers below IC or are intrinsic to IC cells. Second, tuning to SFM rate and specializations for SFM appear to be created in the IC.

3.3.4.6 Tuning to the Rate of Amplitude Modulation

It has been known for some time that a subset of IC units are tuned to the rate of sinusoidally amplitude-modulated sounds (for reviews see Langner 1992; Krishna and Semple 2000). Most studies found the best modulation frequencies to be below 150 Hz, and nearly all were below 300 Hz. This low pass filter, which is characteristic of IC neurons, has been seen in all mammals studied (e.g., cat, Langner and Schreiner 1988; gerbil, Heil et al. 1995; Krishna and Semple 2000; rat: Rees and Møller 1983; squirrel monkey: Muller-Preuss et al. 1994; two species of bats: Burger and Pollak 1998; Reimer 1987; Schuller 1979). In the horseshoe bat, some neurons can phase lock to AM depths as low as 3% (Schuller 1979).

Langner and Schreiner (1988), recording in the cat IC, determined the location of neurons' best modulation frequency (BMF) for SAM within an isofrequency contour and found an orderly arrangement or map of BMF. However, Krishna and Semple (2000) point out that when the overall sound level is changed, the topography of the map would also change.

From the foregoing discussion of how excitation and inhibition temporally offset from one another could shape neural selectivity, it might be proposed that the low-pass characteristics of IC neurons to SAM rate is created in the IC by these mechanisms. Burger and Pollak (1998) tested this hypothesis in the mexican free-tailed bat by asking whether or not phase locking to SAM was altered by the application of $GABA_A$, glycine and $GABA_B$ blockers. Consistent with other studies, most neurons (142 of 195) did not phase lock to SAM rates above 300 Hz. Blocking inhibition did not cause the neurons to phase lock to SAM rates above 300 Hz and thus, inhibition in the IC does not appear to be the mechanism that creats low pass filter characteristics to SAM rate in IC neurons. By blocking NMDA receptors, Burger and Pollak also asked whether the ability of AMPA receptors to follow high stimulation rates might be slowed by the slower action of NMDA receptors. The result here was also negative. Thus, it remains a puzzle as to how the low pass characteristics of IC neurons are generated (Burger and Pollak 1998).

3.3.4.7 Selectivity and Tuning to Sinusoidal Frequency Modulation (SFM)

Most natural sounds, including those generated by animals, contain frequency modulations that repeat over time. Some examples of animal-generated sounds that contain repetitive frequency modulations are: insect communication sounds, amphibian mate attraction sounds, bird song, mammalian communication sounds (for warnings or mate attraction) echolocation sounds, and speech sounds. In the laboratory, it is useful to use simpler sounds that contain some aspects of natural sounds because it is easy to generate them, and because they can be parametrically varied. Sinusoidal frequency modulations (SFM) are often used for this purpose because (1) they contain at least some aspects of natural sounds, (2) it is easy to generate them, and (3) it is easy to vary their parameters in ways that allow quantification of neural responses as a function of a given parameter.

To generate an SFM sound, frequency is continuously modulated up and down with respect to a center or carrier frequency, and the modulation pattern is a sinusoidal function. The depth of modulation refers to the amount of frequency excursion around the center frequency, and the rate of modulation is simply the frequency of the modulating sinusoid. As is the case with simpler sounds, the center frequency and the level of the sound is usually varied.

For the purpose of this chapter, we will be primarily concerned with neurons that show some sort of selectivity for SFM sounds and secondarily with the question of whether this selectivity is created in the IC. First however, we must describe a few basic response properties of IC neurons' responses to SFM. Not all neurons in the IC respond well to an SFM sound. For example, some respond only to the onset of the sound, that is, to the first SFM cycle. Many other IC neurons respond continuously to an SFM sound, and it is these neurons with which we will mainly be concerned. When presented with SFM, the responses of many neurons, both at the IC and lower auditory centers, are tightly locked to the frequency of the SFM. However, the responses of IC neurons (e.g., rat, Rees and Møller 1983; bat, Schuller 1979; Casseday et al. 1997) are typically less tightly locked to the modulating frequency than are auditory cells in the cochlear nucleus (e.g., rat, Møller 1972) or lateral lemniscus (e.g., bat, Huffman et al. 1998). Another difference between lower brainstem auditory neurons and IC neurons is that the latter typically do not respond to SFM rates as high as do those in the cochlear nucleus (e.g., rat, Møller 1972; gerbil, Frisina et al. 1985) or the nuclei of the lateral lemniscus (e.g., bat, Huffman et al. 1998). Thus, the responses of IC neurons to SFM are similar to the responses of IC neurons to SAM in their failure to respond to high modulation rates.

In the big brown bat, some neurons show specialization for SFM sounds. The specialization is manifested by a poor or nonexistent response to pure tones, noise, or single frequency sweeps (Casseday et al. 1997). Examples of other characteristics that distinguish SFM-specialized neurons are shown in Fig. 7.25. These characteristics include sharp tuning to SFM rate (Fig 7.25A and C) and SFM depth (Fig. 7.25B and D). In addition, SFM-specialized neurons respond only once per cycle, and they usually do not respond to the first SFM cycle (Fig. 7.25A and B).

There is evidence that the temporal response characteristics of IC neurons to SFM is generated at the IC. When GABA or glycine is blocked at an IC neuron with bicuculine or strychnine, respectively, the phase-locking response to SFM tuning deteriorats, and the band-pass tuning to SFM rate broadens in many neurons (bat, Koch and Grothe 1998). Thus, tuning to SFM rate, unlike tuning to SAM rate, seems to be modulated by neurotransmitters in the IC.

3.4 Auditory Cortical Control of IC Processing

Stimulation of the auditory cortex can facilitate and inhibit both spontaneous and sound-evoked activity of IC neurons (cat, Mitani et al. 1983; rat, Syka and Poplelar 1984; bat, Sun et al. 1989; Yan and Suga 1996a; Jen and Zhang 1999; guinea pig, Torterolo et al. 1998). Similarly, inactivation of the auditory cortex can produce both facilitation and suppression of IC neurons' responses (bat, Zhang et al. 1997; Jen et al. 1998). These results support the hypothesis that the auditory cortex modulates sensory processing in the IC. The most direct pathway for such modulation is from cells in the auditory cortex that project directly back to most or all the divisions of the IC, as described in Section 2.4 and Chapter 2. This direct projection appears to be excitatory (guinea pig, Feliciano and Potashner 1995; chinchilla, Saint Marie 1996; rat, Saldaña et al. 1996), suggesting that the inhibitory cortical modulation results from connections with inhibitory interneurons within the IC or through even less direct pathways. For example, indirect output from auditory cortex can flow to the IC via (1) the MGB (bat, Kuwabara and Zook 2000), (2) lower brainstem auditory nuclei (rat, Weedman and Ryugo 1996; gerbil, Budinger et al. 2000; see also Spangler and Warr 1991 and Chapter 2), and (3) the globus pallidus and/or the substantia nigra (Section 2.5.2) in conjunction with the striatum (Moriizumi and Hattori 1991b).

3.4.1 What Is the Function of Corticofugal Modulation of the IC?

To date, there is no simple or complete answer to this question. However, an important clue is the finding that cortical stimulation and cortical inactivation affect the responses of an IC neuron differentially depending on whether or not the cortical area and the IC neuron are "matched" in terms of their response to sound (bat, Yan and Suga 1996a; Zhang et al. 1997; see Suga et al. 2000 for review). Matching is determined electrophysiologically by characterizing the best stimulus for a location in cortex and then finding a location in the IC where neurons also respond best to that stimulus. For example, a cortical stimulation site is said to match the IC recording site if both the cortical area and the IC neuron respond optimally to the same frequency or to the same delay between two stimuli. Suga and his colleagues argue that corticofugal activity enhances the responses of IC neurons that are matched to the cortical neurons, whereas it supplies widespread suppression to nonmatched IC neurons (bat, Yan and Suga 1996a; Zhang et al. 1997; see Suga et al. 2000 for review). Within this scheme, corticofugal activity would help enhance and select responses to a particular auditory parameter.

For example, cortical activity can adjust the delay tuning of some IC neurons. Yan and Suga (1996a) stimulated and inactivated small regions of the FM-FM area of the mustached bat's auditory cortex while recording the responses of IC delay-tuned neurons to paired "pulse-echo" stimuli that are

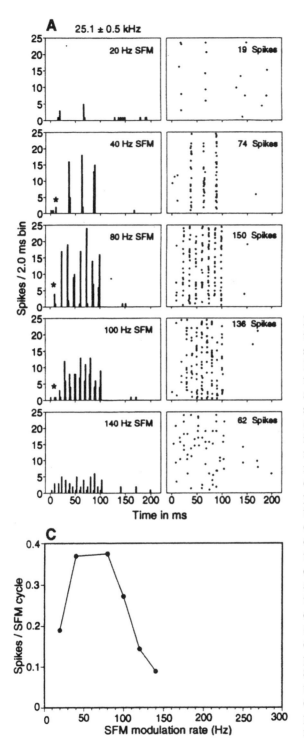

FIGURE 7.25. Example of the responses of an SFM selective neuron in the IC of the big brown bat to changes in SFM rate and SFM depth. *A*. Post-stimulus time histograms (left) and corresponding dot rasters (right) of responses to different rates of SFM. Note that even at SFM rates that yield high spike counts, the number of peaks is one less than the number of cycles, suggesting that the response to the first cycle (*) is diminished or absent. *B*. Post-stimulus time histograms (left) and corresponding dot rasters (right) of responses to different depths of SFM presented at a rate of 80 Hz. At SFM depths that yield high spike counts, the number of peaks is one less than the number of cycles.

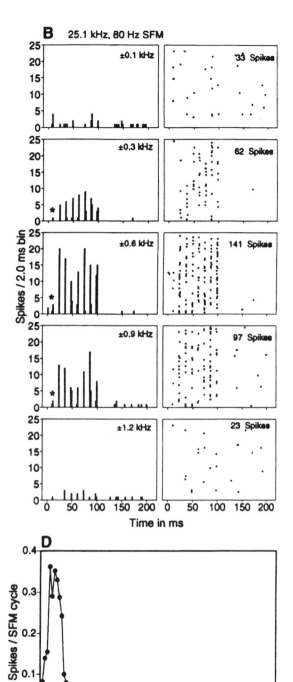

FIGURE 7.25. (*Continued*) C. Number of spikes per SFM cycle for modulation rates from 20 Hz to 140 Hz. The neuron responded maximally to rates of 40 Hz and 80 Hz. D. Number of spikes per SFM cycle for SFM depths from ±0.1 kHz to ±1.2 kHz. The neuron responded best to depths between ±0.4 kHz and ±0.9 kHz. Stimulus duration = 100 ms. (From Casseday et al. 1997.)

meant to mimic the time relationship between the pulse and echo of an echolocating bat. The best delay of units in the area of cortical stimulation or inactivation was either matched (within ± 0.4 ms) or unmatched (difference greater than ± 0.4 ms) to the IC neuron's best delay. When a matched region of the auditory cortex was stimulated, the spike count of the IC neuron increased, the best delay stayed the same, and the delay tuning sometimes sharpened. However, when an unmatched region of the auditory cortex was stimulated, the response rate of the IC neuron decreased around the best delay and the best delay shifted away from the best delay of the auditory cortex region. When the best delay in the cortical area was longer than the best delay of the IC neuron, then the delay tuning of the IC cell shifted to an even shorter delay. When the best delay in the cortical area was shorter than the best delay of the IC neuron, then the delay tuning of the IC cell shifted to an even longer best delay. Inactivation of the auditory cortex had the opposite effects on IC neurons.

Cortical activity can also adjust the frequency tuning of some IC neurons. Suga and his colleagues stimulated and inactivated small regions of the mustached bat's auditory cortex while recording the responses of IC neurons to pure tones, all within the expanded "60 kHz" frequency representation area (Zhang et al. 1997; Zhang and Suga 2000a). The BF of units in the cortex was either matched (within ± 0.2 kHz) or unmatched (greater than ± 0.2 kHz) to the IC neuron's BF. When a matched area of cortex was stimulated, the IC neuron's spike count increased and the BF remained the same. However, when an unmatched area of cortex was stimulated, the IC neuron's spike count decreased around the original BF and the frequency tuning shifted away from the BF of the cortical region. Inactivation of the auditory cortex had the opposite effects on IC neurons. This is the same general pattern of results as described above for delay tuning. Taken together, the changes in delay tuning and frequency tuning in the IC following cortical stimulation or inactivation support the hypothesis that corticofugal modulation, involving both focal excitation and widespread inhibition, occurs for most, if not all, biologically relevant sound parameters.

Excitatory and inhibitory corticofugal modulation of the IC is not limited to the mustached bat. The specific effects of corticofugal modulation on the IC may, however, vary by species. For example, in the big brown bat, stimulation of the auditory cortex produced the same results as in the mustached bat when the BFs of the cortex matched those of the IC; stimulation caused an increase in spike count with no change in BF in the IC neuron (Yan and Suga 1998). However, when the cortical and IC BFs did not match, stimulation of cortex produced a pattern of results different from those found in the mustached bat. Instead of producing a shift in BF away from the BF of cortex, cortical stimulation produced a shift toward the BF of cortex when the IC BF was greater than the cortical BF, and produced no consistent shift when the IC BF was less than the cortical BF. Thus, the specificity of IC corticofugal modulation appears to vary by species.

It is important to note that the changes in spike count, best frequency, and best delay summarized above appear to develop over time, from 2 min to over 30 min, and can last from 30 min to more than 2 hrs. These time courses suggest that one role of corticofugal modulation of the IC is to select, enhance, and maintain processing of specific auditory parameters over long periods of time, perhaps as long as an evening's hunting, in the case of a bat. With this in mind, it is interesting to note that artificial stimulation of the auditory cortex is not the only way to produce frequency tuning changes in the IC. Suga and his colleagues found that frequency tuning in the IC of the big brown bat can change following repetitive presentation of paired auditory and leg shock and that these changes are similar to the changes seen following stimulation of auditory cortex (Gao and Suga 1998, 2000; Yan and Suga 1998).

In addition to the long-term corticofugal modulation of IC that seems to develop over minutes and last for hours, there is also evidence that the auditory cortex can influence processing in the IC on a short-term, stimulus by stimulus basis (Jen et al. 1998; Zhou and Jen 2000a). In order for this to occur, some auditory information must quickly reach the auditory cortex and then return to the IC just prior to, or coincident with, the inputs responding to that same stimulus ascending from the lower brainstem auditory nuclei. In other words, a complex, synaptic delay line must operate, in which at least two different pathways provide input to an IC neuron: one directly from the ascending auditory pathway and one via the descending pathway from auditory cortex. In order for the cortical input to modify the IC neuron's response to the same stimulus that activated the cortex, the ascending pathway to the auditory cortex has to have a shorter latency than the direct ascending auditory pathway to the IC. That is, a short-latency ascending pathway would need to reach the thalamus and cortex so that the descending input from cortex could coincide with a longer-latency ascending pathway to the IC cell. Of course, this timing requirement is much less stringent if the action of the cortex operates only after the first occurrence of the stimulus. In this case, the cortex would still modulate IC processing, but such modulation would be based on recent stimulus history (e.g., what happened 30, 40, or 50 ms ago), and not on the stimulus that the IC neuron is currently responding to.

Mitani et al. (1983) in cats, Syka and Popelár (1984) in rats, and Torterolo et al. (1998) in guinea pigs measured changes in spontaneous or acoustically-evoked activity in IC neurons following ipsilateral stimulation of the auditory cortex. In response to such stimulation, Mitani et al. (1983) reported excitatory and inhibitory latencies in the IC ranging from 1 to 5 ms. Syka and Popelár (1984) reported excitatory and inhibitory latencies ranging from 3 to 15 ms, while Torterolo et al. (1998) reported an average excitatory latency of 7.2 ms. All three studies reported long lasting inhibition, in some cases 70 ms or more. If these results reflect the natural operation of cortical input, then there are at least two important conclusions that can be made concerning how cortical feedback might modulate temporal patterns of sensory input. First, the latencies of excitatory or inhibitory effects on IC cells in response to direct cortical stimulation suggests that cortical feedback to the IC is reasonably fast (as fast as 1–3 ms in the rat and cat, Mitani 1983; Syka and Poplelar 1984), particularly in relation to the shortest latencies of IC neurons to sound (~6 ms in the gerbil, Syka et al. 2000). Thus, the corticofugal IC latency is short enough so that cortical feedback could, in theory, modulate the responses of longer-latency IC neurons evoked by the same stimulus. This idea is further supported by comparing the range of latencies of ascending and descending inputs to the IC with latencies of IC cells. For example, in the big brown bat the range of latencies of a major input to the IC, the nuclei of the lateral lemniscus, is ~2 to 20 ms (Covey and Casseday 1991), the range of latencies in the IC is ~3 to 60 ms (Casseday and Covey 1992), and the range of latencies at auditory cortex is ~6 to 40 ms (Simmons et al. 1995). The fact many IC neurons respond with latencies longer than 10 to 15 ms suggests that they would have ample time to incorporate a cortical contribution that is itself locked to the stimulus. These IC neurons could receive a rapid input from cortex that precedes or overlaps a longer-latency input from the lateral lemniscus or from intrinsic IC projections. Similarly, a cortical input could precede or overlap with an offset, subthreshold, excitatory response, or rebound from inhibition in an IC neuron.

A second conclusion that can be made is that the long-lasting inhibition evoked by cortical stimulation strongly suggests that the cortex may exert effects on IC cells in response to a recently processed stimulus, possibly controlling such temporal properties as the ability of an IC neuron to respond to sounds separated by a specific interval or at a particular repetition rate. In addition, cortical modulation of inhibitory interneurons in the IC might account for the fact that some IC neurons habituate rapidly to the presentation of a sound.

Thus, it is quite possible that the temporal relationship between ascending and descending input plays an important role in shaping an IC neuron's responses on a stimulus by stimulus basis. Whether either of these types of short-lasting cortical modulation occurs in a natural situation is not known.

We do know that artificial cortical stimulation can modulate both frequency tuning (width, minimum threshold, maximum threshold) and perhaps spatial tuning on a stimulus by stimulus basis (Zhou and Jen 2000b,c). Jen has shown that such corticofugal modulation of frequency tuning can be either excitatory or inhibitory regardless of whether the cortex and IC are matched for frequency (Jen et al. 1998). It is possible that short-acting, stimulus by stimulus, corticofugal modulation of the IC does not have the same excitatory and inhibitory pattern as long-term corticofugal modulation. More detailed studies are needed to address short-acting, per stimulus effects of corticofugal modulation of IC processing and how it relates to long-term corticofugal modulation.

3.5 The Inferior Colliculus and Motor Function

3.4.3 IC Activation and Motor Response

What little is known about motor response and IC processing interactions comes mainly from studies investigating the role of midbrain structures in the generation of audiogenic seizures and escape/aversive responses. Audiogenic seizures occur in some rodents (mice for example) due to a genetic mutation. As the name implies, the seizures are brought on by sound. Susceptibility to audiogenic seizures is related to an imbalance of excitation and GABAergic inhibition in the IC (Faingold et al. 1991; Faingold and Naritoku 1992). Two aspects of audiogenic seizures are particularly important concerning the IC and behavior. First, the IC is a crucial structure in the generation of audiogenic seizures. Second, although audiogenic seizures culminate in grand mal seizures, the earlier stage usually begins with "wild running." Wild running has the appearance of escape behavior and suggests a link between aversive/escape behavior and IC processing.

It appears that several structures in the midbrain (including the dorsal pariaqueductual gray, the superior colliculus, the amygdala, the medial hypothalamus, and the IC) form a circuit that is important in mediating fear and escape responses by integrating sensory, autonomic, and behavioral processing components (see Brandão et al. 1994, 1999). The evidence for these midbrain functions comes mainly from studies in which very low levels of electrical or chemical stimulation ("microstimulation") produce reliable, dose dependent escape behaviors (rat, Pandossio and Brandão 1999; see also Brandão et al. 1993, 1999).

The predominant behaviors produced by microstimulation are somewhat dependent on which structure is stimulated. Microstimulation of the dorsal periaqueductal gray (DPAG) mainly produces explosive escape behavior, whereas microstimulation of the deep superior colliculus produces mainly turning behavior and the IC produces freezing followed by escape. The escape behavior following microstimulation of the IC is less explosive than that following microstimulation of the DPAG. In the IC, increasing levels of electrical stimulation generate, in progression, alertness, freezing, and escape. Injection of the $GABA_A$ antagonist bicuculline into the ventral parts of the IC produces a similar behavioral activation that mimics escape behavior. These and other results suggest the hypotheses that GABAergic inhibition, possibly originating in the substantia nigra, serves to modulate the response to threatening sounds (see Brandão et al. 1999). Interestingly, such modulation may also affect the response to nonthreatening sounds as well. Recently, Brandão et al. (2001) found that responses (evoked potentials) in the IC to nonthreatening sounds are altered by fear-evoking stimuli such as sound and light and that such changes are similar to changes caused by glutamate microinjection in the IC.

These microstimulation experiments are beginning to provide a picture of one aspect of behavior in which the IC might participate, namely a response sequence to threatening sounds. However, the full extent of motor reponse and IC processing interaction remains to be determined.

4. Summary and Conclusions

We conclude with some ideas on the significance of some of the currently unresolved issues concerning IC function and consider how they might influence the direction of future studies.

4.1 Structure, Function, and Maps

One unresolved issue has to do with the relationship between commonly used anatomical metrics and the functional organization of the IC. Clearly the fibrodendritic laminae are in some way related to a frequency map, but this finding is not new. To make progress, it is necessary to know whether or not other auditory parameters are mapped relative to the laminae and frequency organization. Despite evidence of a progression of binaural properties along an isofrequency contour (Section 3.2.3), the search for a space map in the mammalian IC has not been a stunning success (although see Ehret 1997). Occasionally, we have seen evidence of other types of maps, such as modulation rate (SAM) organization along an isofrequency contour, but it is not clear how stable such maps are in the face of changing stimulus parameters such as SPL (Section 3.3.4.6). Moreover, there are a number of specialized forms of neural tuning for which a topographical map seems to be missing. Although many IC units are tuned to the delay between two tones, to sound duration, and to FM and SFM, there is no obvious anatomical dimension across which the tuning is organized. If there are topographical maps for these tuned parameters, why have they been so elusive in the laboratory?

Perhaps we should take our failures more seriously and ask whether there is any good reason to expect that the sound parameters we so conveniently produce in the laboratory should be topographically organized in the IC. Sensory receptor surfaces such as the skin, the retina, and the cochlea, have isometric neural surfaces (maps). Furthermore, motor actions in response to sensory input apparently require maps of sensory space to direct motor actions toward or away from the appropriate position in space with respect to the body. If the required map is not a direct representation of a sensory surface, the neural map has to be computed. In the auditory system, the premier example of a map that is computed to guide movement is the auditory space map that is constructed in the auditory midbrain of the barn owl. This map is then transmitted to the optic tectum where it aligns with the visual map; from there the output of the optic tectum engages the motor pathways that control the muscles to move the head and body in the direction of a mouse scampering across the ground at night. Likewise, in the superior colliculus of mammals, there is a computed map of "motor error," in which neurons mark the difference between the actual position of the eyes and how far the eyes must move to center a visual or auditory target in the middle of the neuron's receptive field (e.g., Krauzlis et al. 1997). It is not easy to discover these computed maps because they are based on relationships, not absolute values of a stimulus parameter. Perhaps we should ask if there are not analogous relational maps hidden in the IC of mammals. We know that the spatial receptive field of an IC neuron can be markedly altered by the presence of another sound at a different position (Section 3.2.6). Does this observation tell us that the IC space map is sloppy, or does it tell us to look for a different sort of map? Our auditory space almost always contains multiple sounds. Perhaps what we should be looking for in the IC is not a rigid, point-to-point space map, but a map of spatial distance between two or more sounds in the auditory field.

Is there reason to expect that every sensory parameter will be topographically mapped? Sensory attributes such as color, taste, or smell do not have, as far as we know, parametrically organized topographical maps in the nervous system. Color is represented in a "blob"-like arrangement in the visual cortex. In the chemical senses, it is even less clear how parameters are ordered. Perhaps some sensory parameters do not need to be mapped back on to external space in an isometric way, so that their representation is distributed. This is not to say that there is no point in looking for maps; rather it is to say that there is a point in looking for other forms of organization.

Some maps are difficult to interpret because the tests to discover them may include neurons with very different sensitivities along some untested parameter. Take for example, a threshold map derived from a sample of neurons that includes some quite ordinary neurons responding to pure tones and other sounds and also some very specialized neurons. The unspecialized neurons may have low thresholds to pure tones, but the specialized neurons would most likely have very high thresholds to pure tones. We have seen just this situation for neurons specialized for downward FM sweeps. If the specialized neurons are mainly located at one end of an isofrequency contour and the unspecialized neurons at the other end, then we will find a threshold map, but is threshold really the appropriate parameter of this map? We mention this point because it illustrates that maps must be cautiously evaluated.

4.2 Clustering

If some of the organization is not map-like, what is it? One possibility, analogous to blobs in the visual cortex, is a cluster arrangement. Most of us who look for organization in the IC quickly notice that neurons with similar response properties are often grouped together in clusters along an electrode track. Neurons that respond exclusively to SFM are often found very near one another in an electrode track. Likewise, we very often encounter clusters of band-pass, duration-tuned neurons. Moreover, there are hints of regional distributions of response types. The ICx appears to contains neurons that have different characteristics (e.g., broad-frequency tuning) than do ICc neurons (Section 3.1.1). Duration-tuned cells tend to be encountered more frequently in the caudal part of the IC (section 3.3.4.1). Although this type of general spatial separation may exist, it hardly constitutes a map. Our impression of the dorsal ICc and ICdc seems to fit with other comments in the literature, namely, that this is where one most often finds unusual types of neurons, such as those that quickly habituate, those that respond only to faint broad band noises, or those that simply are unresponsive to the battery of sounds that our computer generates (Section 3.1.1). Although these are mainly anecdotal observations, taken together, they hint at potential new directions for understanding the functional organization of the IC. For example, are there clusters of neural response types along the dorsal-ventral or anterior-posterior planes of an isofrequency contour?

4.3 Gradients

A related idea for IC organization comes from the evidence that there are various connectional and biochemical gradients within the IC. For example, we noted gradients of ascending, descending, and crossed connections, gradients of glycine receptors and terminals, gradients of GABA receptors and terminals, and others (Section 2). It seems likely that these gradients interface in ways that procuce some sort of functional gradient within an isofrequency contour. For example, the ascending and descending connectional gradients suggest that the cells at the ventral stem of an isofrequency contour might have primary-like responses: sustained response patterns, short latency, V-shaped tuning curves, and little or no selectivity for complex sounds. In contrast, cells at the dorsal apex of an isofrequency contour may have complex response patterns, many of which are not seen at lower levels: transient-on or transient-off patterns, long latency, unusual shapes in tuning curves, sensitivity to context, and selectivity for temporally complex sounds. We note that evidence for spike latency fits this idea (Section 3.3.3). Further investigation of this idea might provide new clues to the functional organization of the IC.

4.4 Are Specialized Cells Species Specific?

We reviewed evidence for specialization in IC cells, including delay tuning, band-pass duration tuning, FM direction selectivity, SFM selectivity, and frequency-combination sensitivity. Except for band-pass duration tuning and FM direction selectivity, these specializations were first discovered in the IC of echolocating bats. Therefore, the question arises whether the specializations are a peculiarity of bats and echolocation. On the one hand, if they are bat specializations, it shows how remarkably evolution can change function without altering the basic structural design of the IC. On the other hand, if the specializations are not peculiar to bats, then they reveal the evolution of a general functional design for the IC. It is important to address the issue of species-specific processing because of what it reveals about how plastic the function of the IC has been during evolution. Species comparisons on IC function could answer a fundamental question about the evolution of selectivity for complex sensory stimuli. Have very clear species-specific functions evolved, or have relatively minor revisions been made to a general mammalian plan? Alternatively, does a very minor anatomical revision result in a profound functional difference? The issue is complicated by the fact that most of the data on bats are from unanesthetized preparations, while most of the data from other animals are not. Ideally species comparisons would be made in unanesthetized preparations.

4.5 The Function of Nonascending Auditory Input

A more complete view of what the IC does requires a better understanding of what the descending and IC commissural pathways contribute to function. Recent evidence shows that the auditory cortex modulates the function of some IC neurons, so the IC is not just a "relay" to cortex or motor systems (Section 3.4). Instead it is part of a loop that includes cortex, so it can also be seen as the output stage of cortex.

4.6 The Function of Motor Input to the IC

The IC receives input from motor centers (Section 2.5.2). As yet, we can only speculate about how motor functions affect the function of IC cells. What little is known about the projections from SN and GP suggests that the main input is to ICdc and ICx. Further, the SN projection cells are probably GABAergic and probably tonically active, so that they would exert tonic inhibition on the IC cells that they contact. Therefore, an inhibitory input to the SN cells would in turn release their IC target cells from inhibition, thus providing a window for ascending auditory pathways to be effective. How this affects sensory processing is an important future direction.

References

Adams JC (1979) Ascending projections to the inferior colliculus. J Comp Neurol 183:519–538.

Adams JC (1980) Crossed and descending projections to the inferior colliculus. Neurosci Lett 19:1–5.

Adams JC, Mugnaini E (1984) Dorsal nucleus of the lateral lemniscus: a nucleus of GABAergic projection neurons. Brain Res Bull 13:585–590.

Aitkin L (1986) The Auditory Midbrain: Structure and Function in the Central Auditory Pathways. Clifton, NJ: Humana Press.

Aitkin L, Boyd J (1978) Acoustic input to the lateral pontine nuclei. Hear Res 1:67–77.

Aitkin LM, Phillips SC (1984) The interconnections of the inferior colliculi through their commissure. J Comp Neurol 228:210–216.

Aitkin LM, Fryman S, Blake DW, Webster WR (1972) Responses of neurones in the rabbit inferior colliculus. I. Frequency- specificity and topographic arrangement. Brain Res 47:77–90.

Aitkin LM, Webster WR, Veale JL, Crosby DC (1975) Inferior colliculus. I. Comparison of response properties of neurons in central, pericentral, and external nuclei of adult cat. J Neurophysiol 38:1196–1207.

Aitkin LM, Dickhaus H, Schult W, Zimmermann M (1978) External nucleus of inferior colliculus: auditory and spinal somatosensory afferents and their interactions. J Neurophysiol 41:837–847.

Aitkin LM, Kenyon CE, Philpott P (1981) The representation of the auditory and somatosensory systems in the external nucleus of the cat inferior colliculus. J Comp Neurol 196:25–40.

Aitkin LM, Irvine DR, Nelson JE, Merzenich MM, Clarey JC (1986) Frequency representation in the auditory midbrain and forebrain of a marsupial, the northern native cat (*Dasyurus halluc atus*). Brain Behav Evol 29:17–28.

Aitkin L, Tran L, Syka J (1994) The responses of neurons in subdivisions of the inferior colliculus of cats to tonal, noise and vocal stimuli. Exp Brain Res 98:53–64.

Appell PP, Behan M (1990) Sources of subcortical GABAergic projections to the superior colliculus in the cat. J Comp Neurol 302:143–158.

Bjorkeland M, Boivie J (1984a) The termination of spinomesencephalic fibers in cat. An experimental anatomical study. Anat Embryol 170:265–277.

Bjorkeland M, Boivie J (1984b) An anatomical study of the projections from the dorsal column nuclei to the midbrain in cat. Anat Embryol 170:29–43.

Blauert J (1983) Spatial Hearing: The Psychophysics of Human Sound Localization. Cambridge, MA: MIT Press.

Brand A, Urban R, Grothe B (2000) Duration tuning in the mouse auditory midbrain. J Neurophysiol 84:1790–1799.

Brandão ML, Melo LL, Cardoso SH (1993) Mechanisms of defense in the inferior colliculus. Behav Brain Res 58:49–55.

Brandão ML, Cardoso SH, Melo LL, Motta V, Coimbra NC (1994) Neural substrate of defensive behavior in the midbrain tectum. Neurosci Biobehav Rev 18: 339–346.

Brandão ML, Anseloni VZ, Pandóssio JE, De Araújo JE, Castilho VM (1999) Neurochemical mechanisms of the defensive behavior in the dorsal midbrain. Neurosci Biobehav Rev 23:863–875.

Brandão ML, Coimbra NC, Osaki MY (2001) Changes in the auditory-evoked potentials induced by fear-evoking stimulations. Physiol Behav 72:365–372.

Brückner S, Rübsamen R (1995) Binaural response characteristics in isofrequency sheets of the gerbil inferior colliculus. Hear Res 86:1–14.

Brunso-Bechtold JK, Thompson GC, Masterton RB (1981) HRP study of the organization of auditory afferents ascending to central nucleus of inferior colliculus in cat. J Comp Neurol 197:705–722.

Budinger E, Heil P, Scheich H (2000) Functional organization of auditory cortex in the Mongolian gerbil (*Meriones unguiculatus*). IV. Connections with anatomically characterized subcortical structures. Eur J Neurosci 12:2452–2474.

Burger RM, Pollak GD (1998) Analysis of the role of inhibition in shaping responses to sinusoidally amplitude-modulated signals in the inferior colliculus. J Neurophysiol 80:1686–1701.

Caicedo A, Eybalin M (1999) Glutamate receptor phenotypes in the auditory brainstem and mid-brain of the developing rat. Eur J Neurosci 11:51–74.

Caird D (1991) Processing in the colliculi. In: Altschuler RA, Bobbin RP, Clopton BM, Hoffman DW (eds) Neurobiology of Hearing: The Central Auditory System. pp. 253–292. New York: Raven Press.

Cant NB (1982) Identification of cell types in the anteroventral cochlear nucleus that project to the inferior colliculus. Neurosci Lett 32:241–246.

Casseday JH, Covey E (1992) Frequency tuning properties of neurons in the inferior colliculus of an FM bat. J Comp Neurol 319:34–50.

Casseday JH, Covey E (1995) Mechanisms for analysis of auditory temporal patterns in the brainstem of echolocating bats. In: Covey E, Hawkins H, Port R (eds) Neural Representations of Temporal Patterns. pp. 25–52. New York: Plenum.

Casseday JH, Covey E (1996) A neuroethological theory of the operation of the inferior colliculus. Brain Behav Evol 47:311–336.

Casseday JH, Covey E, Vater M (1988) Connections of the superior olivary complex in the rufous horseshoe bat Rhinolophus rouxi. J Comp Neurol 278:313–329.

Casseday JH, Ehrlich D, Covey E (1994) Neural tuning for sound duration: role of inhibitory mechanisms in the inferior colliculus. Science 264:847–850.

Casseday JH, Covey E, Grothe B (1997) Neural selectivity and tuning for sinusoidal frequency modulations in the inferior colliculus of the big brown bat, *Eptesicus fuscus*. J Neurophysiol 77:1595–1605.

Casseday JH, Ehrlich D, Covey E (2000) Neural measurement of sound duration: control by excitatory-inhibitory interactions in the inferior colliculus. J Neurophysiol 84:1475–1487.

Chen GD (1998) Effects of stimulus duration on responses of neurons in the chinchilla inferior colliculus. Hear Res 122:142–150.

Coleman JR, Clerici WJ (1987) Sources of projections to subdivisions of the inferior colliculus in the rat. J Comp Neurol 262:215–226.

Cooper LL, Dostrovsky JO (1985) Projection from dorsal column nuclei to dorsal mesencephalon. J Neurophysiol 53:183–200.

Covey E, Casseday JH (1986) Connectional basis for frequency representation in the nuclei of the lateral lemniscus of the bat *Eptesicus fuscus*. J Neurosci 6: 2926–2940.

Covey E, Casseday JH (1991) The monaural nuclei of the lateral lemniscus in an echolocating bat: parallel pathways for analyzing temporal features of sound. J Neurosci 11:3456–3470.

Covey E, Casseday JH (1995) The lower brainstem auditory pathways. In: Popper AN, Fay RR (eds) Handbook of Auditory Research, Vol. 5: Hearing and Echolocation in Bats. pp. 235–295. New York: Springer-Verlag.

Covey E, Casseday JH (1999) Timing in the auditory system of the bat. Annu Rev Physiol 61:457–476.

Covey E, Hall WC, Kobler JB (1987) Subcortical connections of the superior colliculus in the mustache bat, *Pteronotus parnellii*. J Comp Neurol 263:179–197.

Covey E, Vater M, Casseday JH (1991) Binaural properties of single units in the superior olivary complex of the mustached bat. J Neurophysiol 66:1080–1094.

Covey E, Kauer JA, Casseday JH (1996) Whole-cell patch-clamp recording reveals subthreshold sound-evoked postsynaptic currents in the inferior colliculus of awake bats. J Neurosci 16:3009–3018.

Dear SP, Suga N (1995) Delay-tuned neurons in the midbrain of the big brown bat. J Neurophysiol 73:1084–1100.

Druga R, Syka J, Rajkowska G (1997) Projections of auditory cortex onto the inferior colliculus in the rat. Physiol Res 46:215–222.

Edgar PP, Schwartz RD (1990) Localization and characterization of 35S-t-butylbicyclophosphorothionate binding in rat brain: an autoradiographic study. J Neurosci 10:603–612.

Edwards SB, Ginsburgh CL, Henkel CK, Stein BE (1979) Sources of subcortical projections to the superior colliculus in the cat. J Comp Neurol 184:309–329.

Ehert G (1997) The auditory midbrain, a "shunting yard" of acoustical information processing. In: Ehert G, Romand R (eds) The Central Auditory System. pp. 259–316. Oxford: Oxford University Press.

Ehret G, Merzenich MM (1988) Complex sound analysis (frequency resolution, filtering and spectral integration) by single units of the inferior colliculus of the cat. Brain Res 472:139–163.

Ehrlich D, Casseday JH, Covey E (1997) Neural tuning to sound duration in the inferior colliculus of the big brown bat, *Eptesicus fuscus*. J Neurophysiol 77: 2360–2372.

Faingold CL, Naritoku DK (1992) The genetically epilepsy-prone rat: neuronal networks and actions of amino acid neurotransmitters. In: Faingold CL, Fromm GH (eds) Drugs for Control of Epilepsy: Actions on Neuronal Networks Involved in Seizure Disorders. pp. 227–308. Boca Raton: CRC Press.

Faingold CL, Gehlbach G, Caspary DM (1989) On the role of GABA as an inhibitory neurotransmitter in inferior colliculus neurons: iontophoretic studies. Brain Res 500:302–312.

Faingold CL, Boersma Anderson CA, Caspary DM (1991) Involvement of GABA in acoustically-evoked inhibition in inferior colliculus neurons. Hear Res 52: 201–216.

Faye-Lund H, Osen KK (1985) Anatomy of the inferior colliculus in rat. Anat Embryol 171:1–20.

Feliciano M, Potashner SJ (1995) Evidence for a glutamatergic pathway from the guinea pig auditory cortex to the inferior colliculus. J Neurochem 65:1348–1357.

Feng AS, Simmons JA, Kick SA (1978) Echo detection and target-ranging neurons in the auditory system of the bat *Eptesicus fuscus*. Science 202:645–648.

FitzPatrick KA (1975) Cellular architecture and topographic organization of the inferior colliculus of the squirrel monkey. J Comp Neurol 164:185–207.

FitzPatrick KA, Imig TJ (1978) Projections of auditory cortex upon the thalamus and midbrain in the owl monkey. J Comp Neurol 177:573–555.

Frisina RD, Smith RL, Chamberlain SC (1985) Differential encoding of rapid changes in sound amplitude by second-order auditory neurons. Exp Brain Res 60:417–422.

Frisina RD, O'Neill WE, Zettel ML (1989) Functional organization of mustached bat inferior colliculus: II. Connections of the FM2 region. J Comp Neurol 284: 85–107.

Fubara BM, Casseday JH, Covey E, Schwartz-Bloom RD (1996) Distribution of $GABA_A$, $GABA_B$, and glycine receptors in the central auditory system of the big brown bat, *Eptesicus fuscus*. J Comp Neurol 369:83–92.

Fuzessery ZM (1994) Response selectivity for multiple dimensions of frequency sweeps in the pallid bat inferior colliculus. J Neurophysiol 72:1061–1079.

Fuzessery ZM, Hall JC (1996) Role of GABA in shaping frequency tuning and creating FM sweep selectivity in the inferior colliculus. J Neurophysiol 76:1059–1073.

Fuzessery ZM, Hall JC (1999) Sound duration selectivity in the pallid bat inferior colliculus. Hear Res 137:137–154.

Gao E, Suga N (1998) Experience-dependent corticofugal adjustment of midbrain frequency map in bat auditory system. Proc Natl Acad Sci U S A 95:12663–12670.

Gao E, Suga N (2000) Experience-dependent plasticity in the auditory cortex and the inferior colliculus of bats: role of the corticofugal system. Proc Natl Acad Sci USA 97:8081–8086.

Garey LJ, Webster WR (1989) Functional morphology in the inferior colliculus of the marmoset. Hear Res 38:67–79.

Gaza WC, Ribak CE (1997) Immunocytochemical localization of AMPA receptors in the rat inferior colliculus. Brain Res 774:175–183.

Geniec P, Morest DK (1971) The neuronal architecture of the human posterior colliculus. A study with the Golgi method. Acta Otolaryngol Suppl 295:1–33.

Glendenning KK, Baker BN, Hutson KA, Masterton RB (1992) Acoustic chiasm V: inhibition and excitation in the ipsilateral and contralateral projections of LSO. J Comp Neurol 319:100–122.

González Hernández TH, Meyer G, Ferres-Torres R (1986) The commissural interconnections of the inferior colliculus in the albino mouse. Brain Res 368:268–276.

González-Hernández T, Mantolán-Sarmiento B, González-González B, Pérez-González H (1996) Sources of GABAergic input to the inferior colliculus of the rat. J Comp Neurol 372:309–326.

Grinnell AD (1963) The neurophysiology of audition in bats: Intensity and frequency parameters. J Physiol 167:38–66.

Groh JM, Trause AS, Underhill AM, Clark KR, Inati S (2001) Eye position influences auditory responses in primate inferior colliculus. Neuron 29:509–518.

Grothe B (1994) Interaction of excitation and inhibition in processing of pure tone and amplitude-modulated stimuli in the medial superior olive of the mustached bat. J Neurophysiol 71:706–721.

Grothe B, Vater M, Casseday JH, Covey E (1992) Monaural interaction of excitation and inhibition in the medial superior olive of the mustached bat: an adaptation for biosonar. Proc Natl Acad Sci U S A 89:5108–5112.

Grothe B, Covey E, Casseday JH (1996) Spatial tuning of neurons in the inferior colliculus of the big brown bat: effects of sound level, stimulus type and multiple sound sources. J Comp Physiol [A] 179:89–102.

Grothe B, Park TJ, Schuller G (1997) Medial superior olive in the free-tailed bat: response to pure tones and amplitude-modulated tones. J Neurophysiol 77: 1553–1565.

Hand PJ, Van Winkle T (1977) The efferent connections of the feline nucleus cuneatus. J Comp Neurol 171:83–109.

Harting KJ, Van Lieshout DP (2000) projections from the rostral pole of the inferior colliculus to the cat superior colliculus. Brain Res 881:244–247.

Hartmann WM (1983) Localization of sound in rooms. J Acoust Soc Am 74: 1380–1391.

Heil P, Schulze H, Langner G (1995) Ontogenetic development of periodicity coding in the inferior colliculus of the Mongolian gerbil. Audit Neurosci 1:363–383.

Huffman RF, Henson OW (1990) The descending auditory pathway and acoustico-motor systems: connections with the inferior colliculus. Brain Res Brain Res Rev 15:295–323.

Huffman RF, Argeles PC, Covey E (1998) Processing of sinusoidally frequency modulated signals in the nuclei of the lateral lemniscus of the big brown bat, *Eptesicus fuscus*. Hear Res 126:161–180.

Hurley LM, Pollak GD (1999) Serotonin differentially modulates responses to tones and frequency-modulated sweeps in the inferior colliculus. J Neurosci 19:8071–8082.

Hurley LM, Pollak GD (2001) Serotonin effects on frequency tuning of inferior colliculus neurons. J Neurophysiol 85:828–842.

Hutson KA, Glendenning KK, Masterton RB (1991) Acoustic chiasm. IV: Eight midbrain decussations of the auditory system in the cat. J Comp Neurol 312: 105–131.

Ingham NJ, Hart HC, McAlpine D (2001) Spatial receptive fields of inferior colliculus neurons to auditory apparent motion in free field. J Neurophysiol 85:23–33.

Irvine DR (1986) The auditory brainstem: A review of the structure and function of auditory brainstem processing mechanisms. In: Ottoson D (ed) Progress in Sensory Physiology. pp. 76–226. Berlin: Springer-Verlag.

Ishii T, Moriyoshi K, Sugihara H, Sakurada K, Kadotani H, Yokoi M, Akazawa C, Shigemoto R, Mizuno N, Masu M, et al. (1993) Molecular characterization of the family of the N-methyl-D-aspartate receptor subunits. J Biol Chem 268: 2836–2843.

Jay MF, Sparks DL (1987a) Sensorimotor integration in the primate superior colliculus. II. Coordinates of auditory signals. J Neurophysiol 57:35–55.

Jay MF, Sparks DL (1987b) Sensorimotor integration in the primate superior colliculus. I. Motor convergence. J Neurophysiol 57:22–34.

Jeffress LA (1948) A place theory of sound localization. J Comp Physiol Psychol 41:35–39.

Jen PH, Zhang JP (1999) Corticofugal regulation of excitatory and inhibitory frequency tuning curves of bat inferior collicular neurons. Brain Res 841:184–188.

Jen PH, Chen QC, Sun XD (1998) Corticofugal regulation of auditory sensitivity in the bat inferior colliculus. J Comp Physiol [A] 183:683–697.

Johnson BR (1993) GABAergic and glycinergic inhibition in the central nucleus of the inferior colliculus of the big brown bat. Ph.D. Dissertation, Duke University, Durham NC.

Kaiser A, Covey E (1997) 5-HT innervation of the auditory pathway in birds and bats. In: Syka J (ed) Acoustical Signal Processing in the Central Auditory System. pp. 71–78.

King AJ, Jiang ZD, Moore DR (1998) Auditory brainstem projections to the ferret superior colliculus: anatomical contribution to the neural coding of sound azimuth. J Comp Neurol 390:342–365.

Kleiser A, Schuller G (1995) Response of collicular neurons to acoustic motion in the horseshoe bat, *Rhinolophus rouxi*. Naturwissenschaften 82:337–340.

Klepper A, Herbert H (1991) Distribution and origin of noradrenergic and serotonergic fibers in the cochlear nucleus and inferior colliculus of the rat. Brain Res 557:190–201.

Klug A, Park TJ, Pollak GD (1995) Glycine and GABA influence binaural processing in the inferior colliculus of the mustache bat. J Neurophysiol 74:1701–1713.

Klug A, Bauer EE, Pollak GD (1999) Multiple components of ipsilaterally evoked inhibition in the inferior colliculus. J Neurophysiol 82:593–610.

Knowlton BJ, Thompson JK, Thompson RF (1993) Projections from the auditory cortex to the pontine nuclei in the rabbit. Behav Brain Res 56:23–30.

Knudsen EI (1982) Auditory and visual maps of space in the optic tectum of the owl. J Neurosci 2:1177–1194.

Koch U, Grothe B (1998) GABAergic and glycinergic inhibition sharpens tuning for frequency modulations in the inferior colliculus of the big brown bat. J Neurophysiol 80:71–82.

Konishi M (1993) Listening with two ears. Sci Am 268:66–73.

Kössl M, Vater M (1985) The cochlear frequency map of the mustache bat, *Pteronotus parnellii*. J Comp Physiol [A] 157:687–697.

Krauzlis RJ, Basso MA, Wurtz RH (1997) Shared motor error for multiple eye movements. Science 276:1693–1695.

Krishna BS, Semple MN (2000) Auditory temporal processing: responses to sinusoidally amplitude-modulated tones in the inferior colliculus. J Neurophysiol 84:255–273.

Kudo M, Nakamura Y, Moriizumi T, Tokuno H, Kitao Y (1988) Bilateral projections from the medial superior olivary nucleus to the inferior colliculus in the mole (*Mogera robusta*). Brain Res 463:352–356.

Kunzle H (1993) Tectal and related target areas of spinal and dorsal column nuclear projections in hedgehog tenrecs. Somatosens Mot Res 10:339–353.

Kuwabara N, Zook JM (2000) Geniculo-collicular descending projections in the gerbil. Brain Res 878:79–87.

Kuwada S, Batra R, Yin TC, Oliver DL, Haberly LB, Stanford TR (1997) Intracellular recordings in response to monaural and binaural stimulation of neurons in the inferior colliculus of the cat. J Neurosci 17:7565–7581.

Langner G (1992) Periodicity coding in the auditory system. Hear Res 60:115–142.

Langner G, Schreiner CE (1988) Periodicity coding in the inferior colliculus of the cat. I. Neuronal mechanisms. J Neurophysiol 60:1799–1822.

Le Beau FE, Rees A, Malmierca MS (1996) Contribution of GABA- and glycine-mediated inhibition to the monaural temporal response properties of neurons in the inferior colliculus. J Neurophysiol 75:902–919.

Li L, Kelly J (1992) Inhibitory influence of the dorsal nucleus of the lateral lemniscus on binaural responses in the rat s inferior colliculus. J Neurosci 12:4530–4539.

Li H, Mizuno N (1997a) Collateral projections from single neurons in the dorsal column nuclei to the inferior colliculus and the ventrobasal thalamus: a retrograde double-labeling study in the rat. Neurosci Lett 225:21–24.

Li H, Mizuno N (1997b) Single neurons in the spinal trigeminal and dorsal column nuclei project to both the cochlear nucleus and the inferior colliculus by way of axon collaterals: a fluorescent retrograde double-labeling study in the rat. Neurosci Res 29:135–142.

Licklider JC (1959) Three auditory theories. In: Koch S (ed) Psychology: A Study of a Science, Study 1. Conceptual and Systematic, Vol. 1. Sensory, Perceptual, and Physiological Formulations. pp. 41–144. New York: McGraw-Hill.

Litovsky RY, Yin TC (1998a) Physiological studies of the precedence effect in the inferior colliculus of the cat. II. Neural mechanisms. J Neurophysiol 80:1302–1316.

Litovsky RY, Yin TC (1998b) Physiological studies of the precedence effect in the inferior colliculus of the cat. I. Correlates of psychophysics. J Neurophysiol 80:1285–1301.

Malmierca MS, Rees A, Le Beau FE, Bjaalie JG (1995) Laminar organization of frequency-defined local axons within and between the inferior colliculi of the guinea pig. J Comp Neurol 357:124–144.

Massopust LC, Hauge DH, Ferneding JC, Doubek WG, Taylor JJ (1985) Projection systems and terminal localization of dorsal column afferents: an autoradiographic and horseradish peroxidase study in the rat. J Comp Neurol 237:533–544.

McAlpine D, Jiang D, Shackleton TM, Palmer AR (2000) Responses of neurons in the inferior colliculus to dynamic interaural phase cues: evidence for a mechanism of binaural adaptation. J Neurophysiol 83:1356–1365.

Meininger V, Pol D, Derer P (1986) The inferior colliculus of the mouse. A Nissl and Golgi study. Neuroscience 17:1159–1179.

Merchán MA, Saldaña E, Plaza I (1994) Dorsal nucleus of the lateral lemniscus in the rat: concentric organization and tonotopic projection to the inferior colliculus. J Comp Neurol 342:259-278.

Merzenich MM, Reid MD (1974) Representation of the cochlea within the inferior colliculus of the cat. Brain Res 77:397–415.

Mitani A, Shimokouchi M, Nomura S (1983) Effects of stimulation of the primary auditory cortex upon colliculogeniculate neurons in the inferior colliculus of the cat. Neurosci Lett 42:185–189.

Mittmann DH, Wenstrup JJ (1995) Combination-sensitive neurons in the inferior colliculus. Hear Res 90:185–191.

Moiseff A (1985) Intracellular recordings from owl inferior colliculus. Soc Neurosci Abstr 11:735.

Møller AR (1972) Coding of amplitude and frequency modulated sounds in the cochlear nucleus of the rat. Acta Physiol Scand 86:223–238.

Monaghan DT, Cotman CW (1985) Distribution of N-methyl-D-aspartate-sensitive L-[3H]glutamate-binding sites in rat brain. J Neurosci 5:2909–2919.

Moore DR (1988) Auditory brainstem of the ferret: sources of projections to the inferior colliculus. J Comp Neurol 269:342–354.

Moore DR, Semple MN, Addison PD (1983) Some acoustic properties of neurones in the ferret inferior colliculus. Brain Res 269:69–82.

Moore DR, Kotak VC, Sanes DH (1998) Commissural and lemniscal synaptic input to the gerbil inferior colliculus. J Neurophysiol 80:2229–2236.

Morest DK, Oliver DL (1984) The neuronal architecture of the inferior colliculus in the cat: defining the functional anatomy of the auditory midbrain. J Comp Neurol 222:209–236.

Moriizumi T, Hattori T (1991a) Pallidotectal projection to the inferior colliculus of the rat. Exp Brain Res 87:223–226.

Moriizumi T, Hattori T (1991b) Pyramidal cells in rat temporoauditory cortex project to both striatum and inferior colliculus. Brain Res Bull 27:141–144.

Moriizumi T, Leduc-Cross B, Wu JY, Hattori T (1992) Separate neuronal populations of the rat substantia nigra pars lateralis with distinct projection sites and transmitter phenotypes. Neuroscience 46:711–720.

Muller-Preuss P, Flachskamm C, Bieser A (1994) Neural encoding of amplitude modulation within the auditory midbrain of squirrel monkeys. Hear Res 80: 197–208.

Musicant AD, Butler RA (1984) The psychophysical basis of monaural localization. Hear Res 14:185–190.

Narins PM, Capranica RR (1980) Neural adaptations for processing the two-note call of the Puerto Rican treefrog, *Eleutherodactylus coqui*. Brain Behav Evol 17: 48–66.

Nelson PG, Erulkar SD (1963) Synaptic mechanisms of excitation and inhibition in the central auditory pathway. J Neurophysiol 26:908–923.

Neuweiler G (1990) Auditory adaptations for prey capture in echolocating bats. Physiol Rev 70:615–641.

O'Neill WE, Suga N (1979) Target range-sensitive neurons in the auditory cortex of the mustache bat. Science 203:69–73.

O'Neill WE, Suga N (1982) Encoding of target range and its representation in the auditory cortex of the mustached bat. J Neurosci 2:17–31.

Olazábal UE, Moore JK (1989) Nigrotectal projection to the inferior colliculus: horseradish peroxidase transport and tyrosine hydroxylase immunohistochemical studies in rats, cats, and bats. J Comp Neurol 282:98–118.

Oliver DL (1984) Neuron types in the central nucleus of the inferior colliculus that project to the medial geniculate body. Neuroscience 11:409–424.

Oliver DL (1987) Projections to the inferior colliculus from the anteroventral cochlear nucleus in the cat: possible substrates for binaural interaction. J Comp Neurol 264:24–46.

Oliver DL, Heurta MF (1992) Inferior and superior colliculi. In: Webster DB, Popper AN, Fay RR (eds) Handbook of Auditory Research, Vol. 1: The Mammalian Auditory Pathway: Neuroanatomy. pp. 168–221. New York: Springer-Verlag.

Oliver DL, Morest DK (1984) The central nucleus of the inferior colliculus in the cat. J Comp Neurol 222:237–264.

Oliver DL, Kuwada S, Yin TC, Haberly LB, Henkel CK (1991) Dendritic and axonal morphology of HRP-injected neurons in the inferior colliculus of the cat. J Comp Neurol 303:75–100.

Oliver DL, Beckius GE, Shneiderman A (1995) Axonal projections from the lateral and medial superior olive to the inferior colliculus of the cat: a study using electron microscopic autoradiography. J Comp Neurol 360:17–32.

Oliver DL, Beckius GE, Bishop DC, Kuwada S (1997) Simultaneous anterograde labeling of axonal layers from lateral superior olive and dorsal cochlear nucleus in the inferior colliculus of cat. J Comp Neurol 382:215–229.

Osen KK (1972) Projection of the cochlear nuclei on the inferior colliculus in the cat. J Comp Neurol 144:355–372.

Paloff AM, Usunoff KG (1992) Projections to the inferior colliculus from the dorsal column nuclei. An experimental electron microscopic study in the cat. J Hirnforsch 33:597–610.

Palombi PS, Caspary DM (1996) GABA inputs control discharge rate primarily within frequency receptive fields of inferior colliculus neurons. J Neurophysiol 75:2211–2219.

Pandóssio JE, Brandão ML (1999) Defensive reactions are counteracted by midazolam and muscimol and elicited by activation of glutamate receptors in the inferior colliculus of rats. Psychopharmacology (Berl) 142:360–368.

Park TJ, Pollak GD (1993a) GABA shapes a topographic organization of response latency in the mustache bat's inferior colliculus. J Neurosci 13:5172–5187.

Park TJ, Pollak GD (1993b) GABA shapes sensitivity to interaural intensity disparities in the mustache bat's inferior colliculus: implications for encoding sound location. J Neurosci 13:2050–2067.

Park TJ, Pollak GD (1994) Azimuthal receptive fields are shaped by GABAergic inhibition in the inferior colliculus of the mustache bat. J Neurophysiol 72: 1080–1102.

Parks TN (2000) The AMPA receptors of auditory neurons. Hear Res 147:77–91.

Pedemonte M, Torterolo P, Velluti RA (1997) In vivo intracellular characteristics of inferior colliculus neurons in guinea pigs. Brain Res 759:24–31.

Peruzzi D, Sivaramakrishnan S, Oliver DL (2000) Identification of cell types in brain slices of the inferior colliculus. Neuroscience 101:403–416.

Petralia RS, Wenthold RJ (1992) Light and electron immunocytochemical localization of AMPA-selective glutamate receptors in the rat brain. J Comp Neurol 318: 329–354.

Petralia RS, Yokotani N, Wenthold RJ (1994) Light and electron microscope distribution of the NMDA receptor subunit NMDAR1 in the rat nervous system using a selective anti-peptide antibody. J Neurosci 14:667–696.

Pollak GD (1997) Roles of GABAergic inhibition for the binaural processing of multiple sound sources in the inferior colliculus. Ann Otol Rhinol Laryngol Suppl 168:44–54.

Pollak GD, Casseday JH (1989) The Neural Basis of Echolocation in Bats. Berlin: Springer-Verlag.

Pollak GD, Park T (1993) The effects of GABAergic inhibition on monaural response properties of neurons in the mustache bat's inferior colliculus. Hear Res 65:99–117.

Poon PW, Sun X, Kamada T, Jen PH (1990) Frequency and space representation in the inferior colliculus of the FM bat, *Eptesicus fuscus*. Exp Brain Res 79:83–91.

Poon PW, Chen X, Cheung YM (1992) Differences in FM response correlate with morphology of neurons in the rat inferior colliculus. Exp Brain Res 91:94–104.

Portfors CV, Wenstrup JJ (2001) Topographical distribution of delay-tuned responses in the mustached bat inferior colliculus. Hear Res 151:95–105.

Rees A, Møller AR (1983) Responses of neurons in the inferior colliculus of the rat to AM and FM tones. Hear Res 10:301–330.

Regenold W, Araujo DM, Quirion R (1989) Quantitative autoradiographic distribution of [3H]AF-DX 116 muscarinic-M2 receptor binding sites in rat brain. Synapse 4:115–125.

Reimer K (1987) Coding of sinusoidally amplitude modulated acoustic stimuli in the inferior colliculus of the rufous horseshoe bat, *Rhinolophus rouxi*. J Comp Physiol [A] 161:305–313.

Riquelme R, Saldaña E, Osen KK, Ottersen OP, Merchán MA (2001) Colocalization of GABA and glycine in the ventral nucleus of the lateral lemniscus in rat: An in situ hybridization and semiquantitative immunocytochemical study. J Comp Neurol 432:409–424.

Rhode WS, Greenberg S (1992) Physiology of the cochlear nuclei. In: Popper AN, Fay RR (eds) The Mammalian Auditory Pathway: Neurophysiology. pp. 94–152. New York: Springer-Verlag.

RoBards MJ (1979) Somatic neurons in the brainstem and neocortex projecting to the external nucleus of the inferior colliculus: an anatomical study in the opossum. J Comp Neurol 184:547–565.

RoBards MJ, Watkins III DW, Masterton RB (1976) An anatomical study of some somesthetic afferents to the intercollicular terminal zone of the midbrain of the opossum. J Comp Neurol 170:499–524.

Roberts RC, Ribak CE (1987) GABAergic neurons and axon terminals in the brainstem auditory nuclei of the gerbil. J Comp Neurol 258:267–280.

Rockel AJ, Jones EG (1973a) The neuronal organization of the inferior colliculus of the adult cat. I. The central nucleus. J Comp Neurol 147:11–60.

Rockel AJ, Jones EG (1973b) Observations on the fine structure of the central nucleus of the inferior colliculus of the cat. J Comp Neurol 147:61–92.

Rouiller EM (1997) Functional organization of the auditory pathways. In: Ehert G, Romand R (eds) The Central Auditory System. pp. 3–96. Oxford: Oxford University Press.

Ruggero MA (1992) Physiology and coding of sound in the auditory nerve. In: Popper AN, Fay RR (eds) The Mammalian Auditory Pathway: Neurophysiology. pp. 34–93. New York: Springer-Verlag.

Ryugo DK, Willard FH (1985) The dorsal cochlear nucleus of the mouse: a light microscopic analysis of neurons that project to the inferior colliculus. J Comp Neurol 242:381–396.

Saint Marie RL (1996) Glutamatergic connections of the auditory midbrain: selective uptake and axonal transport of D-[$_3$H]aspartate. J Comp Neurol 373:255–270.

Saint Marie RL, Baker RA (1990) Neurotransmitter-specific uptake and retrograde transport of [3H]glycine from the inferior colliculus by ipsilateral projections of the superior olivary complex and nuclei of the lateral lemniscus. Brain Res 524:244–253.

Saint Marie RL, Ostapoff EM, Morest DK, Wenthold RJ (1989) Glycine-immunoreactive projection of the cat lateral superior olive: possible role in midbrain ear dominance. J Comp Neurol 279:382–396.

Saint Marie RL, Shneiderman A, Stanforth DA (1997) Patterns of gamma-aminobutyric acid and glycine immunoreactivities reflect structural and functional differences of the cat lateral lemniscal nuclei. J Comp Neurol 389:264–276.

Saint Marie RL, Luo L, Ryan AF (1999) Effects of stimulus frequency and intensity on c-fos mRNA expression in the adult rat auditory brainstem. J Comp Neurol 404:258–270.

Saldaña E, Merchán MA (1992) Intrinsic and commissural connections of the rat inferior colliculus. J Comp Neurol 319:417–437.

Saldaña E, Feliciano M, Mugnaini E (1996) Distribution of descending projections from primary auditory neocortex to inferior colliculus mimics the topography of intracollicular projections. J Comp Neurol 371:15–40.

Sanes DH, Malone BJ, Semple MN (1998) Role of synaptic inhibition in processing of dynamic binaural level stimuli. J Neurosci 18:794–803.

Sato K, Kiyama H, Tohyama M (1993) The differential expression patterns of messenger RNAs encoding non-N-methyl-D-aspartate glutamate receptor subunits (GluR1–4) in the rat brain. Neuroscience 52:515–539.

Schmidt U, Schlegel P, Schweizer H, Neuweiler G (1991) Audition in vampire bats, *Desmodus rotundud*. J Comp Physiol [A] 168:45–51.

Schreiner CE, Langner G (1988) Periodicity coding in the inferior colliculus of the cat. II. Topographical organization. J Neurophysiol 60:1823–1840.

Schreiner CE, Langner G (1997) Laminar fine structure of frequency organization in auditory midbrain. Nature 388:383–386.

Schuller G (1979) Coding of small sinusoidal frequency and amplitude modulations in the inferior colliculus of "CF-FM" bat, *Rhinolophus ferrumequinum*. Exp Brain Res 34:117–132.

Schuller G, Covey E, Casseday JH (1991) Auditory pontine grey: connections and response properties in the horseshoe bat. Eur J Neurosci 3:648–662.

Schwartz IR (1992) The superior olivary complex and lateral lemniscal nuclei. In: Webster DB, Popper AN, Fay RR (eds) The Mammalian Auditory Pathway: Neuroanatomy. pp. 117–167. New York: Springer-Verlag.

Schweizer H (1981) The connections of the inferior colliculus and the organization of the brainstem auditory system in the greater horseshoe bat (*Rhinolophus ferrumequinum*). J Comp Neurol 201:25–49.

Semple MN, Kitzes LM (1987) Binaural processing of sound pressure level in the inferior colliculus. J Neurophysiol 57:1130–1147.

Servière J, Webster WR, Calford MB (1984) Isofrequency labelling revealed by a combined [14C]-2-deoxyglucose, electrophysiological, and horseradish peroxidase study of the inferior colliculus of the cat. J Comp Neurol 228:463–477.

Shammah-Lagnado SJ, Alheid GF, Heimer L (1996) Efferent connections of the caudal part of the globus pallidus in the rat. J Comp Neurol 376:489–507.

Shneiderman A, Henkel CK (1987) Banding of lateral superior olivary nucleus afferents in the inferior colliculus: a possible substrate for sensory integration. J Comp Neurol 266:519–534.

Shneiderman A, Oliver DL (1989) EM autoradiographic study of the projections from the dorsal nucleus of the lateral lemniscus: a possible source of inhibitory inputs to the inferior colliculus. J Comp Neurol 286:28–47.

Shneiderman A, Oliver DL, Henkel CK (1988) Connections of the dorsal nucleus of the lateral lemniscus: an inhibitory parallel pathway in the ascending auditory system? J Comp Neurol 276:188–208.

Simmons JA, Saillant PA, Ferragamo MJ, Haresign T, Dear SP, Fritz J, McMullen T (1995) Auditory computations for biosonal target imaging in bats. In: Hawkins HL, McMullen TA, Popper AN, Fay RR (eds) Auditory Computation. New York: Springer-Verlag 401–468.

Smith PH (1992) Anatomy and physiology of multipolar cells in the rat inferior collicular cortex using the in vitro brain slice technique. J Neurosci 12:3700–3715.

Spangler KM, Warr WB (1991) The descending auditory system. In: Altschuler RA, Bobbin RP, Clopton BM, Hoffman DW (eds) Neurobiology of Hearing: The Central Auditory System. pp. 27–46. New York: Raven Press.

Spitzer MW, Semple MN (1993) Responses of inferior colliculus neurons to time-varying interaural phase disparity: effects of shifting the locus of virtual motion. J Neurophysiol 69:1245–1263.

Spitzer MW, Semple MN (1998) Transformation of binaural response properties in the ascending auditory pathway: influence of time-varying interaural phase disparity. J Neurophysiol 80:3062–3076.

Stanford TR, Kuwada S, Batra R (1992) A comparison of the interaural time sensitivity of neurons in the inferior colliculus and thalamus of the unanesthetized rabbit. J Neurosci 12:3200–3216.

Stiebler I, Ehret G (1985) Inferior colliculus of the house mouse. I. A quantitative study of tonotopic organization, frequency representation, and tone-threshold distribution. J Comp Neurol 238:65–76.

Suga N (1965) Analysis of frequency-modulated sounds by auditory neurones of echo-locating bats. J Physiol 179:26–53.

Suga N (1968) Analysis of frequency-modulated and complex sounds by single auditory neurones of bats. J Physiol 198:51–80.

Suga N (1969) Classification of inferior collicular neurones of bats in terms of responses to pure tones, FM sounds and noise bursts. J Physiol 200:555–574.

Suga N, Simmons JA, Jen PH (1975) Peripheral specialization for fine analysis of doppler-shifted echoes in the auditory system of the "CF-FM" bat *Pteronotus parnellii*. J Exp Biol 63:161–192.

Suga N, O'Neill WE, Manabe T (1978) Cortical neurons sensitive to combinations of information-bearing elements of biosonar signals in the mustache bat. Science 200:778–781.

Suga N, O'Neill WE, Kujirai K, Manabe T (1983) Specificity of combination-sensitive neurons for processing of complex biosonar signals in auditory cortex of the mustached bat. J Neurophysiol 49:1573–1626.

Suga N, Butman JA, Teng H, Yan J, Olsen JF (1995) Neural procesing of target-distance information in the mustached bat. In: Flock A, Ottoson D, Ulfendahl M (eds) Active Hearing. pp. 13–30. Oxford: Pergamon.

Suga N, Gao E, Zhang Y, Ma X, Olsen JF (2000) The corticofugal system for hearing: recent progress. Proc Natl Acad Sci USA 97:11807–11814.

Sun XD, Jen PH, Sun DX, Zhang SF (1989) Corticofugal influences on the responses of bat inferior collicular neurons to sound stimulation. Brain Res 495:1–8.

Syka J, Popelár J (1984) Inferior colliculus in the rat: neuronal responses to stimulation of the auditory cortex. Neurosci Lett 51:235–240.

Syka J, Popelár J, Kvasnák E, Astl J (2000) Response properties of neurons in the central nucleus and external and dorsal cortices of the inferior colliculus in guinea pig. Exp Brain Res 133:254–266.

Thompson AM (1998) Heterogeneous projections of the cat posteroventral cochlear nucleus. J Comp Neurol 390:439–453.

Thompson GC, Thompson AM, Garrett KM, Britton BH (1994) Serotonin and serotonin receptors in the central auditory system. Otolaryngol Head Neck Surg 110:93–102.

Tokunaga A, Sugita S, Otani K (1984) Auditory and non-auditory subcortical afferents to the inferior colliculus in the rat. J Hirnforsch 25:461–472.

Torterolo P, Pedemonte M, Velluti RA (1995) Intracellular in vivo recording of inferior colliculus auditory neurons from awake guinea-pigs. Arch Ital Biol 134:57–64.

Torterolo P, Zurita P, Pedemonte M, Velluti RA (1998) Auditory cortical efferent actions upon inferior colliculus unitary activity in the guinea pig. Neurosci Lett 249:172–176.

Ueyama T, Sato K, Kakimoto S, Houtani T, Sakuma S, Ohishi H, Kase M, Sugimoto T (1999) Comparative distribution of GABAergic and peptide-containing neurons in the lateral lemniscal nuclei of the rat. Brain Res 849:220–225.

Vater M, Feng AS (1990) Functional organization of ascending and descending connections of the cochlear nucleus of horseshoe bats. J Comp Neurol 292:373–395.

Vater M, Habbicht H, Kössl M, Grothe B (1992) The functional role of GABA and glycine in monaural and binaural processing in the inferior colliculus of horseshoe bats. J Comp Physiol [A] 171:541–553.

Vater M, Casseday J, Covey E (1995) Convergence and divergence of ascending binaural and monaural pathways from the superior olives of the mustached bat. J Comp Neurol 351:632–646.

Vater M, Covey E, Casseday JH (1997) The columnar region of the ventral nucleus of the lateral lemniscus in the big brown bat (*Eptesicus fuscus*): synaptic arrangements and structural correlates of feedforward inhibitory function. Cell Tissue Res 289:223–233.

Wada E, Wada K, Boulter J, Deneris E, Heinemann S, Patrick J, Swanson LW (1989) Distribution of alpha 2, alpha 3, alpha 4, and beta 2 neuronal nicotinic receptor subunit mRNAs in the central nervous system: a hybridization histochemical study in the rat. J Comp Neurol 284:314–335.

Wallach H, Newman EB, Rosenzweig MR (1949) The precedence effect in sound localization. Am J Psychol 62:315–336.

Watanabe M, Mishina M, Inoue Y (1994) Distinct distributions of five NMDA receptor channel subunit mRNAs in the brainstem. J Comp Neurol 343:520–531.

Weedman DL, Ryugo DK (1996) Projections from auditory cortex to the cochlear nucleus in rats: synapses on granule cell dendrites. J Comp Neurol 371:311–324.

Wenstrup JJ, Leroy SA (2001) Spectral integration in the inferior colliculus: Role of glycinergic inhibition in response facilitation. J Neurosci 21:RC124.

Wenstrup JJ, Ross LS, Pollak GD (1986) Binaural response organization within a frequency-band representation of the inferior colliculus: implications for sound localization. J Neurosci 6:962–973.

Wenstrup JJ, Larue DT, Winer JA (1994) Projections of physiologically defined subdivisions of the inferior colliculus in the mustached bat: targets in the medial geniculate body and extrathalamic nuclei. J Comp Neurol 346:207–236.

Wiberg M, Blomqvist A (1984) The projection to the mesencephalon from the dorsal column nuclei. An anatomical study in the cat. Brain Res 311:225–244.

Wiberg M, Westman J, Blomqvist A (1987) Somatosensory projection to the mesencephalon: an anatomical study in the monkey. J Comp Neurol 264:92–117.

Wiesendanger R, Wiesendanger M (1982) The corticopontine system in the rat. II. The projection pattern. J Comp Neurol 208:227–238.

Wightman FL, Kistler DJ (1997) Monaural sound localization revisited. J Acoust Soc Am 101:1050–1063.

Willard FH, Martin GF (1983) The auditory brainstem nuclei and some of their projections to the inferior colliculus in the North American opossum. Neuroscience 10:1203–1232.

Willard FH, Rygo DK (1983) Anatomy of the central auditory system. In: Willott JF (ed) The Auditory Psychobiology of the Mouse. pp. 201–304. Springfield, IL: C.C. Thomas.

Wilson WW, O'Neill WE (1998) Auditory motion induces directionally dependent receptive field shifts in inferior colliculus neurons. J Neurophysiol 79:2040–2062.

Winer JA, Larue DT, Pollak GD (1995) GABA and glycine in the central auditory system of the mustache bat: structural substrates for inhibitory neuronal organization. J Comp Neurol 355:317–353.

Winer JA, Larue DT, Diehl JJ, Hefti BJ (1998) Auditory cortical projections to the cat inferior colliculus. J Comp Neurol 400:147–174.

Wright BA, Dai H (1994) Detection of unexpected tones with short and long durations. J Acoust Soc Am 95:931–938.

Wu MI, Jen PH (1995) Responses of pontine neurons of the big brown bat, *Eptesicus fuscus*, to temporally patterned sound pulses. Hear Res 85:155–168.

Yan J, Suga N (1996a) Corticofugal modulation of time-domain processing of biosonar information in bats. Science 273:1100–1103.

Yan J, Suga N (1996b) The midbrain creates and the thalamus sharpens echo-delay tuning for the cortical representation of target-distance information in the mustached bat. Hear Res 93:102–110.

Yan J, Suga N (1999) Corticofugal amplification of facilitative auditory responses of subcortical combination-sensitive neurons in the mustached bat. J Neurophysiol 81:817–824.

Yan W, Suga N (1998) Corticofugal modulation of the midbrain frequency map in the bat auditory system. Nat Neurosci 1:54–58.

Yang L, Pollak GD, Resler C (1992) GABAergic circuits sharpen tuning curves and modify response properties in the mustache bat inferior colliculus. J Neurophysiol 68:1760–1774.

Yasui Y, Nakano K, Kayahara T, Mizuno N (1991) Non-dopaminergic projections from the substantia nigra pars lateralis to the inferior colliculus in the rat. Brain Res 559:139–144.

Yin TC (1994) Physiological correlates of the precedence effect and summing localization in the inferior colliculus of the cat. J Neurosci 14:5170–5186.

Zhang Y, Suga N (1997) Corticofugal amplification of subcortical responses to single tone stimuli in the mustached bat. J Neurophysiol 78:3489–3492.

Zhang Y, Suga N (2000a) Modulation of responses and frequency tuning of thalamic and collicular neurons by cortical activation in mustached bats. J Neurophysiol 84:325–333.

Zhang Y, Wu SH (2000b) Long-term potentiation in the inferior colliculus studied in rat brain slice. Hear Res 147:92–103.

Zhang Y, Suga N, Yan J (1997) Corticofugal modulation of frequency processing in bat auditory system. Nature 387:900–903.

Zhang DX, Li L, Kelly JB, Wu SH (1998) GABAergic projections from the lateral lemniscus to the inferior colliculus of the rat. Hear Res 117:1–12.

Zhou X, Jen PHS (2000a) Brief and short-term corticofugal modulation of subcortical auditory responses in the big brown bat, *Eptesicus fuscus*. J Neurophysiol 84: 3083–3087.

Zhou X, Jen PHS (2000b) Corticofugal inhibition compresses all types of rate-intensity functions of inferior collicular neurons in the big brown bat. Brain Res 881:62–68.

Zhou XM, Jen PHS (2000c) Neural inhibition sharpens auditory spatial selectivity of bat inferior collicular neurons. J Comp Physiol [A] 186:389–398.

Zhou X, Jen PHS (2001) The effect of sound intensity on duration-tuning characteristics of bat inferior collicular neurons. J Comp Physiol [A] 187:63–73.

Zook JM, Casseday JH (1982a) Origin of ascending projections to inferior colliculus in the mustache bat, *Pteronotus parnellii*. J Comp Neurol 207:14–28.

Zook JM, Casseday JH (1982b) Cytoarchitecture of auditory system in lower brainstem of the mustache bat, *Pteronotus parnellii*. J Comp Neurol 207:1–13.

Zook JM, Casseday JH (1985) Projections from the cochlear nuclei in the mustache bat, *Pteronotus parnellii*. J Comp Neurol 237:307–324.

Zook JM, Casseday JH (1987) Convergence of ascending pathways at the inferior colliculus of the mustache bat, *Pteronotus parnellii*. J Comp Neurol 261:347–361.

Zook J, Winer J, Pollak G, Bodenhamer R (1985) Topology of the central nucleus of the mustache bat's inferior colliculus: correlation of single unit properties and neuronal architecture. J Comp Neurol 231:530–546.

8
Location Signaling by Cortical Neurons

John C. Middlebrooks, Li Xu, Shigeto Furukawa, and
Brian J. Mickey

1. Introduction

Localization is a fundamental task of hearing. Identification of the locations
of sound sources permits animals to locate prey or to avoid predators.
Spatial hearing improves signal detection and aids in the segregation of
multiple sources, thus improving identification of sounds. The importance
of sound localization for the auditory system is reflected in the organization of the auditory brainstem, which contains several discrete nuclei that
appear to be specialized for analysis of particular acoustical cues for sound
source location.

The auditory cortex is essential for normal sound localization as is
demonstrated by the localization deficits that result from cortical lesions.
Recognition of the importance of the cortex for localization behavior has
inspired many studies of the spatial sensitivity of cortical neurons. A conventional approach has been to present sounds from various locations and
to characterize location-specific neuronal responses. That is, investigators
have asked: "given a sound source at this location, what is the cortical
response"? More recently, several groups have turned the question around
to ask: "given this cortical response, where is the sound source"? That is,
recent studies have examined the capability of single neurons or neural
populations to signal the location of a sound source. Measures of the accuracy of location signaling serve as an empirical measure of location-specific
information carried by neurons or populations of neurons.

In this chapter, we begin by reviewing studies of localization deficits that
arise from cortical lesions (Section 2), and we review studies of spatial
sensitivity of cortical neurons (Section 3). Then, we present results indicating that single neurons can signal sound source location throughout as
much as 360° of the sound field. Such neural signaling of location has been
tested under conditions in which listeners localize accurately (Section
4) and under conditions in which listeners make systematic localization
errors (Section 5). Finally, we examine several models of sound localization
by small populations of neurons (Section 6). We conclude by considering

the issue of specialization of cortical areas for sound localization (Section 7).

2. Sound Localization Deficits Resulting from Cortical Lesions

2.1 Experimental Animals

Behavioral deficits that follow clinical lesions in humans and cortical ablations in experimental animals implicate the auditory cortex in sound localization behavior. Results from carnivores and monkeys lead to the conclusion that the auditory cortex in each cortical hemisphere represents sound source locations throughout a region that encompasses the contralateral sound hemifield and, possibly, extends a limited distance into the ipsilateral hemifield. Following a unilateral lesion of the auditory cortex, a cat, monkey, or ferret is unable to distinguish the locations of sounds presented in the half of space contralateral to the lesion. Similar deficits result from unilateral lesions in humans although the laterality of the deficit is not as consistent as in the experimental animals that have been tested.

Unilateral cortical ablations in carnivores or monkeys have a clear impact on sound localization only when the behavioral task requires a conditioned response to a brief sound and only when the task requires the animal to discriminate locations within one hemifield. In a task that required discrimination of seven loudspeakers positioned throughout the front half of the horizontal plane, unilateral auditory cortex lesions in cats and monkeys resulted in pronounced localization deficits for targets on the side contralateral to the lesion (Jenkins and Masterton 1982; Thompson and Cortez 1983). Performance on the ipsilateral side was spared. Similarly, ferrets (Kavanagh and Kelly 1987) and monkeys (Heffner and Heffner 1990) showed a contralateral deficit in a task that required discrimination of two locations in one lateral hemifield. Unilateral lesions fail to produce a lasting deficit when the behavioral task requires the animal only to discriminate targets that are positioned on either side of the frontal midline (reviewed by Kavanagh and Kelly, 1987). One interpretation of the left-right discrimination results is that the spatial representation in the surviving cortical hemisphere encompasses source locations that extend across the midline. Left-right discrimination, however, survives even bilateral auditory cortical lesions, albeit with somewhat increased response latencies and reduced accuracy. For instance, Heffner and Heffner (1990) found that monkeys that had sustained bilateral auditory cortical lesions could discriminate between pairs of targets that straddled the frontal midline but not pairs that were restricted to one side. Thus, normal auditory cortical activity apparently is necessary for discrimination of locations within a lateral hemifield but not for discrimination between hemifields.

In contrast to conditioned sound localization, unconditioned (i.e., reflexive) head turns by cats toward a sudden sound are largely unaffected by unilateral auditory cortical lesions (Thompson and Masterton 1978). The effects of bilateral cortical lesions are somewhat greater, resulting in increases in the latencies of head turns, decreases in the duration of movements, and increases in the final error of head orientation toward the source (Beitel and Kaas 1993). Nevertheless, bilateral lesions produce no change in the probability of initial movement toward the correct side.

Jenkins and Merzenich (1984) tested the hypothesis that, of all the auditory cortical areas in the cat, area A1 has a special role in sound localization. They measured cats' ability to localize the sources of pure tones presented from one of seven locations in the frontal half of the horizontal plane. Restricted unilateral lesions were placed within the representations of narrow ranges of the tonotopic frequency representation. Such lesions resulted in localization deficits that were specific to the corresponding tone frequencies presented on the side contralateral to the lesion. Normal localization of tones outside the lesioned frequency range was preserved as was localization of all ipsilateral tones. A complementary lesion was tested in one cat. In that cat, most of area A1 and other auditory areas was ablated in one cerebral hemisphere, sparing only the representation in A1 of a narrow range of frequencies. That cat showed contralateral localization deficits at all but the spared frequency. Based on their results, the authors concluded that area A1 is both necessary and sufficient for sound localization.

One must use some caution in generalizing the Jenkins and Merzenich results to more typical sound localization tasks. A pure tone source, as used in that study, cannot be localized even by a normal animal in a situation in which the source is free to vary in location throughout three dimensions. That is because a tone does not provide the spectral shape cues that are necessary for localization in the vertical and front/back dimensions (reviewed by Middlebrooks and Green 1991); indeed, narrow-band sounds are systematically mislocalized (humans, Blauert 1969/70; Middlebrooks 1992; cats, Populin and Yin 1998; Huang and May 1996). If the sources are constrained to the front half of the horizontal plane, however, source locations can be identified on the basis of individual localization cues, such as interaural level differences. A conservative interpretation of the Jenkins and Merzenich results is that restricted lesions of A1 disrupted cortical processing of localization cues at a particular frequency in A1 and in all areas that receive input from A1. In the one case of the large lesion that spared only a restricted frequency band, such an isolated island of cortex was apparently sufficient to permit processing of a particular spatial cue at a particular frequency. The Jenkins and Merzenich results have been given as a rationale to focus on area A1 in physiological studies of localization coding. Nevertheless, physiological studies have failed to demonstrate any qualitative features that distinguish area A1 from other auditory cortical areas in regard to sound localization (see following sections).

2.2 Lesions in Human Patients

In contrast to results from experimental animals, unilateral cortical lesions in humans do not consistently produce strictly contralateral sound localization deficits. Early neurological studies generally indicated that unilateral damage to the temporal lobe was accompanied by a contralateral localization deficit (Greene 1929; Wortis and Pfeiffer 1948; Sanchez-Longo and Forster 1958; Klingon and Bontecou 1966). More recent studies of hemispherectomy patients have, in contrast, indicated considerable sparing of localization ability bilaterally some years after the cortical excision. Poirier and colleagues (1994) studied three such patients. One of the patients showed no statistically significant effect of the lesion on localization performance. The other two patients showed significantly greater localization errors than did the controls, but still showed localization judgments that shifted systematically further lateral with increasingly lateral target locations. That is, the localization of sounds on the side contralateral to the excised cortex was appreciably better than that demonstrated by animals following an experimental cortical ablation. Similarly, in a study of six hemispherectomy patients, Zatorre and colleagues (1995) found considerable sparing of localization ability, with some subjects performing at near normal levels for targets on ipsi- and contralesional sides. The Zatorre group speculated that the sparing of localization function might be a result of cortical reorganization, with the functions of the absent hemisphere being taken over by the remaining hemisphere. More recently, Zatorre and Penhune (2001) have examined patients that showed temporal lobe lesions that either involved or spared Heschl's gyrus, which is the site of the core auditory fields. Generally, right temporal lobe lesions resulted in bilateral localization deficits. Left lesions produced deficits only when the lesion encroached on Heschl's gyrus, and those lesions produced bilateral deficits.

Parietal cortex lesions in humans also have been reported to cause sound localization deficits although the cortical laterality is controversial. Ruff and colleagues (1981) reported that, in a population of 29 patients with various unilateral cortical lesions, localization errors were greatest among those with lesions on the right side involving parietal, parieto-occipital, or parietotemporal cortex. In contrast, Pinek and colleagues (1989) found severe right- and left-sided sound localization deficits among patients with left parietal lesions. More minor deficits resulting from right parietal lesions were regarded by those authors as secondary to a generalized left hemineglect.

3. Spatial Sensitivity of Auditory Cortical Neurons

The results of lesion studies, at least in carnivores and monkeys, suggest that the auditory cortex in each hemisphere contains a representation of contralateral sound source locations. Influenced by studies of the visual and somatosensory systems, one might believe that a representation of sensory space requires neurons that show sharp spatial tuning and receptive field locations that vary systematically with cortical place to form point-to-point cortical maps of sensory space. In the visual and somatosensory systems, however, spatial receptive fields in the cortex reflect a straightforward cortical representation of the sensory receptor sheets (i.e., representations of the retina or of the body surface). In the auditory system, in contrast, cortical representations of the cochlear receptor sheet amount to maps of sound frequency, not location of sound in space. Indeed, several auditory cortical fields contain systematic frequency maps (e.g., Merzenich et al. 1973; Reale and Imig 1980). Information about sound source locations must be derived through a variety of acoustical cues that result from the interaction of sound with the head and external ears (i.e., the pinnae) (reviewed by Middlebrooks and Green 1991). Given the differences in the mechanisms of spatial processing among the auditory, somatosensory, and visual systems, one might argue that the spatial maps found in the visual and somatosensory cortex are poor models for auditory spatial representation. Somewhat contrary to that argument, however, is the demonstration of a map of auditory space that has been found in the superior colliculus of the midbrain, which is in register with visual and somatosensory space maps (e.g., Palmer and King 1982; Middlebrooks and Knudsen 1984). The superior colliculus results provide a demonstration of the feasibility of a topographic representation of auditory space.

Several investigators have avidly sought spatial receptive fields and maps of space in the auditory cortex. Their work has shown that single auditory cortical neurons can exhibit spatial receptive fields, and that receptive field characteristics are sensitive to stimulus level, stimulus bandwidth, and the positions of the pinnae. Nevertheless, those studies have consistently failed to demonstrate a point-to-point cortical map of auditory space. We review those results in this section, beginning with area A1 in the cat and moving to other cortical areas in cats and monkeys. Readers interested in basic functional organization of the auditory cortex as it relates to coding of acoustical cues for sound location are referred to the review by Clarey and colleagues (1992).

3.1 Area A1 in the Cat

Early studies of the cat auditory cortex indicated that neurons vary in their spatial sensitivity. Evans (1967) reported that about half of the neurons that he studied showed a preference for the location of a transient sound source (generally, snapping fingers), and that the majority of those neurons preferred sounds on the side contralateral to the recording site. Eisenman (1974), using tone and noise stimuli that were presented in the horizontal plane, also found a mixture of spatially sensitive and insensitive neurons with a bias toward best responses to contralateral stimuli. In that study, seven of 43 neurons showed temporal spike patterns that differed according to the sound source location. Sovijarvi and Hyvarinen (1974) studied responses to tones presented from a handheld loudspeaker. About half of the neurons were sensitive to the location of the sound source or to the direction of movement of the source. Again, temporal spike patterns (e.g., on, off, and sustained responses) tended to vary according to the location or movement of the sound source.

Middlebrooks and Pettigrew (1981) presented sounds from a moveable loudspeaker in an anechoic chamber and measured spatial receptive fields of neurons in cortical area A1. Each neuron was tested with tones at its characteristic frequency presented at a level 10dB above the unit's threshold. About half of neurons were "omnidirectional" in the sense that they responded with a spike at least once or twice out of ten presentations of a sound anywhere in a 180° arc in front of the cat. The other half of the units showed either "hemifield" or "axial" receptive fields. Hemifield receptive fields had a border near the frontal midline and extended beyond the contralateral pole of the sound field (i.e., beyond 90° from the frontal midline). Axial receptive fields were completely circumscribed within the front half of the sound field. Hemifield units all had characteristic frequencies (CFs) lower than 12kHz whereas axial units had CFs higher than 12kHz. Separate acoustic measurements of the directionality of the pinna indicated that, at high frequencies, the pinna showed a restricted region of greatest sensitivity in the general region occupied by the axial units. Thus, it was supposed that the restricted axial receptive fields were largely a product of passive acoustics of the pinna whereas hemifield receptive fields probably showed a greater contribution of binaural interactions. Indeed, in a case in which the pinna contralateral to the recording site was deflected medially, an axial receptive field shifted medially.

This early Middlebrooks and Pettigrew study highlighted the importance of pinna acoustics for directionality of cortical neurons. Subsequent quantitative studies of azimuth sensitivity of cortical neurons (Imig et al. 1990; Rajan et al. 1990a) indicated that the restricted axial receptive fields are seen only at near threshold sound levels. Nevertheless, recent studies using virtual auditory space stimuli (Brugge et al. 1996) have demonstrated the imprint of pinna directionality on receptive fields even at levels up to about 50dB above threshold. Moreover, the importance of the pinna for directional filtering of broadband sounds has been increasingly recognized (e.g., Musicant et al. 1990; Rice et al. 1992; Young et al. 1996; Xu and Middlebrooks 2000). In Section 5.2 we consider the impact of such directional filtering on the elevation sensitivity of cortical neurons.

Several groups have generated spike rate versus azimuth plots for neurons in area A1 (Imig et al. 1990; Rajan et al. 1990a; Poirier et al. 1997; Eggermont and Mossop 1998). Those groups found omnidirectional neurons and neurons with hemifield receptive fields similar to those described by Middlebrooks and Pettigrew (1981). They also described a significant number of receptive fields that were centered in front of the cat, and fields that were centered on the side ipsilateral to the recording site. In the studies by the Imig, Rajan, and Eggermont groups, stimuli were presented across broad ranges of sound levels. Figure 8.1 shows examples of the spatial sensitivity of neurons studied by Imig and colleagues (1990). In those plots, the contours represent the normalized activity of neurons as a function of sound source azimuth in the frontal half of space (horizontal axis) and sound level (vertical axis). In that study, 23% of neurons showed an ipsilateral preference (as in Fig. 8.1A), 17% showed a midline preference (Fig. 8.1B), and 60% showed a contralateral preference (Figure 8.1C and D). Middlebrooks and Pettigrew (1981) had illustrated examples of receptive fields that expanded in size with increasing stimulus levels and

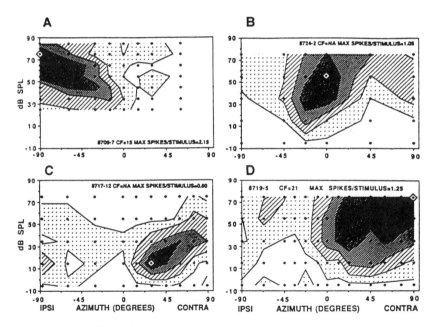

FIGURE 8.1. Amplitude-level response areas for four neurons in cat area A1. Contours represent normalized responses as a function of sound source azimuth (horizontal axis) and sound level (vertical axis). In each panel (A–D), the grid of small diamonds indicates stimulus locations and levels that were tested. The large diamond represents the stimulus that elicited the maximum response. Decreasing densities of shading indicate responses that were greater than 75, 50, 25, and 5% of the maximum. (Adapted from Imig et al. 1990.)

examples of others that maintained their size across a modest range of levels. The Rajan group reported that around 60% of directionally sensitive neurons showed receptive field borders that were constant across a 20 to 40 dB range of sound levels. The Imig group reported that a "minority" of their high-directionality group showed level-invariant spatial sensitivity over a 40 to 60 dB range. The neurons that showed the most restricted spatial tuning also tended to show tuning to particular sound levels (i.e., those neurons had nonmonotonic rate-level functions).

Brugge and colleagues (Brugge et al. 1994, 1996) have applied virtual auditory space techniques to the study of spatial receptive fields of neurons in area A1 of the cat. Impulsive sound sources at up to 1816 virtual locations were simulated by presentation of filtered sounds through headphones. The signals to the headphones were filtered by previously measured transfer functions from 1816 free-field sound sources presented to a cat's left and right tympanic membranes (Musicant et al. 1990); the virtual source locations were distributed through 360° of azimuth and 126° of elevation. At sound levels 10 to 30 dB above neural thresholds, receptive fields of 76% of neurons were largely confined to a quadrant of frontal virtual space. At those sound levels, the spatial medians of receptive fields tended to coincide in virtual location with the axis of greatest sensitivity of one or the other pinna. The great majority of neurons showed receptive fields that increased in size with increasing stimulus level. Of those, some "bounded" receptive fields remained restricted within roughly half of virtual space whereas "unbounded" receptive fields expanded to encompass all locations. The Brugge group also noted that the first spike latencies of neurons tended to show spatial sensitivity. Most receptive fields showed a central core of stimulus virtual locations that elicited neuronal spikes with minimum latencies. Latencies tended to increase monotonically by up to around 5 ms as stimulus virtual locations were increased in eccentricity from the receptive field core. The location of the receptive field core tended to correspond roughly to the axis of the contralateral pinna. First spike latencies for all source locations tended to shorten as sound levels were increased, but the general spatial gradient of latencies was maintained for most neurons. The probability of a particular neuron firing a spike tended to correlate inversely with the neuron's first spike latency as stimulus virtual locations were varied.

One gets a fairly consistent picture of spatial tuning in cat area A1 regardless of whether sound sources are virtual or actual. The great majority of A1 neurons show at least some modulation of spike counts and/or latencies by sound location. The majority of spatially sensitive neurons are tuned to contralateral locations although a minority of neurons show midline or ipsilateral tuning. In most instances, the spatial tuning broadens with increasing sound levels.

A key issue of most studies of auditory spatial tuning has been whether or not the receptive fields of neurons shift systematically with cortical location.

That is, is there a map of auditory space? Studies by the Imig group (Imig et al. 1990; Clarey et al. 1994) and the Rajan group (1990b) confirmed an observation by Middlebrooks and Pettigrew (1981) that neurons with similar spatial sensitivity tended to form multiple clusters within the cortex. Middlebrooks and Pettigrew found that sequences of spatially selective neurons (i.e., neurons showing hemifield or axial receptive fields) along tangential electrode tracks tended to alternate with sequences of omnidirectional neurons. The Rajan group reported that contra-, ipsi- and centralfield neurons tended to form segregated clusters whereas omnidirectional neurons were distributed randomly among the other neurons. The difference between the results from the Rajan group and those obtained by Middlebrooks and Pettigrew in regard to the distribution of omnidirectional neurons can probably be explained by a difference in the spike rate criterion used to define omnidirectional: Middlebrooks and Pettigrew used a criterion that yielded a larger percentage of omnidirectional neurons. The Imig group (Clarey et al. 1994) found that electrode penetrations that were oriented parallel to radial cell columns (i.e., perpendicular to the cortical surface) tended to encounter neurons with similar azimuth preferences and similar sensitivity to sound level. Tangential penetrations, in contrast, tended to encounter clusters of neurons with one type of azimuth preference separated from clusters of neurons showing another type of preference. All these groups agree that the topography of spatial sensitivity in A1 is incompatible with a continuous map of auditory space in the sense of locations in azimuth being mapped systematically onto locations in the cortex.

3.2 Nontonotopic Auditory Areas in the Cat

We have studied the spatial receptive fields of neurons in two areas of the cat auditory cortex outside of area A1: area A2 and the anterior ectosylvian sulcus area (area AES). Both areas are nontonotopic in the sense that neurons show broad frequency tuning and that there is little or no evidence of a systematic progression of frequency tuning as a function of cortical place. Area A2 receives a strong thalamic input primarily from the dorsal division of the medial geniculate body (Andersen et al. 1980); in contrast, area A1, receives input primarily from the ventral division. Area AES receives thalamic auditory input from the posterior nuclear group and the medial division of the medial geniculate body (Roda and Reinoso-Suarez 1983) and sparse auditory cortical input from area A2 and possibly other auditory areas (Reinoso-Suarez and Rodo 1985). Many AES neurons respond both to auditory and visual stimuli (Clarey and Irvine 1990). We chose these nontonotopic areas for study of spatial sensitivity because we reasoned that the broad frequency tuning of neurons might indicate that neurons make use of broadband spectral cues for sound location. Also, AES was of particular interest because it is the only auditory area that is known to send strong projections to the superior colliculus, which contains an auditory space map (Meredith and Clemo 1989). Study of areas A2 and AES, however, produced results that were qualitatively quite consistent with those obtained from area A1.

We measured the mean spike counts elicited by brief noise bursts presented in a free sound field. Stimuli were presented from multiple loudspeakers, one loudspeaker at a time, that were positioned at 20° intervals throughout 360° of azimuth in the horizontal plane (Middlebrooks et al. 1998) and throughout 280° of elevation and front/back location in the vertical midline plane (Xu et al. 1998). One measure of the sensitivity of neurons to source location was given by the depth of modulation of spike count where 0% indicated no sensitivity to source location and 100% indicated that the spike count ranged from some maximum value down to zero spikes as the source location was varied. When sound levels were 20 dB above neural thresholds, the median depths of modulation by azimuth were 86.0% and 75.5% in areas AES and A2, respectively, and the respective median depths of modulation by elevation were 72.7 and 59.6%. The medians for areas AES and A2 dropped to 62.8 and 44.0%, respectively, for modulation by azimuth and to 48.9 and 31.6%, respectively, for modulation by elevation when sound levels were raised to 40 dB above neural thresholds. The depth of modulation by sound source azimuth in our sample of AES neurons was greater than that measured by Korte and Rauschecker (1993) in a sample that included AES neurons and neurons from the tonotopically organized anterior auditory field. That difference probably is due to the narrower range of azimuth that was tested in the Korte and Rauschecker study.

Despite the considerable modulation of spike counts by sound source location, the spatial tuning that we encountered generally was quite broad. Figure 8.2 shows examples of the spatial tuning of area A2 and AES neurons in the horizontal plane (Fig. 8.2A–D) and vertical midline plane (Fig. 8.2E–H). The neuron represented in Figure 8.2A is typical: at lower sound levels it responded with the most spikes to sources in the frontal contralateral quadrant and at higher levels the tuning expanded so that the spike count was about equal for all stimuli in the contralateral hemifield. The spatial tuning in Figure 8.2B showed two (or three) peaks at lower levels that resolved to one peak at higher levels. Figure 8.2C shows a somewhat unusual case in which the strongest responses were elicited by sound sources in the rear. Figures 8.2E through H illustrate the range of sharpness of elevation tuning.

We represented the breadth of spatial tuning by the range of azimuths or elevations over which a stimulus activated a particular neuron to more than half its maximum spike count. At sound levels 20 or 40 dB above neural thresholds, the median widths of azimuth tuning and the median heights of elevation tuning in areas AES and A2 were greater than 180°. As in area A1, the majority of neurons that we studied in areas AES and

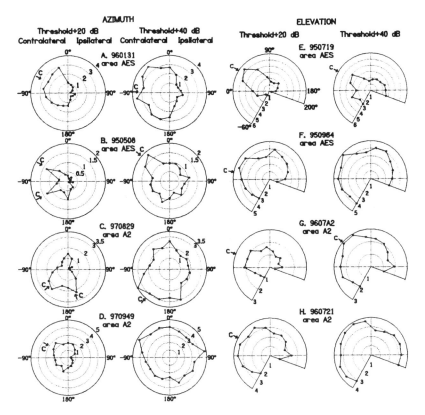

FIGURE 8.2. Spatial tuning of neurons in cat areas AES and A2. Each pair of panels represents spatial tuning of one neuron at 20 dB (left panels) or 40 dB (right panels) levels above the neuron's threshold. In these polar plots, the radial dimension represents the mean spike count of a neuron, and the angular dimension represents the stimulus location in azimuth (panels A–D) or elevation (panels E–H). In the azimuth plots, negative and positive angles are contralateral and ipsilateral, respectively, to the recording site. In the elevation plots, 0° is on the horizontal plane in front of the cat, 90° is overhead, and 180° is on the horizontal plane behind the cat. Arrows labeled "C" indicate Centroids, defined in the text. (Adapted from Middlebrooks et al. 1998 and Xu et al. 1998.)

A2 responded most strongly to contralateral sounds, and the steepest slopes in profiles of spike count versus azimuth tended to lie near the front and rear midline. Figure 8.3 shows the percent of recorded neurons that were activated to 25, 50, and 75% of their maximum spike counts by sounds at various azimuths. Noise bursts at levels 20 dB or more above neural thresholds that were presented at nearly any contralateral location activated nearly all neurons to at least half of their maximum spike counts.

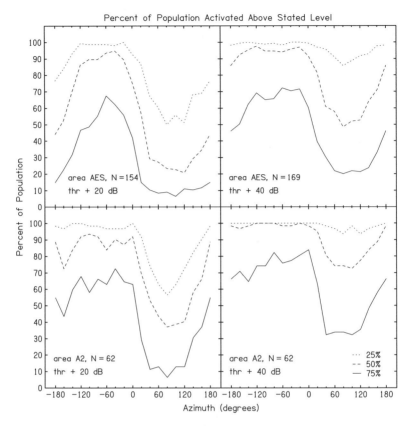

FIGURE 8.3. Percentage of the neural population activated by sound sources at various locations in azimuth. Populations consisted of 154 or 169 neurons in area AES (top) or 62 neurons in area A2 (bottom). (From Middlebrooks et al. 1998.)

At sound levels 20 dB above neural thresholds, profiles of spike count versus azimuth or elevation showed, for many neurons, a well defined "centroid" of maximum response. Some neurons showed bimodal profiles with 2 centroids. Centroids are indicated in Fig. 8.2 by the arrows labeled "C." The distribution of centroids in the azimuth was limited almost entirely to the contralateral half of space with the majority in the frontal contralateral quadrant. Centroids in elevation were limited mostly to upper frontal locations. As stimulus levels were increased to 40 dB above neural thresholds, spatial tuning consistently broadened such that 31% of neurons in area AES and 61% of neurons in area A2 had no discernable centroid in their azimuth profiles. We observed a weak spatial organization of neurons' azimuth centroids in that the centroids of neurons correlated weakly with the centroids of nearby neurons. Nevertheless, there was no indication of a systematic map of auditory space.

3.3 Spatial Sensitivity of Monkey Cortical Neurons

In nonhuman primates, auditory spatial sensitivity has been studied in the primary auditory area and in parietal and frontal association areas. The stimulus sets that have been used in the monkey studies have been much more limited than those used in the cat studies, typically involving only five source locations in the frontal horizontal plane. The advantage of the monkey preparation, however, is the ability to record cortical activity while the animal performs a complex auditory task.

Two groups of investigators have studied the primary auditory cortex in awake macaque monkeys using broadband sound sources located at contralateral 70° and 35°, at 0°, and at ipsilateral 35° and 70°. In a study by Benson and colleagues (1981), 75% of neurons responded with the maximum spike count to contralateral sound sources, and 11% responded best to ipsilateral sources. In that study, neurons often were more responsive to sound when the monkey performed an auditory task than when it was not performing. Nevertheless, fewer than 10% of neurons showed significant differences in responsiveness between conditions in which the monkey was required to localize a sound source and when it was required simply to detect a sound. In a study by Ahissar and colleagues (1992), monkeys were required only to sit quietly in a chair. More than one-half of neurons showed azimuth sensitivity in either their onset, sustained, or offset responses. Of those neurons, about 60% responded best to contralateral sounds and 10% responded best to ipsilateral sounds. In a small number of cases, the correlated firing between pairs of neurons was more azimuth-selective than was the activity of either neuron alone. Some neurons showed sensitivity to the direction of sound movement, but the authors concluded that: "the fact that movement-sensitive units were nearly all azimuth sensitive, and that their movement sensitivity was related to their static preference, implies that location and motion of sound are not encoded by separate mechanisms." (Ahissar et al. 1992, pg. 231). We can conclude from these two studies that, as in the cat, many neurons in the monkey's primary auditory area are sensitive to sound source azimuth and that most respond most strongly to contralateral sounds. Nevertheless, those studies do not permit any conclusions about the dimensions of auditory spatial receptive fields, about any fine differences in tuning among neurons, or about the influence of sound level on spatial sensitivity.

Andersen and colleagues (see references below) have described neurons in the lateral interparietal area (area LIP) of macaque monkeys that respond in relationship to an eye movement to an auditory target. In one report by that group (Mazzoni et al. 1996), about half of light-sensitive neurons in area LIP also responded to sound, and most of the sound-sensitive neurons responded differentially depending on whether the source was 10° to the right or 10° to the left of the midline. When presented with sound sources at five locations in azimuth, most sound-sensitive neurons showed peaked spike count versus azimuth profiles (Stricanne et al. 1996). Nearly half of the sound-sensitive neurons showed azimuth sensitivity in oculocentric coordinates, meaning that the azimuth profile shifted according to the visual fixation point. The remainder of sound-sensitive neurons expressed azimuth sensitivity in head-centered coordinates (i.e., no sensitivity to visual fixation point) or in an intermediate form. More recent reports by the Andersen group indicate that area LIP neurons show sound-evoked responses only in monkeys that have been trained to perform an auditory task (Grunewald et al. 1999) and that the spatial sensitivity of LIP neurons is greater in conditions in which a monkey must distinguish source locations than in a condition in which it must maintain visual fixation (Linden et al. 1999). That group concludes that " . . . these results imply that auditory responses in LIP are best considered supramodal (cognitive or motor) responses, rather than modality-specific sensory responses." (Linden et al. 1999, pg. 354).

Spatially sensitive responses to sounds also have been studied in the frontal cortex. In the frontal eye fields, neurons that show presaccadic activity in response to visual stimuli also tend to do so in response to sound (Russo and Bruce 1994). The magnitude of neural responses tends to correlate with the direction of the ensuing eye movement rather than with the head-centered location of the sound source. Vaadia and colleagues (1986) studied neurons in the periarcuate region of the frontal cortex. Monkeys were required to respond with an arm movement to a visual or auditory target presented from one of five azimuths. Neurons showed similar azimuth sensitivity to visual and auditory targets, generally responding most strongly to targets at one or more contralateral locations. Most neurons responded only in conditions in which an arm movement was required, again suggesting that the spatially sensitive responses to sounds are supramodal rather than modality-specific.

The visual cortex exhibits parallel processing streams that are specialized for processing of visual object information (the ventral, "what" stream) and for processing of visual spatial information (the dorsal, "where" stream) (Ungerleider and Mishkin 1982; Mishkin et al. 1983). Rauschecker (1998) has championed the hypothesis that analogous what and where streams might be present in the primate auditory cortex. The macaque auditory cortex contains, in addition to the core area A1, a core area R and a belt area CM. Rauschecker speculates that area R is specialized for object identification (e.g., identification of communication sounds) and area CM is specialized for localization. In support of that attractive hypothesis, Rauschecker offers results from positron emission tomography (PET) studies in humans and anatomical connection studies in monkeys. The connection studies show that rostral and caudal areas of the lateral auditory belt cortex are connected differentially with areas of the prefrontal cortex that are regarded as nonspatial or spatial, respectively (Romanski et al. 1999). The human PET studies show differential activation of the superior parietal lobule and middle frontal gyrus during auditory localization tasks as compared to auditory frequency discrimination tasks, although there was no task-related, preferential activation of particular auditory core areas of the temporal lobe (Bushara et al. 1999). Rauschecker's model has not yet been tested in any published study of spatial sensitivity of single neurons in areas R and CM. A recent study by Recanzone and colleagues (2000) has shown that spatially selective neurons are somewhat more prevalent in area CM than in area A1.

3.4 Spatial Receptive Fields: Summary and Conclusions

Details of reported auditory spatial sensitivity vary quantitatively among laboratories, species, and cortical fields. Nevertheless, the following properties generally are consistent across studies: (1) A sizeable majority of neurons is spatially sensitive in that neurons respond with higher probability or shorter latency to sounds at particular locations. (2) Most spatial receptive fields increase in size with increases in sound level. (3) At moderate sound levels, most receptive fields occupy a hemifield or more. (4) The majority of spatial receptive fields are centered in the contralateral hemifield, often near the axis of greatest sensitivity of the contralateral pinna. The borders of receptive fields most often lie near the front and rear midlines. (5) In the cat, receptive field properties in areas A1, A2, and AES are qualitatively similar. In the monkey, certain cortical fields are hypothesized to show specialization for sound localization, but no qualitative differences among fields have been published. (6) Nearby neurons tend to show similar spatial sensitivity. (7) Available data are inconsistent with any point-to-point cortical representation of sound source locations.

In a traditional view of stimulus coding, individual neurons are specialized to detect particular stimuli, and the spike rates of such neurons are determined by the probability that the particular stimulus is present. Such a view was formalized by Barlow (1972) as a "neuron doctrine for perceptual psychology." In the case of sound localization and the auditory cortex, such a view would predict that the location of a sound source would be signaled by activity in a restricted subset of cortical neurons. Reported studies of the receptive fields of auditory cortical neurons do not support that prediction. The generally large size of neural receptive fields indicates that the representation of any particular sound source location could involve a sizable majority of the auditory cortical neurons in one or both cortical hemispheres. Also, the spatial sensitivity of auditory neurons can change with changes in stimulus level, thus resulting in changes in the population of maximally activated neurons. A further problem is that the majority of receptive fields are centered well away from the midline even though behaving human and animal subjects show greatest localization accuracy and spatial acuity for stimuli near the midline (Mills 1958; Brown et al. 1982; Heffner and Heffner 1988; Makous and Middlebrooks 1990; May and Huang 1996; Recanzone et al. 1998). One might argue that acuity is greatest on the midline because so many receptive field borders lie near the midline, but then one must invoke a separate explanation to account for localization of off-midline targets. For instance, spatial acuity is still quite good around 45°, yet almost no receptive fields are bordered near contralateral 45°.

We have explored an alternative view, in which broad spatial receptive fields are an asset rather than a liability. We have tested the hypothesis that location-specific characteristic responses of individual neurons might signal the locations of sound sources throughout broad ranges of locations (Middlebrooks et al. 1998; Xu et al. 1998). Results of that investigation are presented in the next section.

4. Panoramic Space Coding by Single Neurons

Figure 8.4 represents the responses of a neuron in area AES to noise bursts presented from various azimuths in the horizontal plane. Each row of dots represents the pattern of spikes recorded during one stimulus presentation; eight presentations at each azimuth are shown. Negative azimuths indicate the cat's left side, contralateral to the recording site. This neuron responded

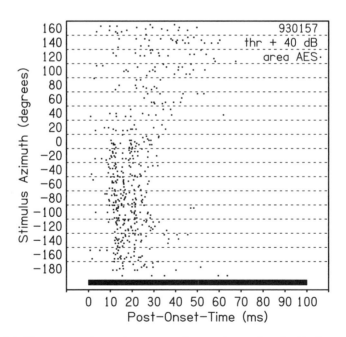

FIGURE 8.4. Responses of a neuron to sounds at various azimuths. Each row of dots represents the spike pattern in response to one presentation of a 100 ms noise burst. The sound source azimuth is indicated on the vertical axis. (From Middlebrooks et al. 1998.)

with high probability to sound sources located throughout the contralateral hemifield and with lower probability to ipsilateral sources. Conventionally, this neuron would be classified as showing a hemifield receptive field. Nevertheless, one can see differences in the response patterns elicited from various locations. For instance, within both the contralateral and ipsilateral hemifields, the first spike latency and the dispersion of spike latencies was systematically less for frontal sound source locations (around 0°) than for rear locations.

We measured the accuracy with which location-specific spike patterns could signal sound source locations. The procedure for analysis of each neuron is summarized here and is detailed elsewhere (Middlebrooks et al. 1998). We recorded responses to 40 or more stimulus presentations at each of 18 azimuths at increments of 20° in the horizontal plane. Spike patterns recorded during odd and even numbered trials were separated into "training" and "test" sets. Within each set, multiple bootstrapped averages were formed by averaging multiple sets of eight responses drawn randomly with replacement (Efron and Tibshirani 1991). An artificial neural network was used as a computer pattern recognition algorithm; the network architecture is diagrammed in Fig. 8.5. The network was configured to take the bootstrapped spike patterns represented with 1 ms time resolution as input and to produce an estimate of the sound source azimuth as output. Patterns from the training set were used to train the artificial neural network, then the trained network was used to classify the patterns from the test set.

Figure 8.6 shows the neural network estimates of sound source locations based on spike patterns from the neuron represented in Fig. 8.4. Each symbol indicates the estimate based on one bootstrapped spike pattern, and the solid line indicates the mean of the estimates of each source location. The mean estimate line generally follows the dashed line that indicates perfect performance. This indicates that the spike patterns of the cortical neuron signaled source locations more or less accurately throughout 360° of azimuth. The progression of estimates shows a discontinuity from 0 to +20°. This was seen in the analysis of the majority of neurons and presumably reflects the high level of neuronal sensitivity to sound source locations around the midline. The median value of the difference between actual and estimated sound source locations, computed on a spike-pattern by spike-pattern basis, was 24.7°. That is, half of the spike patterns estimated sound source locations that were within 24.7° of the correct location. Across all the neurons that were studied, median errors averaged 38.4 and 37.5° in areas AES and A2 in a condition in which sound levels were fixed at 20 dB above neural thresholds. Median errors of neurons in both cortical areas increased by about 6° in conditions in which sound levels were fixed at 40 dB above neural thresholds or in which levels roved in 5 dB steps between 20 and 40 dB above threshold. Note that median errors incorporate pattern by pattern variability in individual network estimates as well as the mean error in network estimates. Mean network estimates based on

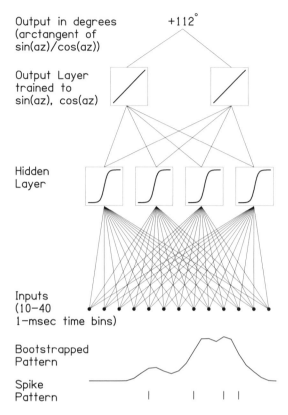

FIGURE 8.5. Artificial neural network architecture. Input to the network consisted of spike density functions that were averaged across eight stimulus presentations and were expressed in 1 ms resolution. The four units in the hidden layer had hyperbolic tangent transfer functions, and the two units in the output layer had linear transfer functions. The network was trained with supervision to estimate the sine and cosine of the stimulus azimuth. The two outputs were represented with a single azimuth term by forming the arctangent of the two outputs. (From Middlebrooks et al. 1998.)

many spike patterns from any given neuron tended to show errors considerably smaller that the medians of trial by trial errors.

Spike patterns of neurons in areas AES and A2 also could signal sound source elevations (Xu et al. 1998). We measured responses to noise sources in the vertical midline in 20° increments around the interaural axis: from −60° (60° below the horizon in front), through +90° (straight overhead), to 200° (20° below the horizon in the rear). Across all elevations and across both cortical areas, median errors in neural network estimation of elevation averaged 47.9° in the condition in which sound levels roved in 5 dB steps from 20 to 40 dB above neural thresholds. When errors were analyzed

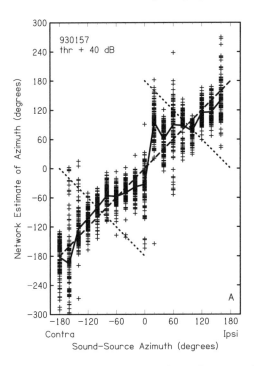

FIGURE 8.6. Artificial neural network estimates of sound source azimuth. Each plus sign represents the network estimate of azimuth based on recognition of one spike pattern. Horizontal and vertical axes represent the actual sound source location and the network estimate of the azimuth. The solid curve represents the circular centroid (i.e., the average) of the estimates at each source azimuth. The dashed line with positive slope represents the loci of perfect estimates, and the dotted lines with negative slope represent loci of perfect front/back confusions. (From Middlebrooks et al. 1998.)

on a location by location basis, errors were smallest for sound sources around 40 to 60° and increased for low frontal and for rear sound sources. Overall, errors were roughly equal for neurons in area A2 as compared to area AES, but neurons in area A2 showed a better balance of accuracy across the entire range of elevations.

It is common practice in studies of sensory coding to summarize the responses of neurons only by mean spike counts and to eliminate any measure of spike timing. We tested the degree to which spike counts might capture all the stimulus-related information in spike patterns (Middlebrooks et al. 1998). We compared the accuracy of sound localization based on recognition of full spike patterns (as described above) with recognition based only on spike counts. Median errors obtained in those conditions are shown in Fig. 8.7. In several instances, the symbols lay near the diagonal line that

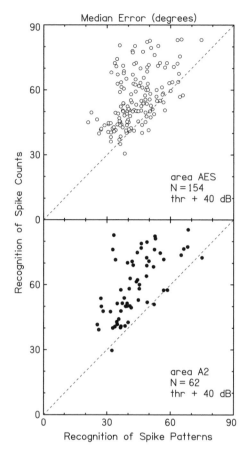

FIGURE 8.7. Accuracy of azimuth coding by spike counts and by complete spike patterns. Accuracy is represented by the median errors of network estimates of azimuth. Each symbol represents results from one neuron. (From Middlebrooks et al. 1998.)

indicates equal performance. In those instances, our analysis procedure was unable to detect any additional stimulus-related information carried by spike timing. In the substantial majority of cases, however, median errors in the full pattern condition were appreciably smaller than in the spike count only condition, indicating that some feature(s) of spike timing carried information about sound source location.

Under the conditions of animal preparation and anesthesia that we used, cortical neurons typically responded to a noise burst with one or a few spikes at the onset of the sound. The sparseness of spike patterns made it difficult to estimate sound source locations on the basis of responses of single neurons to single sound presentations. That is the reason that we chose to analyze averages of responses to 8 sound presentations in the work presented in the previous paragraphs. In analysis of single trials (i.e., no across-trial averaging) in a roving sound level condition, we obtained neural network median errors in azimuth that averaged around 70° but ranged as low as 40° for the best single neurons (Furukawa et al. 2000). We consider in Section 6.3 the increase in accuracy that might be obtained by combining information across multiple neurons.

We found that neural spike patterns signal sound source azimuth with about equal accuracy for sources both contra- and ipsilateral to the recording site (Middlebrooks 1998). Our physiological results seem to predict that, following a unilateral cortical lesion, one surviving cortical hemisphere could signal sound-source locations throughout 360° degrees. That prediction is not supported by the ablation behavioral literature since unilateral cortical lesions tend to produce contralateral localization deficits (see section 2.1). Nevertheless, recordings of cortical neuronal activity indicate that most neurons respond most strongly to contralateral sounds. Our artificial neural networks were able to distinguish ipsilateral sound source locations on the basis of latencies of the infrequent spikes that were elicited by ipsilateral sources. Such low levels of activity, however, are apparently in sufficient to support normal localization of sound sources that are ipsilateral to the surviving hemisphere in a unilaterally lesioned cat or monkey. It is interesting that hemispherectomized human patients described by Poirier and colleagues (1994) and those described by Zatorre and colleagues (1995) showed considerable auditory spatial acuity both contra- and ipsilateral to their surviving hemispheres. In those patients, the onset of severe unilateral cortical pathology occurred early in life, many years prior to localization testing. Perhaps over the course of years the surviving hemispheres in those patients underwent a rebalancing of synaptic weights that permitted a more robust cortical response to ipsilateral sound sources.

5. Responses to Sounds that Produce Spatial Illusions

The panoramic space coding by single neurons that was described in the previous section indicates that the responses of neurons code sound locations accurately under a condition in which human and feline listeners localize accurately: when broadband sounds are presented in an anechoic environment. In certain other conditions, listeners make systematic errors in localization. That is, certain sounds or combinations of sounds produce spatial illusions. If we hypothesize that the cortical neurons that we study contribute to sound localization behavior, it follows that the neurons should respond to certain sounds by signaling erroneous locations that parallel the illusions reported by listeners.

We have used our artificial neural network procedures to study neuronal responses in two conditions in which human listeners report spatial illusions. One condition is the response to paired clicks in which subjects experience "the precedence effect" (Blauert 1983). The second is the response to filtered sounds that results in systematic mislocalization in the vertical and front/back dimensions (Blauert 1969, 1970; Hebrank and Wright 1974; Middlebrooks 1992). In both conditions, we initially recorded the responses of a neuron to broadband sounds that were presented from a broad range of locations in azimuth or elevation, then used those spike patterns to train an artificial neural network. That procedure provided a means by which to associate particular neuronal spike patterns with specific locations. We then recorded the response of the neuron to a sound that, in listeners, produces a spatial illusion, and we presented that spike pattern to the trained network. The output of the network was taken as the location signaled by the neuron in response to the illusion-producing sound.

5.1 The Precedence Effect

Human listeners can localize accurately single, brief, broadband stimuli (i.e., clicks) in the horizontal dimension. Nevertheless, when two sound sources at different locations emit clicks separated by a brief delay, listeners experience a localization illusion. If the interstimulus delay (ISD) is less than about 5 ms, listeners identify only a single sound. When ISDs are between about 1 and 5 ms, the location of the single image is determined mainly by the location of the leading source; that is, the lagging sources does not contribute to the localization judgement. Click pairs with an ISD shorter than about 1 ms produce an image at a location intermediate to the two source locations. For example, if the two sources are placed symmetrically about the midline, the image of the click lies directly in front at zero ISD. As one of the clicks is advanced in time, the image moves toward the source of the earlier click. The relative intensity of the two clicks also influences the image location: an interstimulus level difference biases the image toward the more intense source. The family of phenomena in which the leading sources dominates localization judgements is known as the "precedence effect" and is important to hearing in reverberant spaces (reviewed by Litovsky et al. 1999). The subset of phenomena at delays less than 1 ms has been called "summing localization" and is important in stereophony (reviewed by Blauert 1983; also demonstrated in cats, Populin and Yin 1998).

We have tested correlates of the precedence effect in the cat's auditory cortex (Mickey and Middlebrooks, 2001). Analysis of a neuron from area A2 is shown in Figure 8.8. We first recorded neuronal responses to single clicks presented from various azimuths in the horizontal plane. We used those responses to train and test an artificial neural network, as described above. Localization of single clicks was fairly accurate, with a median error of 27° (Fig. 8.8A). For the same neuron, we also recorded responses to click pairs presented from −50° (left) and +50° (right) with variable ISD and

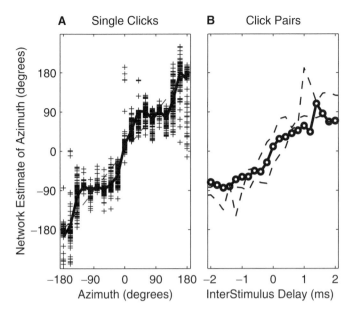

FIGURE 8.8. Network estimates of azimuth based on responses to single clicks or click pairs. The left panel shows estimates based on responses to single clicks presented from various azimuths. Other conventions are as in Figure 8.6. The right panel shows estimates based on responses to click pairs presented from contralateral and ipsilateral 50°. The curves represent circular centroids at each interstimulus delay. The solid curve with open circles represents the condition in which the clicks were presented at equal sound levels. The two dashed curves represent the condition in which the sound levels differed by +9 or −9 dB. (Mickey and Middlebrooks, unpublished.)

interstimulus level difference. The previously trained neural network was then used to produce estimates of azimuth from the responses to click pairs (Fig. 8.8B). In the absence of a level difference, the unit signaled locations near the midline at zero ISD and locations to the left or to the right of the midline as the ISD was shifted towards left-leading or right-leading values, respectively (Fig. 8.8B, solid curve). When the ISD was beyond about −1 or +1 ms, the unit signaled azimuths near one or the other actual source location. When a level difference of 9 dB was introduced, the locations signaled by the neuron shifted systematically toward the more intense source. That is, when the left or right source was made more intense, the azimuth versus ISD curve shifted downward or upward, respectively (Fig. 8.8B, dashed curves). Thus, the locations signaled by this neuron in response to click pairs closely paralleled the locations that would be reported by a listener.

Over 100 neurons from areas A1 and A2 have been analyzed in a similar way. Generally, results were qualitatively similar to those shown in Figure 8.8, and no obvious differences were observed between cortical areas A1 and A2. The range of ISD over which summing localization occurred varied widely among neurons (roughly ±0.1 ms to ±3.0 ms). Analysis of the population of neurons, however, showed a predominant range of about ±0.3 ms. That range corresponds well with the range of ISDs that produced summing localization in cats in a psychophysical study by Populin and Yin (1998). Populin and Yin reported a range of around ±0.3 ms in cats, as compared to the ranges of around ±1 ms that are typically reported for humans (Blauert 1983). In addition to the neurons that mirrored psychophysical performance, a sizeable minority of neurons in our study showed accurate signaling of the locations of single clicks but produced irregular results in the analysis of responses to click pairs. Those neurons were clearly sensitive to localization cues, but since their responses to click pairs disagreed with psychophysical predictions, we speculate that they were not neurons that contribute to localization judgments.

5.2 Elevation Illusions Produced by Filtered Sounds

In a typical listening situation, the folds and cavities of the pinna act to filter a broadband sound as it travels from a free field source to the tympanic membrane, so the spectrum at the tympanic membrane contains peaks and notches that were not present in the spectrum of the source (reviewed by Middlebrooks and Green 1991). The filter characteristics of the pinna (directional transfer functions; DTFs) vary according to the angle of incidence of the sound, so the spectrum at the tympanic membrane contains cues to sound source location. Those spectral-shape cues are most important in the vertical and front/back dimensions in which interaural difference cues are ambiguous. When certain filters are applied to the sound source spectrum, the spectral features introduced by the filters confound the direction-specific spectral features introduced by the pinna. As a result, human and feline listeners make systematic errors in vertical and front/back localization (e.g., Blauert 1969/70; Hebrank and Wright 1974; Populin and Yin 1998; Huang and May 1996).

We have tested localization of narrowband sounds by human listeners (Middlebrooks 1992). The bandwidths were one-sixth octaves, and the center frequencies were 6, 8, 10, and 12 kHz. Listeners tended to localize the narrowband sounds accurately in the horizontal dimension, but made conspicuous errors in the vertical dimension. Each listener tended to associate each center frequency with a particular vertical and front/back location regardless of the actual location of the source. Location judgements varied according to center frequency, and the location associated with a particular center frequency varied among subjects. We measured the DTFs of individual subjects by presenting broadband sounds from various locations while recording from subjects' ear canals with miniature microphones. We found that when a listener localized a narrowband stimulus to a particular location, the listener's DTF for that location tended to resemble the spectrum of the stimulus. An acoustical model was successful in identifying likely location judgements of individual subjects.

We tested a correlate of this elevation illusion in the activity of neurons in the cat's area A2 (Xu et al. 1999). We measured responses of neurons to broadband and one-sixth octave noise bursts presented at various elevations in the vertical midline plane. The responses to broadband sounds were used to train an artificial neural network, and the trained network was used to classify the neural responses to narrowband sounds. The predictions from the human psychophysical results were that neuronal responses to a particular narrowband center frequency would produce a constant elevation estimate, regardless of source elevation, and that elevation estimates would vary with stimulus center frequency. Also, based on the human acoustical model, we predicted that the elevation estimates would fall at elevations at which the DTF of the cat's pinna resembled the narrowband stimulus.

Figure 8.9A shows the network estimates of the elevation of a broadband sound based on the responses of one neuron. The responses of the neuron signaled stimulus elevations fairly accurately across that range. This neuron was typical of many in our sample in that it localized sources above the

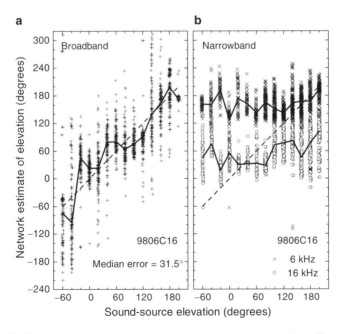

FIGURE 8.9. Network estimates of azimuth based on responses to broadband or narrowband noise bursts. The left panel shows estimates based on responses to broadband noise bursts. The right panel shows estimates based on responses to narrowband noise bursts that were classified using a neural network trained with broadband bursts. Other conventions are as in Figure 8.6. (From Xu et al. 1999. Copyright 1999 by Nature. Reprinted with permission.)

horizon (20 to 160°) more accurately than sources at or below the horizon. When the stimulus was filtered to one-sixth octave bandwidth, the neuron produced responses that were largely independent of source elevation (Fig. 8.9B). Network estimates of elevation were all up and to the front when the stimulus band was centered at 16 kHz, and all were around the rear horizon when the stimulus band was centered at 6 kHz.

We had no behavioral measure of narrowband sound localization by the cats in our physiological study. Nevertheless, it was possible to simulate the behavioral location judgements by adapting the acoustical model that successfully modeled human psychophysical performance (Middlebrooks 1992). We measured the DTFs of each cat's pinnae for sounds in the vertical midline. Then, we computed a spectral difference between each DTF and each narrowband stimulus spectrum Figures 8.10A and B show the distribution in elevation of spectral differences for sounds at various center frequencies. The light shades indicate the elevations at which the spectral difference was low, predicting by hypothesis a high probability of a location judgement at that elevation. The differences in the distributions of shading in Figures 8.10A and B reflect the differences in DTFs between the two cats that are represented (Xu and Middlebrooks 2000).

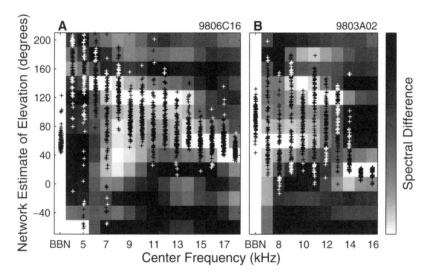

FIGURE 8.10. Neuronal signaling of elevation and an acoustical model. The shaded squares represent the spectral difference between narrowband stimuli of various center frequencies (horizontal axis) and directional transfer functions at various elevations (vertical axis). Light shading indicates a small spectral difference and a model prediction of a high likelihood of response at that elevation. The plus signs indicate the network estimates of elevation in response to narrowband noise bursts and broadband noise (BBN) bursts. (From Xu et al. 1999. Copyright 1999 by Nature. Reprinted with permission.)

The symbols in Figures 8.10A and B indicate the network estimates of elevation based on the responses of single neurons to narrowband sounds of various center frequencies which were presented from a fixed location. The symbols tend to fall in regions of low spectral difference. That is, the neuronal responses to narrowband stimuli tended to signal elevations that corresponded to the elevations predicted by the acoustical model. Across a sample of 146 neurons, network estimates of elevation tended to correspond well with the prediction of the acoustical model for center frequencies between 11 and 16 kHz; the correspondence was less accurate at other frequencies. Behavioral studies show that cats can localize well in elevation when stimulus spectra are restricted to a 10–14 kHz band but localize poorly when spectra are restricted to above or below that band (Huang and May 1996).

5.3 Responses to Sound that Produce Spatial Illusions: Summary and Conclusions

We tested the responses of cortical neurons to two classes of stimuli that produce spatial illusions (illusions in the sense that listeners consistently localize a sound to a position that does not correspond to the actual position of the source). The precedence effect is a well known phenomenon in human psychophysics, and it has been recently demonstrated in cat behavioral trials (Populin and Yin 1998). We find that the locations that are signaled by the responses of A1 and A2 neurons closely follow the behavior of human and feline listeners (Mickey and Middlebrooks, 2001). Psychophysical studies in humans have identified a variety of filter shapes that produce an illusory location judgement, i.e., a reported elevation that is determined more by the characteristics of the filter than by the actual source location. We tested a particular narrowband filter for which we have detailed results in humans and an acoustical model that successfully predicts human localization judgements (Middlebrooks 1992). One study in cats tested vertical localization of narrowband sounds (Populin and Yin 1998). That study showed that normal vertical localization was disrupted but that localization judgements did not show any predictable pattern. Most of the center frequencies that were tested in that study, however, fell below the range of frequencies in which we would expect an illusory elevation judgement. At the highest frequency that was tested, we would have expected the elevation judgement to fall outside of the range in which the cat could respond with eye movements as was required by the experimental design. Huang and May (1996) tested the ability of cats to localize one-half octave-wide noise bursts, which are somewhat wider than the bands that we have used in human psychophysics (Middlebrooks 1992) or cat physiology (Xu et al. 1999). That group found that the cats showed localization errors that were at least qualitatively predicted by an acoustical model similar to the one that we have used.

Given that the neurons that we have studied appear to be capable of signaling sound-source locations with some accuracy, we can entertain the hypothesis that they have some role in sound localization behavior. A necessary prediction of that hypothesis is that accurate location signaling by the neurons would fail under the same condition that accurate localization by listeners fails. That prediction was satisfied in our results. An extension of that prediction is that the neurons not only signal incorrect locations but that the incorrect locations correspond to the incorrect location judgements of listeners. In the case of summing localization, the locations signaled by neurons formed sigmoid azimuth versus interstimulus delay functions like those seen in human and feline psychophysical experiments. In the case of judgements of the elevations of filtered sounds, we have no cat behavioral data with which to compare the physiology. Nevertheless, an acoustical model fit the human behavior closely and the same model adapted to the cat fit the cat physiology closely.

Failure of the physiological results to conform to the behavioral results would have been strong evidence against the hypothesis that these neurons contribute to sound localization behavior. Our success in demonstrating parallels between physiology and behavior supports the hypothesis but does not necessarily confirm it. One might imagine, for instance, that a population of neurons supports some aspect of hearing unrelated to localization yet is sensitive to spectral features that also serve as localization cues. Nevertheless, our results permit the conclusion that the cortical responses that we have described follow the same computational principles that govern listeners' localization judgements. That is, the azimuth tuning of neurons appears to be sensitive to the same manipulations of interstimulus delay and level differences as are listeners, and the elevations signaled by the responses of neurons are predicted by the same computational model that predicts human elevation judgements.

6. Location Coding by Populations of Neurons

We have found that single cortical neurons can signal the locations of real or illusory sound sources with considerable accuracy. Nevertheless, the location coding that we have demonstrated is considerably less accurate than the localization accuracy of a behaving cat. The disparity between our cortical results and the cat behavioral results is not at all surprising when we consider that our results are based on the responses of single neurons, whereas a cat presumably bases its localization judgment on a neural population several orders of magnitude larger. In our work, we have noted that the responses of many individual neurons appear to signal sound locations throughout broad ranges of location. For that reason, we have speculated that an animal's accurate judgement of the location of any particular sound source is distributed across large, possibly widespread, cortical populations (Middlebrooks et al. 1998).

We do not yet know, in any formal way, how the nervous system combines information across multiple neurons. Nevertheless, several investigators have attempted to incorporate small populations of neurons in computational models and to explore ways in which information about sound source location might be carried by interactions among neurons. In the following sections, we review a model based on spatially tuned neurons and one based on relative latencies among multiple neurons. Then, we present extensions of our artificial neural network analysis to encompass multiple neurons.

6.1 A Population Model Based on Spatial Tuning of Spike Counts

One hypothetical mechanism for cortical coding of sound source locations might involve spatially tuned neurons that have well-defined receptive fields. As discussed in Section 3, such a hypothesis is challenged by the large receptive fields of auditory cortical neurons. In the motor cortex, however, neurons show large movement fields, yet a population model based on movement fields has been quite successful in modeling the coding of arm movements (Georgopoulos et al. 1986). In that model, each neuron is assigned a preferred movement direction. The neuron's activity during a particular movement is represented as a vector that has a direction given by the neuron's preferred direction and a length determined by the neuron's rate of activity. The direction that is signaled by a population of neurons is computed by forming a vector sum of the vectors that represent the individual neurons. The direction of the resultant vector is taken as the direction that is signaled by the neuronal population.

Eggermont and Mossop (1998) attempted to adapt such a model to area A1 of the cat auditory cortex. Their implementation of the model used responses from 102 neurons averaged across 50 stimulus presentations. For the purpose of optimizing model performance, each neuron was assigned to one of nine preferred-azimuth classes, and response rates were derived from the mean response in each group. The performance of the model was rather disappointing. Across all frontal stimulus locations, model estimates of the location gave a root-mean-squared error no better than 32°. The model estimates varied dramatically across changes in stimulus levels. For a sound source located straight ahead at 0° azimuth, the model output showed an ipsilateral bias of 20° that increased to more than 90° for sounds at lower sound levels. Possible explanations for the poor performance of this population model include the absence of radial symmetry of auditory spatial receptive fields, the nonuniform distribution of preferred azimuths, and the pronounced tendency of spatial tuning to shift and broaden at high sound levels. We have informally attempted to apply a similar model to our results from cortical areas A2 and AES and obtained similarly disappointing results.

6.2 A Population Model Based on Spatial Tuning of First Spike Latencies

As discussed in Section 3.1, Brugge and colleagues (1996) showed that the first-spike latencies of neurons in area A1 tend to form a systematic gradient, most often increasing with increasing separation of sound sources from the acoustical axis of the contralateral ear. Jenison and colleagues (1998) have modeled those data using spherical approximation functions. Jenison then derived a maximum likelihood estimation procedure to attempt to estimate sound source locations based on simulated latencies within populations of up to 65 cortical neurons (Jenison 1998). He calculated the Cramer-Rao lower bound, which is an estimate of the best possible estimation error of an ideal observer. The estimation error decreased with the inverse of the square root of the number of neurons, approaching errors comparable to those of cat and human behavior as the number of neurons approached 65. Jenison states that "absolute comparisons could be unwarranted given the number of parametric assumptions" (Jenison 1998, pg. 172), but the generally high level of performance of a model based on large spatial receptive fields is encouraging.

Jenison did not explicitly test the sensitivity of his model to changes in stimulus sound level. In the data set on which the model was based, Brugge and colleagues (1996) reported that latencies tended to shorten as sound levels increased but that the general structure of the spatial gradients were preserved. The level dependence of latencies could be added as another parameter to the model, but the need to estimate an additional parameter presumably would add uncertainty to the location estimate. Another limitation of the current form of Jenison's model is that spike latencies are measured relative to the onset of the stimulus. Of course, the stimulus onset time would not be known precisely by a behaving animal, so the nervous system would need to estimate a time reference from within the distribution of spike times within the neural population. That estimate would contribute an additional source of uncertainty.

Spike latencies present an attractive alternative to mean spike counts as a source of stimulus-related information. Plots of mean spike counts averaged over 20 or more trials often show systematic azimuth sensitivity. However most auditory cortical neurons tend to fire with only one or a few spikes in response to the onset of a stimulus, so the spike count from a single neuron on a single trial could distinguish no more than about two sound source locations. The latency of one or more spikes provides a continuously variable term that could potentially signal a continuum of sound source locations. Even in our artificial neural network classification of spike patterns averaged across eight trials (Section 4), we found that the identification of sound source locations was substantially more accurate when spike timing information was preserved than when responses were represented only by mean spike counts.

6.3 Artificial Neural Network Classification of Responses of Neural Ensembles

We have extended our analysis of stimulus coding by temporal spike patterns to include small populations of neurons, which we refer to as "neural ensembles" (Furukawa et al. 2000). We record neural ensembles with multichannel recording probes that permitted simultaneous recording of differentiated spike activity at up to 16 cortical sites. The results that we present here are based on recordings of five to 19 distinct neurons or small clusters of neurons. In our analysis, we have concentrated on information that was available on single trials. That is, there was no averaging across trials. All analysis was based on conditions in which sound levels varied in 5 dB steps from 20 to 40 dB above neural threshold.

In one analysis, we measured the accuracy with which an artificial neural network could identify sound source locations based on recognition of the spike patterns recorded at five to 19 sites. Multiple neuron spike patterns were recorded simultaneously and were represented as five to 19 sets of 25 2 ms time bins. The network architecture did not attempt to model any physiological specialization for coordinated coding by neural populations. For that reason, the results that we obtained probably underestimate the stimulus-related information contained in these neural ensembles. Nevertheless, sound localization on individual trials was remarkably accurate. Figure 8.11 shows the median errors in azimuth identification. The bars indicate the ensemble performance, and the plus signs indicate the performance of the single neurons that constituted each ensemble. Performance of the ensembles ranged from near chance levels to a median error of 22.9° produced by one ensemble. The mean and standard error of the median errors of 34 neural ensembles on single trials was 49.2 ± 11.9°, which compares with the mean performance of responses of single neurons averaged across eight trials (46.0 ± 10.3°).

We also explored the accuracy of single trial azimuth identification by larger neural ensembles (Furukawa 2000). For that purpose, we compiled spike patterns recorded nonsimultaneously across multiple neurons in multiple cats. In that condition, trial by trial median errors averaged around 20° for ensembles of 128 neurons drawn randomly from our sample population. For the purpose of comparison with a cat behavioral study (Huang and May 1996), we tested the accuracy of localization of targets in the frontal half of

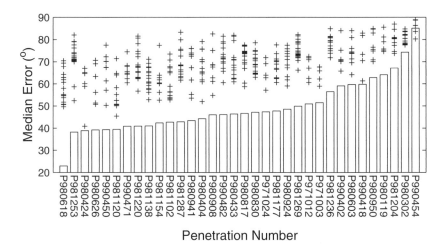

FIGURE 8.11. Median errors of azimuth signaling by neural ensembles. Each bar represents the median error of a network estimate of azimuth based on simultaneously recorded spike patterns from five to 19 neurons. The plus signs above each bar indicate median errors obtained from the constituent neurons when they were analyzed individually. The bars are arranged in ascending order for convenience of presentation. (From Furukawa et al. 2000.)

the horizontal plane by an ensemble of the 128 neurons that were most accurate in single neuron tests. In that condition, the average error of the neural ensemble (i.e., the unsigned error of the mean network estimate for each target azimuth, averaged across target azimuth was 8.9°, which was roughly half the average error of the behaving cats in the Huang and May study. The trial by trial standard deviation of the ensemble estimates was about double that of the behavior. One should not overinterpret these comparisons because our neural network recognition of multiple neuron spike patterns does not incorporate any specific comparisons between neurons that might be used by the brain and because the specific patterns of localization errors by the network differ from the errors made by the behaving cats. Nevertheless, it is encouraging that neural ensembles of modest size can signal sound source locations with accuracy comparable to behavioral accuracy.

As was the case for single neurons (Section 4), sound localization by neural ensembles was more accurate when the analysis preserved spike timing information than when responses were represented only by spike counts. Nevertheless, performance could be quite good in some cases even when only spike counts were used. The neural ensembles that produced the most accurate sound localization tended to be those in which the constituent single neurons showed the greatest diversity in their spatial sensitivity. In terms of spike counts alone, one can think of activity shifting from one subset of neurons to another as the sound source location was varied.

Considerable location-related information was apparently carried by the distribution of spike counts across the neural ensemble. When we manipulated ensemble spike patterns in various ways, performance was significantly better when the profile of the cross population distribution of spike counts was preserved than when the profile was obliterated and only the cross population spike count mean was preserved.

Our analysis of ensemble spike patterns is subject to same criticism that was applied to Jenison's model in Section 6.2. That is, spike times were timed related to the stimulus onset, which would not be known precisely to a behaving animal. We tested the effect of expressing spike times relative to the first spike in the ensemble, effectively eliminating direct knowledge of the stimulus onset time. The result varied among neural ensembles. In some cases, there was no apparent loss of information carried by spike times whereas in other cases some or all of the temporal information was lost by expressing spike times relative to the ensemble minimum. In another manipulation, we expressed each spike time relative to the first spike recorded from each neuron on each trial. That procedure eliminated absolute latencies, eliminated relative timing among neurons, and preserved only interspike intervals within single neuron spike patterns. In that condition, we saw no evidence of stimulus-related information carried by interspike intervals. It is difficult to abandon the idea that information is carried by interspike intervals, and the result might reflect a limitation of our analysis procedure or of the experimental conditions. Nevertheless, the conclusion that interspike intervals carry little or no location-related information also has been reached following a rather different type of analysis by Eggermont (1998).

7. Summary and Conclusions

We can make several generalizations about coding of sound source locations by neurons in the auditory cortex: (1) Unilateral cortical lesions in carnivores and monkeys result in contralateral localization deficits. Localization deficits are observed in humans, but the deficits are not always strictly contralateral to the lesion. (2) The majority of neurons in auditory cortical areas that have been studied in cats and monkeys are sensitive to sound source location, most often preferring the sounds in the contralateral hemifield. Receptive fields typically are large (a hemifield or more) and enlarge as sound levels are increased. There is no evidence of cortical maps of auditory space that are based on systematic cortical progressions of receptive fields. (3) The spike patterns of neurons vary with sound source location. Recognition of spike patterns by artificial neural networks can identify source locations with varying degrees of accuracy throughout up to 360° of auditory space. (4) Sounds that produce spatial illusions in feline and human listeners produce corresponding errors in the locations signaled by the spike patterns of cortical neurons. (5) Populations on the order of 100 neurons can signal sound source locations with accuracy comparable to the accuracy measured in behavioral trials.

It is somewhat surprising that across all cortical areas that have been studied in cats, no distinct population of neurons has stood out as specialized for sound localization. Various investigators have found it convenient to distinguish neurons that were more or less directionally sensitive (e.g., the "high-directionality" and "low-directionality" neurons defined by the Imig group (Imig et al. 1990)). Such distinctions, however, have generally amounted to setting an arbitrary boundary within a unimodal distribution of some measure of directionality. The types of spatial sensitivity described in cortical areas A1, A2, and AES are qualitatively similar although areas might differ in the distributions of quantitative measures of spatial sensitivity. In studies of panoramic space coding by the spike patterns of neurons in areas A2, AES, and in our unpublished observations in area A1, we have seen quantitative differences in the distributions of median errors of azimuth coding and the uniformity of elevation coding. Nevertheless, the general characteristics of panoramic coding are very similar between cortical areas. One might argue that the apparent absence of cortical neurons specialized for spatial coding is due to the use of anesthesia. Limited studies in awake monkeys, however, have produced results that are quite consistent with the observations from anesthetized cats.

Results from lesion studies have led to the general belief that the auditory cortex has a special role in spatial sensitivity. Nevertheless, we wish to consider the hypothesis that the special role of the cortex is only in distributing preprocessed information about sound source location to appropriate perceptual and motor stations, not in actual computation of source locations. The likely sites for analysis of specific localization cues are in the brainstem; examples include the dorsal cochlear nucleus and the superior olivary complex. Output from those nuclei converges in the central nucleus of the inferior colliculus (ICC). Not many synapses removed from the ICC is the superior colliculus, which contains a map of auditory space. The superior colliculus receives descending input from the cortex, but its auditory responses show short latencies, which are consistent with a midbrain source (e.g., Middlebrooks and Knudsen 1984). The form of the auditory spatial representation in the superior colliculus is very different from anything reported in the cortex, but the presence of a space map indicates that the superior colliculus has access to processed auditory spatial information. Ascending auditory input takes multiple parallel routes through the thalamus on its way to the auditory cortex. Imig and colleagues have recorded from two divisions of the auditory thalamus and report spatial sensitivity that is essentially the same as they record from cortical area A1 (Barone et al. 1996). It is reasonable to speculate that computation of sound source location is essentially complete by the level of the midbrain (perhaps by one or two synaptic levels beyond the level of the ICC) then is distributed to the superior colliculus and to multiple thalamocortical systems.

These considerations lead to the idea that the pathways to discrete cortical areas might be specialized for various aspects of hearing, but many or all of those pathways might exhibit at least some degree of spatial sensitivity. One could imagine that some cortical areas might utilize auditory spatial information for an overtly spatial task, such as for directing eye movements. Other cortical areas might perform functions that are not overtly spatial, such as identification of communication sounds, but spatial information might assist those functions, such as by helping to segregate multiple sound sources. In this view, one might predict that differences in the spatial sensitivity of neuronal responses between cortical areas might appear during appropriately designed behavioral tasks, but that such differences might not be evident in an anesthetized preparation or in a behavioral context that did not emphasis localization. We are optimistic that use of awake behaving recording preparations in animals or functional imaging techniques in humans will reveal cortical areas that are specialized for the coupling of auditory spatial information to systems for spatial behavior and perception.

References

Ahissar M, Ahissar E, Bergman H, Vaadia E (1992) Encoding of sound-source location and movement-activity of single neurons and interactions between adjacent neurons in the monkey auditory cortex. J Neurophysiol 67:203–215.

Andersen P, Knight L, Merzenich MM (1980) The thalamocortical and corticothalmic connections of AI, AII, and the anterior auditory field (AFF) in the cat: Evidence for two largely segregated systems of connections. J Comp Neurol 194: 663–701.

Barlow HB (1972) Single units and sensation: A neuron doctrine for perceptual psychology? Perception 1:371–394.

Barone P, Clarey JC, Irons WA, Imig TJ (1996) Cortical synthesis of azimuth-sensitive single-unit responses with nonmonotonic level tuning: A thalamocortical comparison in the cat. J Neurophysiol 75:1206–1220.

Beitel RE, Kaas JH (1993) Effects of bilateral and unilateral ablation of auditory cortex in cats on the unconditioned head orienting response to acoustic stimuli. J Neurophysiol 70:351–369.

Benson DA, Hienz RD, Goldstein Jr MH (1981) Single-unit activity in the auditory cortex of monkeys actively localizing sound sources: Spatial tuning and behavioral dependency. Brain Res 219:249–267.

Blauert J (1969–1970) Sound localization in the median plane. Acustica 22:205–213.

Blauert J (1983) Spatial Hearing. Cambridge, MA: MIT Press.

Brown CH, Schessler T, Moody D, Stebbins W (1982) Vertical and horizontal sound localization in primates. J Acoust Soc Am 72:1804–1811.

Brugge JF, Reale RA, Hind JE, Chan JCK, Musicant AD, Poon PWF (1994) Simulation of free-field sound sources and its application to studies of cortical mechanisms of sound localization in the cat. Hear Res 73:67–84.

Brugge JF, Reale RA, Hind JE (1996) The structure of spatial receptive fields of neurons in primary auditory cortex of the cat. J Neurosci 16:4420–4437.

Bushara KO, Weeks RA, Ishii K, Catalan M, Tian B, Rauschecker JP, Hallett M (1999) Modality-specific frontal and parietal areas for auditory and visual spatial localization in humans. Nat Neurosci 2:759–766.

Clarey JC, Irvine DRF (1990) The anterior ectosylvian auditory field in the cat: I. An electrophysiological study of its relationship to surrounding auditory cortical fields. J Comp Neurol 301:289–303.

Clarey JC, Barone P, Imig T (1992) Physiology of Thalamus and Cortex. In: Popper AN, Fay RR (eds) The Mammalian Auditory Pathway: Neurophysiology. New York: Springer-Verlag, pp. 232–334.

Clarey JC, Barone P, Imig TJ (1994) Functional organization of sound direction and sound pressure level in primary auditory cortex of the cat. J Neurophysiol 72:2383–2405.

Efron B, Tibshirani R (1991) Statistical data analysis in the computer age. Science 253:390–395.

Eggermont JJ (1998) Azimuth coding in primary auditory cortex of the cat. II. Relative latency and interspike interval representation. J Neurophysiol 80:2151–2161.

Eggermont JJ, Mossop JE (1998) Azimuth coding in primary auditory cortex of the cat. I. Spike synchrony versus spike count representations. J Neurophysiol 80:2133–2150.

Eisenman LM (1974) Neural encoding of sound location: An electrophysiological study in auditory cortex (AI) of the cat using free field stimuli. Brain Res 75:203–214.

Evans EF (1967) Cortical representation. In: de Reuck AUS, Knight J (eds) Symposium on Hearing Mechanisms in Vertebrates. London: CIBA, Churchill, pp. 272–295.

Furukawa S, Xu L, Middlebrooks JC (2000) Coding of sound-source location by ensembles of cortical neurons. J Neurosci 20:1216–1228.

Georgopoulos AP, Schwartz AB, Kettner RE (1986) Neuronal population coding of movement direction. Science 233:1416–1419.

Greene TC (1929) The ability to localize sound: a study of binaural hearing in patients with tumor of the brain. Arch Surg 18:1825–1841.

Grunewald A, Linden JF, Andersen RA (1999) Responses to auditory stimuli in macaque lateral intraparietal area. I. Effects of training. J Neurophysiol 82:330–342.

Hebrank J, Wright D (1974) Spectral cues used in the localization of sound sources on the median plane. J Acoust Soc Am 56:1829–1834.

Heffner HE, Heffner RS (1990) Effect of bilateral auditory cortex lesions on sound localization in Japanese macaques. J Neurophysiol 64:915–931.

Heffner RS, Heffner HE (1988) Sound localization acuity in the cat: Effect of azimuth, signal duration, and test procedure. Hear Res 36:221–232.

Huang AY, May BJ (1996) Sound orientation behavior in cats. II. Mid-frequency spectral cues for sound localization. J Acoust Soc Am 100:1070–1080.

Imig TJ, Irons WA, Samson FR (1990) Single-unit selectivity to azimuthal direction and sound pressure level of noise bursts in cat high-frequency primary auditory cortex. J Neurophysiol 63:1448–1466.

Jenison RL (1998) Models of direction estimation with spherical-function approximated cortical receptive fields. In: Brugge JF, Poon PWF (eds) Central Auditory Processing and Neural Modeling. New York: Plenum, pp. 161–174.

Jenison RL, Reale RA, Hind JE, Brugge JF (1998) Modeling of auditory spatial receptive fields with spherical approximation functions. J Neurophysiol 80: 2645–2656.

Jenkins WM, Masterton RB (1982) Sound localization: Effects of unilateral lesions in central auditory system. J Neurophysiol 47:987–1016.

Jenkins WM, Merzenich MM (1984) Role of cat primary auditory cortex for sound-localization behavior. J Neurophysiol 52:819–847.

Kavanagh GL, Kelly JB (1987) Contribution of auditory cortex to sound localization by the ferret (Mustela putorius). J Neurophysiol 57:1746–1766.

Klingon GH, Bontecou DC (1966) Localization in auditory space. Neurology 16: 879–886.

Korte M, Rauschecker JP (1993) Auditory spatial tuning of cortical neurons is sharpened in cats with early blindness. J Neurophysiol 70:1717–1721.

Linden JF, Grunewald A, Andersen RA (1999) Responses to auditory stimuli in macaque lateral intraparietal area. II. Behavioral modulation. J Neurophysiol 82: 343–358.

Litovsky RY, Colburn HS, Yost WA, Guzman SJ (1999) The precedence effect. J Acoust Soc Am 106:1633–1654.

Makous JC, Middlebrooks JC (1990) Two-dimensional sound localization by human listeners. J Acoust Soc Am 87:2188–2200.

May BJ, Huang AY (1996) Sound orientation behavior in cats. I. Localization of broadband noise. J Acoust Soc Am 100:1059–1069.

Mazzoni P, Bracewell RM, Barash S, Andersen RA (1996) Spatially tuned auditory responses in area LIP of macaques performing delayed memory saccades to acoustic targets. J Neurophysiol 75:1233–1241.

Meredith MA, Clemo HR (1989) Auditory cortical projection from the anterior ectosylvian sulcus (field AES) to the superior colliculus in the cat: An anatomical and electrophysiological study. J Comp Neurol 289:687–707.

Merzenich MM, Knight PL, Roth GL (1973) Cochleotopic organization of primary auditory cortex in the cat. Brain Res 63:343–346.

Mickey BJ, Middlebooks JC (2001) Responses of auditory cortical neurons to pairs of sounds: correlates of fusion and localization. J Neurophysiol 86:1333–1350.

Middlebrooks JC (1992) Narrow-band sound localization related to external ear acoustics. J Acoust Soc Am 92:2607–2624.

Middlebrooks JC (1998) Location coding by auditory cortical neurons. In: Poon PWF, Brugge JF (eds) Central auditory processing and neural modeling. New York: Plenium Press, pp. 139–148.

Middlebrooks JC, Green DM (1991) Sound localization by human listeners. Ann Rev Psychol 42:135–159.

Middlebrooks JC, Knudsen EI (1984) A neural code for auditory space in the cat's superior colliculus. J Neurosci 4:2621–2634.

Middlebrooks JC, Pettigrew JD (1981) Functional classes of neurons in primary auditory cortex of the cat distinguished by sensitivity to sound location. J Neurosci 1:107–120.

Middlebrooks JC, Xu L, Eddins AC, Green DM (1998) Codes for sound-source location in nontonotopic auditory cortex. J Neurophysiol 80:863–881.

Mills AW (1958) On the minimum audible angle. J Acoust Soc Am 30:237–246.

Mishkin M, Ungerleider LG, Macko KA (1983) Object vision and spatial vision: two cortical pathways. TINS 414–417.

Musicant AD, Chan JCK, Hind JE (1990) Direction-dependent spectral properties of cat external ear: New data and cross-species comparisons. J Acoust Soc Am 87:757–781.

Palmer AR, King AJ (1982) The representation of auditory space in the mammalian superior colliculus. Nature 299:248–249.

Pinek B, Duhamel JR, Cave C, Brouchon M (1989) Audio-Spatial deficits in humans: Differential effects associated with left versus right hemisphere parietal damage. Cortex 25:175–186.

Poirier P, Lassonde M, Villemure J-G, Geoffroy G, Lepore F (1994) Sound localization in hemispherectomized patients. Neuropsychologica 32:541–553.

Poirier P, Jiang H, Lepore F, Guillemot JP (1997) Positional, directional and speed selectivities in the primary auditory cortex of the cat. Hear Res 113:1–13.

Populin LC, Yin TC (1998) Behavioral studies of sound localization in cat. J Neurosci 18:2147–2160.

Rajan R, Aitkin LM, Irvine DRF, McKay J (1990a) Azimuthal sensitivity of neurons in primary auditory cortex of cats. I. Types of sensitivity and the effects of variations in stimulus parameters. J Neurophysiol 64:872–887.

Rajan R, Aitkin LM, Irvine DRF (1990b) Azimuthal sensitivity of neurons in primary auditory cortex of cats. II. Organization along frequency-band strips. J Neurophysiol 64:888–902.

Rauschecker JP (1998) Parallel processing in the auditory cortex of primates. Audiol Neurootol 3:86–103.

Reale RA, Imig TJ (1980) Tonotopic organization in auditory cortex of the cat. J Comp Neurol 192:265–291.

Recanzone GH (1998) Rapidly induced auditory plasticity: The ventriloquism aftereffect. Proc Nat Acad Sci USA 95:869–875.

Recanzone GH, Guard DC, Phan ML, Su T-IK (2000) Correlation between the activity of single auditory cortical neurons and sound-localization behavior in the macaque monkey. J Neurophysiol 83:2723–2739.

Reinoso-Suarez F, Roda JM (1985) Topographical organization of the cortical afferent connections to the cortex of the anterior ectosylvian sulcus in the cat. Exp Brain Res 59:313–324.

Rice JJ, May BJ, Spirou GA, Young ED (1992) Pinna-based spectral cues for sound localization in cat. Hear Res 58:132–152.

Roda JM, Reinoso-Suarez R (1983) Topographical organization of the thalamic projections to the cortex of the anterior ectosylvian sulcus in the cat. Exp Brain Res 49:131–139.

Romanski LM, Tian B, Fritz J, Mishkin M, Goldman-Rakic PS, Rauschecker JP (1999) Dual streams of auditory afferents target multiple domains in the primate prefrontal cortex. Nat Neurosci 2:1131–1136.

Ruff RM, Hersh NA, Pribram KH (1981) Auditory spatial deficits in the personal and extrapersonal frames of reference due to cortical lesions. Neuropsychologica 19:435–443.

Russo GS, Bruce CJ (1994) Frontal eye field activity preceding aurally guided saccades. J Neurophysiol 71:1250–1253.

Sanchez-Longo LP, Forster FM (1958) Clinical significance of impairment of sound localization. Neurology 8:119–125.

Sovijarvi ARA, Hyvarinen J (1974) Auditory cortical neurons in the cat sensitive to the direction of sound source movement. Brain Res 73:455–471.

Stricanne B, Andersen RA, Mazzoni P (1996) Eye-centered, head-centered, and intermediate coding of remembered sound locations in area LIP. J Neurophysiol 75:1233–1241.

Thompson GC, Cortez AM (1983) The inability of squirrel monkeys to localize sound after unilateral ablation of auditory cortex. Behav Brain Res 8:211–216.

Thompson GC, Masterton RB (1978) Brainstem auditory pathway involved in reflexive head orientation to sound. J Neurophysiol 41:1183–1202.

Ungerleider LG, Mishkin M (1982) Two cortical visual systems. In: Ingle DJ, Goodale MA, Mansfields RJW (eds) Analysis of Visual Behaviour. Cambridge: MIT Press, pp. 549–586.

Vaadia E, Benson DA, Hienz RD, Goldstein JR, Moise H (1986) Unit study of monkey frontal cortex: Active localization of auditory and visual stimuli. J Neurophysiol 56:934–952.

Wortis SB, Pfeiffer AZ (1948) Unilateral auditory-spatial agnosia. J Nerv Ment Dis 108:181–186.

Xu L, Middlebrooks JC (2000) Individual differences in external-ear transfer functions of cats. J Acoust Soc Am 107:1451–1459.

Xu L, Furukawa S, Middlebrooks JC (1998) Sensitivity to sound-source elevation in non-tonotopic auditory cortex. J Neurophysiol 80:882–894.

Xu L, Furukawa S, Middlebrooks JC (1999) Auditory cortical responses in the cat to sounds that produce spatial illusions. Nature 399:688–691.

Young ED, Rice JJ, Tong SC (1996) Effects of pinna position on head-related transfer functions in the cat. J Acoust Soc Am 99:3064–3076.

Zattore RJ, Penhume VB (2001) Spatial localization after excision of human auditory cortex. J Neurosci 21:6321–6328.

Zattore RJ, Ptito A, Villemure J-G (1995) Preserved auditory spatial localization following cerebral hemispherectomy. Brain 118:879–889.

9
Feature Detection by the Auditory Cortex

ISRAEL NELKEN

1. Introduction

The view of neurons in sensory cortices as performing feature detection on the peripheral input gained acceptance to a large extent from studies of cortical processing of visual information. In this view of the visual system, information from simple feature detectors in the retina (photoreceptors and retinal ganglion cells) converges at the level of the primary cortex to give rise to more complex, orientation-tuned neurons. These, in turn, send their outputs to higher cortical areas where more complex features are detected. For example, in some parts of the Infero-Temporal cortex (IT) there are neurons that can be considered to be medium-level feature detectors: they are sensitive to features that are more complex than oriented bars but less complex than full visual objects (Fujita et al. 1992; Ito et al. 1995). These neurons presumably converge further up to produce object-sensitive neurons such as the face-sensitive cells (Perrett et al. 1982).

The information we have on the auditory system is partially consistent with such a view. For example, the auditory system starts with simple feature detectors, hair cells, which respond to acoustic events occurring within their tuning curve. Next, auditory nerve fibers transfer information to the cochlear nucleus, which corresponds, in terms of synaptic distance from the receptors, to the retinal ganglion cells of the retina. In the cochlear nucleus, information diverges to several parallel, multisynaptic ascending pathways that are reminiscent of the X, Y, and W cells of the cat retina, or to the M and P pathways in the monkey visual system.

Whereas visual information flows largely unchanged through the thalamus to the primary visual cortex (where the oriented line detectors are first found), auditory information is intensely processed in the brainstem and the midbrain. In terms of number of synapses, the distance between photoreceptors and primary visual cortex is the same as the distance between hair cells and the inferior colliculus. Thus, the information reaching the thalamocortical segment of the auditory system is, at least potentially, much farther along a hierarchical processing scheme than the visual information reaching the primary visual cortex. If this is indeed so, it should be expected that feature detection in the primary auditory cortex would reflect this added complexity.

Possibly due to these additional layers of processing and possibly because acoustic stimuli are so complex, the picture of signal processing in the primary auditory cortex is still rather murky. There is no accepted account of feature extraction analogous to the hierarchical scheme of the visual cortex, both within V1 and across visual fields. Instead, the field is very fragmented, with a substantial number of different approaches that investigate different facets of neuronal responses to sounds. The experimental paradigms chosen by different researchers depend on methodological decisions that are to a large extent a matter of personal preferences and beliefs, without direct experimental support. Thus, this chapter starts with a list of what I consider to be the principal methodological issues in the study of the auditory cortex. Then, one major dichotomy is used to group studies of the auditory cortex neurons into two sets: detailed studies of the responses of single neurons on the one hand, and mapping studies using a local population signal such as cluster activity on the other hand. The emphasis in this chapter is on the cat primary auditory cortex. The bat auditory cortex, which is also extensively studied, has been reviewed recently in a previous volume of this series (O'Neil 1995).

2. Methodological Issues in the Study of the Auditory Cortex

This section provides a brief overview of some of the methodological issues that have plagued studies of the auditory cortex. These issues are presented dichotomously; reality is probably more nuanced than indicated here.

2.1 Anesthetized versus Awake Animals

The influence of anesthesia on activity in the auditory system is great, even at subcortical levels. Under barbiturate anesthesia, principal neurons in the dorsal cochlear nucleus (DCN) have no spontaneous activity and are excited by pure tones at their best frequencies (BF). In unanesthetized, decerebrate animals, DCN principal cells have a high level of spontaneous activity, and many of them are inhibited at medium and high levels by pure tones at their best frequencies (Young and Brownell 1976). In the thalamus and cortex, pentobarbital and ketamine reduce activity relative to nitrous oxide alone, increase sharpness of tuning, and may modify temporal response patterns (Zurita et al. 1994). Under pentobarbital, neurons in the cortex tend to respond with a short burst of spikes at the onset of the stimulus; this presumably has led to the widespread use of very short stimuli in

studies of auditory cortex. In awake animals, neurons have more spontaneous activity and more sustained responses (Frostig et al. 1983; Shamma
and Symmes 1985; Ahissar et al. 1992). Under the lighter halothane anesthesia, properties of neurons are somewhat intermediate, with more sustained responses to pure tones (Nelken et al. 1999).

In spite of these well known effects, most studies of the auditory cortex
are performed in anesthetized animals. Although this poses a problem for
interpreting data, it is widely believed that feature detection and integration performed under anesthesia will also be performed in the awake
animal. It is presumed that, in the awake animal, more complex integrative
mechanisms may be expressed and that these mechanisms are superimposed on the simpler (but still not well understood) mechanisms apparent
in the anesthetized animals. Whether this view is correct requires experimental tests in awake animals. In practice, the use of anesthesia requires
comparisons between studies using different anesthetics to be done with
care.

2.2 Single Neurons versus Cluster Activity

Often, adjacent or closely spaced single neurons in the auditory cortex have
similar properties, reflecting the columnar organization of cortex (Abeles
and Goldstein 1970) and the local continuity of feature maps (Schreiner
1992, 1995; Eggermont 1996). Thus, it is possible to speak about the
response properties of a recording location that are common to many
neurons in the vicinity of a recording electrode and, presumably, to most
neurons in the same column. In fact, many papers report properties of
single neurons and nonseparated clusters of neurons together without
distinguishing between them.

That this may be problematic can be seen in Fig. 9.1, which shows the frequency response areas (FRAs) of two neurons recorded from the same
electrode. The spikes of these two neurons were of widely different shape.
In spite of being so close together, these two simultaneously recorded
neurons have FRAs of different shapes. The fact that they seem to be
complementary to one another suggests that there might be an inhibitory
interaction between the two. This pair of neurons was recorded in a
halothane-anesthetized cat. It is possible that the presence of such pairs,
which have not been described in the literature, has to do with the lighter
anesthetic state of the animal.

In a number of studies, Schreiner and his co-workers (Schreiner and
Sutter 1992; Sutter and Schreiner 1995) measured cluster responses and
single neurons from the same recording locations in barbiturate-
anesthetized cats. They showed that a given property of cluster activity, such
as a wide tuning curve, could result from a different combination of units
in different parts of cortical area A1. For example, clusters with wide bandwidths in dorsal A1 are often composed of broadly tuned individual

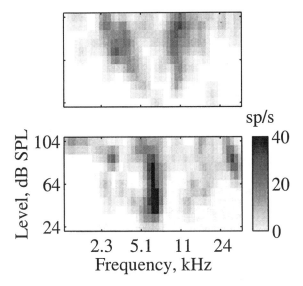

FIGURE 9.1. Frequency response areas of two well-separated neurons recorded simultaneously from the same electrode in the primary auditory cortex of a halothane-anesthetized cat. The two neurons vary in best frequency and the width of their tuning curve. The relative position of their tuning curves suggest that there are inhibitory interactions between the two.

neurons whereas clusters with wide bandwidths in ventral A1 may be composed of a number of narrowly tuned single neurons with somewhat different best frequencies.

Thus, spike separation reveals that neighboring neurons can have different response properties, and that homogeneity and heterogeneity exist simultaneously in the auditory cortex. Whether more emphasis is put on homogeneity, leading to the use of cluster activity as the measure of neuronal signaling, or on heterogeneity, emphasizing single neuron properties, is at the moment at the discretion of the experimenter. These different approaches can lead to different interpretations of experiments.

2.3 Maps versus Detailed Characterizations of Single Neurons

If the important facets of neuronal activity are encoded in the common, averaged response at the recording location, then we would expect that the responsiveness to features that are processed at the level of A1 will vary between cortical locations. In addition, there may be some gradients of feature sensitivity on the cortical surface. This reasoning has led to mapping studies in which the responses to a (usually) small set of sounds are studied using cluster activity in a large number of cortical locations. Alternatively, if the differences between single neurons are important, then single neurons must be studied to a much greater extent, and it is difficult to collect enough data to reveal the systematic representation of features as function of cortical location.

2.4 Simple versus Complex Sounds

The stimuli used to study the response properties of neurons in the primary visual cortex are almost invariably quite simple, such as stationary or drifting bars and gratings. These stimuli are nonetheless more complex than the stimuli thought to best activate lower stations of the visual system: mostly just dots of light. In the auditory cortex, a large number of studies have used pure tones or simple modifications of pure tones, such as amplitude-modulated (AM) tones, frequency-modulated (FM) tones, pairs of tones in some definite temporal arrangement, and so on. A minority of studies has used more complex sounds, such as multitone complexes (Nelken et al. 1994a,b,c; deCharms et al. 1998), ripple spectra (Schreiner and Calhoun 1994; Shamma and Versnel 1995; Shamma et al. 1995; Kowalski et al. 1996a,b; Calhoun and Schreiner 1998; Versnel and Shamma 1998), or natural sounds of various kinds (Smolders et al. 1979; Creutzfeldt et al. 1980; Glass and Wollberg 1983a,b; Olsen and Suga 1991; Wang et al. 1995).

In barbiturate-anesthetized animals, the activity evoked by the more complex stimuli consists mostly of short bursts of spikes immediately following the sound onset and therefore does not differ substantially from the activity evoked by simple pure tones or wide band noise bursts. This statement has to be qualified in that periodic stimuli, such as amplitude-modulated sounds or sinusoidally frequency-modulated sounds, often cause cortical neurons to be excited in a repetitive manner, increasing the evoked spike counts. However, the nature of each response phase is still qualitatively similar to the response that would be evoked by a simple stimulus with the corresponding parameters. Thus, in terms of their ability to evoke responses, simple and complex stimuli do not seem to be very different from one another. To find the differences between responses to simple and complex sounds, it therefore becomes necessary to go beyond counting spikes. In halothane-anesthetized cats, or in awake cats, responses are more tonic. As a result, the difference between responses to simple and complex sounds becomes much more apparent. Nevertheless, under halothane, it is still true that for many neurons the largest number of action potentials evoked by pure tones is comparable to that evoked by complex sounds.

The use of complex stimuli made it possible to discover phenomena with longer time scales than those described using the simple stimuli (Calford and Semple 1995; Brosch and Schreiner 1997; Nelken et al. 1999). The use of complex stimuli also made it possible to predict the responses of cortical neurons to stimuli outside the set measured directly. For example,

Shamma and his co-workers used ripple spectra to predict the responses to other broadband spectra (Shamma and Versnel 1995; Kowalski et al. 1996b; Versnel and Shamma 1998). Such predictions are better than predictions based only on the excitatory tuning curves of the neurons (Versnel and Shamma 1998). Similarly, Nelken et al. (1994c) showed that population responses to simultaneous combinations of four tones can be predicted better by the responses to pairs of tones than by the responses to single tones.

Thus, the utility of the use of complex sounds in studying the auditory cortex has become well established. Complex sounds make it possible to study the integrative mechanisms that are not engaged in responses to pure tones and to create predictive models for the responses of auditory neurons to complex sounds.

2.5 Specialized versus General-Purpose Processing

To what extent is the auditory cortex adapted for processing of sounds in the natural environment? Although it is commonly assumed that the auditory cortex is the result of an evolutionary process that optimized neural processing to the properties of natural sounds, different researchers interpret the implications of this assumption differently. The actual sounds found in natural environments, whether animal-made or not, form a very small part of all possible sounds. Thus, it seems unreasonable to assume that the auditory cortex is general-purpose machine for analyzing any possible input. As a result, the use of artificial sounds, whether simple or complex, is always problematic because such sounds may miss some crucial integrative mechanisms that are specific for the sounds against which auditory cortex was optimized by evolution.

However, the relevant subset of all possible sounds is not well defined at the moment for many mammals. Thus, species-specific vocalizations are often invoked as the "right" stimulus to use in studies of the auditory cortex. The enormous advances in understanding the bat echolocation system by using echolocation calls is an example of the successful use of this principle. In other animals, however, the results of studies using species-specific calls are interesting but are hard to generalize because the acoustic structure of species-specific vocalizations is so rich. Thus, this research direction does not seem to be as useful in the study of the cat or monkey auditory cortex as it was in the study of the bat echolocation system. In fact, the few studies published on responses of neurons in the bat auditory cortex to social communication calls (in contrast to echolocation calls, e.g., Ohlemiller et al. 1994) show a picture that is probably as complex as that found in other mammals.

Furthermore, it is unclear whether the auditory cortex is indeed specialized for processing species-specific vocalizations. Animals live in a far richer acoustic environment than just their own vocalizations. For example, species-specific vocalizations often appear in mixture with other sounds and not in an acoustically clean form. In addition, animals such as cats are very successful in environments that are very different from those in which they evolved (for example, cats are successful both in cities and on farms, two environments with highly divergent acoustic environments). Thus, it may be possible that the auditory cortex is performing other forms of high level auditory processing that is still specialized for the general statistics of natural sounds but is not specifically tailored for processing species-specific vocalizations (e.g., Nelken et al. 1999).

2.6 Spike Rate Codes vs. Spike Time Codes

In the lower stations of the auditory system, it is common to distinguish between place/rate codes and temporal codes of neural information. For example, low frequencies can be coded by the rate of firing relative to the other neurons in the tonotopic array (neurons whose BFs are close to the tone frequency would fire maximally) and by a time code (the tendency of neurons to fire at a fixed phase relative to the acoustic input). Both of these codes, however, reflect properties of the external input.

In the cortex, there is a different meaning for time code. It is possible for a neuron to generate different temporal response patterns that may be used to distinguish between different acoustic inputs, but, in this case, the temporal response patterns are internally generated and need not be directly related to the dynamics of the acoustic input (Middlebrooks et al. 1994, see also Chapter 8). A related feature of cortical neurons is the degree of synchronization in the times of action potentials of pairs of neurons driven by the acoustic input (Ahissar et al. 1992; deCharms and Merzenich 1996). In this case, again, the temporal response features of the neurons change internally as a function of the auditory input, and these changes cannot be directly attributed to the dynamics of the acoustic input.

It is still an open question whether differences in temporal response patterns or changes of neuronal synchronization are important for the coding of sounds in auditory cortex. Under barbiturate anesthesia, most of the response occurs as a short burst of spikes at the onset of sound, and the difference between temporal patterns and spike counts are not very large. Under other kinds of anesthetic agents, such as chloralose or halothane, temporal response patterns do seem to exist and provide a significant amount of information about stimuli.

2.7 Primary versus Secondary Auditory Fields

What auditory cortical areas are involved in feature detection? In the mustached bat (*Pteronotus parnellii*, Suga 1990) response properties in secondary auditory fields seem to be simpler to interpret than response properties in the primary auditory cortex. Taking this approach in the auditory cortices of nonspecialized mammals such as cats and monkeys is problematic, in that it is not clear what special parameters are coded in the higher cortical areas. The overwhelming majority of studies of the cat auditory cortex were performed in the primary auditory cortex. In the cat, the most intensively studied auditory fields beyond A1 are the anterior and posterior auditory fields AAF and PAF A2, the large ventral auditory field, is essentially uncharted. In the monkey, Rauschecker et al. (1995) started the task of charting secondary auditory fields, hinting that some rather sophisticated auditory processing may be performed in those fields.

One secondary, nontonotopically organized field in the cat cortex which may perform a special function is the auditory area in the posterior third of the anterior ectosylvian gyrus (AES), which is known as field AES (FAES, Meredith and Clemo 1989) or the anterior ectosylvian auditory area (AEA, in analogy to the partially overlapping visual field AEV, Rauschecker and Korte 1993). FAES has been implicated in spatial processing because of its strong projection to deep layers of the superior colliculus (Meredith and Clemo 1989). Neurons in FAES show spatial selectivity expressed as a temporal code (Middlebrooks et al. 1994) and as a rate code (Korte and Rauschecker 1993; Nelken et al. 1997a). These neurons are exquisitely sensitive to features of sounds that are spatially informative while being less sensitive to sound features that are independent of spatial direction (Nelken et al. 1997a). Thus, studying spatial sensitivity in FAES seems to give reasonably simple answers.

The case of FAES would suggest that in cat (and presumably in other mammals), it would be worthwhile to search for feature detection in nonprimary fields. In fact, it may be easier to understand a more specific class of signal processing operations first. In spite of this, the majority of studies of the auditory cortex are performed in the primary auditory cortex.

2.8 Exclusive versus Inclusive Criteria for Auditory Areas

The auditory cortex is best charted in the cat where it is usually considered to include the tonotopically organized fields A1, AAF and PAF, and the nontonotopically organized fields VP and A2. However, auditory responses have been reported over a much wider area of cortex. The FAES is a case in point: it was recognized as a separate auditory field less than 20 years ago. In addition, reports of auditory activity, some of it sharply tuned, exist for Somatonsensory field S2 (Toldi and Feher 1984), and association areas (Irvine and Huebner 1979). In the monkey, bimodal auditory-visual neurons are found in the LIP, which is involved in sensorymotor coordination (Grunewald et al. 1999; Linden et al. 1999). Similar auditory activity exists in the monkey frontal cortex (Vaadia 1989). Thus, auditory activity can be recorded in many cortical areas that are not considered part of auditory cortex.

2.9 Pure Auditory Responses versus Set-related Activity

Finally, there are indications that the auditory cortex in behaving animals codes not only the pure sensory signal, but also shows set-related activity: response components which change according to the behavior required from the animal. For example, Vaadia et al. (1982) showed that neurons in the auditory cortex of macaques responded differently to the same sounds depending on whether the sound instructed the monkey to move a lever to the right or to the left. Similarly, lesions of the supratemporal field that left A1 intact caused monkeys to fail in memory tasks while pure sensory discrimination remained intact, at least for the tasks used in that study (Colombo et al. 1990, 1996). Using functional magnetic resonance imaging (fMBI), Grady et al. (1997) showed that the MRI signal in A1 is modulated by attention, suggesting that neural activity is also different according to the amount of attention paid to auditory stimuli. Thus, the auditory cortex may not be purely sensory. The influence of such nonsensory factors will probably be an important field of research once studies in awake, behaving animals are routinely performed in auditory cortex.

3. Responses Properties of Single Neurons in the Auditory Cortex

3.1 Responses of Single Neurons to Pure Tones

The simplest sound used to study the auditory cortex is the pure tone. Pure tones have been used to define best frequencies and frequency response areas (FRA) of neurons and to characterize their binaural properties. Pure tones are defined by frequency, amplitude, and phase; however, since tones are almost invariably presented as tone bursts, an additional factor in their use is the shape of the onset and offset windows. Until recently, this has been considered a nuisance parameter, whose exact shape is unimportant. In the last few years, however, a number of studies (Phillips et al. 1995; Heil 1997a,b) have shown that on the contrary, the onset window is extremely important in shaping the responses of neurons in the the auditory cortex to pure tones.

The following properties are generally associated with for the responses of auditory cortical neurons to pure tones in anesthetized animals (Fig. 9.2): (1) Neurons respond to pure tones with short bursts of spikes. Thus, it usually makes more sense to talk about the probability of responding to any specific tone than about firing rate. (2) Neurons tend to respond to tones over a restricted frequency range (at least close to threshold). This property is usually called narrow tuning. (3) Neurons have a rather small dynamic range, often occupying less than 20 dB. (4) Most neurons are non-monotonic to some degree, showing a drop in their response probability at high levels. Usually, neurons are considered as nonmonotonic only if the reduction in the response probability at the highest sound level, relative to the maximum response probability, is sufficiently large. Values of 50% or 75% of the maximum are often used (Sutter and Schreiner 1995). Such a distinction is, however, to a large degree arbitrary, and by manipulating some seemingly nonessential parameters in the stimuli, neurons may pass from being monotonic to being nonmonotonic (Phillips 1988; Heil 1997b). Within these generalities, there is a large amount of variability (Eggermont 1996). For example, the sharpness of tuning covers a large range of values (Q10 ranging from less than 1 to over 10 in a narrow BF range, Schreiner and Mendelson 1990); minimum threshold can vary between −10 and over 30 dB SPL (Schreiner et al. 1992); and minimum latency can vary from less than 10 to over 30 ms (Mendelson et al. 1997).

The most important recent advance in understanding the responses of A1 neurons to pure tones was the realization that their response properties depend on the shape of the tone burst onset window. The first indications that this is indeed the case were published by Phillips (1988), who showed that changing the rise time of the linear onset ramp resulted in a change in the shape of the rate-level function of cortical neurons, with shorter rise times leading to more nonmonotonic responses. He interpreted this finding as evidence for lateral inhibitory input, which is activated by spectral splatter resulting from the shorter rise time.

The main breakthrough, however, came with the publication of studies by Heil and Irvine (1996) and Heil (1997a,b) which showed that the dependence of the response on the onset window follows simple rules (Fig. 9.3). For linear ramps, Heil showed that the latency from tone onset to the first spike is a universal function of the rate of rise with two free parameters that fix the location of this curve. Furthermore, the probability of spike occurrence is a function of the sound level at the time of first spike generation; the shape of this function may be specific for each neuron but is independent of the details of the onset window. Since the time of the first spike generation is a function of the rate of rise of the linear ramp, the spike count is approximately a function of the rate of rise. For cosine-shaped ramps, Heil showed that the latency is a universal function of the maximum acceleration (second derivative) of the ramp shape. In addition, the probability of spike occurrence is a function of the sound level at the time of first spike generation and is therefore approximately a function of the maximum acceleration of the ramp. Furthermore, there are indications that these results hold for off-BF tones too. Thus, it seems that neural processing leading to cortical spikes can be separated into two components: a frequency filtering stage followed by a circuit that is sensitive to the shape of the onset ramp. These results have been duplicated by Phillips in A1 and PAF (Phillips et al. 1995; Phillips 1998).

The rough features of these results can be explained by a model with a fixed threshold, where the spike is elicited at the moment that the onset

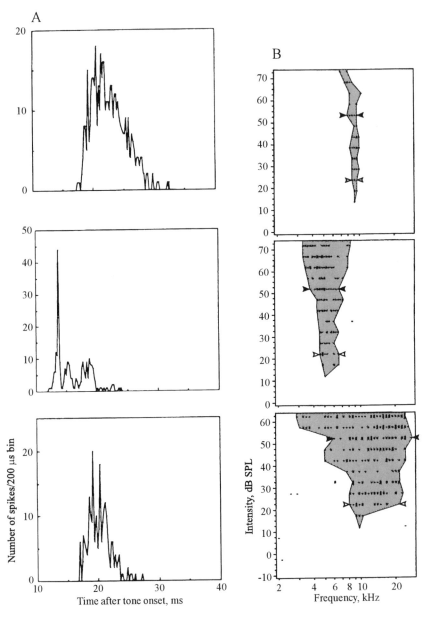

FIGURE 9.2. Responses of single neurons to pure tones in auditory cortex of barbiturate-anesthetized cats. A: PSTHs of three neurons. The responses occur as a short burst of spikes (adapted from Phillips and Sark (1991) with permission from Elsevier Science). B: Frequency response areas (FRAs) of three representative single neurons (adapted from Schreiner and Sutter 1992). C: Dynamic range and non-monotonic behavior to BF tones of representative neurons. (Adapted from Sutter and Schreiner 1995.)

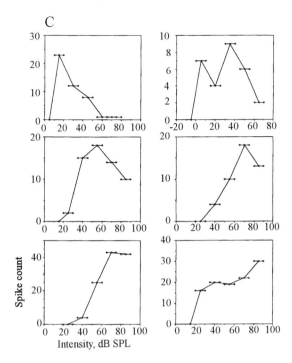

FIGURE 9.2. *Continued*

window crosses the threshold. However, the detailed structure of the results of Heil and Phillips argue against this explanation (Heil and Irvine 1996; Heil 1998). For linear ramps, for example, such a model would predict that onset latency is a linearly increasing function of ramp duration; however, experimentally measured values show a strong compression, with onset latencies for longer ramps being shorter than expected (Heil and Irvine 1996). Recently, Fishbachital (2001) noticed that all the properties of the latency could be explained by a preprocessing stage in which the log-compressed envelope is lowpass filtered by a neuronal membrane. When a constant threshold is imposed over the resulting waveform, Heil's latencies results are reproduced. Thus, the properties of the latency (but not the properties of the spike counts) are simple consequences of the neuronal circuitry of the auditory system and can indeed be demonstrated in the auditory nerve (Heil and Irvine 1997), the inferior colliculus (IC) (Phillips and Burkard 1999), and the medial geniculate body (MGB) (Fishbach et al. 2001).

The recent work reviewed above suggests that even when tested with pure tones, neurons in A1 cannot be considered as simple frequency filters. They behave much more like detectors of fast amplitude changes, provided these occur within their tuning curves.

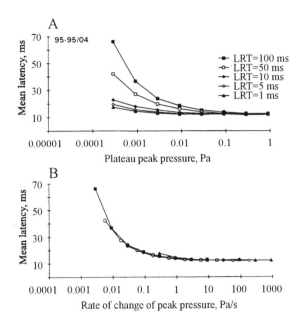

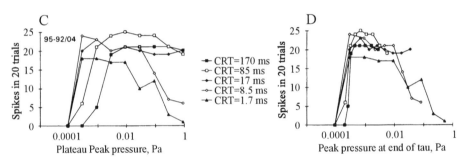

FIGURE 9.3. The effects of onset dynamics on responses to BF tones in the primary auditory cortex of barbiturate-anesthetized cats. In A and B, linear onset ramps have been used. A: First spike latency as a function of the peak tone level, showing that response latency is not an invariant function of peak level (LRT—linear rise time). B: The same data plotted as function of linear ramp velocity (adapted from Heil 1997a). In C and D, cosine-shaped onset ramps have been used. C: Spike count as function of peak tone level, showing that spike count is not an invariant function of peak level (CRT—cosine rise time). D: Spike count as a function of ramp level at the moment of first spike generation. Spike counts for this neuron are monotonic functions of peak level for rise times of 17 ms and slower, and nonmonotonic for faster rise times.

Binaural properties of auditory cortex neurons have also been studied in depth using BF tones. Neurons in A1 are often characterized as receiving excitation from the two ears (EE) or as excited by one ear (usually the contralateral) and inhibited by the other one (EI for contralateral excitation,

IE for ipsilateral excitation). Most neurons are excited by the contralateral ear, and are therefore either EE or EI, or, if they are unaffected by stimulation of the ipsilateral ear, they are classified as EO.

The exact definition of EE and EI neurons is not uniform across studies. One standard definition (Matsubara and Phillips 1988) is as follows: neurons that are excited by BF tones presented to the contralateral ear alone and to the ipsilateral ear alone are considered as EE; neurons that are excited only by contralateral presentation are then tested with a low-level BF tone contralaterally, varying the level of an ipsilateral BF tone. If the response to the contralateral tone is suppressed by the ipsilateral stimulation, the neuron is considered EI. Otherwise, it is considered as EO. This definition does not deal with interaction between simultaneous presentations of tones to the two ears in EE neurons. Such interaction can be facilitatory (response to the binaural stimulus is greater than to either monaural stimuli) or suppressory (response to the binaural stimulus is smaller than the larger of the two monaural stimuli) (Phillips and Irvine 1983). In the first studies of binaural interactions (e.g., Imig and Adrian 1977), the last category was considered as EI.

The problem in defining the two classes consistently shows that in A1, binaural interactions are complex and cannot be easily defined in terms of two classes. Early studies avoided these complexities by defining the binaural interaction using near-threshold BF tone bursts. However, this led to the obvious question about the relationships between binaural interactions at near-threshold levels and at suprathreshold levels. Characterizing neurons as EI, for example, presupposes that they "listen" to interaural level differences (ILD). Therefore, the responses of these neurones would be suppressed at sufficiently extreme ILDs, independent of the absolute level of the stimuli and their nature. Conversely, these neurons would respond at the same level for e.g. an ILD of 0 dB, independent of the absolute level of the sounds (provided they are suprathreshold).

That this is incorrect has been shown by a number of researchers. By using tone bursts with different frequencies in the two ears, Mendelson et al. (1992) demonstrated that the nature of the stimuli presented to the two ears may change the binaural interactions. Such results suggest that binaural interactions using broadband noise, for example, may be different from those elicited with BF tones.

Reale and Kettner (1986), using BF tones in the two ears, described a significant fraction of EE neurons as showing suppression once the ipsilateral tone level increased. They concluded that the binaural interactions changed as a function of average binaural level (ABL, the average of the levels in the two ears). The definitive study in this respect was performed by Semple and Kitzes (Semple and Kitzes 1993a,b). By varying independently the sound levels in the two ears, they showed that nearly all A1 neurons are influenced both by the ABL and the ILD. They concluded, therefore, that characterizing binaural interactions in terms of the EE and EI classes is not useful because the binaural sensitivity of cortical neurons is not a simpler function of ILD and ABL than a function of the absolute levels in the two ears. In particular, they described a significant class of neurons that were nonmonotonically sensitive to both ABL and ILD (Fig. 9.4). These neurons had focal sensitivity to sound level in the two ears; they were most highly activated by a specific combination of sound levels in the two ears. As a

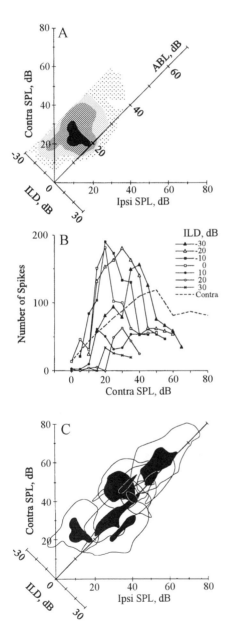

FIGURE 9.4. Responses of single neurons in the auditory cortex of barbiturate-anesthetized cats to BF tones with different levels in the two ears. A: responses plotted simultaneously as function of the level at the two ears (Cartesian axes) and as function of ABL and ILD (diagonal axes). Note that this neuron is maximally activated by a restricted range of levels in the two ears. B: The same data represented as spike counts as function of level in the contralateral ear, with the ILD as a parameter (equivalent to cuts along the ABL diagonal in A). The responses to tones presented to the contralateral ear alone are plotted as a dashed line. The addition of an ipsilateral tone can have mixed effects: at low ILDs, it causes facilitation of the responses, at high level it causes suppression. C: The focus of activation for ten selected neurons. These foci cover a large range of level combinations in the two ears. (Adapted from Semple and Kitzes 1993b.)

population, their foci of activation covered a large number of combinations of sound levels. Thus, a population of such neurons could represent both overall level and interaural level differences in a highly sensitive manner.

3.2 From Simple to Complex Stimuli, or on the Importance (or Lack Thereof) of the Tuning Curve

Characterizing a single cortical neuron by its best frequency, or even by its full FRA, presupposes that its responses to other sounds will depend only (or to a significant level) on its pure tone responses. This assumption holds, to a large degree, in the auditory nerve and in some neurons of the cochlear nuclei. However, even at the level of the cochlear nuclei, there are important classes of neurons whose responses to pure tones and to broadband stimuli differ dramatically (Nelken and Young 1994). Thus, whether tuning curves are relevant in predicting cortical responses to more complex stimuli is an important open question. There are several indications that, indeed, tuning curves and FRAs are not good tools for predicting the responses of cortical neurons to complex sounds.

The tuning curve or even the full FRA measured by pure tones is an insufficient characterization of auditory cortex neurons in two respects. First, spectral integration is not predicted well by the tuning curve. Second, the tuning curve does not describe important temporal factors which may strongly influence the responses of cortical neurons to complex, time-varying sounds.

Considering first spectral factors, Versnel and Shamma (1998) studied the representation of vowels in the activity of neurons in ferret auditory cortex. They predicted the relative size of the responses to the vowels either by the spectral level of the vowels at the BF of the neuron, or by using a broadband characterization of the spectral integration properties of the neuron called the response field (calculated by using ripple spectra, see below). By using broadband stimuli to characterize the neuron, it was possible to identify inhibitory sidebands, which are hard or impossible to measure using pure tones. They showed that although a reasonable prediction of neural responses by the stimulus sound level at BF was possible, the predictions improved significantly by considering the full response field.

Ehret and Schreiner (1997), who looked for correlates of the psychophysical phenomenon of critical bands in the auditory cortex, found further evidence for the importance of inhibitory sidebands in the processing of wideband stimuli. Critical bands are a measure of the bandwidth of the basic frequency channels of the auditory system. Psychophysically, they can be measured in a number of ways, including measures of tone threshold in wideband noise and changes in tone thresholds as a function of the width of narrowband maskers. Ehret and Schreiner measured BF tone thresholds in broadband and narrowband maskers in order to estimate the neuronal critical bands, and also measured the full FRAs of the same neurons. Measures of frequency bandwidth based on the tuning curve correlated poorly with measures of bandwidth based on masking in narrowband or broadband noise. Thus, the integrative mechanisms that sum noise energy for masking are not represented well by the neuronal tuning curve.

In a more direct way, Bar Yosef et al. (in preparation) used FRAs in the cat auditory cortex to predict the responses to natural sounds and some modifications of the same natural sounds. Predictions were made by counting the number of bins inside the neuronal tuning curve in the frequency-level plane that were crossed by the sound power spectrum. Although some neurons showed a positive correlation between this prediction and the actual responses, these correlations were weak (with the large majority below 0.5), and in many cases they were negative. In fact, weak background noise often elicited responses that were as large as those elicited by tonal components well inside the tuning curve.

The second reason for which tuning curves and frequency response areas are an insufficient characterization of neuronal responses in the auditory cortex is the fact that they do not describe the dynamic response properties of cortical neurons. Cortical neurons tend to be sluggish relative to their counterparts in the lower auditory centers. For example, whereas auditory nerve fibers can follow amplitude modulation of their input up to rates of hundreds of Hz, and even neurons of the inferior colliculus can follow amplitude modulation up to tens of Hz (and in some cases even significantly higher), neurons in the auditory cortex often fail to follow AM at rates higher than 10 Hz (Schreiner and Urbas 1988), although in awake animals, best AM rates may be higher.

It is still unclear how best to characterize the temporal response properties of cortical neurons. To measure spectrotemporal characteristics of neurons in the auditory cortex, deCharms et al. (1998) used a version of the reverse correlation method (Eggermont 1993), using random tonal complexes as stimuli. deCharms and colleagues found that many neurons had a clear excitatory patch at short delays before spikes and around the best frequency of the neuron. However, in addition to this "classical" patch, they found in many cases a number of excitatory and inhibitory patches surrounding it. They interpreted some of these findings as evidence for auditory edge detectors for temporal and spectral edges, although the evidence for this interpretation is weak. An alternative way for measuring the spectrotemporal characteristics of neurons is to use moving ripple stimuli, as described below. Interpretation aside, these results again emphasize the need for characterizing the neuron beyond its tuning curve.

Brosch and Schreiner (1997) and Calford and Semple (1995) studied yet another manifestation of temporal processing by neurons in auditory cortex, using delayed two-tone paradigms similar to forward masking. The two groups independently found that previous tones suppressed the responses to a later BF tone over a large temporal and frequency extent (Fig. 9.5). This suppression occurred even at frequencies at which the first tone did not elicit responses at all. Thus, long-lasting inhibitory inputs strongly affect the responses of cortical neurons; such effects are not captured by the tuning curve.

3.3 Artificial Complex Sounds

Upon finding that the tuning curve gives a limited characterization of the neuronal responses to complex sounds, it became necessary to use more complex stimuli in order to engage and study those more complicated integrative mechanisms. A variety of families of artificial sounds have been used in such attempts.

3.3.1 Multiple Tone Complexes

The low spontaneous firing rate of cortical neurons makes it difficult to locate their inhibitory sidebands. A common procedure used to find those sidebands is to measure the suppression of a response to a low-level BF tone by a family of tones that sweep over a wide frequency range. Such manipulations have been used, for example, by Sutter and Schreiner (1991) and very extensively by Shamma and his co-workers (1993). Figure 9.6 shows results for neuronal clusters in the ferret auditory cortex. Here, the BF tone was delayed slightly relative to the other tone. As a result, it is possible to observe both the frequency dependence of the neuronal responses and the suppression of the responses to the BF tone. When the earlier tone was within the neuron tuning curve, the late BF tone did not elicit any responses. This suppression of the responses to the late BF tone also extended slightly beyond the boundaries of the tuning curve. The suppression of the responses to the BF tone by off-BF tones is interpreted by Shamma as evidence for the occurrence of inhibitory sidebands.

However, there are a number of reasons why a neuron might not respond to the delayed BF tone, not all of them having to do with inhibition. For example, Heil's results (Heil 1997a,b) may be used to give an alternative explanation (Fishbach, per com). Consider the situation where the first, swept tone is at BF too. In this case, the addition of the delayed BF tone would result in a step increase in the level of the stimulus. However, the step increase is small: if the two tones are of the same level, the step increase would be at most 6 dB. Since the onset ramp of the delayed tone is of constant duration, the result is a very shallow slope of the amplitude step, which may not activate the neuron. If the first tone is far off-BF, the presumed amplitude step might be larger but over a range of frequencies the delayed BF tone will still create slopes which are too shallow to evoke activity. An alternative explanation for the lack of response is synaptic depression: once the synapses driving the neuron have been depleted of transmitter by the

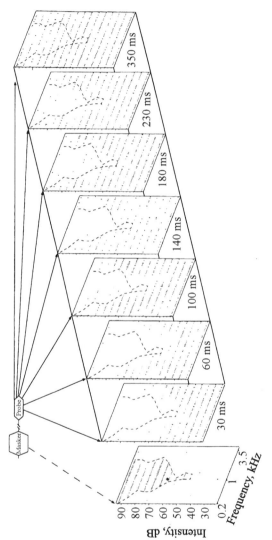

FIGURE 9.5. Forward masking in the auditory cortex of barbiturate-anesthetized cats. The masker varies in frequency and level whereas the probe is a constant level BF tone. The responses to the masker alone are presented in the leftmost panel. The probe was presented at various times following the masker. The responses to the probe are strongly suppressed both within and outside the excitatory tuning curve of the neuron following masker offset, and this suppression lasts for over 250 ms. (From Brosch and Schreiner 1997.)

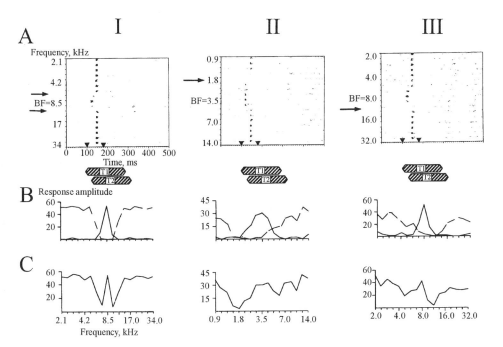

FIGURE 9.6. Suppressive sidebands of three neuronal clusters measured by using a delayed two-tone suppression in the auditory cortex of barbiturate-anesthetized ferrets. A: The raster plots of the responses. Below the raster plots, a schematic representation of the stimulus time course is shown. The earlier responses are produced by the first tone and are generated only when it is inside the excitatory tuning curve of the clusters. The late responses are produced by the second tone, which is always at BF. They are suppressed in a range of frequencies inside and flanking the excitatory region. B: Spike counts of the responses to the first tone (continuous lines) and the second tone (dashed lines). C: The total spike counts evoked by the two-tone stimulus, showing symmetric suppressory sidebands (in I), a strong below-BF suppression (in II), and a strong above-BF suppression (in III). (From Shamma et al. 1993.)

first stimulus, the delayed tone is incapable of reaching threshold. Indeed, some thalamocortical synapses in brain slice preparations have been found to be depressing (Markram et al. 1997).

To support the claim that suppressive sidebands are indeed the effect of inhibition, Shamma and his co-workers showed that there is a correlation between the position of the major sideband relative to BF and the directional preference for FM chirps. This correlation is weak, however (Kowalski et al. 1995), and is sensitive to the exact shape of the frequency trajectory of the sweep (Nelken and Versnel 2000).

3.3.2 FM Sensitivity of Single Neurons

FM sweeps have been used quite extensively to study neurons in auditory cortex. Most studies in nonspecialized mammals use a large-extent, single direction sweep going from low frequencies to high frequencies or vice versa. A minority of studies (e.g., Gaese and Ostwald 1995) used also sinusoidally modulated FM sounds. FM sweeps are also among the few complex sounds which have been studied outside A1 (in AAF, Tian and Rauschecker 1994, 1998; in PAF, Heil and Irvine 1998).

Large-extent sweeps are usually defined in terms of their velocity and their direction. However, there are many different ways of generating FM sweeps. For example, frequency may change linearly with time (linear FM) or exponentially with time so that equal frequency intervals are swept in equal time intervals (exponential FM). These are considered "nuisance parameters" that should not affect properties such as directional selectivity. There is, however, evidence that the details of the frequency trajectory influence dramatically the directional selectivity of cortical neurons (Nelken and Versnel 2000).

Some general features of the responses to FM sweeps are consistently reported (Heil et al. 1992a; Mendelson et al. 1993; Tian and Rauschecker 1994; Ricketts et al. 1998; Nelken and Versnel 2000). In deeply anesthetized animals, the responses to large-extent sweeps occur usually as short bursts of spikes when the instantaneous frequency is close to BF (Fig. 9.7). Using multiunit responses, Heil et al. (1992a) and Nelken and Versnel (2000) showed that the response occurs as the sweep reaches a trigger frequency that is below BF for upward sweeps and above BF for downward sweeps and is independent of sweep velocity. In recordings from single neurons, there is some controversy about this, with the results in PAF being interpreted differently by Heil and Irvine (1998) and by Tian and Rauschecker (1998); these differences may however result from different analyses, rather than from real differences in the data. All researchers agree that neurons are activated once the sweep reaches the approximate border of their tuning curves. In that respect, the responses to large-extent sweeps can be inferred from the responses to pure tones.

These responses depend, however, on the velocity of the sweep and on its direction. Directional selectivity (DS) is commonly measured by using the contrast between the responses to upward and downward sweeps:

$$DS = (R_{up} - R_{down})/(R_{up} + R_{down})$$

where R_{up} and R_{down} are the firing rates for upward and downward sweeps, respectively. In some papers, the DS is defined with the opposite sign (Shamma et al. 1993; Nelken and Versnel 2000). In some studies, strong directional selectivity is considered to consist of DS > 0.15 (Mendelson et al. 1993), or DS > 0.33 (Heil et al. 1992a). In the first case, the preferred direction evokes just 1.35 more spikes than the opposite direction while in

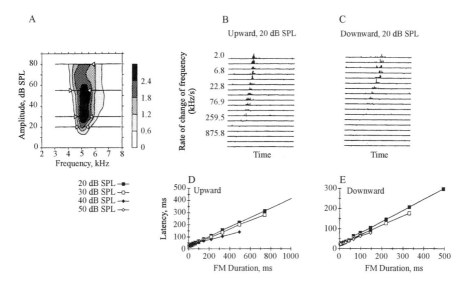

FIGURE 9.7. Responses to large-extent FM sweeps in the posterior field of the auditory cortex of barbiturate-anesthetized cats. A: Frequency response area; the arrows point to the triggering frequencies of FM sweeps at different levels. B: PSTHs of the responses to FM sweeps at 20dB SPL, as a function of the sweep velocity. Response latency increases as sweep velocity decreases, because it takes a longer time for the sweep to reach the triggering frequency. C: Plots of response latency as a function of FM sweep duration. When the triggering frequency is independent of sweep velocity, these plots form straight lines (which is the case here). The triggering frequency can be extracted from the slope of these plots. (Adapted from Heil and Irvine 1998.)

the second case, the preferred direction evokes twice as many spikes. Although these differences may be significant, they are not very large. Thus, directional selectivity is a matter of degree rather than of kind of responses.

Similar results were found regarding velocity preferences. There too, neurons tend to respond to many FM velocities, but there is often a velocity to which the neurons are more responsive than to others. In the cat A1, neurons often prefer very high velocities. In the ferret A1, preferences to various velocities are more equally distributed. In cat PAF, sensitivity to slow modulation rates is more common (Heil and Irvine 1998; Tian and Rauschecker 1998).

In conclusion, in spite of the importance of FM sweeps in animal communication, and in spite of their appealing analogy to bars or moving bars in visual cortex, FM sweeps do not seem to elicit responses dramatically different from pure tones. Furthermore, selectivity to FM parameters, although it exists, is weak.

3.3.3 Ripple Analysis

Recently, a new family of stimuli called spectral ripples has been used by two groups (Schreiner and Calhoun 1994; Shamma et al. 1995; Calhoun and Schreiner 1998). These stimuli are wideband with a spectral envelope that is sinusoidal on a logarithmic frequency scale. Many variations of these stimuli have been reported, such as spectrotemporal ripples (Kowalski et al. 1996a) and sums of spectrotemporal ripples (TORCS, see Klein et al. 2000).

The basic idea behind the use of ripple stimuli is that they are dual to pure tones in the same sense that sine waves (in time) are dual to temporal impulses (clicks): the spectral envelope of a pure tone (a delta function) can be generated as a sum of the spectral envelopes of ripple stimuli, or alternatively, the spectral envelopes of the ripple stimuli are the Fourier components of the spectral envelope of a pure tone. Thus, if auditory neurons process spectral envelopes linearly, any statement about pure tone responses can be immediately transformed into a statement about the responses of a neuron to ripple spectra, and vice versa. Since the system is not exactly linear, however, the duality does not exactly hold. Thus, in order to study responses of neurons to broadband stimuli, it makes sense to characterize the cells by using broadband stimuli, and for this purpose, the ripple spectra are natural candidates.

In practice, ripple spectra are defined by their spatial frequency (in units of cycles/octave) and their phase; they include a large number of nuisance parameters, such as the nature of the carrier (in most studies, logarithmically spaced sine waves are used), the depth of the modulation, the absolute level of the ripple, and so on. With the appropriate choice of these parameters, the response to the ripple becomes a function of its spatial frequency and phase. For each spatial frequency, the response is periodic in the phase of the ripple, and often it can be approximated by a clipped sine wave (Fig. 9.8). Thus, for each spatial frequency, it is possible to define a gain factor and a phase shift. Plotting these as a function of spatial frequency gives rise to the ripple transfer function. By using an inverse Fourier transform, the ripple transfer function can be used to calculate a frequency response function, which in many cases is rather similar to response functions measured directly with pure tones.

Thus, in cases such as those shown in Fig. 9.8, the duality of pure tones and ripple spectra holds reasonably well, and the system is rather linear. This is achieved by setting the overall level low enough that responses do not saturate, and by using an appropriate modulation depth (Calhoun and Schreiner 1998).

Spectrotemporal ripples are spectral ripples in which the modulating envelope also changes sinusoidally in time. The responses to these ripples, after a short transient, are periodic in time. It is therefore possible to extract not only a ripple transfer function but also a temporal transfer function and a temporal impulse response. Using a characterization based on downward-moving ripples, Kowalski et al. (1996b) showed that they could predict the responses to stimuli composed of sums of downward-moving ripples of varying temporal and spatial frequencies.

It remains to be seen whether this methodology will be able to give a general characterization of auditory neurons, independent of the nuisance parameters. Evidence presented below suggests (apparently contrary to the results presented here) that auditory neurons are in fact highly nonlinear, at least when they process sounds containing both narrowband and broadband components. This discrepancy will be discussed below.

3.4 Single Neurons and Species-specific Vocalizations

At the other extreme from the artificial sounds described above, a number of studies approached feature detection by testing neurons with natural sounds, most often by species-specific vocalizations. These studies are somewhat sparse, and their results, although highly intriguing, require duplication and controls.

Coding of species-specific vocalizations in the auditory cortex of squirrel monkeys have been studied by Wollberg and his co-workers over a period of 20 years (Wollberg and Newman 1972; Newman and Wollberg 1973a,b; Glass and Wollberg 1983a; Pelleg-Toiba and Wollberg 1991). They tested neurons with a standard set of vocalizations and some variants of those, pure tones, time reversed vocalizations (llacs), and sequences of vocalizations . Their first findings pointed to a set of neurons with an exquisite selectivity to a small number of vocalizations. This finding was interpreted as evidence for responses that are not generated by simple integration of acoustic features (Wollberg and Newman 1972). However, later studies raised a series of problems with this interpretation. For example, calls and llacs gave rise to a similar level of responses (Pelleg-Toiba and Wollberg 1991). In the final publication of the series, Pelleg-Toiba and Wollberg (1991) used a larger number of controls and more sophisticated data analysis and concluded that most neurons in auditory cortex are sensitive to a set of acoustic features and that the natural calls are in fact coded by a distributed array of such elementary feature detectors.

Wang et al. (1995) also tested natural calls in marmosets. They used "twitter calls," which consist of a succession of extremely fast, wideband transients at a rate of about 10 Hz. They manipulated these calls by expanding or compressing them in time and by reversing them. They described a class of neurons that are best activated by the natural parameter of the calls and are activated less by reversed calls or time-expanded and time-reversed calls (Fig. 9.9). They concluded that these results are indications for specializations for processing of natural calls, but that there are no species-specific call detectors. However, these initial results require more controls. For example, in the spectrograms presented in Wang et al. (1995), each

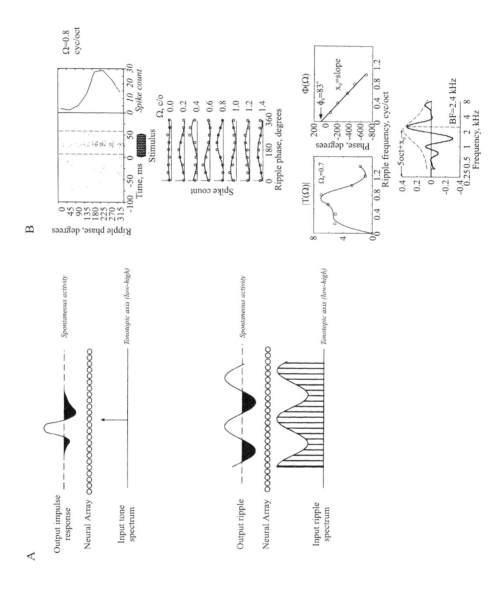

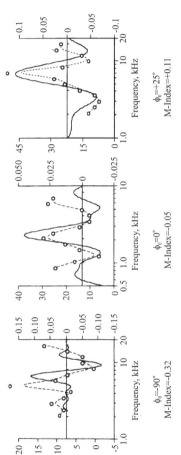

FIGURE 9.8. Theory and practice of ripple spectra. A: Schematic responses of neurons along the tonotopic axis to a pure tone (upper plot) and to a sinusoidal spectral pattern (lower plot). The responses to a pure tones are analogous to the impulse response of a linear filter, with the time axis replaced by the logarithmic tonotopic array. The plot of firing rate as a function of tone frequency is therefore interpreted as the "impulse response" of a system analyzing spectral profiles (upper plot). If this system were linear, It would be expected that a sinusoidal spectral pattern would cause a sinusoidal response pattern along the same axis (lower plot). For linear systems, the two representations are dual to each other in the sense that the amplitude and phase of the sinusoidal response patterns evoked by the ripple stimuli should be equal to the amplitude and phase of the Fourier coefficient of the single tone responses at the same spatial frequency. [(Adapted from Shamma et al. 1995, Copyright OPA (Overseas Publishers Association) N.V., with permission from Gordon and Breach Publishers)]. B: Measurements of a ripple transfer function. The raster plot presents the responses of a neuron in the auditory cortex of a ferret to ripple spectra of frequency 0.8cyc/oct and with different phases. Below, spike counts are plotted as function of ripple phase for eight different ripple frequencies. These plots are, by necessity, periodic with the ripple period, and they can often be fitted by one cycle of a sine wave. The amplitude and phase of the sine wave at each ripple frequency are collected to form the ripple transfer function, |T(W)| and Φ(W). By using an inverse Fourier transform, the expected responses to pure tones can be computed, (the response field) (from Versnel and Shamma 1998). C: Response fields (continuous lines) and measured two-tone responses (dashed lines with symbols) for three neurons in the ferret primary auditory cortex. [Adapted from Shamma et al. 1995, Copyright OPA (Overseas Publishers Association) N.V., with permission from Gordon and Breach Publishers.] Note the close similarity between the two, supporting the hypothesis that the processing of spectral envelopes by auditory cortical neurons is to a large extent linear.

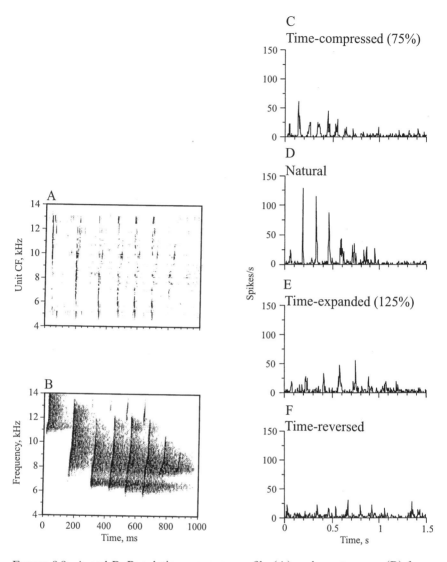

FIGURE 9.9. A and B: Population response profile (A) and spectrogram (B) for a twitter call of a marmoset. Only neurons which significantly phase locked to the call temporal envelope are included here. Note that the neuronal population gives a rough spectrotemporal representation of the call. C–F: Compound PSTHs for the class of neurons which were excited most strongly by the natural call relative to the manipulated versions. The compound PSTH to the natural call itself is shown in D; it is largest because of the way the participating neurons have been selected. For this class of neurons, responses are reduced for time-compressed (C), time-expanded (E) and reversed (F) calls. (Adapted from Wang et al. 1995.)

transient in the twitter call was accompanied by echoes, which decay slowly. When reversing such sounds, the sharp onset, slow decay structure of the sound is transformed into a slow onset, fast decay structure. Thus, the preference for calls relative to reverse calls may result for nonspecific factors such as those studied by Heil (1997b). Such mechanisms, although they would generate sensitivity to the twitter call, are not specific for the communication calls and could play a role in processing a much larger family of sounds. Thus, there is only negative evidence today for the presence of special detectors for species-specific vocalizations in the primary auditory cortex of monkeys.

One family of mammals has well-known specific detectors for its own vocalizations: the bats. Bat echolocation is one of the best understood auditory submodalities. In terms of feature selectivity in the auditory cortex, the main specialization of the bat auditory system for echolocation seems to be the presence of combination-sensitive neurons: neurons that are sensitive to a combination of a feature in the call and a feature of the echo (e.g., in the mustached bat, *Pteronotus parnellii*, Suga 1990). For example, FM-FM neurons are sensitive to the time delay between an FM component in the call of the bat and an FM component of the echo. Such neurons are believed to be involved in measuring the distance between the bat and the source of the echoes. In the mustached bat, which was studied by Suga and his co-workers over many years, these neurons are arranged in regular maps in various secondary auditory fields. In another bat, the big brown bat (*Eptesicus fuscus*), on the other hand, there is no regular map of delay tuning, although delay-tuned neurons exist and their tuning covers the relevant echo delay range for the animal (Dear et al. 1993).

Echolocation, however, is different from the auditory modality discussed here in that it is geared to measure specific physical parameters of the call-echo pair, in contrast with the less well-defined computational goals of what is usually considered as hearing. The mustached bat has also a large repertoire of communication calls (Kanwal et al. 1994). Interestingly, it seems that when tested with communication calls, the auditory cortex of bats appears to be as complex as the auditory cortices of other mammals. For example, Ohlemiller et al. (1994) reported that neurons, which were classified as FM-FM when tested with echolocation calls, were also sensitive to FM combinations in communication calls, but with different frequency and delay specificities. They also found neurons that were tuned to intersyllable intervals (Ohlemiller et al. 1996); such neurons resemble the neurons in the marmoset's cortex that are specific to the natural rate of twitter calls (Wang et al. 1995). Finally, Esser et al. (1997) found neurons that were highly selective to specific sequences of pairs of different syllables. They interpreted these findings as indications for the processing of syntax in the bat auditory cortex. Such neurons, obviously, cannot be considered as specific detectors of species-specific vocalizations. Rather, they are implicated in some high level parsing of the acoustic input.

3.5 Single Neurons, Signals in Noise, and Auditory Scene Analysis

In nature, species-specific calls are rarely encountered in isolation. In most natural conditions, calls will appear on the background of other calls and nonanimal made sounds. Thus, even before the auditory system can address the issue of the meaning of the vocalizations, it must be able to separate them from their background. The separation of a relevant signal from noise is part of a more general problem which has been called auditory scene analysis (ASA) (Bregman 1990). Auditory scene analysis in its full complexity has not been addressed in electrophysiological research. Here only the special case of signals in noise will be addressed.

In a series of papers, Phillips and his co-workers described the way in which white noise affects the responses of cortical neurons to BF tones. They described two phenomena that underlie much of the research in this field. When the background noise is continuous, the main finding is that the neuronal response shifts to higher level by essentially one dB for each increase of one dB in the noise level (Phillips et al. 1985). This exact compensation is significant, because at the level of the auditory nerve and the cochlear nucleus, rate-level functions shift less and are degraded more in the presence of continuous white noise than in the cortex (for example by loss of dynamic range, see Costalupes et al. 1984; Gibson et al. 1985). Thus, the degradation in the peripheral representation of tones caused by the addition of continuous noise is partially corrected at the level of the auditory cortex.

On the other hand, when the noise and the tone are gated together, Phillips and Cynader (1985) described a phenomenon they called "strong-signal capture." In strong-signal capture, the neuron responds to the combination in the same way it would respond to the single stimulus, which, by itself, would give rise to the stronger response. An example of strong signal capture is shown in Fig. 9.10 (left column). In this figure, the PSTHs of the responses of a neuron to a BF tone in white noise are presented as functions of time and of tone level. At low tone level, the response is essentially the response to the noise alone. Once the tone is strong enough, the response it generates occurs earlier than the noise response. At this point, the neuron shows a response that occurs at the same latency with essentially the same strength as the tone response alone.

White noise, however, is uncommon in natural scenes. Most natural background sounds are much more structured than white noise. For example, Nelken et al. (1999) have shown that many background sounds can be well approximated as colored, gaussian noise that is multiplied by a temporal envelope. This is interesting because such modulated noise bands have an important place in human auditory psychophysics. When pure tones are masked by narrowband gaussian noise, the masked threshold increases with bandwidth up to a critical bandwidth. However, when the noise bands are

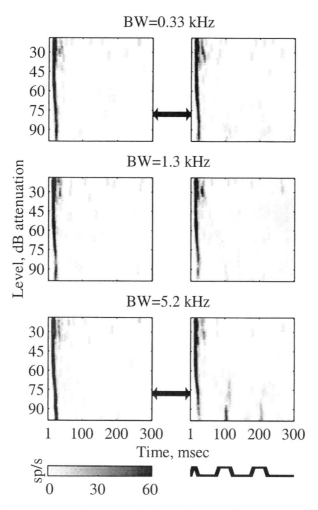

FIGURE 9.10. Responses of a neuron in the primary auditory cortex of halothane-anesthetized cat to tones in noise. All the stimuli consisted of a fixed-level masker to which BF tones of varying levels were added. Each panel shows the responses to tones in the presence of one type of masker. Maskers were either unmodulated noise bands (left column), or trapezoidally-modulated noise bands with the same bandwidth (right column, the modulating waveform is schematically shown on the bottom). Each row corresponds to a different bandwidth of the masker. The responses are presented as functions of time after stimulus onset (abscissa) and tone level (ordinate). Thus, each horizontal cut through a response plane is the PSTH of the responses of the neuron to a combination of a masker with a tone at one specific level; these PSTHs are stacked to form the response plane. For unmodulated noise, the responses show strong signal capture as described by Phillips and Cynader (1985): at low tone levels, the response is identical to the response to the noise alone. At higher tone levels, the response is identical to the response to the tone alone. On the other hand, modulated maskers (right column) elicit additional response components which are phase locked to the temporal envelope, provided the bandwidth of the noise band is large enough. Adding a tone suppresses this envelope-following response, increasing the contrast between the responses to the modulated noise alone and to the modulated noise with added low-level tones.

multiplied by a slowly fluctuating temporal envelope, thresholds actually decrease with increases in bandwidth at the appropriate range of bandwidths (Hall et al. 1984; Schooneveldt and Moore 1989). Thus, an increase in bandwidth, which introduces more noise energy into the auditory system, actually results in decrease in thresholds. This phenomenon is called comodulation masking release (CMR). The results of Nelken et al. (1999) show, therefore, that natural background sounds have the right structure to elicit CMR. It is appealing to speculate that CMR is a specialization of the auditory system for detecting weak sounds over natural backgrounds. Klump and Langemann (1995) have shown that birds have CMR, suggesting that the phenomenon might be general to higher vertebrates.

Nelken et al. (1999, 2001) tested CMR electrophysiologically in the auditory cortex of halothane-anesthetized cats (Fig. 9.10). They found a class of neurons that show neural correlates to CMR. The CMR-like behavior is caused by the conjunction of two phenomena. First, neurons tend to follow the slow amplitude fluctuations of the maskers, and they do so better when the masker bandwidth is wider. Second, this envelope following is weak and is suppressed by the addition of BF tones. Thus, when comparing the responses to tone in noise and the responses to the noise alone, there are distinct changes in the temporal pattern of the responses and in the total number of spikes evoked after the masker onset with and without tones.

An interesting facet of this result is the fact that when the tone suppresses the phase locking to the temporal envelope of the masker, the temporal pattern of the responses becomes similar to the responses to the tone alone although the tone level may be weak relative to the masker level. Thus, the cortical neurons seem to respond to the acoustically weak component in the mixture.

A similar phenomenon was found by Bar Yosef et al. (in preparation) when testing the responses in the auditory cortex of halothane-anesthetized cats to bird calls recorded in nature. They digitally separated the bird call from its noise background, which included both echoes and unrelated acoustic components (Fig. 9.11). Then, they presented to neurons the natural sound, the clean bird call by itself, and the background noise by itself. In many cases, the responses to the natural sound by itself had a similar temporal pattern to that of the response to the noise presented alone, although the noise component was significantly weaker than the cleaned bird call. In this case again, it seems that neurons in the auditory cortex responded to a mixture in the same way that they would have responded to the acoustically weak component when it was presented alone.

The differences between these results and the strong-signal capture reported by Phillips and Cynader (1985) probably have to do both with the differences in stimuli and in the differences in the anesthetic state of the animals. When white noise was used as the masker, strong-signal capture

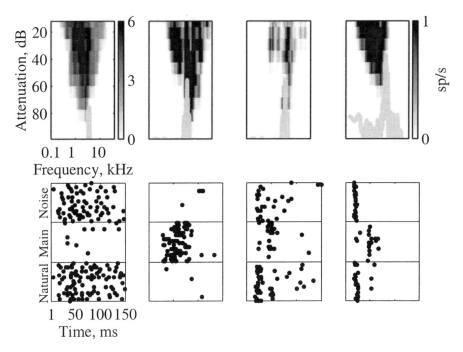

FIGURE 9.11. Responses of four neurons in the auditory cortex of halothane-anesthetized cats to natural sounds consisting of bird calls in noise. The upper row shows the frequency response areas of these neurons, with the power spectrum of the natural sound superimposed in light gray. The raster plots present the responses to the full natural sound (Natural), to the bird call after background noise was removed (Main), and to the background noise alone (Noise). For these neurons, the responses to the original call is essentially identical to the response to the background noise alone, in spite of the fact that their FRAs contain the frequency range of the call.

was present in the auditory cortex of halothane-anesthetized cats, as shown in Fig. 9.10, in the same way that it existed in the auditory cortex of barbiturate-anesthetized cats used by Phillips and Cynader. However, the richer temporal response structure that appears in the lighter, halothane anesthesia makes it possible to use a richer set of stimuli, and to observe a richer set of phenomena as found by Nelken et al. (1999) and Bar Yosef et al. (in preparation).

These two studies suggest that auditory cortical neurons are able to separate foreground and background sounds, and that, in a mixture, they would be able to respond to the weaker component. Thus, it is tempting to speculate that the auditory cortex actually performs auditory scene analysis, and that, in fact, it performs the harder part of this task in the sense that cortical neurons respond to the weaker components in a mixture.

4. Feature Maps in Auditory Cortex: Maps of Cluster Activity

This section will survey the information in mapping studies gathered by using cluster activity. As discussed above, the use of cluster activity is usually accompanied by the following assumptions: (1) that the cortex is locally homogeneous so that nearby neurons share most or all of their important characteristics, (2) that the average activity of groups of neurons has functional significance, and (3) that, since the stimuli used in these experiments are often rather simple, knowing the responses of neurons to simple sounds is at least relevant, if not fully sufficient, for predicting their responses to more complex stimuli.

4.1 Maps of Features of Simple Sounds

The basic principle of organization in the auditory system is the tonotopic order. The presence of tonotopic order in the primary auditory cortex of mammals is well established. Its details have been best studied in cats. Historically, there was a controversy about the amount of order and disorder in the tonotopic map of the primary auditory cortex. Goldstein and Abeles (Goldstein et al. 1970; Goldstein and Abeles 1975) claimed that the tonotopic order, while certainly there, is only a rough indication for the frequency preference of neurons at each location in the auditory cortex. However, this conclusion was based partially on pooled data from many cats, which was based on anatomical landmarks. Merzenich et al. (1975) then presented detailed maps from the cortices of single cats and showed that anatomical landmarks are unreliable for the position of the tonotopic map. In the maps published by Merzenich et al., tonotopic order is very precise along the anterior-posterior axis.

Part of the discrepancy between these two studies results from the pooling procedure used by Abeles and Goldstein. However, it seems today that Merzenich et al. may have gone too far in the opposite direction. In barbiturate-anesthetized cats, the tonotopic map is most precise and easiest to demonstrate in a central strip of A1. Dorsally, single neurons tend to be more widely tuned, making it more difficult to determine their BF whereas ventrally, approaching the border of A2, neurons with different BFs can coexist in the same column (Schreiner and Sutter 1992). Furthermore, isofrequency lines are only roughly oriented along the medio-lateral axis and tend to be oriented somewhat diagonally (Reale and Imig 1980; Matsubara and Phillips 1988), which contrasts with the maps of Merzenich et al. (Merzenich et al. 1975).

Another part of the discrepancy between the conclusions drawn by Abeles and Goldstein and those of Merzenich et al. may have to do with the fact that, whereas Abeles and Goldstein used awake, muscle-relaxed cats, Merzenich et al. used barbiturate anesthesia. Under halothane anesthesia, neurons simultaneously recorded on the same electrode can have quite different BFs (e.g., Fig. 9.1). In spite of these reservations, it is clear that the tonotopic map is the clearest and strongest map in the primary auditory cortex.

Various other parameters related to the pure tone responses are reported to have nonrandom distribution on the cortical surface. In the center of A1, there is a region in which neurons have narrow tuning curves (Schreiner and Mendelson 1990; Heil et al. 1992b; Schreiner 1995). Ventrally and dorsally, neuronal clusters show a tendency for broader tuning (although narrow tuning curves are still found in these regions, Fig. 9.12A). This increase in the bandwidth is accompanied with a tendency to respond better to a broadband transient (Schreiner and Mendelson 1990). The increase in bandwidth of tuning curves of clusters is caused by different mechanisms in ventral and in dorsal A1 (Schreiner and Sutter 1992).

Cluster activity also shows a nonrandom distribution of response thresholds: there is often a central strip of low-threshold neurons surrounded by regions with a tendency to have higher thresholds. Here too, there is a substantial amount of randomness within this tendency. For example, in a set of data for single penetrations (Schreiner et al. 1992) the region with low thresholds (−5 to −10 dB SPL) also has penetrations with some substantially higher thresholds (up to 20 dB SPL, Fig. 9.12B). The spatial distribution of thresholds is consistently correlated with the distribution of best response levels (the level at which the neuronal cluster responded at maximum), which is therefore also nonrandom.

The nonmonotonicity of the cluster responses at BF also show a similar pattern, with a central strip of strongly nonmonotonic neurons flanked by regions of lesser nonmonotonicity. There is, possibly, another region of nonmonotonic cluster responses more dorsally in A1 (Schreiner et al. 1992; Sutter and Schreiner 1995). Finally, response latencies seem also to be shortest at a strip in central A1, and to increase ventrally and dorsally (Mendelson et al. 1997).

These tendencies are correlated with one another to some extent, although rather weakly. The central strip with narrow tuning curves also contains (in some cats) the low-threshold clusters and the nonmonotonic clusters (Heil et al. 1992b; Schreiner et al. 1992). Thus, these three strips are sometimes, but not always, overlapping. Furthermore, the penetration-by-penetration correlations of these parameters are mostly weak and inconsistent across animals (Schreiner et al. 1992; Mendelson et al. 1997). Thus, it seems that these parameters are mostly distributed in a combinatorial fashion across the cortical surface, with any single penetration combining an almost random set of properties.

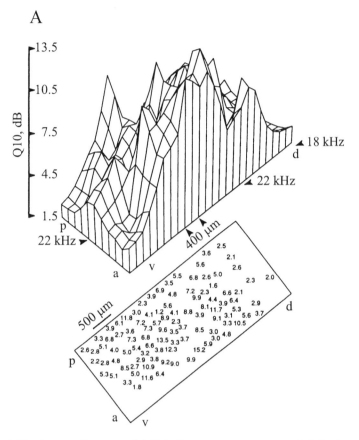

FIGURE 9.12. Parametric maps of Q10 (A) and response threshold (B) of neuronal clusters in the primary auditory cortex of barbiturate-anesthetized cats. There is a tendency for more sharply tuned neurons to be in the middle of A1 (A), as shown by the smoothed map; however, in the same region there are also neurons with wider tuning curves, as can be seen in the raw data presented below. Similarly, dorsally and ventrally to this strip, Q10 values tend to be smaller, although there are rather sharply tuned clusters in those areas too. The same general phenomena can be seen for the threshold map (B). There is a central strip where clusters tend to have low thresholds, but there are also clusters with higher thresholds in the same area (A, Adapted from Schreiner and Mendelson 1990; B, Adapted from Schreiner et al. 1992 with permission from Springer-Verlag).

As a consequence of the wide distribution of thresholds, nonmonotonicity, and best response levels, the representation of suprathreshold tones may be rather complicated on the cortical surface. Phillips et al. (1994) studied this question by constructing distributed response maps for single neurons in mapping experiments (Fig. 9.13). They found that for low-level tones,

B

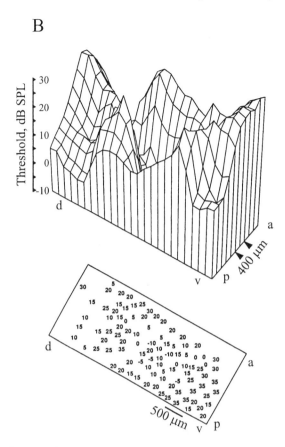

FIGURE 9.12. *Continued*

most responses occurred on the isofrequency contour corresponding to the stimulus frequency. However, only the low-threshold parts of the isofrequency contour were activated. At higher levels, low-threshold, nonmonotonic neurons stopped responding. On the other hand, high-threshold and monotonic neurons on the isofrequency contour, and neurons outside the isofrequency contour, started responding to the same tones. As a result, the cortical regions which were maximally activated at suprathreshold levels had very little resemblance to the threshold isofrequency contours. Similar results were described also by Heil et al. (1994) and by Schreiner (1998).

In this sense, the incorrect BF maps published by Abeles and Goldstein in 1970 are closer to reality than the correct near-threshold BF maps of later studies. For suprathreshold stimulation, the near-threshold tonotopic

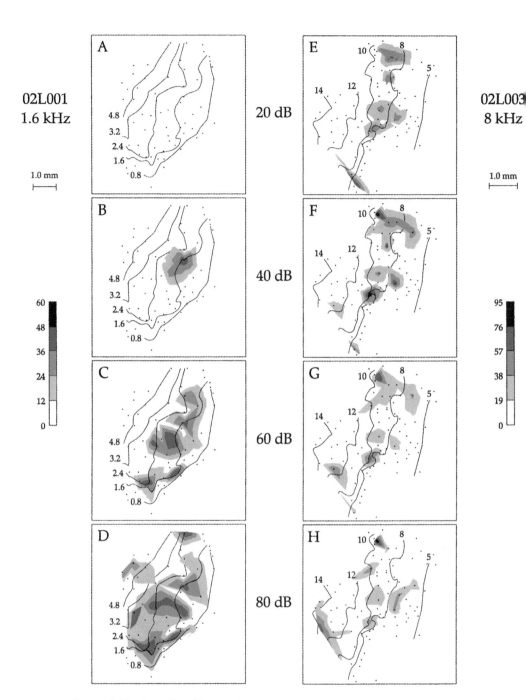

FIGURE 9.13. Areas in which single neurons respond to a tone of the specified frequency, as function of level, in the auditory cortex of barbiturate-anesthetized cats. A–D: tone frequency is 1.6 kHz. E-H, tone frequency is 8 kHz. At low levels, response patches occur mostly on the isofrequency line corresponding to the stimulation frequency. At high levels, the response patches occur over a wide cortical territory, without any obvious correlation to the isofrequency line. (Adapted from Phillips et al. 1994, with permission from Springer-Verlag.)

map of A1 is of very limited usefulness, and responses occur in a complex network of patches both on and outside the isofrequency contour.

All the maps mentioned above are usually measured by using sounds presented to the dominant (almost always the contralateral) ear. However, most neurons in A1 are influenced by the two ears. Two main interaction classes for sounds presented to the two ears have been defined. Most neurons are defined as either EE or EI, where the first epithet describes responses to the contralateral ear and the second to the ipsilateral ear. The problems in defining these categories have been discussed above.

Imig and Adrian (1977) demonstrated the existence of columns of binaural interaction in A1, and then Middlebrooks et al. (1980) presented evidence that A1 has binaural interaction bands which run parallel to the tonotopic axis. These findings led to the assumption that the binaural interaction bands are a direct analog of the visual ocular dominance bands in V1.

Later studies again refined the picture and made it less neat. All studies agree that binaural properties tend to cluster in A1. However, anatomical work (Imig and Brugge 1978) and mapping studies (Reale and Kettner 1986; Matsubara and Phillips 1988) did not find that the binaural interaction classes were organized consistently in bands. Rather, the general picture is that there are about twice as many EE responses than EI responses on the average, and that recording locations with EE responses form a background in which clusters of EI and other, less common response types (IE, EO) are interspersed. In some cats, the EI clusters merge to form continuous bands, but in other cats, they do not.

4.2 Maps Based on Complex Sounds

4.2.1 Two-tone Complexes

Two-tone complexes have been used in the auditory cortex to study the organization of sideband inhibition. Shamma et al. (1993) used the delayed two-tone paradigm described above to characterize the inhibitory sidebands of cortical neurons. The asymmetry of the inhibitory sidebands was quantified by the contrast between the number of spikes evoked above BF and those evoked below BF. They mapped the primary auditory cortex of ferrets by using these sounds and showed that the asymmetry index tends to be small in absolute value in the center of the ferret A1; it tends to become large and positive on one side of this strip and large and negative on the other side.

In the cat, there is only one study of the same phenomenon (Rotman et al. 2001). They used an objective criterion for the presence of clustering in the sideband asymmetry to conclude that, at the resolution of their map, no such clustering could be found. Whether the differences between ferrets and cats are real or are caused by the use of somewhat different methodologies remains to be seen.

4.2.2 Frequency-modulated Sweeps

FM sweeps have been used to map the auditory cortex by a number of authors (in cats, Heil et al. 1992b; Mendelson et al. 1993; in ferrets, Shamma et al. 1993; Nelken and Versnel 2000). Most studies used large sweeps, but different groups chose different nuisance parameters, such as linear or exponential frequency trajectory.

In general, the responses of neuronal clusters to FM sweeps in mapping experiments are similar to the responses of single neurons as described above. The response usually occurs as a short burst of spikes when the frequency trajectory approaches the BF. There are effects of velocity and direction, typically with a weak selectivity to some range of values in both. What is of interest here is whether this selectivity forms a coherent parametric map. The results are highly variable between animals and between laboratories, and the sources of these differences are only partially understood.

The two animals in which the most work has been done are the cat and the ferret. In cats, Mendelson et al. (1993), using exponential FM sweeps, reported the existence of maps of directional selectivity, roughly orthogonal to the isofrequency bands. They did not perform objective tests for non-randomness of the distribution of directional selectivity. Their results are largely reported as smoothed maps. Smoothing in displaying maps is problematic because with enough smoothing, which is equivalent to lowpass filtering of the two-dimensional map, slowly varying features are certain to appear. In the case of the study of Mendelson et al. (1993), inspection of the raw data also shows some apparent order in the distribution of the data. However, there is substantial variability around the reported selectivity gradients. Thus, in a region in which the smoothed map predicts selectivity for upward FM sweeps, there are penetrations with strong downward preference. It is therefore more conservative to conclude that there may be a patchy organization of FM directional selectivity, with statistically more probable selectivity to upward or downward sweeps in some parts of the cortex.

Heil et al. (1992b) also measured maps in cat auditory cortex, but by using linear FM sweeps. They mapped a single isofrequency contour with a large number of closely spaced penetrations. With all the penetrations along a single (somewhat winding) line, they could apply run tests to check whether there is any clustering of directional selectivity. The runs (sets of nearby penetrations with the same directional selectivity) were not longer than expected by a random assignment of directional selectivity, leading Heil and his coauthors to cautiously conclude that there is no map of directional selectivity, although they discuss possible reasons why their experiment could miss such organization had it existed. Thus, the results in the cat are inconsistent.

In the ferret, Shamma et al. (1993) reported that directional selectivity is mapped orthogonal to the isofrequency lines. In their hands, clusters in the center of A1 respond equally to the two FM directions, but on one side of this strip clusters prefer upward FM sweeps, and on the other side they tend to prefer downward FM sweeps. Beyond these central bands, directional selectivity could change sign or reverse the progression, but this usually occurred in the margins of A1, where auditory activity was weaker.

The FM sweeps used by Shamma et al. (1993) were different from those used by Mendelson et al. (1993) and Heil et al. (1992b). The sweeps started with an interval of a fixed frequency at the lowest frequency used. They then swept to the high frequency, remained there for a while, and then swept back to the low frequency. This choice of FM stimuli could be the source of some of the discrepancies with the results from cat studies. Thus, Nelken and Versnel (2000) studied a large number of FM paradigms in the ferret, including linear and exponential sweeps, sweeps structured like those of Shamma et al. (1993), and upward and downward sweeps separated in time, as used by Mendelson et al. (1993) and Heil et al. (1992b). The results of the study demonstrated that directional selectivity depended strongly on the exact way that the FM sweep was generated. Maps of directional selectivity of linear FM sweeps were different from maps of directional selectivity of exponential FM sweeps, and maps for the continuous-tone paradigm as used by Shamma et al. (1993) were different from maps for separate up and down sweeps. In some, but not all, of the animals, clusters of nearby penetrations with similar directional selectivity could be found using an objective test.

The current picture is that directional selectivity in A1 depends strongly on the paradigm used to measure it. It is, therefore, highly sensitive to the nuisance parameters. As a result, it is difficult to attribute any special place in the population coding of sounds in A1 to the FM directional selectivity.

The picture is completely different for velocity preference. All studies agree that preference to FM velocity is robustly clustered in A1, both in cats and in ferrets, and this clustering is independent of the details of the frequency trajectory in the different studies. The previously discussed study by Nelken and Versnel showed that velocity preferences for linear and exponential FM sweeps are correlated. Maps of velocity preference are also more robust and similar to each other in different paradigms than maps of directional selectivity. Thus, FM velocity preference seems a more basic and robust feature detected and processed by A1 population activity than directional selectivity.

The studies reviewed above in the cat and the ferret, combined with studies in the rat, also show some differences that may be interpreted as real species differences. For example, studies in A1 of cats tend to find more clusters that are sensitive to downward sweeps than to upward sweeps. In the ferret, such bias is absent in some studies (Kowalski et al. 1995) and is actually reversed in others (Nelken and Versnel 2000). In the rat, there seems to be a preference for upward sweeps, as in the ferret. For velocity preferences, cat A1 and AAF seem to prefer very fast FM sweeps (Heil et al. 1992a; Mendelson et al. 1993; Tian and Rauschecker 1994); sweeps as fast as those are very rare in nature. In PAF, neurons prefer slower FM sweeps (Heil and Irvine 1998; Tian and Rauschecker 1998), which leads to the suggestion that PAF is more suitable for processing natural communication sounds than neurons in A1 of AAF (Tian and Rauschecker 1998). This suggestion is also consistent with the close ties of the posterior fields with limbic structures such as the amygdala (Reale and Imig 1983) and the cingulate gyrus (Rouiller et al. 1990). In the ferret A1, velocity preferences are more evenly distributed (Nelken and Versnel 2000) but again, neurons in rat auditory cortex seem to prefer faster sweeps (Ricketts et al. 1998). These differences may be related to differences in the important acoustic events that must be processed by each of the three species, but it is difficult to imagine what differences in nature could give rise to such different adaptations in the brains of the three animals.

4.2.3 Pitch and the Missing Fundamental

Feature maps tend to be measured by using simple sounds. There have been few efforts to measure maps using sounds more complicated than those discussed above. One exception is the percept of pitch. Pitch is an interesting phenomenon, in that it is a common feature of many sounds with wildly different spectral structure. Determining the pitch of a stimulus therefore requires high level processing.

Periodic sounds elicit the sensation of pitch, and their pitch is essentially equal to their fundamental frequency. Such sounds can be expressed as a Fourier series, that is, as a sum of sine waves at the fundamental frequency and its harmonics, where the phase and level of each component may vary. An important constraint on theories of pitch formation is the phenomenon of the missing fundamental: the fact that the fundamental of the harmonic series may be removed under many conditions without changing its pitch. Thus, to elicit a certain pitch, sounds do not need spectral energy at the pitch frequency.

The basic auditory organization is tonotopic: it is based on the frequency content of sounds. Thus, pitch is not explicitly represented, for example, at the level of the auditory nerve (although there are features of auditory nerve activity from which pitch can be extracted, e.g., Cariani and Delgutte 1996a,b). One question that has been raised repetitively is whether pitch is represented explicitly in higher brain centers. For example, it may be that the cortical tonotopic map is really a map of pitch, which appears as tonotopic map when tested with pure tones. Alternatively, it may be that a pitch map exists in A1, orthogonal to or in general independent of the tonotopic map.

One way of testing whether a neuron that is sensitive to a certain frequency really codes for pitch at that frequency is to test it with sounds that have this pitch but from which the fundamental (or several low harmonics) has been removed. Such tests have been published in the behaving monkey (Schwarz and Tomlinson 1990). None of the tested neurons responded to pitch; all of them responded only to the actual energy present within their integration bandwidth, which could include both excitatory and inhibitory subfields. Similar negative results, using MUA recordings and field potentials, have been also reported (Fishman et al. 1998; Steinschneider et al. 1998; Fishman et al. 2000).

In contrast with these negative results, there are two reports of magnetoencephalogram (MEG) recordings in humans wherein periodic stimuli with missing fundamentals were used. Pantev et al. (1989) reported that the localization of the source of the magnetic field was similar for a pure tone at 250 Hz and a harmonic complex whose missing fundamental was 250 Hz. In contrast, a tone at 1600 Hz had a source in a different location. They produced a similar result by using stimuli that required binaural fusion to create the right pitch (Pantev et al. 1996). They concluded that the tonotopic map in A1 is in reality a pitch map. This conclusion is, however, somewhat premature, given the small number of stimuli used in this study and some problems in its design. For example, Pantev et al. (1989) used a narrow noise band at the fundamental frequency in order to mask cochlear combination tones. However, the noise approximate location of the fundamental in A1, giving rise to a mistaken localization of the activity in response to the harmonic complex.

Using MEG, Langner et al. (1997) employed a much larger number of stimuli in an attempt to measure both the tonotopic gradient and, if possible, the pitch gradient. They found that, indeed, pure tones of different frequencies are mapped in a regular way in the human A1, and that complex sounds with different pitches are mapped in a regular way. The two axes were, however, essentially orthogonal to each other. Thus, Langner et al. concluded that timbre (as measured by the spectral content) and pitch (as measured by the period of the sound) are processed and mapped independently of each other in the auditory cortex. The results of all MEG studies are still limited by methodological considerations (e.g., the assumption of a single equivalent dipole, see Lutkenhoner et al. 2001), and therefore require support using more direct measurements at the single neuron level.

4.3 Feature Maps and (Other) Complex Sounds

The crucial test of feature maps is whether, when presented with a complex sound, the activity in the map represents the actual feature in the stimulus. In the auditory cortex, as discussed above, feature maps are superimposed on each other. It is therefore expected that the activity evoked by a stimulus would peak at a location corresponding to the intersection of all features present in the stimulus, with possible weaker activity in other parts of the map.

Some obvious candidates for feature maps actually fail this test rather dramatically. The tonotopic map is a case in point. As discussed above, at different tone levels, different patches of cortex become active, both on and off the isofrequency contours, depending on the stimulus frequency and level (Phillips et al. 1994). Thus, the tonotopic map fails to represent the frequency of high-level tones.

So, do the parametric maps make sense? In limited settings, the answer seems to be positive. For example, in their study of the width of frequency response areas, Schreiner and Mendelson (1990) actually performed a generalization test by using a broadband transient to test their neuronal clusters. They showed that neurons with wider bandwidth respond, on the average, more strongly to the broadband transient. Thus, in this limited sense, by measuring FRA bandwidth, they did measure a parameter which is related to the bandwidth over which a cluster integrates its spectral input.

Shamma et al. (1993) showed some correspondence between maps of sideband asymmetry in the ferret auditory cortex and maps of FM direction selectivity. The underlying idea is that neurons with a strong inhibitory sideband below BF would respond better to downward FM sweeps, which hit their excitatory tuning curve first and only later hit the inhibitory sidebands, than to upward FM sweeps, which hit the inhibitory sideband before reaching the excitatory region. However, correlations between sideband asymmetry and FM directional selectivity of the same clusters are weak. Nelken and Versnel (2000) found that this correlation was significant only in the special FM paradigm used by Shamma et al. (1993). In the other paradigms they tested, directional selectivity was not correlated significantly with sideband asymmetry. Shamma et al. (1993) also showed that sideband asymmetry was correlated with strength of responses to noisebands positioned just above BF or just below BF. These correlations show that sideband asymmetry as measured by Shamma indeed has some of the expected consequences; however, the weakness of the correlations show that at best, the way sideband asymmetry was measured is only a partial characterization of these mechanisms.

Rotman et al. (2001) measured responses to six short natural sounds in the auditory cortex of cats under barbiturate anesthesia and plotted the neuronal responses as a function of the cluster's BF. This resulted in a clear population profile resembling the power spectra of the stimuli, even at rather high presentation levels. This representation is permissive in the sense that some clusters whose BF is close to the spectral peak of a stimulus responded strongly to that stimulus but other clusters with similar BF responded rather weakly; clusters whose BF lies far from the spectral peak mostly responded rather weakly to the same sounds. These results are similar to those of Phillips et al. (1994) in their single tone representation study. The results of Rotman et al. suggest that in spite of the problems with the representation of pure tones, natural sounds with "natural" spectral peaks (in terms of their width and signal to noise ratio) are represented reasonably well by the cortical population. Thus, this study validates to some extent the expected meaning of the tonotopic map, at least for the class of sounds used there. It should be remembered, however, that this representation is a rather coarse one and loses details both spectrally and temporally.

The general pattern of the results presented here indicates that the parametric feature maps which are currently measured in A1 are not very useful for the purpose of representing general sounds. This conclusion is consistent with, and reinforces, the conclusions on the place of frequency response areas in the characterization of single neurons.

5. Synthesis and Speculations

5.1 Are There Feature Detectors in the Auditory Cortex?

There are two, almost opposite, views of feature detectors. On the one hand, a perfect feature detector neuron may be defined as a neuron whose response to a sound depends only on the presence of the specific feature it codes. A perfect feature detector would respond only when the feature it detects is present in the sound, and its response level may then depend on the amount, level, or strength of the feature in the sound. We intuitively expect that a feature detector will be activated rather sparsely, only when its specific feature is actually present in the sound. Thus, we expect that feature detectors will be highly nonlinear in the sense that if the feature is further subdivided, the detectors will respond weakly or not at all to each of the components, but will respond strongly when the whole feature is present. For example, a species-specific call detector would respond to the presence of that call, but not to any of its acoustical components when presented in isolation. Furthermore, we expect that such a call detector would be insensitive to differences between variations of the same call, but would differentiate between the specific call and other, maybe acoustically similar, calls.

A pure feature detector of this kind has not been described, but examples of neurons that may approximate feature detectors exist in the research literature. Perhaps the best examples in the auditory cortex are the combination-sensitive neurons in the bat auditory cortex. These neurons respond to specific combinations of the echolocation call and its echo, without responding much, if at all, to either the call or echo by themselves. Delay-tuned neurons respond only to specific time intervals between the call and its echo (Suga 1990); such neurons are probably used for detecting the distance between the bat and the source of the echo. CF/CF neurons are activated by a combination of the first harmonic of the bat call and a specific doppler-shifted higher harmonic of the echo; such neurons are probably used for detecting relative velocity between the bat and the source of the echo.

However, even these almost perfect feature detectors turn out to be more complicated than that. When tested with species-specific communication sounds, these neurons seem to respond to different sets of features. Thus, these neurons are not pure feature detectors.

The alternative view of feature detectors is that of linear filter functions. Under some circumstances, the responses of auditory cortical neurons to complex sounds can be explained linearly (or almost linearly) by their responses to simpler sounds (Nelken et al. 1994b,c; Shamma and Versnel 1995; Kowalski et al. 1996b; Versnel and Shamma 1998), making it possible to describe these neurons as approximate linear systems. Neurons such as these would respond to a large variety of sounds, essentially measuring the similarity of the stimulus to their filter function. Thus, the shape of the filter function may be considered as the feature detected by such neurons.

So, is the auditory cortex really an array of nearly linear filter functions? The weight of the evidence presented above is that it is not. It seems that for some families of sounds, such as sounds of similar bandwidth, similar overall level, and similar temporal structure, responses depend linearly on a suitably chosen stimulus representation. This is the case for ripple spectra, for example. However, when the nuisance parameters of the ripple spectra are manipulated, as in the study of Calhoun and Schreiner (1998), it appears that the responses may depend strongly on factors other than just the filter function. Therefore, the linear filter functions are "local": they can be used for sounds which are near the sounds used in their characterization, but the farther one moves away from those sounds, the worse the predictive ability of the linear approximation becomes. Similar conclusions hold even in a subcortical structure (Nelken et al. 1997b; Nelken and Young 1997).

Thus, it seems that at the current level of knowledge, it is hard to discuss the auditory cortex in terms of feature detectors. Rather, neurons in the auditory cortices of cats, ferrets and monkeys are promiscuous: they respond to a large number of different features. In some cases, their sensitivity to some classes of sounds can be explained by their responses to simpler sounds. In other cases, they may not be explained more simply. To understand this neuronal promiscuity, we need an increased understanding of the fundamental processing tasks subserved by the auditory cortex.

5.2 Single Neurons Versus Cluster Activity Revisited

Studies of maps gave rise to some important insights into the organizational principles of the auditory cortex. For example, they gave rise to the hypothesis that features other than low-level BF are mapped in rather disordered way, with clear patches in which clusters have similar response properties. The size and organization of these patches may have different spatial scales for different parameters (Schreiner 1995, 1998). They also gave rise to a functional parcellation of A1 into dorsal and ventral parts, with different functional properties both at the level of clusters and at the level of single neurons (Schreiner et al. 1992; Schreiner and Calhoun 1994; Sutter and Schreiner 1995; Mendelson et al. 1997). On the other hand, having realized that order and disorder both play a role in feature maps of auditory cortex, we need some objective means for testing randomness and order. A start in this direction was done by Heil et al. (1992b) and by Nelken and Versnel (2000), but their methods require more testing and refinements.

The main problem with maps, as they are currently measured, is the fact that most are measured with simple sounds. It is unclear whether these maps have any relevance to the processing of real-world signals. To understand the issues involved better, it would be required on the one hand to validate feature maps by testing them with a rich collection of test sounds, to see whether the map is activated as expected by such sounds. On the other hand, studies of single neurons may be able to come up with new features, for which it would be possible to measure maps.

New technologies make it easier to perform mapping studies. In particular, advances in the use of optical imaging, both of intrinsic signals and with dyes, will increase the speed with which such maps can be acquired. With such technologies, it will be possible to measure activation by a larger number of stimuli. Best-frequency maps using intrinsic signals have already been demonstrated in the cat (using electrical stimulation of the cochlea, Dinse et al. 1997a,b), chinchilla (Harrison et al. 1998), and guinea pig and rat (Bakin et al. 1996). Recordings with voltage-sensitive dyes have been also performed (Horikawa et al. 1996; Hosokawa et al. 1997; Taniguchi et al. 1997). In addition, mapping studies are the only studies that can be performed in humans by using imaging techniques such as fMRI and position emission tomography (PET).

Do single neurons teach us more about auditory processing then just measuring functional maps? There are again two issues here: one is whether responses of single neurons show phenomena that cannot be explained by their location in feature maps, and the other one is whether these differences have functional significance. Given the research discussed in section 2, the answer to the first question is an obvious yes. The second question is, however, harder to answer since it depends on the readout of the neuronal activity in the "next stage" of the auditory system.

5.3 Who Reads the Auditory Cortex?

It is probable that at least some of the complexity found at the single neuron level (and often not tested at the cluster level) is averaged out when cluster activity is recorded. Does this averaging reduce noise, or does it erase the crucial signal? This question, of course, has to be answered by the next stage: those brain structures that are affected by the neuronal activity in A1. The next stage in this case can be both higher auditory areas and subcortical targets of the auditory system, including the strong feedback connections to the MGB and the IC.

There is very little information about the way inputs from the auditory cortex influence its targets. In the bat, Suga and his co-workers (Zhang and Suga 1997; Zhang et al. 1997; Gao and Suga 1998; Yan and Suga 1999) suggested that auditory feedback acts as a selection signal, increasing the sharpness of tuning of the subcortical centers to the parameters which are of interest to the auditory cortex. In the cat, Villa and his collaborators (Villa et al. 1991, 1999) showed that cortical inactivation by cooling had a multiplicity of effects on MGB function, and concluded that the cortical feedback adaptively modifies the thalamic activity, adjusting sensory filtering by the thalamus as a function of cortical activity. In the superior colliculus, Meredith and his co-workers have shown that removing cortical input from field AES reduces general excitability (Meredith and Clemo 1989) and decreases behavioral multimodal facilitation (Wilkinson et al. 1996). The information passing to higher auditory centers is even less understood. For example, in barbiturate-anesthetized cats, Kitzes and Hollrigel (1996) found that ablating the primary auditory cortex does not change the responses of neurons in PAF to pure tones at all, suggesting that the input from A1 is not used in PAF, at least for processing pure tones.

Within this general lack of understanding, it should be remembered that different cortical layers give rise to projections to different targets: for example, the supragranular layers are involved in corticocortical and commissural projections whereas the infragranular layers also project to subcortical targets (Winer 1992). It is reasonable to assume that different projection targets may use different kinds of information from the cortical activity. If this hypothesis is true, then it would be expected that neurons projecting to different targets would process information in different ways; in this case, differences between single neurons in the auditory cortex would be of extreme relevance. Alternatively, it is possible that all target structures receive some kind of an averaged signal from the cortex, and that the differences between layers that give rise to the different projections have only developmental importance. Under this hypothesis, averaged cluster activity would be the relevant output signal of the cortex. Thus, a better understanding of the nature of auditory cortical signaling at its targets may give important clues to the kind of processing performed by the auditory cortex itself.

5.4 Auditory Psychophysics and the Auditory Cortex

What is the role of the auditory cortex in the processing leading to known psychophysical phenomena? In the pyschoacoustical literature, there is a tendency to explain most phenomena by properties of the activity in the auditory nerve. Obviously, any information processed in the cortex should exist at the level of the auditory nerve. In order to assign a processing task to the auditory cortex, the question is whether responses in the auditory cortex appear to be more processed and more easily read out than the activity in the auditory nerve, and whether such activity does not already appear in lower stations of the auditory system.

We don't know the answers to these questions in any specific case. Some recent studies suggest that the auditory cortex is involved in the parsing of auditory scenes. Two of these findings have been discussed above: (1) the responses to tones masked by AM noise, in which phase locking to the noise envelope is suppressed by low-level tones, giving rise to a response that has the temporal characteristics of responses to pure tones (Nelken et al. 1999), and (2) the tendency of neurons to respond to a mixture of bird song in background noise in the same way they respond to the background noise alone (Bar Yosef et al., in preparation). In both cases, auditory cortical neurons respond to the mixture in the same way they would respond to the acoustically weak component of the mixture presented alone.

A third case in which activity in the auditory cortex can be interpreted as related to auditory scene analysis follows from the rules Heil found for onset responses in the auditory cortex (Heil 1997a,b). Independently, Fishbach et al. (2001) studied the perception of auditory discontinuities. They used somewhat different stimuli, consisting of tone or noise bursts that begin at a fixed level; after a while, the level is increased with a linear ramp to a higher level. Fishbach found that the perception of discontinuities did not depend solely on the difference between the levels or on the transition time by themselves; rather, it depended on the ratio of the two, that is, the slope of the ramp. Thus, the perception of discontinuities follows the same rules as those governing onset responses of the auditory cortex. Although the stimuli are not identical, this correspondence is highly suggestive and makes it tempting to hypothesize that onset responses in the auditory cortex signal the appearance of a new auditory object.

Fishbach et al. (2001) developed a computational model that can reproduce the results of a large array of experiments in which rise time was manipulated. These experiments include single neuron data (Phillips et al. 1995; Heil 1997a,b), evoked potential responses in the IC (Barth and Burkard 1993; Phillips and Burkard 1999), psychophysical data from experiments on the detection of amplitude transients (Bregman et al. 1994; Fishbach et al. 2001), and the effects of probe rise time on threshold in forward masking (Turner et al. 1994). The model interprets the cortical onset responses as the result of temporal edge detection. It consists of a standard auditory preprocessing stage that includes demodulation and logarithmic compression, followed by an array of neurons that vary in their latencies (Fig. 9.14). Standard receptive field operators on the output of this array, such as first or second derivatives, can reproduce all of the above data. Thus, a single computational model can reproduce measurements of the psychoacoustic effects of onset dynamics as well as measurements of

A

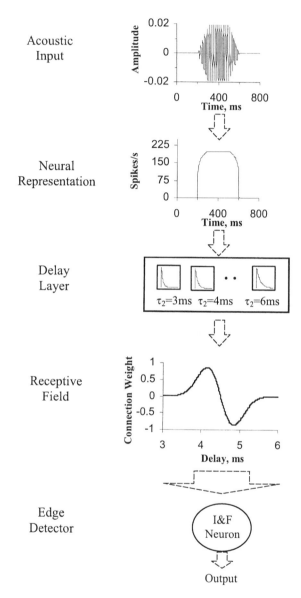

FIGURE 9.14. A: A model for onset responses in the auditory cortex. The acoustic input is demodulated, and its envelope is logarithmically compressed. The envelope is then processed through a layer of neurons with varying delays. Finally, a receptive field with inhibitory sidebands, mimicking first or second order derivative, operates on the outputs of the delay layer, computing an approximate temporal derivative of the amplitude envelope.

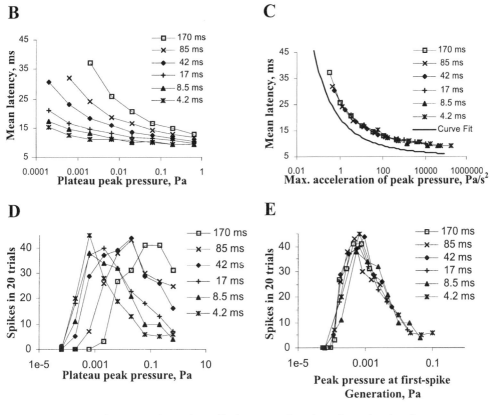

FIGURE 9.14. (*Continued*) B: First spike latency as function of tone level at the output of the model, for cosine-shaped onset ramps. The latency is not an invariant function of tone level. C: The same data plotted as function of the maximum acceleration of the ramp, as suggested by Heil (1997a). D: Spike count as function of the tone level as computed by the model. The spike counts depend on the rise time of the onset ramps, with the model neuron becoming more monotonic at longer rise times. E: The same data plotted as function of the ramp level at the time of first spike generation, as suggested by Heil (1997b). Thus, the model reproduces both the latency and the spike-count rules described by Heil. (Courtesy of A. Fishbach.)

physiological effects of onset dynamics, strongly suggesting that the physiology of onset responses in the auditory cortex underlies psychoacoustic parsing of amplitude transients in auditory scenes.

6. Summary

The picture emerging here is of the primary auditory cortex performing some sort of middle-level auditory processing. There are probably no species-specific call detectors that are sensitive to the meaning of calls without reference to their acoustic composition: on the contrary, auditory cortex probably represents species-specific calls by a spectrotemporal analysis of their structure. There may be specializations related to the processing of species-specific vocalizations, but these specializations may have to do with the low-level analyzers' adaptations to the general spectral and temporal features of the communication calls or even to features of the general auditory world.

Thus, species-specific vocalizations are probably not the most important sounds processed by the auditory cortex; they are useful for studying neurons only in the sense that they contain lower-level acoustic events that are detected and processed by the auditory cortex. But, in the same vein, pure tones, in spite of their utility in mapping the auditory cortex, are able to only partially engage the processing mechanisms in the auditory cortex.

I suggest that the middle-level processing performed by neurons in the auditory cortex has the flavor of auditory scene analysis, of detecting the occurrence of new auditory sources, and of performing the harder parts of scene parsing by listening to the weak acoustic components. This hypothesis leads to some specific experimental choices: the use of complex sounds, the studying of single neurons rather than measuring maps, the close interaction between physiological and psychophysical research, and the necessity to better understand those features in the external auditory world that can be used for parsing auditory scenes. This experimental approach will need, however, to interact with all the other approaches described in this chapter in order to fully elucidate the function of the auditory cortex.

References

Abeles M, Goldstein MH, Jr. (1970) Functional architecture in cat primary auditory cortex: columnar organization and organization according to depth. J Neurophysiol 33:172–187.

Ahissar M, Ahissar E, Bergman H, Vaadia E (1992) Encoding of sound-source location and movement: activity of single neurons and interactions between adjacent neurons in the monkey auditory cortex. J Neurophysiol 67:203–215.

Bakin JS, Kwon MC, Masino SA, Weinberger NM, Frostig RD (1996) Suprathreshold auditory cortex activation visualized by intrinsic signal optical imaging. Cereb Cortex 6:120–130.

Barth C, Burkard R (1993) Effects of noise bursts rise time and level on the human stem auditory evoked response. Audiology 32:225–233.

Bregman AS (1990) Auditory Scene Analysis: The Perceptual Organization of Sound. Cambridge, MA: MIT Press.

Bregman A, Ahad P, Kim J, Melnerich L (1994) Resetting the pitch-analysis system. 1. Effects of rise times of tones in noise backgrounds or of harmonics in a complex tone. Perception and Psychophysics 56:155–162.

Brosch M, Schreiner CE (1997) Time course of forward masking tuning curves in cat primary auditory cortex. J Neurophysiol 77:923–943.

Calford MB, Semple MN (1995) Monaural inhibition in cat auditory cortex. J Neurophysiol 73:1876–1891.

Calhoun BM, Schreiner CE (1998) Spectral envelope coding in cat primary auditory cortex: linear and non-linear effects of stimulus characteristics. Eur J Neurosci 10:926–940.

Cariani PA, Delgutte B (1996a) Neural correlates of the pitch of complex tones. I. Pitch and pitch salience. J Neurophysiol 76:1698–1716.

Cariani PA, Delgutte B (1996b) Neural correlates of the pitch of complex tones. II. Pitch shift, pitch ambiguity, phase invariance, pitch circularity, rate pitch, and the dominance region for pitch. J Neurophysiol 76:1717–1734.

Colombo M, Rodman HR, Gross CG (1996) The effects of superior temporal cortex lesions on the processing and retention of auditory information in monkeys (*Cebus apella*). J Neurosci 16:4501–4517.

Colombo M, D'Amato MR, Rodman HR, Gross CG (1990) Auditory association cortex lesions impair auditory short-term memory in monkeys. Science 247: 336–338.

Costalupes JA, Young ED, Gibson DJ (1984) Effects of continuous noise backgrounds on rate responses of auditory nerve fibers in cat. J Neurophysiol 51:1326–1344.

Creutzfeldt O, Hellweg FC, Schreiner C (1980) Thalamocortical transformation of responses to complex auditory stimuli. Exp Brain Res 39:87–104.

Dear SP, Fritz J, Haresign T, Ferragamo M, Simmons JA (1993) Tonotopic and functional organization in the auditory cortex of the big brown bat, Eptesicus fuscus. J Neurophysiol 70:1988–2009.

deCharms RC, Merzenich MM (1996) Primary cortical representation of sounds by the coordination of action-potential timing. Nature 381:610–613.

deCharms RC, Blake DT, Merzenich MM (1998) Optimizing sound features for cortical neurons [see comments]. Science 280:1439–1443.

Dinse HR, Godde B, Hilger T, Reuter G, Cords SM, Lenarz T, von Seelen W (1997a) Optical imaging of cat auditory cortex cochleotopic selectivity evoked by acute electrical stimulation of a multi-channel cochlear implant. Eur J Neurosci 9: 113–119.

Dinse HR, Reuter G, Cords SM, Godde B, Hilger T, Lenarz T (1997b) Optical imaging of cat auditory cortical organization after electrical stimulation of a multichannel cochlear implant: differential effects of acute and chronic stimulation. Am J Otol 18:S17–18.

Eggermont JJ (1993) Wiener and Volterra analyses applied to the auditory system. Hear Res 66:177–201.

Eggermont J (1996) How homogeneous is cat primary auditory cortex? Evidence from simultaneous single-unit recordings. Aud Neurosci 2:79–96.

Ehret G, Schreiner CE (1997) Frequency resolution and spectral integration (critical band analysis) in single units of the cat primary auditory cortex. J Comp Physiol (A) 181:635–650.

Esser KH, Condon CJ, Suga N, Kanwal JS (1997) Syntax processing by auditory cortical neurons in the FM-FM area of the mustached bat *Pteronotus parnellii*. Proc Natl Acad Sci U S A 94:14019–14024.

Fishbach

Fishman YI, Reser DH, Arezzo JC, Steinschneider M (1998) Pitch vs. spectral encoding of harmonic complex tones in primary auditory cortex of the awake monkey. Brain Res 786:18–30.

Fishman Y, Reser D, Arezzo J, Steinschneider M (2000) Complex tone processing in primary auditory cortex of the awake monkey. II. Pitch versus critical band representation. J Acoust Soc Am 108:247–262.

Frostig RD, Gottlieb Y, Vaadia E, Abeles M (1983) The effects of stimuli on the activity and functional connectivity of local neuronal groups in the cat auditory cortex. Brain Res 272:211–221.

Fujita I, Tanaka K, Ito M, Cheng K (1992) Columns for visual features of objects in monkey inferotemporal cortex [see comments]. Nature 360:343–346.

Gaese BH, Ostwald J (1995) Temporal coding of amplitude and frequency modulation in the rat auditory cortex. Eur J Neurosci 7:438–450.

Gao E, Suga N (1998) Experience-dependent corticofugal adjustment of midbrain frequency map in bat auditory system. Proc Natl Acad Sci U S A 95:12663–12670.

Gibson DJ, Young ED, Costalupes JA (1985) Similarity of dynamic range adjustments in auditory nerve and cochlear nuclei. J Neurophysiol 53:940–958.

Glass I, Wollberg Z (1983a) Auditory cortex responses to sequences of normal and reversed squirrel monkey vocalizations. Brain Behav Evol 22:13–21.

Glass I, Wollberg Z (1983b) Responses of cells in the auditory cortex of awake squirrel monkeys to normal and reversed species-specific vocalizations. Hear Res 9:27–33.

Goldstein MH Jr., Abeles M, Daly RL, McIntosh J (1970) Functional architecture in cat primary auditory cortex: tonotopic organization. J Neurophysiol 33: 188–197.

Goldstein MH Jr., Abeles M (1975) Note on tonotopic organization of primary auditory cortex in the cat. Brain Res 100:188–191.

Grady CL, Van Meter JW, Maisog JM, Pietrini P, Krasuski J, Rauschecker JP (1997) Attention-related modulation of activity in primary and secondary auditory cortex. Neuroreport 8:2511–2516.

Grunewald A, Linden JF, Andersen RA (1999) Responses to auditory stimuli in macaque lateral intraparietal area. I. Effects of training. J Neurophysiol 82: 330–342.

Hall JW, Haggard MP, Fernandes MA (1984) Detection in noise by spectro-temporal pattern analysis. J Acoust Soc Am 76:50–56.

Harrison RV, Harel N, Kakigi A, Raveh E, Mount RJ (1998) Optical imaging of intrinsic signals in chinchilla auditory cortex. Audiol Neurootol 3:214–223.

Heil P (1997a) Auditory cortical onset responses revisited. I. First-spike timing. J Neurophysiol 77:2616–2641.

Heil P (1997b) Auditory cortical onset responses revisited. II. Response strength. J Neurophysiol 77:2642–2660.

Heil P (1998) Further observations on the threshold model of latency for auditory neurons [comment]. Behav Brain Res 95:233–236.

Heil P, Irvine DR (1996) On determinants of first-spike latency in auditory cortex. Neuroreport 7:3073–3076.

Heil P, Irvine DR (1997) First-spike timing of auditory-nerve fibers and comparison with auditory cortex. J Neurophysiol 78:2438–2454.

Heil P, Irvine DR (1998) Functional specialization in auditory cortex: responses to frequency-modulated stimuli in the cat's posterior auditory field. J Neurophysiol 79:3041–3059.

Heil P, Rajan R, Irvine DR (1992a) Sensitivity of neurons in cat primary auditory cortex to tones and frequency-modulated stimuli. I: Effects of variation of stimulus parameters. Hear Res 63:108–134.

Heil P, Rajan R, Irvine DR (1992b) Sensitivity of neurons in cat primary auditory cortex to tones and frequency-modulated stimuli. II: Organization of response properties along the "isofrequency" dimension. Hear Res 63:135–156.

Heil P, Rajan R, Irvine DR (1994) Topographic representation of tone intensity along the isofrequency axis of cat primary auditory cortex. Hear Res 76:188–202.

Horikawa J, Hosokawa Y, Kubota M, Nasu M, Taniguchi I (1996) Optical imaging of spatiotemporal patterns of glutamatergic excitation and GABAergic inhibition in the guinea-pig auditory cortex in vivo. J Physiol (Lond) 497:629–638.

Hosokawa Y, Horikawa J, Nasu M, Taniguchi I (1997) Real-time imaging of neural activity during binaural interaction in the guinea pig auditory cortex. J Comp Physiol A 181:607–614.

Imig TJ, Adrian HO (1977) Binaural columns in the primary field (A1) of cat auditory cortex. Brain Res 138:241–257.

Imig TJ, Brugge JF (1978) Sources and terminations of callosal axons related to binaural and frequency maps in primary auditory cortex of the cat. J Comp Neurol 182:637–660.

Irvine DR, Huebner H (1979) Acoustic response characteristics of neurons in nonspecific areas of cat cerebral cortex. J Neurophysiol 42:107–122.

Ito M, Tamura H, Fujita I, Tanaka K (1995) Size and position invariance of neuronal responses in monkey inferotemporal cortex. J Neurophysiol 73:218–226.

Kanwal JS, Matsumura S, Ohlemiller K, Suga N (1994) Analysis of acoustic elements and syntax in communication sounds emitted by mustached bats. J Acoust Soc Am 96:1229–1254.

Kitzes LM, Hollrigel GS (1996) Response properties of units in the posterior auditory field deprived of input from the ipsilateral primary auditory cortex. Hear Res 100:120–130.

Klein D, Depireux D, Simon J, Shamma S (2000) Robust spectrotemporal reverse correlation for the auditory system: optimizing stimulus design. J Comput Neurosci 9:85–111.

Klump GM, Langemann U (1995) Comodulation masking release in a songbird. Hear Res 87:157–164.

Korte M, Rauschecker JP (1993) Auditory spatial tuning of cortical neurons is sharpened in cats with early blindness. J Neurophysiol 70:1717–1721.

Kowalski N, Versnel H, Shamma SA (1995) Comparison of responses in the anterior and primary auditory fields of the ferret cortex. J Neurophysiol 73:1513–1523.

Kowalski N, Depireux DA, Shamma SA (1996a) Analysis of dynamic spectra in ferret primary auditory cortex. I. Characteristics of single-unit responses to moving ripple spectra. J Neurophysiol 76:3503–3523.

Kowalski N, Depireux DA, Shamma SA (1996b) Analysis of dynamic spectra in ferret primary auditory cortex. II. Prediction of unit responses to arbitrary dynamic spectra. J Neurophysiol 76:3524–3534.

Langner G, Sams M, Heil P, Schulze H (1997) Frequency and periodicity are represented in orthogonal maps in the human auditory cortex: evidence from magnetoencephalography. J Comp Physiol (A) 181:665–676.

Linden JF, Grunewald A, Andersen RA (1999) Responses to auditory stimuli in macaque lateral intraparietal area. II. Behavioral modulation. J Neurophysiol 82:343–358.

Lutkenhoner B, Lamertmann C, Ross B, Steinstrater O (2001) Tonotopic organization of the human auditory cortex revisited: High-precision neuromagnetic studies. In: Houtsma A, Kohlrausch A, Prijs V, Schoonhoven R (eds) Physiological and Psychophysical Bases of Auditory Function. Maastricht: Shaker Publishing BV, p. 13.

Markram H, Lubke J, Frotscher M, Roth A, Sakmann B (1997) Physiology and anatomy of synaptic connections between thick tufted pyramidal neurones in the developing rat neocortex. J Physiol Lond 500:409–440.

Matsubara JA, Phillips DP (1988) Intracortical connections and their physiological correlates in the primary auditory cortex (AI) of the cat. J Comp Neurol 268: 38–48.

Mendelson JR (1992) Neural selectivity for interaural frequency disparity in cat primary auditory cortex. Hear Res 58:47–56.

Mendelson JR, Schreiner CE, Sutter ML, Grasse KL (1993) Functional topography of cat primary auditory cortex: responses to frequency-modulated sweeps. Exp Brain Res 94:65–87.

Mendelson JR, Schreiner CE, Sutter ML (1997) Functional topography of cat primary auditory cortex: response latencies. J Comp Physiol (A) 181:615–633.

Meredith MA, Clemo HR (1989) Auditory cortical projection from the anterior ectosylvian sulcus (Field AES) to the superior colliculus in the cat: an anatomical and electrophysiological study. J Comp Neurol 289:687–707.

Merzenich MM, Knight PL, Roth GL (1975) Representation of cochlea within primary auditory cortex in the cat. J Neurophysiol 38:231–249.

Middlebrooks JC, Dykes RW, Merzenich MM (1980) Binaural response-specific bands in primary auditory cortex (AI) of the cat: topographical organization orthogonal to isofrequency contours. Brain Res 181:31–48.

Middlebrooks JC, Clock AE, Xu L, Green DM (1994) A panoramic code for sound location by cortical neurons. Science 264:842–844.

Nelken I, Versnel H (2000) Responses to linear and logarithmic frequency-modulated sweeps in ferret primary auditory cortex. European J Neurosci 12:549–562.

Nelken I, Young ED (1994) Two separate inhibitory mechanisms shape the responses of dorsal cochlear nucleus type IV units to narrow-band and wide-band stimuli. J Neurophysiol 71:2446–2462.

Nelken I, Young ED (1997) Linear and Nonlinear Spectral Integration in Type IV Neurons of the Dorsal Cochlear Nucleus. I. Regions of Linear Interaction. J Neurophys 78:790–799.

Nelken I, Prut Y, Vaadia E, Abeles M (1994a) In search of the best stimulus: an optimization procedure for finding efficient stimuli in the cat auditory cortex. Hear Res 72:237–253.

Nelken I, Prut Y, Vaadia E, Abeles M (1994b) Population responses to multifrequency sounds in the cat auditory cortex: one- and two-parameter families of sounds. Hear Res 72:206–222.

Nelken I, Prut Y, Vaddia E, Abeles M (1994c) Population responses to multifrequency sounds in the cat auditory cortex: four-tone complexes. Hear Res 72: 223–236.

Nelken I, Bar Yosef O, Young ED (1997a) Responses of field AES neurons to virtual space stimuli. In: Palmer AR, Rees A, Summerfield AQ, Meddis R (eds) Psychophysical and Physiological Advances in Hearing. London: Whurr Publ. pp. 504–512.

Nelken I, Kim PJ, Young ED (1997b) Linear and Nonlinear Spectral Integration in Type IV Neurons of the Dorsal Cochlear Nucleus. II. Predicting Responses With the Use of Nonlinear Models. J Neurophys 78:800–811.

Nelken I, Rotman Y, Bar Yosef O (1999) Responses of auditory-cortex neurons to structural features of natural sounds. Nature 397:154–157.

Nelken I, Jacobson G, Ahdut L, Ulanovsky N (2001) Neural correlates of comodulation masking release in auditory cortex of cats. In: Houtsma A, Kohlrausch A, Prijs V, Schoonhoven R (eds) Physiological and Psychophysical Bases of Auditory Function. Maastricht: Shaker Publishing BV.

Newman JD, Wollberg Z (1973a) Multiple coding of species-specific vocalizations in the auditory cortex of squirrel monkeys. Brain Res 54:287–304.

Newman JD, Wollberg Z (1973b) Responses of single neurons in the auditory cortex of squirrel monkeys to variants of a single call type. Exp Neurol 40:821–824.

Ohlemiller K, Kanwal J, Butman J, Suga N (1994) Stimulus design for auditory neuroethology: synthesis and manipulation of complex communication sounds. Aud Neurosci 1:19–38.

Ohlemiller KK, Kanwal JS, Suga N (1996) Facilitative responses to species-specific calls in cortical FM-FM neurons of the mustached bat. Neuroreport 7:1749–1755.

Olsen JF, Suga N (1991) Combination-sensitive neurons in the medial geniculate body of the mustached bat: encoding of relative velocity information. J Neurophysiol 65:1254–1274.

O'Neil WE (1995) The Bat Auditory Cortex. In: Popper AN, Fay RR (eds) Hearing by Bats. New York: Springer-Verlag pp. 416–180.

Pantev C, Hoke M, Lutkenhoner B, Lehnertz K (1989) Tonotopic organization of the auditory cortex: pitch versus frequency representation. Science 246:486–488.

Pantev C, Elbert T, Ross B, Eulitz C, Terhardt E (1996) Binaural fusion and the representation of virtual pitch in the human auditory cortex. Hear Res 100:164–170.

Pelleg-Toiba R, Wollberg Z (1991) Discrimination of communication calls in the squirrel monkey: "call detectors" or "cell ensembles"? J Basic Clin Physiol Pharmacol 2:257–272.

Perrett DI, Rolls ET, Caan W (1982) Visual neurones responsive to faces in the monkey temporal cortex. Exp Brain Res 47:329–342.

Phillips DP (1988) Effect of tone-pulse rise time on rate-level functions of cat auditory cortex neurons: excitatory and inhibitory processes shaping responses to tone onset. J Neurophysiol 59:1524–1539.

Phillips DP (1998) Factors shaping the response latencies of neurons in the cat's auditory cortex [see comments]. Behav Brain Res 93:33–41.

Phillips DP, Burkard R (1999) Response magnitude and timing of auditory response initiation in the inferior colliculus of the awake chinchilla. J Acoust Soc Am 105:2731–2737.

Phillips DP, Cynader MS (1985) Some neural mechanisms in the cat's auditory cortex underlying sensitivity to combined tone and wide-spectrum noise stimuli. Hear Res 18:87–102.

Phillips DP, Irvine DR (1983) Some features of binaural input to single neurons in physiologically defined area AI of cat cerebral cortex. J Neurophysiol 49:383–395.

Phillips DP, Sark

Phillips DP, Orman SS, Musicant AD, Wilson GF (1985) Neurons in the cat's primary auditory cortex distinguished by their responses to tones and wide-spectrum noise. Hear Res 18:73–86.

Phillips DP, Semple MN, Calford MB, Kitzes LM (1994) Level-dependent representation of stimulus frequency in cat primary auditory cortex. Exp Brain Res 102:210–226.

Phillips DP, Semple MN, Kitzes LM (1995) Factors shaping the tone level sensitivity of single neurons in posterior field of cat auditory cortex. J Neurophysiol 73: 674–686.

Rauschecker JP, Korte M (1993) Auditory compensation for early blindness in cat cerebral cortex. J Neurosci 13:4538–4548.

Rauschecker JP, Tian B, Hauser M (1995) Processing of complex sounds in the macaque nonprimary auditory cortex. Science 268:111–114.

Reale RA, Imig TJ (1980) Tonotopic organization in auditory cortex of the cat. J Comp Neurol 192:265–291.

Reale RA, Imig TJ (1983) Auditory cortical field projections to the basal ganglia of the cat. Neuroscience 8:67–86.

Reale RA, Kettner RE (1986) Topography of binaural organization in primary auditory cortex of the cat: effects of changing interaural intensity. J Neurophysiol 56: 663–682.

Ricketts C, Mendelson JR, Anand B, English R (1998) Responses to time-varying stimuli in rat auditory cortex. Hear Res 123:27–30.

Rotman Y, Bar Yosef O, Nelken I (2000) Relating Cluster and Population Responses to Natural Sounds and Tonal Stimuli in Cat Primary Auditory Cortex. Hear Res.

Rouiller EM, Innocenti GM, De Ribaupierre F (1990) Interconnections of the auditory cortical fields of the cat with the cingulate and parahippocampal cortices. Exp Brain Res 80:501–511.

Schooneveldt GP, Moore BC (1989) Comodulation masking release (CMR) as a function of masker bandwidth, modulator bandwidth, and signal duration. J Acoust Soc Am 85:273–281.

Schreiner CE (1992) Functional organization of the auditory cortex: maps and mechanisms. Curr Opin Neurobiol 2:516–521.

Schreiner CE (1995) Order and disorder in auditory cortical maps. Curr Opin Neurobiol 5:489–496.

Schreiner CE (1998) Spatial distribution of responses to simple and complex sounds in the primary auditory cortex. Audiol Neurootol 3:104–122.

Schreiner CE, Calhoun BM (1994) Spectral Envelope Coding in Cat Primary Auditory Cortex: Properties of Ripple Transfer Functions. Auditory Neurosci 1:39–62.

Schreiner CE, Mendelson JR (1990) Functional topography of cat primary auditory cortex: distribution of integrated excitation. J Neurophysiol 64:1442–1459.

Schreiner CE, Sutter ML (1992) Topography of excitatory bandwidth in cat primary auditory cortex: single-neuron versus multiple-neuron recordings. J Neurophysiol 68:1487–1502.

Schreiner CE, Urbas JV (1988) Representation of amplitude modulation in the auditory cortex of the cat. II. Comparison between cortical fields. Hear Res 32: 49–63.

Schreiner CE, Mendelson JR, Sutter ML (1992) Functional topography of cat primary auditory cortex: representation of tone intensity. Exp Brain Res 92: 105–122.

Schwarz DW, Tomlinson RW (1990) Spectral response patterns of auditory cortex neurons to harmonic complex tones in alert monkey (*Macaca mulatta*). J Neurophysiol 64:282–298.

Semple MN, Kitzes LM (1993a) Binaural processing of sound pressure level in cat primary auditory cortex: evidence for a representation based on absolute levels rather than interaural level differences. J Neurophysiol 69:449–461.

Semple MN, Kitzes LM (1993b) Focal selectivity for binaural sound pressure level in cat primary auditory cortex: two-way intensity network tuning. J Neurophysiol 69:462–473.

Shamma SA, Symmes D (1985) Patterns of inhibition in auditory cortical cells in awake squirrel monkeys. Hear Res 19:1–13.

Shamma SA, Versnel H (1995) Ripple Analysis in Ferret Primary Auditory Cortex. II. Prediction of Unit responses to Arbitrary Spectral Profiles. Auditory Neurosci 1:255–270.

Shamma SA, Fleshman JW, Wiser PR, Versnel H (1993) Organization of response areas in ferret primary auditory cortex. J Neurophysiol 69:367–383.

Shamma SA, Versnel H, Kowalski N (1995) Ripple Analysis in Ferret Primary Auditory Cortex. I. Response Characteristics of Single Units to Sinusoidally Rippled Spectra. Auditory Neurosci 1:233–254.

Smolders JW, Aertsen AM, Johannesma PI (1979) Neural representation of the acoustic biotope. A comparison of the response of auditory neurons to tonal and natural stimuli in the cat. Biol Cybern 35:11–20.

Steinschneider M, Reser DH, Fishman YI, Schroeder CE, Arezzo JC (1998) Click train encoding in primary auditory cortex of the awake monkey: evidence for two mechanisms subserving pitch perception. J Acoust Soc Am 104:2935–2955.

Suga N (1990) Cortical Computational maps for Auditory Imaging. Neural Networks 3:3–21.

Sutter ML, Schreiner CE (1991) Physiology and topography of neurons with multipeaked tuning curves in cat primary auditory cortex. J Neurophysiol 65:1207–1226.

Sutter ML, Schreiner CE (1995) Topography of intensity tuning in cat primary auditory cortex: single-neuron versus multiple-neuron recordings. J Neurophysiol 73:190–204.

Taniguchi I, Horikawa J, Hosokawa Y, Nasu M (1997) Optical imaging of neural activity in auditory cortex induced by intracochlear electrical stimulation. Acta Otolaryngol Suppl 532:83–88.

Tian B, Rauschecker JP (1994) Processing of frequency-modulated sounds in the cat's anterior auditory field. J Neurophysiol 71:1959–1975.

Tian B, Rauschecker JP (1998) Processing of frequency-modulated sounds in the cat's posterior auditory field. J Neurophysiol 79:2629–2642.

Toldi J, Feher O (1984) Acoustic sensitivity and bimodal properties of cells in the anterior suprasylvian gyrus of the cat. Exp Brain Res 55:180–183.

Turner C, Relkin E, Doucet J (1994) Psychophysical and physiological foward masking probe duration and rise-time effects. J Acoust Soc Am 96:795–800.

Vaadia E (1989) Single-unit activity related to active localization of acoustic and visual stimuli in the frontal cortex of the rhesus monkey. Brain Behav Evol 33:127–131.

Vaadia E, Gottlieb Y, Abeles M (1982) Single-unit activity related to sensorimotor association in auditory cortex of a monkey. J Neurophysiol 48:1201–1213.

Versnel H, Shamma SA (1998) Spectral-ripple representation of steady-state vowels in primary auditory cortex. J Acoust Soc Am 103:2502–2514.

Villa AE, Rouiller EM, Simm GM, Zurita P, De Ribaupierre Y, De Ribaupierre F (1991) Corticofugal modulation of the information processing in the auditory thalamus of the cat. Exp Brain Res 86:506–517.

Villa AE, Tetko IV, Dutoit P, De Ribaupierre Y, De Ribaupierre F (1999) Corticofugal modulation of functional connectivity within the auditory thalamus of rat, guinea pig and cat revealed by cooling deactivation. J Neurosci Methods 86:161–178.

Wang X, Merzenich MM, Beitel R, Schreiner CE (1995) Representation of a species-specific vocalization in the primary auditory cortex of the common marmoset: temporal and spectral characteristics. J Neurophysiol 74:2685–2706.

Wilkinson LK, Meredith MA, Stein BE (1996) The role of anterior ectosylvian cortex in cross-modality orientation and approach behavior. Exp Brain Res 112: 1–10.

Winer JA (1992) The Functional Architecture of the Medial Geniculate Body and the Primary Auditory Cortex. In: Webster DB, Popper AN, Fay RR (eds) The Mammalian Auditory Pathway: Neuroanatomy. New York: Springer-Verlag pp. 222–409.

Wollberg Z, Newman JD (1972) Auditory Cortex of Squirrel Monkey: Response Patterns of Single Cells to Species-Specific Vocalizations. Science 175:212–214.

Yan J, Suga N (1999) Corticofugal amplification of facilitative auditory responses of subcortical combination-sensitive neurons in the mustached bat. J Neurophysiol 81:817–824.

Young ED, Brownell WE (1976) Responses to tones and noise of single cells in dorsal cochlear nucleus of unanesthetized cats. J Neurophysiol 39:282–300.

Zhang Y, Suga N (1997) Corticofugal amplification of subcortical responses to single tone stimuli in the mustached bat. J Neurophysiol 78:3489–3492.

Zhang Y, Suga N, Yan J (1997) Corticofugal modulation of frequency processing in bat auditory system. Nature 387:900–903.

Zurita P, Villa AE, de Ribaupierre Y, de Ribaupierre F, Rouiller EM (1994) Changes of single unit activity in the cat's auditory thalamus and cortex associated to different anesthetic conditions. Neurosci Res 19:303–316.

Index

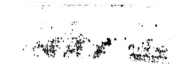